Benchmark Papers in Geology

Series Editor: Rhodes W. Fairbridge
Columbia University

Published Volumes and Volumes in Preparation

ENVIRONMENTAL GEOMORPHOLOGY AND LANDSCAPE CONSERVATION, VOLUME I: Prior to 1900 / Donald R. Coates
RIVER MORPHOLOGY / Stanley A. Schumm
SPITS AND BARS / Maurice L. Schwartz
TEKTITES / Virgil E. Barnes and Mildred A. Barnes
GEOCHRONOLOGY: Radiometric Dating of Rocks and Minerals / C. T. Harper
SLOPE MORPHOLOGY / Stanley A. Schumm and M. Paul Mosley
MARINE EVAPORITES: Origin, Diagenesis, and Geochemistry / Douglas W. Kirkland and Robert Evans
ENVIRONMENTAL GEOMORPHOLOGY AND LANDSCAPE CONSERVATION, VOLUME III: Non-Urban / Donald R. Coates
GLACIAL ISOSTASY / John T. Andrews
WEATHERING AND SAPROLITE / Dan Yaalon
EARTHQUAKES: Geological Effects / Clarence R. Allen
ROCK CLEAVAGE / Dennis S. Wood
ENVIRONMENTAL GEOMORPHOLOGY AND LANDSCAPE CONSERVATION, VOLUME II: Urban / Donald R. Coates
BARRIER ISLANDS / Maurice L. Schwartz
VERTEBRATE PALEONTOLOGY / J. T. Gregory
PALEOMAGNETISM / Kenneth Creer
PALEOECOLOGY / John Chronic, Jr.
SOLID STATE GEOPHYSICS / Thomas J. Ahrens
OCEANOGRAPHY / Takashi Ichiye
PERMAFROST / J. D. Ives
PLAYAS AND CLAY-PANS / James L. Neal
GEOCHEMISTRY OF SULFUR / Isaac Kaplan
MODERN CARBONATES / Gerald Friedman
PHILOSOPHY OF GEOLOGY / Claude C. Albritton

Benchmark Papers in Geology

A *BENCHMARK*™ Books Series

MARINE EVAPORITES:

Origin, Diagenesis, and Geochemistry

Edited by
DOUGLAS W. KIRKLAND and ROBERT EVANS
Mobil Research and Development Corporation

Dowden, Hutchinson & Ross, Inc.
Stroudsburg, Pennsylvania

Library of Congress Catalog Card Number: 73-11151
ISBN: 0-87933-043-0

Library of Congress Cataloging in Publication Data

Kirkland, Douglas W comp.
Marine evaporites.

(Benchmark papers in geology)
1. Evaporites--Addresses, essays, lectures.
I. Evans, Robert, 1939- joint comp.
II. Title.
QE471.K496 552'.5 73-11151
ISBN 0-87933-039-2

Manufactured in the United States of America.

Exclusive distributor outside the United States and Canada: John Wiley & Sons, Inc.

Acknowledgments and Permissions

ACKNOWLEDGMENTS

Mineralogical Society—*Mineralogical Magazine*
"The Mineralogy of the British Permian Evaporites"

Society of Economic Paleontologists and Mineralogists
Dolomitization and Limestone Diagenesis
"Dolomitization of Recent and Plio-Pleistocene Sediments by Marine Evaporite Waters on Bonaire, Netherlands Antilles"

Journal of Sedimentary Petrology
"An Occurrence of Gypsum in Southwest Texas"
"Modern Evaporite Deposition and Geochemistry of Coexisting Brines, the *Sabkha,* Trucial Coast, Arabian Gulf"
"Origin of the Salina Salt of Michigan"
"Origin and Diagenesis of Gypsum and Anhydrite"

PERMISSIONS

The following papers have been reprinted with the permission of the authors and the copyright owners.

American Association of Petroleum Geologists, Inc.
The American Association of Petroleum Geologists Bulletin
"Modern Evaporite Deposition in Peru"
"A Modern Evaporite Deposit in Mexico"
"Origin of Nodular and Bedded Anhydrite in Permian Shelf Sediments, Texas and New Mexico"
"Modes of Formation, Sedimentary Associations, and Diagnostic Features of Shallow-Water and Supratidal Evaporites"
"Sedimentation in Permian Castile Sea"
"Deposition of Evaporites"

Fluids in Subsurface Environments—A Symposium
"Sulfur-Isotope Geochemistry of Petroleum, Evaporites and Ancient Seas"

Izvestiya of the Academy of Sciences of the USSR, Geological Series
"Geologic Conditions of Formation of Bottom Sediments in Karabogaz-Gol in Connection with Fluctuations of the Caspian Sea Level"
"Formation of Halogenic Zones in Marine Basins Illustrated by the Example of the Lower Cambrian Evaporite Basin of the Siberian Platform"

Geochemistry
"Geochemistry of Bromine in the Processes of Salt Deposition and the Use of the Bromine Content as a Genetic and Prospecting Criterion"

Consultants Bureau, Plenum Publishing Corporation—*Lithology and Mineral Resources*
"The Problem of Kara-Bogaz-Gol"

Elsevier Publishing Company—*Chemical Geology*
"Isotope Geochemistry of Sedimentary Sulfates"

The Geological Society of America—*Geological Society of America Bulletin*
"Halite–Anhydrite Seasonal Layers in the Middle Devonian Prairie Evaporite Formation, Saskatchewan, Canada"
"Intrabasin Varve Correlation"

Institution of Mining and Metallurgy—*Institution of Mining and Metallurgy, Transactions*
"Origin of Marine Evaporites by Diagenesis"
"Recent Halite Rock, Baja California, Mexico"

Macmillan (Journals) Ltd.—*Nature*
"Association of Dolomite and Anhydrite in the Recent Sediments of the Persian Gulf"

The Mineralogical Society of America—*American Mineralogist*
"Diagenetic Polyhalite in Recent Salt from Baja California"

Northern Ohio Geological Society, Inc.
Second Symposium on Salt
"Bromide Geochemistry of Salt Rocks"

Third Symposium on Salt
"Shelf Carbonate Sedimentation Controlled by Salinity in the Paradox Basin, Southeast Utah"
"Wind and Water Depth and Their Bearing on the Circulation in Evaporite Basins"

Springer-Verlag—Berlin–Heidelberg–New York—*Salt Deposits: Their Origin and Composition*
"Rhythmic Bedding and Influxes"

American Journal of Science (Yale University)—*American Journal of Science*
"Dolomite Sedimentation in the South-East of South Australia"
"Shallow-Water and Deep-Water Evaporite Deposition"

Yorkshire Geological Society—*Proceedings of the Yorkshire Geological Society*
"Evaporite Diagenesis in the Lower Purbeck Beds of Dorset"

Series Editor's Preface

The philosophy behind the "Benchmark Papers in Geology" is one of collection, sifting, and rediffusion. Scientific literature today is so vast, so dispersed, and, in the case of old papers, so inaccessible for readers not in the immediate neighborhood of major libraries that much valuable information has been ignored by default. It has become just so difficult, or so time consuming, to search out the key papers in any basic area of research that one can hardly blame a busy man for skimping on some of his "homework."

This series of volumes has been devised, therefore, to make a practical contribution to this critical problem. The geologist, perhaps even more than any other scientist, often suffers from twin difficulties—isolation from central library resources and immensely diffused sources of material. New colleges and industrial libraries simply cannot afford to purchase complete runs of all the world's earth science literature. Specialists simply cannot locate reprints or copies of all their principal reference materials. So it is that we are now making a concerted effort to gather into single volumes the critical material needed to reconstruct the background of any and every major topic of our discipline.

We are interpreting "Geology" in its broadest sense: the fundamental science of the Planet Earth, its materials, its history, and its dynamics. Because of training and experience in "earthy" materials, we also take in astrogeology, the corresponding aspect of the planetary sciences. Besides the classical core disciplines such as mineralogy, petrology, structure, geomorphology, paleontology, and stratigraphy, we embrace the newer fields of geophysics and geochemistry, applied also to oceanography, geochronology, and paleoecology. We recognize the work of the mining geologists, the petroleum geologists, the hydrologists, the engineering and environmental geologists. Each specialist needs his working library. We are endeavoring to make his task a little easier.

Each volume in the series contains an Introduction prepared by a specialist (the volume editor)—a "state of the art" opening or a summary of the objects and content of the volume. The articles, usually some thirty to fifty reproduced either in their entirety or in significant extracts, are selected in an attempt to cover the field, from the key papers of the last century to fairly recent work. Where the original works are in foreign languages, we have endeavored to locate or commission translations. Geologists, because of their global subject, are often acutely aware of the oneness of our world. The selections cannot, therefore, be restricted to any one country, and whenever possible an attempt is made to scan the world literature.

To each article, or group of kindred articles, some sort of "Highlight Commentary" is usually supplied by the volume editor. This should serve to bring that article into historical perspective and to emphasize its particular role in the growth of the field. References, or citations, wherever possible, will be reproduced in their entirety—for by this means the observant reader can assess the background material available to that particular author, or, if he wishes, he too can double check the earlier sources.

A "benchmark," in surveyor's terminology, is an established point on the ground, recorded on our maps. It is usually anything that is a vantage point, from a modest hill to a mountain peak. From the historical viewpoint, these benchmarks are the bricks of our scientific edifice.

Rhodes W. Fairbridge

Preface

What are evaporites? The product of neither magmatic activity nor high-temperature reactions deep within the crust, they are a residue of deposition in seawater. They are not the carbonate residue of abundant life within the seas, not the accumulated debris of uplift and inevitable denudation, but the unique precipitates that result from the concentration of seawater by evaporation. To the geologist concerned with describing and understanding the history of processes on the earth's surface, they reflect extreme and unusual conditions of deposition and alteration, and it is toward an appreciation of these conditions that we hope to lead the reader of this collection of papers.

Aware of this objective, our audience should not then be disappointed to discover that we have omitted papers that deal with the formation, alteration, or geochemistry of potash salts. Such works comprise the majority of papers written on the geology of evaporite rocks, but they deal with deposits that are of great rarity.

Accordingly, our attention has been concentrated on three aspects of the non-potash marine evaporites: origin and sedimentation; petrology and diagenesis; equilibrium, trace-element, and isotope geochemistry. In the part on deposition we consider recent examples of evaporite formation in lagoon, estuary, and supratidal flat; we also review papers on models of sedimentation in deep water, on a shelf, and in environments analogous to present-day occurrences. That part of the book labeled "petrology" deals with problems of "varves" and postdepositional changes; and the final part introduces some complexities of the gypsum–anhydrite equilibrium, the substitution of Br^- for Cl^- in evaporitic chlorides, and the significance of differences in the isotopic composition of sulfur in sedimentary sulfates. The general reader may wonder about the inclusion of material on trace elements and isotopes, extremely specialized aspects of the geology of evaporites. These chemical data, however, give us a useful insight into the processes of formation and the processes of alteration of the rocks we seek to understand.

A collection of papers can serve only to introduce the stranger to the subject at hand. Carefully chosen, however, such a collection can both arm and protect the unknowing reader when he journies into the wilderness of the literature on any aspect of any science. Our intent at the outset of this venture was to make a judicious and useful selection; time and the reader will tell if we have succeeded.

Douglas W. Kirkland
Robert Evans

Contents

II. EVAPORITE PETROLOGY

III. GEOCHEMISTRY OF EVAPORITES

Contents by Author

Introduction

When Lot's wife turned to view the destruction of Sodom and Gomorrah her life may have ended, but her value to society increased, for the salt of Mt. Sodom (near the Dead Sea) has played an important part in local trade for thousands of years. It is not surprising, therefore, to realize that speculation on the origin of salt springs had occupied the thoughts of many of the early naturalists. More than nineteen centuries ago, Pliny the Elder discussed the salt springs of Germany, France, and Spain, but it is to Georgius Agricola that credit must go for explaining their origin. In *De Natura Fossilium,* published in 1558, he suggested that they depended on subterranean salt strata for their existence.

The first test of Agricola's theory was not made until the eighteenth century, when the discovery of a small rock salt occurrence in Bottendorf was made. At the beginning of that century, von Borlach wrote "Salzgebirge müssen sich überall gleich sein, und, wo Solquellen gefunden werden, muss auch Steinsalz vorkommen" (Fulda, 1938—Salt formations must be similar everywhere, and where salt springs are found, rock salt must also occur). Both he and Agricola were proved correct, for in the early nineteenth century, in response to the demand for more salt than could be supplied by salt springs, a search was made for subsurface salt beds. Borings at various localities throughout Germany were successful, discoveries being made at Lothringen (1819), Gotha (1825), Hannover (1831), Thüringen (1831), Artern (1837), and Salzungen (1842) (Fulda, 1938).

As a result of drilling in the area of Stassfurt during the period 1839–1851, information that stimulated exploration for evaporites elsewhere in northern Germany became available. Impetus was provided by two things: first, an enormous thickness of salt was discovered, more than 300 m at Stassfurt; second, large quantities of potassium and magnesium salts were found associated with halite. Because, at that time, these bitter salts were regarded as contaminants in the salt of interest, they were called "Abraumsalze" (waste or rubbish salts) and were discarded in large heaps on the surface.

However, in 1840, Baron Justus von Liebig had shown that potassium was vital for the growth of plants and that "since growing plants assimilated potassium, this element must necessarily be resupplied to the soil." The "Abraumsalze" accumulating in mounds near the mine shafts at Stassfurt soon became valuable agricultural commodities and were, therefore, granted the more respectful title of "Edelsalze" (noble or vital salts). Some potash was sold as early as 1857, the first plant for the extraction of potash was built in 1861, and for more than 50 years thereafter the bulk of the world's potash needs were supplied by the mines near Stassfurt. Many early ideas on the origin of evaporites derived from studies conducted at those same mines.

Students of the history of science devote a great deal of attention to the differences of opinion, the conflicts of personality, the factionalism that marked, and still marks, the supposedly objective investigation of the natural world. To their delight, but to the chagrin of some geologists, the study of the earth has been a battleground on which many a champion has risen, to fall later to another, perhaps better armed. We may note, for example, that the name of Abraham Gotlob Werner survives, but not his Neptunist philosophy. And is not Alfred Wegener, so long the subject of amused scorn, now treated with the reverence, perhaps even awe, that is given to a major figure in the development of our science?

The study of evaporites has produced some notable differences of opinion, perhaps because no one worked at the level of detail that we now know is necessary. The first controversy, and this is no surprise, involved the origin of the rocks themselves. Were they caused by deposition from volcanic vapors arising from the depths of the earth, as Friedrich von Alberti proposed in 1852? His two-volume work *Halurgische Geologie* included this hypothesis, which was accepted by many of his contemporaries. Or were they precipitated from seawater in the manner described by Bischof in 1854? Two chapters of Bischof's influential work *Elements of Chemical and Physical Geology* were devoted to rock salt and gypsum, and he considered the possibility that these rocks formed in a basin behind a bar over which the sea poured only during periods of high tide. How pleased Werner would have been with this "Neptunian" exposition!

A most significant event in the history of our understanding of evaporites occurred in 1849 with the publication of the results of Usiglio's classic work with seawater. This Italian chemist performed experiments that have never been repeated, or rather, it is more accurate to say that results of any repetition of Usiglio's work have never been reported. With primitive equipment, Usiglio took an early step into the field of experimental petrology by trying to determine the nature of, and the order of, the precipitates from seawater progressively concentrated by evaporation. In an atmosphere dried by calcium oxide and maintained at 40°C, 5-liter aliquots of seawater were evaporated in porcelain dishes. Periodically, the brine was cooled to 21°C, filtered, and the precipitates weighed and identified. The sequence of precipitates he identified, from calcium carbonate to calcium sulfate to sodium chloride to potassium and magnesium compounds, has since been recognized in many major evaporite deposits. If for nothing else, the experiments of Usiglio deserve to be remembered for the role they played in inspiring van't Hoff to conduct his detailed studies of the chemistry of the Stassfurt potash deposits.

Another great difference of opinion developed during the period 1900–1904 and involved Carl Ochsenius and the great stratigrapher Johannes Walther. In 1877,

Ochsenius, elaborating on the earlier Bischof model, proposed that the salt deposits at Stassfurt had been formed in a lagoon cut off from the ocean by a bar of such dimension as to allow only as much water to enter the lagoon as evaporated from its surface. This model differed from that of Bischof by being "open," allowing for the development of an evaporite section as thick as the basin was deep. In order to explain the rarity of potash deposits, he considered that the dense brines, rich in magnesium and potassium as a result of deposition of carbonate, calcium sulfate, and sodium chloride, would fill the basin to the level of the bar and then flow outward beneath the inflowing water of normal salinity. In the event of the relative level of the bar being raised from time to time, the dense brine would be retained in the closed basin and would, as a result of continued evaporation, release its load of magnesium and potassium compounds. Ochsenius used Kara-Bogaz-Gol as an extant example of his bar theory, and this remarkable lagoon on the eastern side of the Caspian Sea is described in this volume in papers by Dzens-Litovskiy and Vasil'yev (1962) and Dzens-Litovskiy (1967).

The conflict arose in 1900 when Walther published his "desert theory" as an explanation for the origin of evaporites. A professor at the University of Halle, he had become impressed by the paucity of marine faunal remains in evaporites and noted that if fossils were found, they were exclusively nonmarine, as is the case for the mammals described by Cuvier from the gypsiferous strata in the Paris basin. Clearly, he reasoned, this would not be the case had the rocks been deposited in a coastal lagoon.

Furthermore, he noted that natural salt sequences only rarely matched the sequences predicted after the experiments of Usiglio and van't Hoff. He concluded that salts were deposited in a large landlocked basin; those exposed at the margins were dissolved and transported into the center of the depression by meteoric waters, there to be reprecipitated after further evaporation. Walther stated his arguments against the theory of Ochsenius in a bitter polemic in 1903.

Although the "desert theory" appeared in the later editions of "Das Gesetz der Wüstenbildung in der Gegenwart und Vorzeit" until 1924, and was championed by Grabau in his *Principles of Salt Deposition* (1920), the idea was largely ignored in favor of the bar theory. Recently, however, the concept of a landlocked basin has been resurrected by Hsü (1972) in his surprising explanation of the origin of the evaporites of the western Mediterranean, an explanation derived from discoveries made in the *Glomar Challenger* deep-sea drilling program.

From time to time new theories of the origin of evaporites have been presented, but few have been paid the compliment of survival. Among them are the "saline lava theory" of Lichtenberger (1927), the "infiltration theory" of Fulda (1927), the "great tidal flood theory" of Wilfarth (1933), and, most recently, the curious "replacement theory" of Kudryavtsev (1971).

The Bischof–Ochsenius "bar theory" (1877, 1888, 1893) became widely accepted as a reasonable explanation of thick accumulations, providing for a continuous supply of salts and persistent subsidence of a basin, which in an arid region would lead inevitably to such deposits.

In an intriguing theory, Branson (1915) attempted to explain the monomineralic composition characteristic of many evaporites, for he and others realized that the unmodified "bar theory" would allow for no segregation of the major evaporite minerals in marine basins. Branson's model consisted of a series of evaporating bodies of water, within each of which an individual salt precipitated. The first would be characterized by calcium carbonate, the second by calcium sulfate, the next by halite, and so on. Water flowing from one basin to another would be "preconcentrated" and would have divested itself of the least-soluble salts by precipitation in the first restricted body of water in the series imagined by Branson. The fact that neither ancient nor modern examples of his "modified bar theory" could be found resulted in the passing of Branson's hypothesis into well-deserved obscurity.

Another modification of the bar theory was introduced in 1947 by Ralph H. King, in an attempt to explain the thick anhydrite of the Castile Formation. His "theory of reflux" has survived and been supported by observations in modern estuaries. This significant contribution to the understanding of evaporites is included in this volume (Paper 13).

The early 1950s saw the students of evaporites arm themselves with details of modern evaporite depositional processes, details of which were able to buttress the purely theoretical models prevalent in the literature. Papers by Scruton (1953), Masson (1955), and Morris and Dickey (1957), all included in this volume, heralded this new approach to the problems of evaporitic sedimentation, an approach that has produced useful and stimulating studies. One need consider only the exciting discoveries in the tidal flats of Abu Dhabi (Persian Gulf), to appreciate the significance of this new approach. Almost half of the papers selected for this volume deal with the results of detailed examination of modern evaporitic environments.

Unfortunately, this work on the formation of gypsum, halite, and the other evaporites is limited by the paucity of suitable sites, and also by the fact that many of the places where evaporites are forming have been so altered by the needs of commerce that they are unsuitable for further geologic work. Nevertheless, what has been learned from the Holocene may now be applied to ancient rocks, and it appears that the detailed sedimentological approach to investigations of evaporites will dominate future work on the subject. In a review of the translation of Braitsch's book *Entstehung und Stoffbestand der Salzlagerstätten,* Eugster (1971) prophesied:

> Like no other branch of earth sciences, evaporite research has been dominated by chemists, foremost among them van't Hoff. It looks as if it is now the sedimentologists' turn. A detailed analysis of the depositional environments and sedimentary structures of evaporites promises to yield more insight than further phase diagrams. . . .

There are many deposits of evaporites that have not been studied at all; there are even more that have not been studied in the detail necessary for a complete explanation of their origin. This collection of papers will, therefore, be of value if it stimulates some of the work mentioned in Eugster's prophecy.

Origin of Evaporites

I

Editors' Comments on Papers 1 Through 5

1 **Dzens-Litovskiy and Vasil'yev:** *Geologic Conditions of Formation of Bottom Sediments in Karabogaz-Gol in Connection with Fluctuations of the Caspian Sea Level*

2 **Dzens-Litovskiy:** *The Problem of Kara-Bogaz-Gol*

3 **Morris and Dickey:** *Modern Evaporite Deposition in Peru*

4 **Phleger:** *A Modern Evaporite Deposit in Mexico*

5 **Alderman and Skinner:** *Dolomite Sedimentation in the South-East of South Australia*

Holocene Evaporites: Lagoon

With recognition of the fact that many deposits of gypsum, anhydrite, halite, and potash salts were the result of evaporation of seawater, and after the subsequent publication of the Bischof–Ochsenius lagoonal bar model to explain the deposition of thick accumulations of these evaporites, attention was directed toward those areas where such lagoons were to be found. The choice of Kara-Bogaz-Gol as the "type example" of the lagoonal model of evaporite deposition has been well justified and supported by all investigations of that inhospitable but important embayment on the eastern shore of the Caspian Sea.

Kara-Bogaz-Gol embodies many features of the classic evaporite lagoon: it is very shallow, it exhibits a marked salinity gradient away from the inlet, and there is a persistent inflow of water from the neighboring Caspian Sea, with which it has a narrow restricted connection. Over the past 45 years the amount of water entering Kara-Bogaz-Gol each year has fluctuated between a maximum of 25.6 km^3 in 1929 and a minimum of 6.04 km^3 in 1939, and although the salinity of the Caspian (at 12.8 ‰) is only about one third that of normal seawater, the intensity of evaporation in this arid region has produced in the Gulf a brine more saline ($>$300 ‰) than in any similar basin on earth.

The Soviet geologist A. I. Dzens-Litovskiy is the leading authority on this modern evaporite basin, and in the two papers included here, he and his colleague G. V. Vasil'yev summarize the sedimentation of the various salts (mirabilite, astrakhanite, epsomite, gypsum, and halite), describe the underlying succession of salts, and review the recorded sequence of events. It should be noted that the waters of the Caspian Sea differ from normal seawater in that they are richer in calcium and sulfate and poorer in sodium and chloride. The actual mineralogy of the deposits of Kara-Bogaz-Gol is, therefore, not typical of marine evaporites.

Of particular significance in the following papers on this evaporite lagoon is the observation of the effect that sea-level variations have had on salt deposition. The level of the Caspian Sea is now −28 m, the level of the Kara-Bogaz-Gol −33 m (Garbell, 1963), and over the past few centuries the level of the major water body has varied over more than 5 m (Fairbridge, 1968, fig. 2). The salt succession revealed in boreholes in the gulf records the transgressions and regressions of the Caspian and shows the stratigraphic complexities that can result from sea-level variations. It is instructive to make a comparison with the work of Hite (1970) (Paper 17) in which he describes the effects of sea-level changes on the evolution of a major *ancient* evaporite deposit.

The lagoonal model of evaporite deposition is not intended to explain only small, localized, monomineralic bodies. Kara-Bogaz-Gol at the beginning of this century occupied 18,000 km^2, but is progressively shrinking. Three different evaporite facies can be recognized within it, for corresponding with the horizontal salinity gradient across the gulf, there is a lateral segregation of evaporite precipitates that form simultaneously within the facies belts, in a pattern identified in ancient salt basins.

Dzens-Litovskiy has been a student of the gulf since the early 1950s; he and his colleague Vasil'yev work at the renowned All-Union Institute of Halurgy in Leningrad.

Another example of the formation of Holocene marine evaporites in a "lagoonal" setting is the Bocana de Virrila, a shallow estuary that extends into the Sechura Desert of Peru. Over a distance of about 30 km and behind a series of sand bars and low islands, salinity in the channel increases from 3.5 percent to nearly 35.5 percent. Calcium carbonate, calcium sulfate, and sodium chloride are precipitated within the salinity ranges predicted by the results of Usiglio's evaporation experiments in 1849.

The paper by Morris and Dickey is the report of a 1955 reconnaissance of the Bocana de Virrila. It is one of the tragedies of evaporite geology that this natural laboratory has been dammed in the interest of salt production and is no longer available for the detailed chemical and sedimentological study that is needed (W. E. Dean, Jr., Syracuse University, personal communication, 1972).

A similar fate has befallen Ojo de Liebre lagoon, a large hypersaline inlet on the Pacific coast of Baja California. Huge crystallization pans have been constructed for the purpose of halite extraction, so that the environments described by Holser (1966) and Phleger (1969) have been greatly modified. The papers are of special interest because they describe this environment before the commercial alteration took place.

The sedimentology and oceanography of the Ojo de Liebre Lagoon have been described by Phleger and Ewing (1962), and in a later article, reproduced here, Phleger adds details on the formation of the evaporites associated with the margins of the lagoon. Supratidal flats about 275 km^2 in area mark the edge of the inner part of the main lagoon, and almost one-quarter of that area is covered by shallow brine pans that receive periodic influxes of wind-driven water from the lagoon proper.

Of great interest is the organic productivity of the lagoon and associated salt pans. It is high enough to provide sufficient food for the herds of California gray whales that use the outer lagoons as their winter calving grounds. Several other evaporitic environments are reported to have similarly high organic productivity:

Laguna Madre, in Texas (Hedgpeth, 1947), and artificial salt pans (Carpelan, 1957; Caspers, 1952). The abundance of algae in the salt pans of Ojo de Liebre decreases the carbon dioxide concentration within the brines and apparently causes local precipitation of calcium carbonate.

F. B. Phleger is Professor of Oceanography and Director of the Marine Foraminifera Laboratory at Scripps Institute of Oceanography in San Diego, California. Though trained as a geologist and paleontologist, he became interested in oceanography before World War II, and the paper included here is a natural adjunct of oceanographic studies made in several lagoons associated with Ojo de Liebre.

In many evaporite basins, for example Castile, Paradox, Zechstein, and Todilto, carbonates are a characteristic feature of the lower part of the chemical sequence. Until recently, there were no known modern occurrences of carbonates in an evaporitic setting, so the question of their organic or inorganic origin remained unsolved. Our understanding of the origin of such carbonates has been improved by studies in Coorong Lagoon and an associated series of lakes in southeast South Australia.

Since the publication of the original paper by Alderman and Skinner in 1957, at least ten works have appeared dealing with the Coorong and nearby lakes. Bathurst (1971) gives an excellent summary of the current state of knowledge of the carbonate sediments in this area and concludes: "As an open air laboratory on a grand scale, the region rivals the much studied Trucial Coast."

In the paper that follows, Alderman and Skinner establish the presence of Holocene dolomite–calcite sediments in the Coorong and show that there is a relationship among plant growth, hydrogen-ion concentration, and carbonate precipitation. As in the Laguna Ojo de Liebre and in portions of the Persian Gulf (Illing and Wells, 1964), it appears that photosynthesis can promote carbonate formation.

A. R. Alderman, petrologist and geochemist, became interested in dolomite as a result of these studies in the Coorong, although much of his earlier work had been in the petrology of ceramics. He is now Professor Emeritus at the University of Adelaide in South Australia.

His colleague during the Coorong work, Catherine Skinner, has turned her geological training and mineralogical and geochemical experience to rather unique use since the South Australia studies. Married to B. J. Skinner, Professor of Geology at Yale, she is now a Research Associate in the Department of Surgery and Molecular Biophysics and Biochemistry at the same university.

More studies of evaporite deposition in restricted lagoons are urgently needed before commercial modification destroys any remaining natural laboratories. In addition to the alteration of the Bocana de Virrila and Ojo de Liebre, a ship channel has been cut into and through Laguna Madre, and a regulating sluice has been proposed for the strait in Kara-Bogaz-Gol. In the Western Hemisphere, Rio Lagartos, a narrow coastal lagoon on the northern coast of Yucatan, Mexico, is still in a natural state and deserves intensive study (Wilson et al., 1970). Gypsum is precipitated from water at the distal end of the lagoon in an area of abundant algal growth.

American Geological Institute translation of *Izvestiya Acad. Sci. USSR, Geol. Ser.*, **3**, 79–86 (1962)

GEOLOGIC CONDITIONS OF FORMATION OF BOTTOM SEDIMENTS IN KARABOGAZ-GOL IN CONNECTION WITH FLUCTUATIONS OF THE CASPIAN SEA LEVEL[1]

1

by

A. I. Dzens-Litovskiy, and G. V. Vasil'yev

The Karabogaz-Gol (Figure 1), located on the east Caspian coast, is known in world literature as the classic example of a marine lagoon where huge masses of salt deposits and other highly concentrated saline deposits have been and still are accumulating.

The area of the gulf is about 18,000 km^2. Deeply incised into the land, it is separated from the sea by northern and southern sand spits extended meridionally. They are separated by Karabogaz Strait which connects the gulf and the sea.

Not a single river flows to the Karabogaz-Gol. Extending to the north and east of it is the Ust-Urt desert plateau; the sandy desert of Kara Kum touches it in the southeast. Its coast is an arid desert.

Overlain by the sands at the base of the spits are dense Ancient Caspian limestones changing downward to calcareous shell-bearing sandstones, with a total thickness up to 10 m. They are underlain by dense gypsiferous Oligocene claystone. The two horizons dip west; in many shallow places they form numerous reefs and submarine ledges.

The surface of these spits is hilly to barchan-like. Present among the barchans, in the axial parts of the spits are saline troughs, dry and issueless, but still preserving evidence of a recent connection between the sea and the bay.

Exposed in isolated places of both the northern and southern spits are ledges of ancient Caspian limestones and sandstones, the former submarine reefs. Well expressed is the ancient beach zone and a single terrace standing 4 to 5 m above the present sea level, a testimony to the last Caspian transgression.

Karabogaz-Gol is fed by Caspian waters, through the strait, and its hydrologic and hydrochemical regimen is closely related to the sea-level regimen. Waters of the Caspian, because of the considerable difference in their level as compared with the bay, continuously flow to the gulf where they are evaporated because of the arid climate. Brine accumulates in the bay, and salts are deposited as the result of complex physicochemical processes.

Thus, Karabogaz-Gol can be compared to a vast evaporation vat where sea water gains in density. The volume of inflow, and consequently the water balance of the bay, are determined on the whole solely by the height of the Caspian level.

The amount of precipitation falling on the water area of the bay does not exceed 10% of the inflow.

The Caspian sea level is subject to wide fluctuations. It has fallen almost three meters in the last 25 to 30 years. Most students explain that by climatic causes [3, 5, 13, 16]. The recent warming up of the climate has caused a reduction in the Volga discharge, and consequently a drop in the Caspian Sea level; the latter, in turn, has caused a drop in the Karabogaz-Gol level.

Another reason for the Caspian level drop is human activity in connection with the building of water storage basins for the Volga and its tributaries [2].

Because of the drop in the Caspian Sea level, the flow of sea water into the bay has been cut sharply; the volume of its surface brines has been reduced also. As early as 1929, the annual flow of sea water into the bay was 25.8 km^3, as against the present figure of about 8 km^3. There have been temporary peaks on the generally downward discharge curve. Between 1938 and 1956, the bay depth has been reduced from ten to three meters; its water surface area from 18,000 to almost 10,000 km^2; and the brine volume from 103 to 28 km^3. The present volume of surface brine is about 25 km^3.

This contraction of the bay water surface has

---

[1] **Geologicheskiye usloviya formirovaniya donnykh otlozheniy Kara-Bogaz-Gola v svyazi s kolebaniyami urovnya Kaspiyskogo morya.**

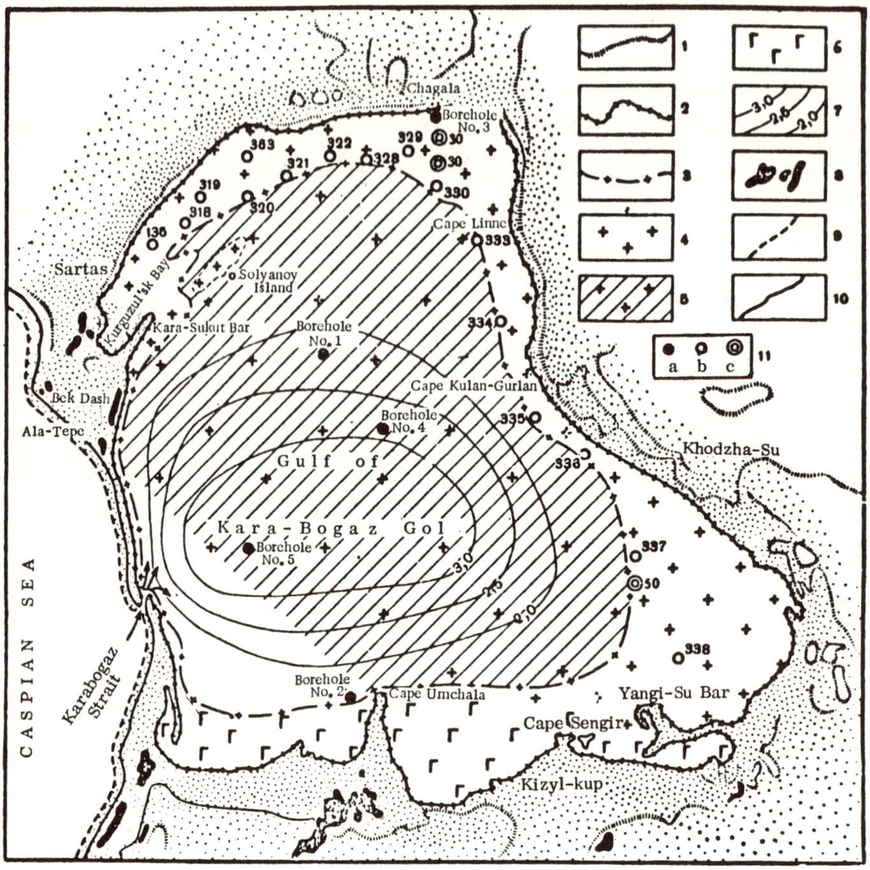

FIGURE 1. Generalized map of Karabogaz-Gol

1 - ancient coastal swells and terraces; 2 - generalized 1930 boundary of brines; 3 - generalized 1956 brine boundary; 4 - exposed part of the upper salt layer; 5 - surface salt layer below the brine; 6 - gypsiferous saline (dry); 7 - 1956 brine isobaths; 8 - dry salines; 9 - generalized 1956 boundary of Caspian waters; 10 - generalized 1930 boundary of Caspian waters; 11 - exploration boreholes: a - 1934-1935; b - 1954; c - 1951.

led to the formation of dry salines, from 30 to 50 km wide, along its shores. Karabogaz Strait itself has changed much in recent years.

In the thirties, it was 5.6 km long, 200 to 700 m wide, and up to 6 m deep, allowing the passage of boats to load mirabilite (Glauber salt) and thenardite and to carry the minerals to the Caspian ports [1]. The present Karabogaz Strait is about 10.5 km long and 100 to 150 m wide; it forms a peculiar "sea waterfall" about 4.0 m high.

Under the arid desert climate, the change in the Karabogaz-Gol water balance has lead to a concentration of its surface brines [12, 14]. The rise in this concentration since 1897 is shown in Table 1.

Present in the surface brine are sodium chloride, magnesium sulfate, magnesium chloride, potassium chloride, and other salts.

Only Glauber salt was deposited from Karabogaz-Gol surface brines, prior to the thirties. The brine concentration has been rising since 1934, and the first precipitation of NaCl was observed in the summer of 1939.

Published in 1869 was a work by K. Behr [4], in German; based on erroneous data, it told of halite crystallization in Karabogaz-Gol and of the formation of salt deposits at its bottom. That communication has become part of all geologic textbooks and is cited by many even now.

Up to 1938, the bay brine had about the same

Table 1

Maximum concentration of salts in brines of the gulf (autumn composition)

Year	Density	Content of salts in %	Year	Density	Content of salts in %	Year	Density	Content of salts in %
1897	1.136	16.40	1936	1.208	25.20	1947	1.290	31.16
1909	1.149	18.10	1937	1.222	26.60	1948	1.290	31.73
1922	1.172	20.90	1938	1.230	28.20	1949	1.290	31.75
1930	1.175	21.30	1939	1.240	28.80	1950	1.292	31.48
1935	1.198	23.50	1945	1.275	31.77	1954	1.312	31.90
			1946	1.280	31.26			

composition throughout its area. The post-1939 summer precipitation of halite did not take place everywhere in the bay. In the western and southern parts of the bay, where the brine was diluted by the incoming sea water, a precipitation of carbonates and gypsum has been observed; however, there is no halite precipitation, as of now.

The highest brine concentration prevails in the northern and eastern parts of the gulf, the farthest removed from the strait. As a result of halite crystallization here, the brine composition, in the summer, is quite different from that in the southern and western parts: the brine is low in Na and Cl ions.

A further rise in the brine concentration has led to a simultaneous precipitation of halite and astrakhanite [11]; in recent years, halite and epsomite have been deposited in certain northern parts of the bay.

At the present time, three zones can be identified in the gulf by their surface brine and its physicochemical processes:

1. The dilution zone embracing the western and southern parts of the bay (adjacent to the coast), from Ala-Tepe to the Yangi-Su spit. Here, the brine is not halite saturated, with mirabilite precipitated in winter months.

2. The intermediate zone of metamorphism embraces the eastern area of the bay, from the Yangi-Su spit to Point Linne, and the northwestern coastal zone from Ala-Tepe to Solyánoy Island (it also includes the central part of the bay, according to the Expedition of the Institute of Organic and Inorganic Chemistry, the U. S. S. R. Academy of Sciences). This zone is characterized by summer precipitation of halite with an addition of astrakhanite ($MgSO_4 \cdot Na_2SO_4 \cdot 4H_2O$) and epsomite ($MgSO_4 \cdot 7H_2O$) and winter precipitation of mirabilite ($Na_2SO_4 \cdot 10H_2O$) with an occasional addition of epsomite.

3. The zone of high metamorphism occurs along the north shore of the bay, in the area of Solyanoy Island-Chagala-Malyy Linne. Taking place here is summer and winter precipitation of halite and epsomite (mixed) and autumn precipitation of epsomite. A slight solution of salts deposited during the year takes place in the spring.

It should be noted that there are no clean-cut boundaries between these zones. Their boundaries are unstable and dependent on the season, the volume of inflow, and the direction and intensity of prevailing winds which displace sizable masses of brine.

The determination of simultaneous deposition of different salt in the same salt basin but in different localities is of exceptional scientific and practical value. The brine composition in the gulf tends to change in the direction of eutonics.

Recent exploration in the Karabogaz-Gol has revealed three additional layers of mixed salts buried underneath the upper salt layer; they consist of halite, glauberite ($Na_2SO_4 \cdot CaSO_4$) and astrakhanite (Figure 2). The bottom salt layers are separated by horizons of gypsiferous carbonate oozes with a Caspian fauna. They are in turn underlain by ancient Caspian and Oligocene clays [7-9].

Thus, the Karabogaz-Gol has lived through four inundation stages and as many stages of dessication, undoubtedly related to periodic transgressions and regressions of the Caspian after the Würm glaciation of the Russian platform [10, 15, 16].

The geologic past and present of the Karabogaz-Gol are clósely related to the Caspian. Its transgressions are marked by deposition of gypsiferous carbonate oozes in the bay; during the regressions, the oozes became more saline and became an accumulation site for mixed salts, depending on the concentration of the brines.

Table 2

Composition of buried interstitial brines in Karabogaz-Gol[1]

Data	Collection date	Sampling locality	Brine horizon	Sampling depth	Specific Weight
1	22/IX—1953	Test No. 306, southern part of the area of detailed survey (Kurguzul Cove)	II	8.0	1.153
2	"	Same	II	9.5	1.172
3	9/VII—1952	Test No. 106, central part of the detailed survey area (Sartas Bay)	II	4.5	1.203
4	"	Same	II	6.0	1.219
5	"	"	II	8.0	1.227
6	"	"	II	10.0	1.233
7	"	"	II	12.0	1.237
8	5/IX—1956	Test No. 150	II	6.0	1.223
9	"	Same	II	7.15	1.221
10	"	"	II	9.0	1.233
11	"	"	II	11.5	1.226
12	"	"	II	13.0	1.245
13	27/X—1954	Test No. 366, northern part of the detailed survey area (Sartas Bay)	II	4.5	1.231
14	"	"	II	6.5	1.296
15	"	"	II	8.0	1.246
16	"	"	II	10.0	1.254
17	3/IX—1954	Test No. 328, Chagala	II	6.5	1.267
18	"	"	II	8.0	1.270
19	"	"	II	10.5	1.273
20	18/X—1954	Test No. 338, Khodzha-Su	II	4.8	1.266
21	"	"	II	8.0	1.269
22	"	"	II	12.0	1.273
23	"	"	II	14.0	1.273
24	15/VII—1952	Test No. 106, central part of the detailed survey area (Sartas Bay)	III	18.5	1.237
25	"	"	III	20.5	1.243
26	7/VII—1956	Test No. 189	III		1.260

[1] Analyses performed at the chemical laboratory of the Karabogaz Geologic Exploration Party, the All-Union Institute of Halurgy (N.S. Garkavi and M.F. Kritskaya, Analysts).

The last 10,000 to 12,000 years have witnessed four transgressive-regressive cycles of the Caspian, corresponding to the four ooze horizons and to as many salt layers in the bay.

The first (surface) salt layer has been forming since 1939, as a result of the present dessication of the bay. Deposition is still going on, with the present thickness up to 38 m.

(Table 2 continued)

Ion content in %							Salt content in %						
HCO_3'	SO_4''	Cl'	$Ca^{\cdot\cdot}$	$Mg^{\cdot\cdot}$	$Na^{\cdot}$		$Ca(HCO_3)_2$	$CaSO_4$	$Mg(HCO_3)_2$	$MgSO_4$	$MgCl_2$	NaCl	Total ions
0.11	3.62	8.36		1.19	4.96	18.24	—	—	0.13	4.54	0.98	12.58	18.23
0.11	4.20	9.18		1.15	5.83	20.47	—	—	0.13	5.26	0.27	14.80	20.46
0.05	2.64	13.87	0.02	0.91	8.53	26.02	0.07	0.01	—	3.29	0.97	21.68	26.02
0.04	2.69	15.00	0.01	1.03	9.07	27.84	0.05	—	--	3.37	1.37	23.05	27.84
0.04	3.46	14.93		1.51	8.50	28.44	—	—	0.05	4.34	2.43	21.63	28.45
0.03	4.05	14.78		2.15	7.47	28.48	—	--	0.04	5.07	4.40	18.96	28.48
0.04	4.36	14.55		2.34	7.11	28.39	—	—	0.05	5.46	4.82	18.07	28.40
0.04	2.86	14.48	0.02	1.17	8.54	27.10	0.05	0.01	—	3.57	1.76	21.71	27.10
0.04	3.01	14.43	0.02	1.26	8.41	27.16	0.05	0.01	—	3.76	1.95	21.39	27.16
0.04	3.70	14.40		1.75	7.80	27.69	—	—	0.05	4.63	3.17	19.84	27.69
0.05	3.52	14.16		1.73	7.59	27.05	—	—	0.06	4.40	3.29	19.30	27.05
0.08	5.21	14.08		3.04	5.88	28.28	—	—	0.09	6.52	6.71	14.96	28.28
0.04	3.65	14.47		1.45	8.40	28.01	—	—	0.05	4.57	2.04	21.35	28.01
0.05	4.42	14.30		2.09	7.45	28.31	—	—	0.06	5.54	3.76	18.96	28.32
0.08	5.30	14.04		2.66	6.64	28.72	—	—	0.1	6.64	5.09	16.90	28.73
0.08	5.82	13.85		3.04	6.05	28.84	—	—	0.1	7.29	6.07	15.38	28.84
0.14	7,44	13.32		3.44	5.75	30.09	—	—	0.17	9.32	5.99	14.61	30.09
0.11	7.94	12.77		3.06	6.34	30.22	—	—	0.13	9.95	4.03	16.11	30.22
0.09	8.43	12.49		2.84	6.80	30.65	—	—	0.11	10.56	2.70	17.28	30.65
0.14	7.44	13.25		3.51	5.56	29.90	—	—	0.17	9.32	6.26	14.11	29.91
0.15	7.58	13.55		3.88	5.13	30.29	—	—	0.18	9.50	7.56	13.06	30.30
0.14	7.73	13.46		3.89	5.13	30.35	—	—	0.17	9.68	7.48	13.01	30.34
0.14	7.71	13.48		3.90	5.11	30.34	—	—	0.17	9.66	7.52	12.99	30.34
0.03	4.75	14.61		2.76	6.54	28.69	—	—	0.04	5.95	6.07	16.53	28.69
0.06	5.62	14.23		2.93	6.40	29.24	—	—	0.07	7.04	5.87	16.26	29.24
0.02	6.69	13.49		2.58	7.07	29.85	—	—	0.02	8.38	3.47	17.98	29.85

This layer is represented by three principal minerals: halite, epsomite, and astrakhanite [6]. It covers about 75% of the bottom area. Its surface dips gently from shore to the center of the gulf and gradually plunges under the brine.

Its porosity is low (about 10%) and it is

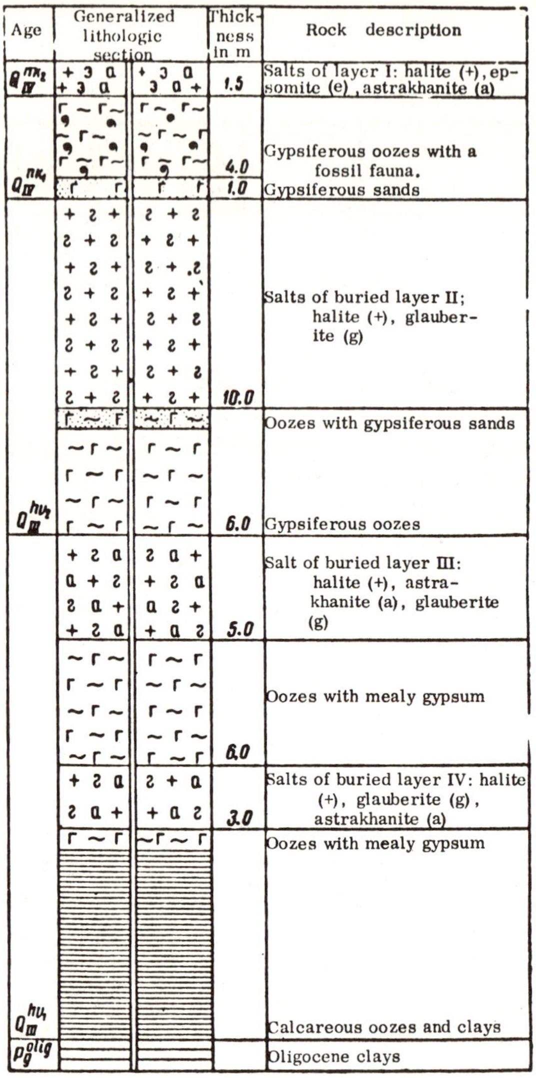

FIGURE 2. Generalized lithologic section of bottom sediments at the northwestern shore of Karabogaz-Gol, from data by the Karabogaz geologic exploratin party of the All-Union Institute of Halurgy.

saturated by interstitial brines which generate it.

The second salt layer was penetrated in 1951-1954 by numerous exploratory tests, below the Neo-Caspian deposits in Sartas Bay, Kurguzul Cove, and along the northwestern, northern, and eastern bay shore, at an average depth of 5.5 m.

This layer is separated from the overlying and the underlying ones by about 5 m of gypsiferous carbonate oozes with a Caspian fauna.

The second salt layer usually consists of two halite and three glauberite beds and is about 10 m thick, on the average. It thickens gradually from the shore to the center. It is highly porous and is everywhere saturated with fossil brines.

Present along with pores in this layer are larger cavities; they have been observed in cores and have been detected by sudden penetrations of the drilling tool.

The fossil brines which saturate this salt layer and a thin layer of gypsiferous sands at its top are at the second horizon of high-pressure buried brines.

This brine layer is characterized by its extremely high permeability and brine-producing capacity, determined by its high porosity and the presence of cavities. Its percolation rate usually exceeds 2000 m/day, and the flow rate in tests is 400 to 500 m^3/hr.

The composition of the brines varies laterally and with depth. Sodium chloride and magnesium sulfate and chloride (Table 2) are prominent among the constituent salts. Present in relatively lower concentrations are salts of potassium, boron, bromine, etc. As a rule, these brines have a higher sulfate-ion concentration offshore than farther away from it and have a correspondingly higher mirabilite yield upon cooling. They become richer in sulfate and magnesium ions, with depth, and their density rises from 1.19 to 1.27, with a corresponding increase in total salts, from 25 to 30%. In the annual and secular cycles, buried brines, unlike the surface ones, are marked by stability of composition and temperature (15 to 17°C).

Brines of the second horizon are particularly important as raw material for sodium chloride as well as for the combined production of a number of other magnesium and potassium salts and microcomponents.

Beginning in 1955, the "Karabogazsulfate" combine has been completely converted for exploiting a new king of halurgic raw material, buried brines of the second salt layer.

The third salt layer occurs at an average depth of 18 m. Its maximum penetrated thickness is 18.2 m, being about 5 m on the average. This layer is represented by halite, glauberite, and astrakhanite and is underlain by gypsiferous carbonate oozes which pass gradually to clay. The salt layer is porous and thoroughly saturated with brines at the third horizon of high-pressure buried brines.

Compared with the second, this brine horizon is characterized by low permeability and yield. Its brines are marked by a somewhat higher mineralization and comparatively uniform composition. Their temperature is 16.4 to 17°C.

The fourth salt layer has been penetrated in a limited area, in the Kurguzul Cove and Sartas Bay. Like the third layer, it is represented by halite, glauberite, and astrakhanite. Its brines have not been studied.

Because of the Caspian Sea level drop, Karabogaz-Gol is now in a transition stage from a lagunal-lacustrine salt deposit to a deposit of fossil salts.

In order to maintain its function as an important source of raw material (surface brines) it is necessary to regulate the inflow of sea waters by a lock gate. The minimum inflow for an optimum composition of surface brines should be established by additional hydrogeologic and hydrochemical studies.

The study of Karabogaz-Gol is going on. An understanding of its geologic past and present is of exceptional theoretical and practical interest.

REFERENCES

1. Andrusov, N.I., Kara-Bogaz-Gol i yego promyshlennoye znacheniya. [KARA-BOGAZ-GOL AND ITS COMMERCIAL VALUE]: Materialy dlya izucheniya yestestv. proizv. sil Rossii, vyp. 7, 1916.

2. Apollov, B.A., Kaspiyskoye more i yego basseyn. [THE CASPIAN SEA AND ITS BASIN]: Izd-vo Akad. Nauk SSSR, 1956.

3. Berg, L.S., Uroven' Kaspiyskogo morya za istoricheskoye vremya. [THE CASPIAN SEA LEVEL IN HISTORIC TIMES]: Ocherki po fizicheskoy geografii. Izd-vo Akad. Nauk SSSR, 1949.

4. Behr, K.M., Uchenyye zametki o Kaspiyskom more i yego okrestnostyakh. [ACADEMIC OBSERVATIONS ON THE CASPIAN SEA AND ITS VICINITY]: Zap. Geogr. ob-va, t. 2, 1856.

5. Bruyevich, S.V., Izmeneniye klimata i urovnya Kaspiyskogo morya za posledneye tysyacheletiye po khimicheskim dannym. [CHANGES IN THE CLIMATE AND LEVEL OF THE CASPIAN SEA DURING THE LAST MILLENNIUM, FROM CHEMICAL DATA]: Tr. Okeanogr. in-ta, t. 4, (16), 1948

6. Vakhrameyeva, V.A., K mineralogii i petrografii solyanykh otlozheniy zaliva Kara-Bogaz-Gol. [MINERALOGY AND PETROGRAPHY OF SALT DEPOSITS IN KARABOGAZ-GOL]: Tr. Vses. n.-i. in-ta galurgii, vyp. 32, 1956.

7. Dzens-Litovskiy, A.I., Geologiya i gidrogeologicheskiye usloviya novykh vidov

sul'fatnogo syr'ya zaliva Kara-Bogaz-Gol. [GEOLOGY AND HYDROGEOLOGIC CONDITIONS OF A NEW KIND OF SULFATE RAW MATERIAL IN KARABOGAZ-GOL BAY]: Tr. Okeanogr. komissii, t. 5, 1959.

8. Dzens-Litovskiy, A. I., M. Yu. Garkavi, and L. V. Yelovskaya, Geologiya donnykh otlozheniy i gidrokhimicheskiy rezhim Kurguzul'skoy bukhty i Sartasskogo zaliva Kara-Bogaz-Gola. [GEOLOGY OF BOTTOM DEPOSITS AND THE HYDROCHEMICAL REGIMEN IN THE KURGUZUL COVE AND SARTAS BAY OF KARA-BOGAZ-GOL]: Tr. In-ta geol. nauk Akad. Nauk TurkmSSR, t. 1, 1956.

9. Dzens-Litovskiy, A. I., G. A. Vasil'yev, and M. Yu. Garkavi, Donnyye solevyye otlozheniya i pogrebennyye mezhkristal'nyye rassoly Kara-Bogaz-Gola kak syr'yevaya baza galurgicheskoy promyshlennosti. [BOTTOM SEDIMENTS AND BURIED INTERSTITIAL BRINES OF KARABOGAZ-GOL AS A RAW MATERIAL BASE OF THE HALURGIC INDUSTRY]: In the book, Problemy kompleksnogo ispol'zovan. mineralog. bogatstv Kara-Bogaz-Gola. Izd-vo Akad. Nauk Turkm-SSR, 1959.

10. Leont'yev, O. K., and P. V. Fedorov, K istorii Kaspiyskogo morya v pozdne i poslekhvalynskoye vremya. [HISTORY OF THE CASPIAN SEA IN POST-KHVALYN TIME]: Izv. Akad. Nauk SSSR, ser. geogr., no. 4, 1953.

11. Pel'sh, A. D., Sostav rapy i kristallizatsiya astrakhanita v zalive Kara-Bogaz-Gole 1943—1945 gg. [THE COMPOSITION OF BRINE AND THE CRYSTALLIZATION OF ASTRAKHANITE IN KARABOGAZ-GOL, 1953-1945]: Tr. Vses. n.-i. in-ta galurgii, v. 23, 1952.

12. Sedel'nikov, G. S., Gidrokhimicheskiy rezhim zaliva Kara-Bogaz-Gol i perspektivy yego promyshlennogo ispol'zovaniya. [HYDROCHEMICAL REGIMEN IN KARABOGAZ-GOL AND PROSPECTS FOR ITS COMMERCIAL UTILIZATION]: Tr. Okeanogr. komis., t. 5, 1959.

13. Tugolesov, D. A., O prichinakh transgressiy i regressiy Kaspiyskogo morya. [THE CAUSES OF TRANSGRESSIONS AND REGRESSIONS OF THE CASPIAN SEA]: Izv. Akad. Nauk SSSR, ser. geol., no. 6, 1948.

14. Urazov, G. G., and V. D. Polyakov, Isledovaniye Kara-Bogaz-Gola. [A STUDY OF THE KARABOGAZ-GOL]: Priroda, no. 9, 1956.

15. Fedorov, P. V., Stratigrafiya chetvertichnykh otlozheniy i istoriya razvitiya Kaspiyskogo morya. [STRATIGRAPHY OF QUATERNARY DEPOSITS AND THE HISTORY OF DEVELOPMENT OF THE CASPIAN SEA]: Tr. Geol. in-ta Akad. Nauk SSSR, vyp. 10, 1957.

16. Shnitnikov, A. V., Izmenchivost' obshchey uvlazhnennosti materikov Severnogo polushariya. [GENERAL CHANGES IN HUMIDITY OF THE NORTHERN HEMISPHERE CONTINENTS]: Zap. Geograf. ob-va, t. 16, nov. ser., 1957.

All-Union Institute of Halurgy
Leningrad

Received, 11 May 1959

Erratum: Page 15, eighteenth line from the bottom of the page, column 1 should read "ploiting a new kind of halurgic raw material,"

Reprinted from *Lithol. Miner. Resour.*, **1**, 70–72, 76 (1967)

THE PROBLEM OF KARA-BOGAZ-GOL

[UDC 553.631(262.8.04K-B]

2

A. I. Dzens-Litovskii

All-Union Scientific Research Institute of Halurgy, Leningrad

Translated from Litologiya i Poleznye Iskopaemye, No. 1,
pp. 88-96, January-February, 1966
Original article submitted June 23, 1964

The Gulk of Kara-Bogaz-Gol, on the eastern shore of the Caspian Sea, is widely known from the literature as a classic example of a basin in which huge masses of salts and highly concentrated brines have formed and accumulated in the recent geologic past and are continuing to form and accumulate at the present time.

During the Soviet regime, extensive investigations concerning the geology, saline processes, and general physical chemistry of the basin have been made, and a huge store of data has been gathered, all of which must be of great interest to anyone concerned with the theory of salt accumulation in arid basins as well to anyone studying the use of such saline deposits in the economy.

1. Geologic History of Development of the Kara-Bogaz-Gol Basin

During the Middle Pliocene huge areas of the Transcaspian region and the northeastern part of the floor of the present Caspian Sea were dry land. A huge stream ran through the present Kara-Bogaz-Gol, running from the mountainous Mangyshlak to the southern Caspian basin, where a marine basin was then preserved (Fig. 1). According to Rikhter [1955], the internal basins of Chagaly-Sor and Karyn-Zharyk, lying northeast of Chagaly Gulf of Kara-Bogaz-Gol, represent this river valley as altered by later processes.

Drilling has revealed a deeply buried erosional depression north of the village of Karshi, at the opposite end of Kara-Bogaz-Gol, extending from southwest to northeast and being filled with Akchagylian sediments. A deep submarine canyon has been found on the continental slope of the Caspian, in the open sea opposite the bay bar [Leont'ev, 1961].

Thus, this ancient valley of the large Karyn-Zharyk stream, with its left and right tributaries, was flooded by waters of the Akchagylian transgression, and deeply indenting, branching embayments were formed in its place.

According to Leont'ev [1961], at the end of Akchagylian time the shoreline in the northern and eastern parts of Kara-Bogaz-Gol lay inside the present line (farther from shore).

We have fewer data regarding the paleogeography of Kara-Bogaz-Gol during Apsheronian and Bakinian (Baku) times. The absence of Bakinian deposits along the shores of the gulf indicates that the ancient shore was farther out than the present shore. The Bakinian sea was broadly open to the larger sea on the west.

The general outlines of Kara-Bogaz-Gol during the second half of Khazarian time were clearly more like those of the present gulf. Intensive abrasion occurred along the shores. At the broad entrance to the gulf, a huge bar of shelly, oölitic, abraded material began to form, completely cutting the gulf off from the sea. Remnants of the bar of this Khazarian lagoon have been preserved within the bounds of the beach and the present spits (Fig. 2).

The post-Khazarian regression, as suggested by Shnitnikov [1956], Fedorov [1937],* and Leont'ev and others [1959], was very prolonged and was characterized by a lowering of sea level to at least 20 m below the present level.

In all probability, the connection between Kara-Bogaz-Gol and the sea was first cut off during the Khazanian regression, and the gulf was converted to a salt lake. Salt deposits must have surely formed on the floor of the basin at that time. However, investigations during the expedition of VNIIG (All-Union Scientific Research Institute of Halurgy) along the margins of Kara-Bogaz-Gol uncovered no Khazarian salt deposits. They were either too soluble, and were eroded by waters of the transgressing Early Khazarian sea, or they were preserved only in the central part of the Kara-Bogaz-Gol basin [Dzens-Litovskii, 1956_1, 1956_2].

* This date differs from the one in the literature citation—Publisher.

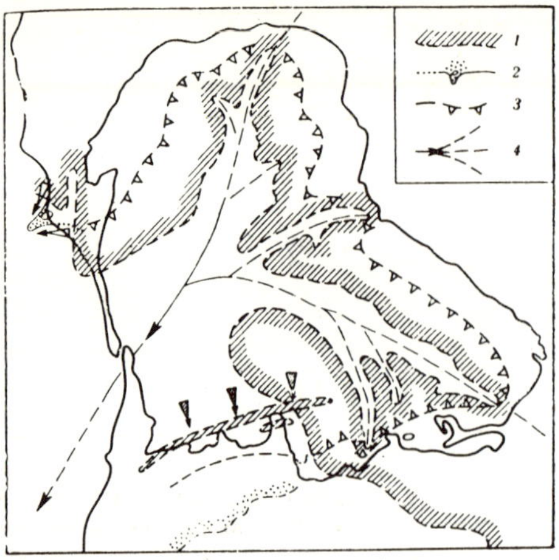

Fig. 1. Kara-Bogaz-Gol in Late Pliocene and Bakinian times (after Leont'ev, with modifications and additions by Dzens-Litovskii). 1) Hypothetical shoreline during maximal Akchagylian transgression, 2) shore and depositional shore forms in the Bekdash region during Late Akchagylian time, 3) shore of the Bakinian sea (teeth indicate abraded segments of the shore), 4) large hypothetical Karyn-Zharik stream with its tributaries.

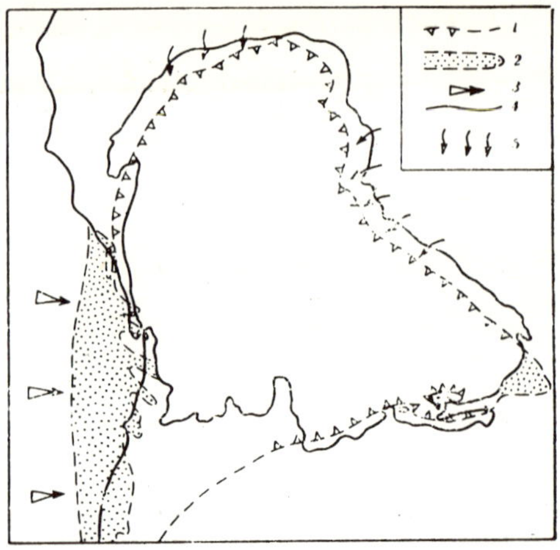

Fig. 2. Kara-Bogaz-Gol in Late Khazarian time (after Leont'ev, with modifications and additions by Dzens-Litovskii). 1) Shore of the Late Khazarian sea (teeth indicate abraded segments), 2) Late Khazarian bar at entrance to gulf and assumed buried depositional terrace, 3) transverse drift of detritus, 4) present shoreline, 5) shoreline slides.

Fig. 3. Kara-Bogaz-Gol in Early Khvalynian time (after Leont'ev). 1) Shoreline of the gulf in Early Khvalynian time (teeth indicate abraded shore), 2) shoreline depositional features of Early Khvalynian time, 3) longitudinal shifting of detritus, 4) transverse shifting of detritus.

The Early Khvalynian transgression of the Caspian strongly eroded the bay bar that had developed during the Khazarian regression. Abrasion of the shores recurred strongly, and the coastal outline was smoothed. During all of Lower Khvalynian time, Kara-Bogaz-Gol was a deep open gulf. High waves strongly attacked the shore, giving rise to slumps and huge slides along the eastern and northern shores of the basin (Fig. 3).

During the Lower Khvalynian, an archipelago, consisting of a complex network of island, peninsulas, and bays, existed in the Bekdash region [Leont'ev, 1961]. Salt lakes on the drainage divide between Bekdash Gulf on the Caspian and Kurguzul Bay of Kara-Bogaz-Gol are relicts of the Lower Khvalynian epoch.

During the Middle Khvalynian regression, Kara-Bogaz-Gol was again separated from the sea, and salt deposits again accumulated (the fourth horizon from the top) on its floor. These were cut in a number of holes drilled during the Kara-Bogaz-Gol expedition of VNIIG [Dzens-Litovskii, 1959, 1962_1, 1962_2].

During the first half of Late Khvalynian time, Kara-Bogaz-Gol again became an open gulf of the Caspian, and muds accumulating in this sea covered a lens of Middle Khvalynian salts.

Regression at the end of Late Khvalynian time converted Kara-Bogaz-Gol to a salt lake, and halite and glauberite accumulated in the bottom deposits (third horizon from the top).

Maximal Concentration of Salts in Brine of the Gulf
(Autumn composition)

Year	Spec. gravity	Total salts	Year	Spec. gravity	Total salts
1897	1,136	16,40	1945	1,275	31,77
1909	1,149	18,10	1946	1,280	31,86
1922	1,172	20,90	1947	1,290	31,96
1930	1,175	21,30	1948	1,290	31,93
1935	1,198	23,50	1949	1,290	31,95
1936	1,208	25,20	1950	1,292	31,98
1937	1,222	26,60	1954	1,312	33,90
1938	1,230	28,20	1955	1,318	34,02
1939	1,240	28,80	1961	1,362	34,53

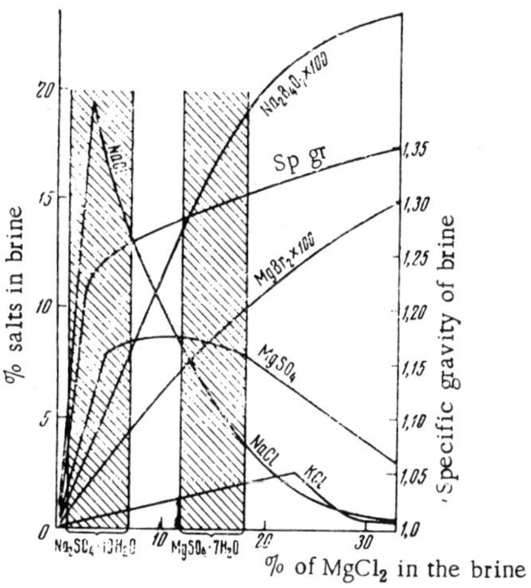

Fig. 4. Change in chemical composition of the brine in the gulf depending on decrease in influx of sea water and amount of diminution in volume of brine.

The New Caspian transgression again restored the connection between Kara-Bogaz-Gol and the Caspian and permitted strong development of spits on both sides of the bay bar, on the west toward the sea and on the east toward the gulf. Several lagoons were formed at these spits at that time. The dry salt lake Ala-Tepe is a remnant of the lagoon at the northern spit, and the Shakh-Nefes and Karshi salt marshes are remnants at the southern spit.

The basin of the former Kukurt-Ala lagoon on the southern shore of the gulf was flooded at that time, and planation of the shoreline continued. Depositional shore features developed.

Later, Kara-Bogaz-Gol went through two stages of saline development, during which salt beds III (third from the bottom) and IV (surface) were formed.

* * * * * * *

LITERATURE CITED

Dzens-Litovskii, A. I., Geology and hydrogeologic conditions of new species of sulfate raw material in Kara-Bogaz-Gol. Tr. Okeanografich. komis., 5, Izd. AN SSSR, 1959.

Dzens-Litovskii, A. I., Garkavi, M. Yu., and Elovskaya, L. V., Geology of the bottom deposits and the hydrochemical conditions in Kurguzul Bay and Sartas Gulf of Kara-Bogaz-Gol. Tr. Inst. geol. nauk AN Turkm. SSR, Ashkhabad, 1956_1.

Dzens-Litovskii, A. I., The Geological Past and Present Kara-Bogaz-Gol. Berg Readings, 1 and 2 [in Russian], Izd. AN SSSR, 1956_2.

Dzens-Litovskii, A. I., Vasil'ev, G. A., and Garkavi, M. Yu., Bottom Salt Deposits and Brines of Kara-Bogaz-Gol as Raw Material for the Salt Industry. Problems of Composite Use of the Mineral Wealth of Kara-Bogaz-Gol [in Russian], Izd. AN Turkm. SSSR, Ashkhabad, 1959.

Dzens-Litovskii, A. I., The past and present Caspian and its gulfs. Tr. Leningr. obsh. estestvoispytatelei, 73, No. 1, Izd. LGU, 1962_1.

Dzens-Litovskii, A. I., The Caspian Sea and Kara-Bogaz-Gol. Izv. VGO, No. 1, 1962_2.

Fedorov, P. V., Stratigraphy of the Quaternary deposits and developmental history of the Caspian Sea. Tr. Geol. inst. AN SSSR, No. 10, 1957.

Kurnakov, N. S., et al., The problem of Kara-Bogaz-Gol. Uspekhi khimii, 8, No. 2, 1938.

Leont'ev, O. K., History of shore development on Kara-Bogaz-Gol. Tr. Inst. okeanologii, 68, 1961.

Leont'ev, O. K. and Fedorov, P. V., History of the Caspian Sea in Late Khvalynian and post-Khvalynian time. Izv. AN SSSR, ser. geogr., No. 4, 1959.

Rikhter, V. G., History of the Kara-Bogaz-Gol spit. Dokl. AN SSSR, 105, No. 3, 1955.

Sedel'nikov, G. S., Hydrochemical conditions in Kara-Bogaz-Gol and the potential of the gulf for industrial use. Tr. Okeanografich. komis., 5, Izd. AN SSSR, 1959.

Sedel'nikov, G. S. and Lazareva, A. I., Bromine in the brines of Kara-Bogaz-Gol. Zh. "Neorg. khimiya," 9, No. 1, 1964.

Shnitnikov, A. V., Rhythms of the Caspian in post-Valdai time [in Russian], Report at the annual readings in memory of L. S. Berg, 1-3, 1956.

Reprinted from *Amer. Assoc. Petroleum Geologists Bull.*, **41**, 2467–2474 (1957)

MODERN EVAPORITE DEPOSITION IN PERU[1]

ROBERT C. MORRIS[2] AND PARKE A. DICKEY[3]
Lima, Peru, and Tulsa, Oklahoma

3

ABSTRACT

The Bocana de Virrila, located near Bayovar in the central part of the Sechura, illustrates principles of evaporite deposition. The Bocana, a marine estuary which normally lacks fresh-water inflow, extends about 20 kilometers inland and is about 2 kilometers wide.

Gypsum is being precipitated near the head, and halite at the extreme end of the Bocana. Black muds are found on the bottom. Life in the upper reaches of the Bocana includes red and green algae and insect larvae.

The high rate of evaporation of the lagoonal waters causes an increase in the degree of concentration of the various salts found in normal sea water. Chemical studies have shown a horizontal salinity gradient in which the concentration of total salts increases to more than 350,000 parts per million at the head of the estuary. A vertical salinity stratification was also noted near the head, where warmer, more saline waters are found on the bottom, below a cooler, less saline surface layer. It is believed that normal marine waters enter the Bocana at the surface to replace the lagoonal waters lost by evaporation, become more concentrated by evaporation and mixing, sink, and tend to escape seaward near the bottom. Physical and dynamic barriers inhibit the escape of the bottom brines.

INTRODUCTION

The Bocana de Virrila is located on the coast of northwestern Peru in the Sechura Desert (Fig. 1). It is a relict river channel which forms an estuary with an open connection to the Pacific Ocean. The topographic features of this ancient river bottom and the excessive aridity of the region have combined to form an interesting example of the deposition of evaporites in an arid climate. It is the purpose of this report to describe where and how these processes are taking place and to confirm some of the hypotheses postulated by other workers concerning the deposition of salts in a restricted estuary where evaporation exceeds precipitation plus runoff.

The report summarizes a reconnaissance investigation of this interesting area made by A. G. Fischer and the writers in August, 1955. No detailed study has ever been made.

DESCRIPTION OF CHANNEL

Beginning near the village of Sechura in the Sechura Desert near Paita, Peru, a broad area slightly lower than the surrounding flat Pleistocene beach deposits extends southward about 30 kilometers, encounters a ridge of flat Miocene shale beds, then curves westward out to the Pacific Ocean. During exceptional years when rain falls on the western slopes of the Andes Mountains, this whole lowland is covered by large sheets of water derived from the spill-over of the Piura River, which normally enters the sea west of Sechura. For example, during the rainy

[1] Manuscript received, May 6, 1957.

[2] International Petroleum Company.

[3] Carter Research Laboratory.

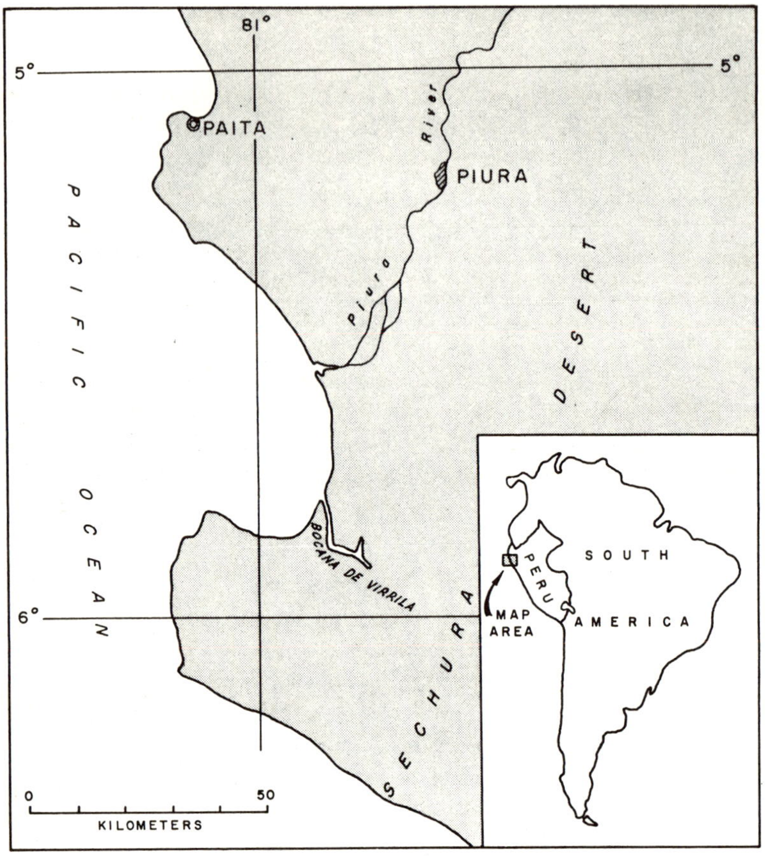

FIG. 1.—Location of Bocana de Virrila, Peru.

year of 1925, geological maps made at that time show the Piura River flowing along this course. However, during normal years, the upper part of this low area is generally dry except for a few shallow evaporating pans. Thin salt and gypsum crusts surround these areas. In the subsurface, gypsum has been observed in numerous shot holes at depths of 30 feet, presumably formed during recent time. Small mining operations for salt and gypsum have been carried out for years throughout this area. No vegetation is present. Although it is some distance from the sea, the area is very near sea-level.

Closer to the sea, the broad area becomes more constricted to form a channel that gradually falls below sea-level. This channel forms an estuary called the Bocana de Virrila, about 20 kilometers long and 2 kilometers maximum width. As this lower part of the channel contains a year-round supply of sea water, considerable quantities of salt and gypsum are being deposited at the present time.

DEPOSITIONAL ENVIRONMENTS

In accord with the classification used by Sloss (1953), the depositional environments of the Bocana have been classified as normal marine, penesaline, and saline. Each of these environments is quite distinct geographically and is represented by characteristic mineral associations and water compositions. The nature and location of the environments may change seasonally, and certainly change in those occasional years when there is abnormal rainfall on the coast of Peru. The present paper describes the condition as it was in August, 1955.

The area extending from somewhere just inside the mouth of the Bocana to a sill near the upper reaches of the lagoon has been designated as penesaline and is intermediate in salinity between normal marine and saline waters. According to Sloss (1953), the penesaline environment is characterized by deposition of evaporitic carbonates, interbedded with anhydrite. The waters from the Bocana are a milky, yellowish green. Water depths probably do not exceed 2 meters. Surface waters had a temperature of about 23°C. Throughout the penesaline environment, no indications of sodium chloride or calcium sulphate being deposited were to be found. Black muds covered the floor of the estuary near shore. Farther from shore, a white marl $\frac{1}{4}$–$\frac{1}{2}$ inch thick covered the floor of the estuary. Directly below were found soft black muds approximately 1 foot thick. This mud in turn overlies gypsum crystals. This gypsum might indicate that higher salinities had previously prevailed much closer to the mouth.

The saline environment of the Bocana occurs at the extreme upper end of the estuary. A small sill composed of several small islands and shallow sand bars greatly increases the physical restrictions of the dense, outflowing brines. This sill was located approximately where the two forks of the upper part of the Bocana meet. Just inside the sill, black muds are found on the bottom. The margins of small islands which occur in the area contain small gypsum crystals. Many of these crystals have a leached, indistinct appearance, as if they have partly gone into solution several times. This is probably due to fluctuations in the horizontal salinity gradient caused by tides, changes in rate of evaporation, or possibly some other reason.

Most of the halite and gypsum deposits are found at the extreme margins of the saline environment. Water depths do not exceed 1 foot. Large interlocking crystals of gypsum up to $\frac{1}{2}$ inch in length have been observed. The growth of these interlocking crystals has bowed up the floor of the estuary, in some places thrusting large folds of the crystals completely above water level. The transition zone between the principal gypsum and halite deposits is very indistinct as both halite and gypsum crystals appear the same when seen under shallow depths of water. The color of the water in the saline environment, especially in the upper reaches where the vast majority of the evaporites are being deposited, is a bright pink. This pink color is thought to be due to red algae in the water. The same color has been noted in other evaporating pans on the western coast of Peru.

The only forms of life noted in the saline environment are red and green algae and insect larvae. The temperatures in this environment ranged between 25° to 27°C., somewhat warmer than in the lower reaches of the Bocana.

COMPOSITION OF WATER SAMPLES

Analyses of six water samples taken from the different environments are shown in Table I. Sample locations are shown on the map (Fig. 2). An analysis of normal sea water taken from Clarke's "Data of Geochemistry" is also presented for comparison (Table I). The columns are arranged in order of increasing salinity from left to right. The absolute concentration of the ions is given in parts per million of total water sample. The relative amount of each ion is given in percentage of total solids.

Although a detailed sampling project of the entire estuary would be much more desirable, these preliminary figures readily bring certain facts to light concerning relative ion concentrations in the different environments.

Calcium ions show an increase in concentration in the penesaline environment and a decrease in the saline environment. Percentage-wise, however, calcium ion concentration drops steadily from normal marine to saline conditions. This is because the calcium is precipitated in the lower reaches of the estuary in the normal and penesaline environments, which increases the relative percentage of the more soluble ions, such as sodium, remaining in solution.

Sodium ions show a gradual increase in concentration until the innermost samples are reached; then a sharp decrease is noted. Sodium is not precipitated until the highly saline environment is reached.

TABLE I. CHEMICAL ANALYSES OF BOCANA DE VIRRILA WATER SAMPLES

Ion	*Normal Sea Water*	*Loc. E Surface*	*Loc. D Surface*	*Loc. C Surface*	*Loc. C Bottom*	*Loc. B Surface*	*Loc. A Surface*
	Ion Concentration Expressed as Parts Per Million of Water Sample						
K^+	380	2,314	2,600	1,426	1,912	6,645	9,962
Na^+	10,566	26,110	30,150	40,515	58,550	94,176	67,500
Ca^{++}	400	960	1,040	1,000	1,040	300	300
Mg^{++}	1,272	3,446	4,316	5,580	8,340	22,390	39,612
Cl^-	18,980	48,990	57,155	74,195	107,920	190,200	190,200
$SO_4^{=}$	2,649	6,868	7,828	9,567	13,489	31,216	46,417
HCO_3	140	175	173	171	245	551	948
Total	34,387	88,863	103,262	132,454	191,496	345,478	354,939
	Ion Concentration Expressed as Percentage of Total Solids						
K^+	1.1	2.6	2.5	1.1	1.0	1.9	2.8
Na^+	30.6	29.4	29.2	30.6	30.6	27.2	19.0
Ca^{++}	1.2	1.1	1.01	0.75	0.54	0.1	0.1
Mg^{++}	3.7	3.9	4.17	4.2	4.4	6.5	11.1
Cl^-	55.3	55.1	55.3	56.0	56.3	55.1	53.6
$SO_4^{=}$	7.7	7.7	7.6	7.2	7.0	9.0	13.1
HCO_3	0.40	0.20	0.17	0.13	0.13	0.16	0.27
Total	100.0	100.0	100.0	100.0	100.0	100.0	100.0
Sp. gr.		1.067	1.077	1.094	1.135	1.227	1.243
pH		7.4	7.6	7.8	6.65	7.3	7.1
Temperature		23°C.	25°C.	25°C.	27°C.	26°C.	27°C.

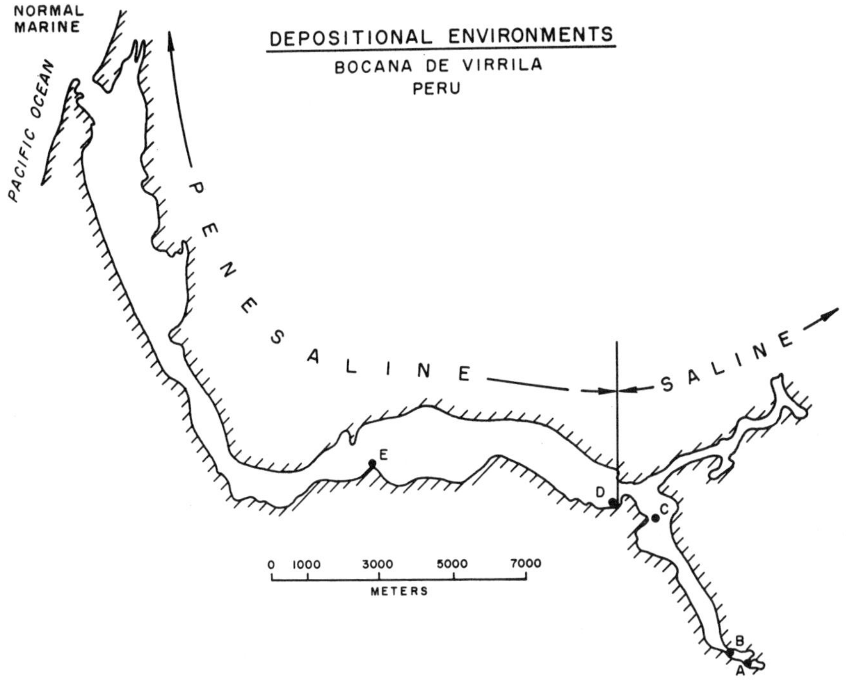

FIG. 2.—Depositional environments, Bocana de Virrila, Peru.

Magnesium ions show a slight increase in concentration until the innermost saline environment is reached, where a sharp increase is noted both in absolute concentration and relative percentage of total solids. Potassium ion concentration is fairly constant until the innermost end of the estuary is reached, where a sharp increase is apparent. The relative percentage is about the same in all samples. Thus neither of these ions is being precipitated appreciably under present conditions, but both are being concentrated by loss of less soluble ions and by evaporation. Still more restricted conditions would be necessary before they could be deposited.

Among the anions, the chloride ions increase in absolute concentration from the lower part of the estuary to the extreme upper end. However, their relative percentage first increases slightly as a result of the precipitation of calcium sulphate, and finally decreases markedly in the place where sodium chloride is precipitated. This comes also as a result of concentration of magnesium and sulphate ions in the last stages.

The sulphate ions show a steady rise in absolute concentration, increasing in the upper reaches of the estuary. There is a slight decrease in relative amount in the lower reaches where gypsum is precipitating. The bi-carbonate ions increase in concentration toward the upper end, but decrease in relative per cent.

Summarizing these results, we find that at first there is a simple increase in concentration of the individual ions above that of normal sea water. As precipitation of calcium carbonate, calcium sulphate, and sodium chloride takes place successively, the relative concentration of these ions decreases while other ions, such as magnesium, potassium, and sulphate increase. Should further concentration of the salts take place due to evaporation, magnesium sulphate and magnesium chloride would probably be precipitated, with the last bitterns being rich in potassium and chloride ions.

LITHOLOGIC DESCRIPTION OF CORES FROM BOCANA VIRRILA

At locality A in the upper reaches of the estuary, the stratified evaporite deposits were sampled to a depth of about 3 feet by driving 3-inch pipes into the sediments. The cores contain interstratified beds of halite, gypsum, and unconsolidated sandstone, indicating that both water level and salinity have fluctuated in the past. Following is a description of the cores.

Unit	*Thickness (Centimeters)*	*Description*
11	7.0	Halite, very light gray, with minor amounts (±5%) of silt (quartz and feldspar), dark organic material, and gypsum; halite is euhedral to subhedral and ranges in size from 0.3 to 2.0 mm.; some larger halite crystals contain inclusions of silt or organic material; organic material is unevenly distributed, concentrated in places along bedding or in isolated spots; bedding shows considerable disturbance due to downward thrust of core tube
10	1.0	Mixed layer of fine sand, gypsum and halite; dark gray color due to 15–25% content of organic material; sand consists mainly of quartz with minor plagioclase feldspar
9	11.0	Gypsum, light greenish gray, anhedral to subhedral grains 0.5–3.0 mm. in diameter; contains several 1-cm. layers of gypsum with 15–20% dark organic material and several pockets of soft greenish gray microcrystalline (.01 mm.) halite; X-ray shows trace quartz
8	2.0	Gypsum, solid crust of crystals, 1–5 mm. in length
7	2.0	Gypsum, acicular crystals 0.5–2 mm. in length with massive gypsum in center of unit; layers, 2 mm. in length with massive gypsum in center of unit; layers, 2 mm. thick, of greenish gray, microcrystalline halite at top and bottom of unit
6	6.0	Fine sand, yellowish brown; with 15–20% acicular gypsum crystals 0.5–1.8 mm. in length, and several pockets of soft, greenish gray; microcrystalline halite near base of unit; microcrystalline halite is also disseminated throughout with total content possibly 20–25%; sand is chiefly quartz with minor plagioclase and orthoclase feldspars; X-ray analysis shows trace illite
5	1.0	Gypsum, white, acicular crystals up to 1.0 mm. in length
4	8.0	Gypsum, greenish gray, acicular crystals 1.0–4.0 mm. in length, oriented largely in vertical position; lower 4 cm. is solid layer of vertical gypsum crystals 2–2.5 cm. in length
3	7.5	Fine sand, yellowish brown, well sorted, with microcrystalline halite distributed throughout (10–15% approx.); sand composed chiefly of quartz (subangular) —65–70%, gypsum (cleavage fragments)—10%, and plagioclase feldspar—5–10%; 5 cm. below top of unit is 3-mm. layer of fine sand with 40–50% acicular gypsum crystals 0.5–1.5 mm. in length

Unit	Thickness (Centimeters)	Description
2	7.5	Gypsum, greenish gray to white, acicular crystals 0.5–6.0 mm. in length; contains 10–15% subangular sand and coarse silt grains (mainly quartz-minor plagioclase) distributed throughout; contains several pockets of soft greenish gray microcrystalline halite with considerable silt (10–20%), less amounts of gypsum, illite, and trace of feldspar
1	17.0	Sand, yellowish brown, fine- to medium-grained, well sorted, subangular to subrounded grains; composed mainly of quartz grains (70–80%) with minor basic plagioclase feldspar (15–20%), halite (5–10%), and gypsum (1–5%)
		Bottom of core

PHYSICAL AND CHEMICAL PROCESSES

The order of the precipitation of the evaporites in the estuary conforms closely to Usiglio's order of precipitation. He has shown that calcium carbonate will be precipitated from solutions with salinities ranging from 72 to 199 parts per thousand. The gray marls from the floor of Bocana Virrila were found to be underlying waters with salinities ranging from 88 to 103 parts per thousand. Usiglio showed that calcium sulphate would be precipitated from waters with salinities ranging between 199 and 353 parts per thousand. A water sample taken above gypsum deposits in the estuary totalled 354 parts per thousand. Finally, Usiglio showed that sodium chloride would be deposited when the salinities reached 457 parts per thousand.* A water sample taken above halite deposits from the estuary contained a salinity of 355 parts per thousand. Thus, a good correlation exists between the laboratory data prepared by Usiglio and actual concentrations found above the various salts from Bocana Virrila.

The presence of black muds covering the floor of the estuary in the lower part of the saline environment is believed to furnish a valuable clue in explaining the frequent association of black shales with evaporites. Black muds from the Bocana Virrila were observed to occur below thin crusts of gypsum in the saline environment. In the transition zone between the penesaline and saline environments, black muds were observed on the floor of the estuary, and presumably are being deposited at the present time. The pH of the surface waters measured 7.8 while the bottom waters directly overlying the black muds were slightly acidic and measured 6.6. The acid conditions may be brought about by sulphate-reducing bacteria which reduce the sulphates to sulphides, thus freeing hydrogen sulphide gas which tends to acidize the water.

COMPARISON WITH OTHER LOCALITIES

Super-saline waters are not uncommon in many places throughout the world. The majority of those found along coasts, such as the Bocana Virrila, contain modified sea water. As the sodium chloride crystallizes out they become enriched in magnesium sulphate like that in the Bocana Virrila. Their actual chemical composition, however, varies rather widely, depending on local circumstances

**Editors' footnote:* Usiglio's experiment showed that halite *first* appeared at 353 parts per thousand, but only in conjunction with gypsum. At 457 parts per thousand, halite precipitates alone (Scruton, 1953, Table I)

and the contribution of water from other sources. The published analyses most closely resembling that of the Bocana Virrila are from the Gulf of Karaboghaz on the east side of the Caspian Sea (Clarke, 1924, p. 169) where a similar inflow from the sea area occurs, although on a much larger scale than here. The same disappearance of calcium, decrease in sodium chloride, and increase in magnesium sulphate is noted. The water of the Dead Sea is quite different, containing practically no sulphate and substantial amounts of calcium. The interior lakes of the Rocky Mountains and Great Basin areas of the United States are still different, containing sulphate and carbonate as the predominant anions and calcium in quantities comparable with sodium among the anions. These lakes were not derived directly from the evaporation of sea water.

The possibility that evaporite deposition could occur as a result of a countercurrent, with less dense brine entering from the sea at the surface, concentrating by evaporation, and returning seaward along the bottom was suggested by P. B. King in 1942 to explain evaporite deposition in the Permian of Texas and New Mexico. Scruton in 1953 pointed out that counterflows of this type had been observed repeatedly in estuaries and river mouths, and might have been a dominant factor in the deposition of evaporites. The counterflow has not yet actually been observed at the Bocana de Virrila, but it may be going on.

REFERENCES

CLARKE, F. W., 1924, "Data of Geochemistry," *U. S. Geol. Survey Bull. 770*, p. 125.

KING, P. B., 1942, "Permian of West Texas and Southeastern New Mexico," *Bull. Amer. Assoc. Petrol. Geol.*, Vol. 26, pp. 535–763.

KRUMBEIN, W. C., AND GARRELS, R. M., 1952, "Origin and Classification of Chemical Sediments in Terms of pH and Oxidation-Reduction Potentials," *Jour. Geol.*, Vol. 60, pp. 1–33.

SCRUTON, P. C., 1953, "Deposition of Evaporites," *Bull. Amer. Assoc. Petrol. Geol.*, Vol. 37, No. 11, pp. 2498–2512.

SLOSS, L. L., 1953, "The Significance of Evaporites," *Jour. Sed. Petrology*, Vol. 23, pp. 143–61.

TWENHOFEL, W. H., 1950, *Principles of Sedimentation*. 641 pp. McGraw-Hill Book Company, Inc.

Erratum: The King reference, above, should read: King, P. B., 1947, "Sedimentation in Permian Castile Sea," *Bull. Amer. Assoc. Petrol. Geol.,* Vol. 31, No. 3, pp. 470–477.

Reprinted from *Amer. Assoc. Petroleum Geologists Bull.*, **53**, 824–829 (1969)

A Modern Evaporite Deposit in Mexico[1]

4

FRED B PHLEGER[2]

La Jolla, California 92037

Abstract Marine evaporites are being deposited in the Ojo de Liebre Lagoon area, Baja California, in shallow diked basins flooded by high tides from an adjacent lagoon. Aragonite is deposited in the brine pans where daytime pH is high because of photosynthesis. Halite is deposited when wind transports brine into inner basins, and most of the gypsum is intermediate in location. Evaporites and brine support a very large standing stock of organisms.

Introduction

Extensive evaporites are present in marine rocks of various ages, and interpretation of these deposits is important to an understanding of the stratigraphic record. Borchert (1965), Scruton (1953), Sloss (1953), and others have advanced hypotheses to account for the accumulation of the salts. Most of the explanations require a constant flow of seawater into a barred basin or series of basins where it evaporates (Branson, 1915). There may be a fractionation of the salt by a return flow to the open ocean of the brines containing salts which have not precipitated (King, 1947); reverse current flow along the bottoms of some estuaries has been demonstrated (Pritchard, 1950, 1951).

A few workers have described evaporite deposits which are forming in the present cycle (*e.g.*, Morris and Dickey, 1957). This report describes some of the major features of a modern evaporite sequence at Ojo de Liebre Lagoon. Several reconnaissance examinations of the salt and brine pans in the area were made in connection with a study of the oceanography and sedimentology of the adjacent lagoons (Phleger and Ewing, 1962; Phleger, 1965). This preliminary report is primarily a general description of the lateral sequence of evaporites, the processes of flooding, the standing stock of organisms, and some of the sedimentary features. It is hoped that more detailed studies will be made in the area, especially of the chemical compositions of the surface and ground brines, of the mineralogy of the evaporites, and of their distribution beneath the surface

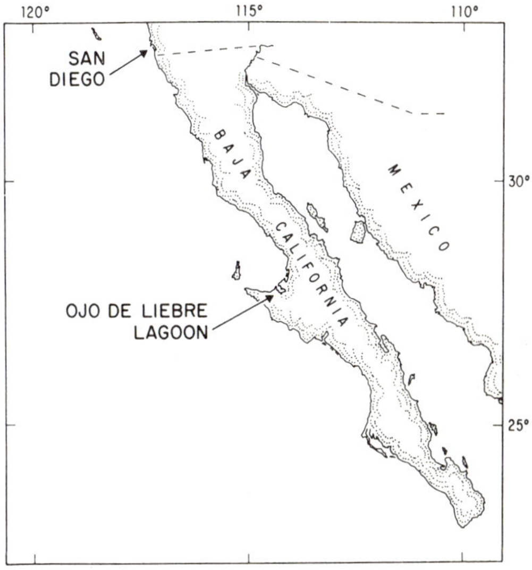

Fig. 1.—Location of Ojo de Liebre Lagoon area, Baja California, Mexico.

Study Area

The Ojo de Liebre Lagoon area, approximately midway on the Pacific coast of Baja California, Mexico, at about 28° N lat. (Fig. 1), consists of three coastal lagoons which have been fragmented from a single larger lagoon (Fig. 2). The lagoons are on the seaward edge of the Vizcaíno basin where the climate is arid, with an average yearly rainfall of about 2 cm, and where temperatures are moderate. Low clouds formed of the moist, relatively cool air of the Pacific generally are present from early evening until late morning, and the air humidity is somewhat higher than that of most deserts. There is no stream runoff in the area and no drainage pattern can be observed. On most days an afternoon wind blows onshore. Rates of evaporation are high.

[1] Manuscript received, September 26, 1968; accepted, December 23, 1968. Contribution from the Scripps Institution of Oceanography.

[2] Professor of Oceanography, Scripps Institution of Oceanography, University of California, San Diego.

This work was supported by the Petroleum Research Fund of the American Chemical Society and by the National Science Foundation.

Ojo de Liebre Lagoon extends inland approximately 65 km and is connected with the ocean by an inlet 12 m deep and more than 3 km wide. A relatively deep channel extends to near the head of the lagoon. Tide ranges average about 1.5 m and have a maximum of about 2.7 m at the lagoon inlet. Some maximum tide ranges appear to be greater in the inner lagoon than at the inlet. Tidally induced current velocities attain a maximum of 200 cm/sec at the inlet and decrease inward in the lagoon. Tides are at least 3 hours later in the inner lagoon than at the inlet, but there is some variation. The water at the head of the lagoon has a salinity almost twice that of the adjacent open ocean.

The evaporite and brine pans are on the border of the inner part of Ojo de Liebre Lagoon. They are separated from the lagoon by a low barrier of sand, shell, and mud built by wave action in the lagoon and assisted by deposition of sediment during high-tide overflow of lagoon water. Tidal channels extend through the barrier and invade the brine pans, and they are bounded by levees of sand and mud formed

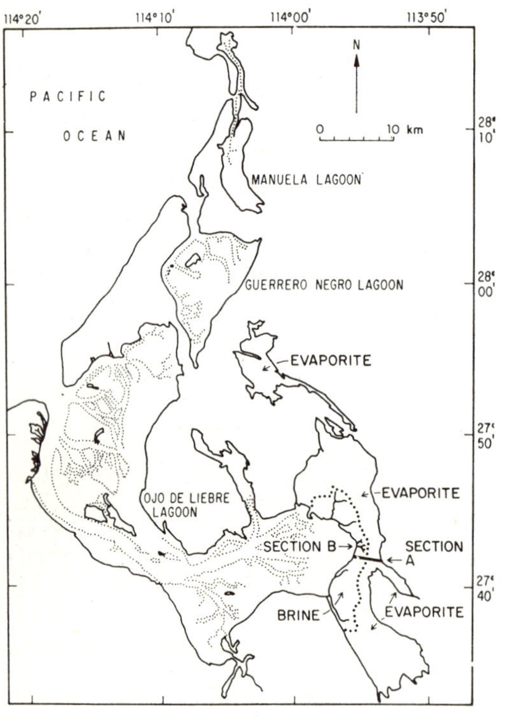

FIG. 2.—Ojo de Liebre Lagoon area. Shows locations of cross-sections A and B of Figure 3.

during overflow at high tide. The wave-built barrier and the channel levees form an effective dike system to contain water that overflows from the lagoon.

The lower part of the basin which is adjacent to the lagoon forms a brine pan that is covered with a layer of rubbery algal "pads" composed of filamentous algae and diatoms. A sparse marsh assemblage of *Salicornia* is present on the channel levees. The brine pans are always covered with brine of varying concentration a few to several centimeters deep. A very gentle upward slope inland restricts the distance which the brine can flood inland under the usual conditions. The brine and evaporite basins cover at least 275 km^2, of which the permanent brine pans occupy about 65 km^2.

The evaporite and brine pans consist of a series of connected old basins at the head of Ojo de Liebre Lagoon which have become essentially filled and are cut off from the main lagoon by the dike system. Fossil lagoon mollusks are present in the substrate of these old basins. Only basins containing the lower parts of the brine pans were formed during the Holocene cycle of the lagoon. The main basins containing modern sodium chloride are more than 30,000 years old and represent an earlier cycle, including one or more Pleistocene interglacials. The substrate of most of the evaporite basins contains an appreciable percentage of silt- and clay-size sediments because of the location within the lagoon system and the resulting low current velocities (Postma, 1965). This muddy substrate forms a relatively impervious foundation for holding brine, although in places it contains an appreciable amount of sand and is permeable, especially around the edges of certain individual evaporite basins.

EVAPORITES

Figure 3 shows the distribution of the principal evaporites in cross section. The algal pads in the brine pans have an actively growing surface 3–5 mm thick. Under the growing surface is a layer of light-colored, mud-size material 2–3 mm thick which was determined by x-ray diffraction to be composed entirely of small acicular crystals of aragonite. No other carbonate material was observed except that in mollusk shells. The aragonite appears to be precipitating from the brine, aided by the algal colonies. The measured density values of the brine within the algal-pad area are within the range of precipitation of calcium carbonate from concentration of seawater (Clarke, 1924).

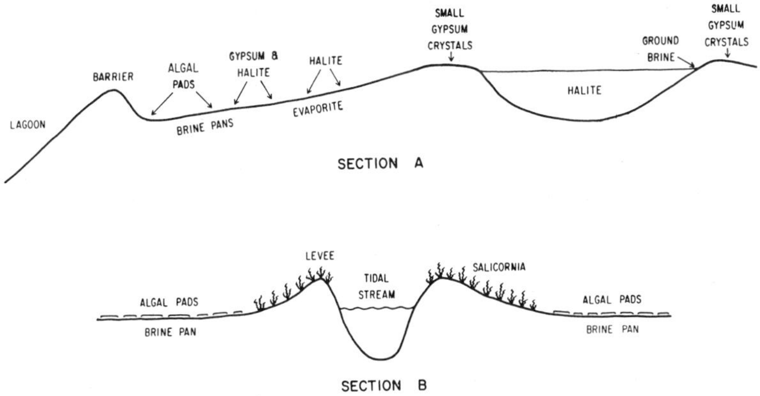

FIG. 3.—Diagrammatic cross sections of brine and evaporite. Traverse A extends from main brine barrier to inner halite basin. Traverse B is section across tidal channel in brine basin. Locations of cross sections shown on Figure 2.

The high pH induced by the growing algae also must affect the calcium carbonate precipitation. A pH as high as 9.2 was measured at the base of the growing algae during the afternoon. The pH of the lagoon water entering the brine pans is about 8.2 and within the normal range of seawater. The high pH within the algal colonies is related to the use of carbon dioxide during photosynthesis and facilitates the precipitation of calcium carbonate in the form of aragonite from the brine. This is believed to explain its concentration in close association with the algae.

The salinity of the brine in the brine pans varies through a wide range of values. The pans are flooded only during the higher spring tides which occur at about 2-week intervals, with one flooding over the barrier each day for 3 or 4 days. The lagoon water entering the pans has a salinity of about 5–6 percent, or almost twice that of open ocean water. This water mixes with residual brine to increase the salinity to about 8–9 percent, depending on the relative volumes of flood water and brine. The brine in the pans becomes further concentrated by evaporation before the next flooding over the dikes. Concentration occasionally must reach at least about one-tenth the volume of seawater before the next flooding, because a thin layer of sodium chloride commonly is precipitated on the algal pads (Clarke, 1924). The sodium chloride is redissolved when flooding next occurs and again becomes a part of the brine.

Inland from the algal pads is a large area where gypsum is common and halite also is present. There is no distinct boundary between this zone and the algal-pad zone, and the boundary varies with very minor elevation irregularities and with tidal phase. On the inner end of the zone halite is dominant, representing the more concentrated inner brine. The inner edge of the zone of gypsum and halite is the limit of the usual flooding by spring tides.

Inland from the gypsum and halite area, and at a slightly higher elevation, the principal halite deposits are in interconnected older lagoon basins. The halite is known to be as much as 2 or 3 m thick in the centers of some of the depressions where it has been penetrated by drilling, and it thins toward the edges of the basins. Deposits of small crystals of gypsum characteristically are present around the edges of these basins and considerable gypsum also is found within the halite. In places small gypsum crystals have been transported by wind to form low mounds. Considerable polyhalite also is reported by Holser (1966) from the halite area, much of it replacing gypsum crystals.

Flooding of Halite Basins

The halite basins are high enough that normal flooding by spring tides does not reach them. They are flooded only when high wind velocities are constant during a period of several days or weeks. If the wind is sufficiently prolonged, it will coincide with a spring tide which aids flooding of the halite basins. The writer observed one such flooding when wind was sustained for a period of about 3 weeks between a minimum velocity of about 30 kph and a maximum of more than 80 kph. Under these conditions the more concentrated brine was pushed up on to the halite basins where it was observed to be trapped; it eventually evaporated and a new layer of halite was precipitated.

The deposition of a 2–3 m thickness of halite with gypsum in the basins requires the equivalent of a column of seawater about 200–300 m deep at normal oceanic salinity. The flooding, which is observed in the brine pans and in the halite basins, is very shallow; the water may reach a maximum depth of about 25 cm, but usually is much shallower. The salt concentration in brine which reaches the halite basins is at least 10 times that of seawater, because of concentration by evaporation and also redissolution of sodium chloride en route to the salt flats. Twenty-five centimeters of brine thus would be the equivalent of about 2.5 m of seawater at normal salinity.

Flooding of the inner halite areas must be infrequent because it is believed that the prolonged wind necessary for transport of the brine is not common. About 100 flooding events with a total delivery of 25 cm of concentrated brine to the basins would be required for the accumulation of 3 m of halite in the inner halite areas. It has been estimated that the sea reached approximately its present level in the Ojo de Liebre area about 2,000 years ago (Phleger, 1965). The average frequency of the storm tides which cause flooding of the halite basins therefore is not more than about once in 20 years.

At that rate, a deposit of halite 300 m thick could accumulate in about 200,000 years. It appears, however, that the inner basins in the Ojo de Liebre area soon will be filled with halite to an elevation at which a significant amount of flooding with brine cannot occur. A marked increase in thickness of the salt would require either a rise in sea level or a depression of the land.

The brine and evaporite areas seem to be essentially a closed system retaining all the salts present in seawater. This is indicated partly by the presence of polyhalite (Holser, 1966). Possibly other magnesium and potassium minerals are present as well. The entrapment of most or all of the bitterns also is indicated clearly by the analysis of the ground brine under the surface of one of the salt flats (Holser, 1966), which shows a marked enrichment of magnesium, potassium, and bromine (Table I).

No return of brines to the lagoon could be detected, although a minor amount of seepage may occur. No evidence was discovered of a possible return flow of saline water out of the lagoon, as postulated to obtain segregation of the evaporite phases. Apparently the rate of exchange of water from the inner part of Ojo de Liebre Lagoon to the open ocean is very slow and is not an important process.

Table I. Composition of Ground Brine under Ojo de Liebre Salt Flat and of Seawater
(percent)

Element	*Ground Brine* (*After Holser, 1966*)	*Seawater* (*After Sverdrup et al., 1942*)
Na	16	30.6
K	4	1.1
Mg	16	3.7
Ca	0.03	1.2
Cl	62	55.3
Br	0.8	0.19

Organic Production

One of the striking features of the Ojo de Liebre brine and evaporite area is the very large standing stock of organisms. It is evidenced by the algal pads which form a continuous surface in the brine pans. The *Salicornia* assemblage which fringes the flood channels contains an unusually large standing stock of Foraminifera. They are mostly forms with calcareous tests, in contrast to those with arenaceous tests in many other marine marshes, a characteristic that probably reflects the high-pH and high-salinity environment in which they are living. Foraminifera are not present elsewhere in the brine and evaporite areas. The water in the brine pans contains a very dense population of chlorophytes (Carpelan, 1957) which color the brine red under proper conditions of concentration. Algae also are abundant within the halite, apparently being concentrated at the upper level of the ground brine within the salt a few centimeters beneath the surface, where they color the halite a bright green or red.

It is now well known that hypersaline lagoons support an unusually great organic productivity (Phleger, 1960; Phleger and Ewing, 1962). It is also well known among commercial producers of solar salt that the salt supports an abundant bloom of algae. Commercial salt must be washed with brine after harvesting to remove the considerable amount of organic debris contained in it.

The great organic productivity in areas where seawater is concentrated by evaporation can be explained, at least in general terms. When seawater is concentrated, the nutrients present within it also are concentrated. In a hypersaline lagoon there is a constant resupply of seawater to replace that lost by evaporation, and the new water contains additional nutrients. The nutrients in a brine are highly concentrated, and replenishment to the evaporite area is by tidal and wind-induced flooding. It seems possible that evaporite areas are important suppliers of source materials for petroleum.

Additional Observations

Certain surface features of the halite deposits are of interest. The surface salt characteristically has an irregular 30–40 cm polygonal pattern. The cause of the pattern is not apparent. The edges of the polygons are raised a few millimeters above the level of the surface, apparently because of upward migration of brine along cracks and deposition of sodium chloride at the surface. Large quantities of the shallow-water plant, *Zostera,* are transported from the inner lagoon into the evaporite area during flooding. Some of the clumps of *Zostera* are distributed in patterns at right angles to the prevailing wind direction. There are also "windrows" on the evaporite of a marsh gastropod, *Melampus,* which is transported by the flood waters for several miles.

Abundant evaporites have been found to a depth of 200 m in the cuttings from a water well near the northern edge of the Ojo de Liebre halite flats. These subsurface evaporites may be deposited by interstitial circulation of brine and precipitation of salt within the sediment. They also may represent older deposits that were precipitated when relative sea level was lower than it is today.

Discussion

The modern Ojo de Liebre evaporite represents one type of marginal marine deposit for which seawater is the source of the salts. The processes and conditions in this area may clarify some problems concerning older deposits. Perhaps the most interesting features are that the salts are precipitated from a thin layer of brine which is resupplied from the lagoon at partly frequent and partly infrequent intervals, and that there is an imperfect lateral segregation of the evaporites.

It has been suggested that some older evaporites were deposited in "deep basins." At the International Conference on Saline Deposits in 1962, evidence for "deep-water basins" was suggested by the Deposition and Geochemistry Work Sessions (Ingerson, 1968). It was the opinion of that group that "bituminous matter" which is associated with some evaporites would be preserved only in deep water because "the indicated reducing conditions could not be maintained in a shallow basin" (Ingerson, 1968, p. 674). The evidence at Ojo de Liebre does not support this suggestion. The muds beneath the brine and evaporite are strongly reducing, as indicated by a very high content of hydrogen sulfide. In addition, a concentrated brine is itself an excellent preservative for organic detritus, as evidenced by the presence of many insects, birds, lizards, and *Zostera* leaves completely preserved within the salt. Both the salt and the substrate contain abundant organic detritus.

It was argued that very thin clay layers almost certainly would be broken and partly destroyed in shallow water before they could be covered and preserved (Ingerson, 1968, p. 674). The finest grained sediments in a lagoon system where tidal currents are the transporting agent are concentrated in the inner part of the lagoon, because velocity of the transporting current is least at that location (Postma, 1965). The brine and evaporite basins are only an inner extension of the lagoon system, and the water furnished them during spring tides contains mostly silt- and clay-size particles. The water from the lagoon invades the basins at a very low velocity and no disturbance of the substrate over which it advances has been observed during flooding. Transport of *Zostera* leaves, *Melampus* shells, and other large materials is by flotation in the slowly advancing water.

Ripple marks, as ordinarily defined, were not observed in the Ojo de Liebre evaporites. There are no loose mineral particles to be transported by waves or currents into ripples, even with appreciable brine on the halite flats. The small crystals precipitating out of the brine appear to become fused with the saline substrate very rapidly. All the observed "windrows" were associated with clumps of *Zostera* or *Melampus* cemented together by the salt.

A considerable amount of dust and sand is transported by wind into the Ojo de Liebre area. The surface of the halite flats becomes covered with sand and dust during the long intervals when they receive no brine. The salt precipitated from the brine when flooding occurs covers this layer of wind-borne sediment, which is preserved within the halite. During very infrequent heavy rains, water accumulates on the surface of the halite basins. It dissolves the surface salt, and the sand and silt sink through the brine thus produced. Evaporation eventually causes precipitation of the salt as a sediment-free surface layer. The production of a thin layer of sediment by either of these processes is infrequent.

Estimates of high rates of evaporite deposition have been based on "varved" sequences (Ingerson, 1968, p. 674). The high rate has been given as a reason for requiring a deep basin to allow a thick series to accumulate in a relatively short time. However, it appears that a "varve" does not necessarily mark a yearly event and may be produced infrequently, and that thick sequences of varves may represent a much longer time than has been estimated.

References Cited

Borchert, H., 1965, Principles of oceanic salt deposition and metamorphism, *in* Chemical oceanography, v. 2: New York and London, Academic Press, p. 205–276.

Branson, E. B., 1915, Origin of thick gypsum and salt deposits: Geol. Soc. America Bull., v. 26, p. 231–242.

Carpelan, L. H., 1957, Hydrobiology of the Alviso Salt Ponds: Ecology, v. 38, p. 375–390.

Clarke, F. W., 1924, The data of geochemistry, 5th ed.: U.S. Geol. Survey Bull. 770, 841 p.

Holser, W. T., 1966, Diagenetic polyhalite in recent salt from Baja California: Am. Mineralogist, v. 51, p. 99–109.

Ingerson, E., 1968, Deposition and geochemistry work sessions, *in* Saline deposits: Geol. Soc. America Spec. Paper 88, p. 671–681.

King, R. H., 1947, Sedimentation in Permian Castile Sea: Am. Assoc. Petroleum Geologists Bull., v. 31, p. 470–477.

Morris, R. C., and P. A. Dickey, 1957, Modern evaporite deposition in Peru: Am. Assoc. Petroleum Geologists Bull., v. 41, p. 2467–2474.

Phleger, F. B, 1960, Foraminiferal populations in Laguna Madre, Texas: Tohoku Univ. Sci. Repts., 2d ser. (Geology), spec. vol. 4, p. 597–605.

——— 1965, Sedimentology of Guerrero Negro Lagoon, Baja California, Mexico, *in* Submarine geology and geophysics, Colston Research Soc. 17th symposium: London, Butterworth (Colston Papers), v. 17, p. 205–237.

——— and G. C. Ewing, 1962, Sedimentology and oceanography of coastal lagoons in Baja California, Mexico: Geol. Soc. America Bull., v. 73, p. 145–182.

Postma, H., 1965, Water circulation and suspended matter in Baja California lagoons: Netherlands Jour. Sea Research, v. 2, p. 566–604.

Pritchard, D. W., 1950, Notes on the dynamics of estuarine waters: Woods Hole Oceanographic Inst., Ref. no. 50–37, p. 49–60.

——— 1951, The physical structure, circulation and mixing in a coastal plain estuary: Ph.D. dissert., California Univ. Scripps Inst. Oceanography, 60 p.

Scruton, P. C., 1953, Deposition of evaporites: Am. Assoc. Petroleum Geologists Bull., v. 37, p. 2498–2512.

Sloss, L. L., 1953, The significance of evaporites: Jour. Sed. Petrology, v. 23, p. 143–161.

Sverdrup, H. U., M. W. Johnson, and R. H. Fleming, 1942, The oceans: New York, Prentice Hall, Inc., 1087 p.

Reprinted from *Amer. Jour. Sci.*, **255**, 561–567 (1957)

5

DOLOMITE SEDIMENTATION IN THE SOUTH-EAST OF SOUTH AUSTRALIA

A. R. ALDERMAN and H. CATHERINE W. SKINNER

ABSTRACT. Fine dolomite-calcite sediments of Pleistocene to Recent age occur over a wide area in the South-East province of South Australia. Similar sediments are being formed in a number of saline lakes and in a shallow inlet of the sea. The origin of the dolomitic sediments is ascribed to elevated pH caused by plant growth in shallow saline water.

INTRODUCTION

The physiography of the South-East province of South Australia is dominated by a series of ranges of consolidated sand dunes. They trend NNW parallel to the present coast and are separated by long flat interdunal areas. The ranges represent former coastal dunes left stranded during Quaternary time by the retreating sea. The latest phase of this is to be seen where the Coorong, a long and narrow stretch of sea water, is separated from the open sea by Younghusband Peninsula, a strip of sand dunes. At its southern end a former extension of the Coorong is now represented by a series of lakes and lagoons from which the sea has retreated and which indicate the similar origin of the interdunal areas further inland.

The surface rocks of the region are predominantly carbonates. Limestones of Early to Middle Tertiary age, which are partly dolomitic, outcrop over wide areas in the south and apparently underlie the Pleistocene and Recent marine and freshwater sediments and dune rocks of the central and northern parts of the province. The widespread and prominent dunes consist of varying proportions of quartz sand and shell fragments and are mostly capped by a layer of kunkar limestone. The geology of the region has been discussed by Hossfeld (1950) and Sprigg (1952).

The main drainage of the whole region is underground. Meteoric waters disappear very rapidly in the dunal areas either by penetration or by surface run-off from the kunkar. The flat interdunal areas are however relatively impervious and lead to the development of swamps and shallow lakes after rain. Some lakes are permanent but others are shallow and generally become dry during the summer season.[1] Water plants, the most abundant being *Ruppia maritima* Linn., and a fauna of which copepods and small molluscs are the most obvious members, regenerate in the intermittent lakes during the winter and populate the permanent lakes throughout the year.

The sediment from one of these intermittent lakes, situated 10 miles north of Kingston and referred to in this paper as "Kingston Lake," was found by Sir Douglas Mawson (1929) to consist largely of dolomite. A recent examination of similar material from the same locality shows that it is a very fine aggregate of dolomite and calcite with some halite and small amounts of quartz and clay minerals. Sediments of this nature which are extremely fine-grained and possess marked plastic properties when wet have been locally

[1] The average annual rainfall varies from 15 inches in the north to 34 inches in the south. Kingston has an average of 24.3 inches. Average maximum temperatures are highest, 80°F., in February and lowest, approximately 40°F., in July.

known as "pipe-clay". During arid summers this carbonate sediment forms a hard white surface on the dry beds of the lakes, especially those which were formerly the southward continuation of the Coorong. Similar sediment also occurs in the southernmost parts of the Coorong itself and is exposed during dry summers when, at the head of this long and narrow inlet of the sea, water-level is lowered by evaporation.

Our investigations were concentrated on the sedimentation in Kingston Lake. Similar sediments have been found between the dune ranges in an area which extends over many hundreds of square miles. Morover, Rogers, Quirk and Norrish (1956) have described dolomite nodules and dolomite-aluminian-sepiolite marl below a surface soil of aluminian-sepiolite, montmorillonite, illite, kaolin, quartz and dolomite from a depression at Tintinara, South Australia, 60 miles NNE of Kingston.

Other examples of plastic dolomitic sediments have been recorded in areas remote from the South-East province. A specimen obtained from Eastern Cove on Kangaroo Island, 140 miles NW of Kingston, has similar composition and properties and seems to have been formed as a shallow water marine sediment. At the base of an evaporite sequence in the bed of Lake Eyre (nearly 600 miles north of Kingston), C. W. Bonython (1956) has recorded beds of dolomite which also contain small amounts of palygorskite, kaolinite, illite, quartz and gypsum. It appears that the occurrence of dolomite in the southeast part of South Australia is not of restricted importance. Direct comparison can be made with beds of similar material formed in comparatively recent times over a very wide area.

COMPOSITION OF THE SEDIMENTS

A typical sample of the dolomitic sediment is white with a grey tinge when dry. Microscopic examination shows an aggregate of particles most of which are so fine that, apart from the fact that they are birefringent, ordinary optical examination gives very little information. Larger particles of sandy and shelly material as well as organic fragments make up varying proportions of each sample. The carbonates are always very fine-grained and were identified by X-ray powder diffraction. Dolomite, even in the surface samples, has the true dolomite structure (space group R3). Dolomite is always accompanied by calcite but calcite may occur without dolomite. Some X-ray powder diffraction lines are very broad. Intensity profiles of these lines are highly asymmetric and work is proceeding on them. The unit cell of the calcite is generally smaller than that of pure calcite, suggesting solid solution of magnesium in the calcite. The dolomite generally has a slightly larger unit cell than pure dolomite. Further X-ray analyses are in progress.

Six samples collected from widely spaced localities and at various depths were selected to illustrate the general mineral and chemical composition of the carbonate-bearing sediments. Compositions are given in table 1 and localities in figure 1. Samples A. 78.1, A. 78.17, D 41 and USE 7-4 can be described as white, extremely fine-grained carbonates cementing quartz grains and organic fragments. D 45 and W 6-4 are dark-gray quartz sands with discrete carbonate

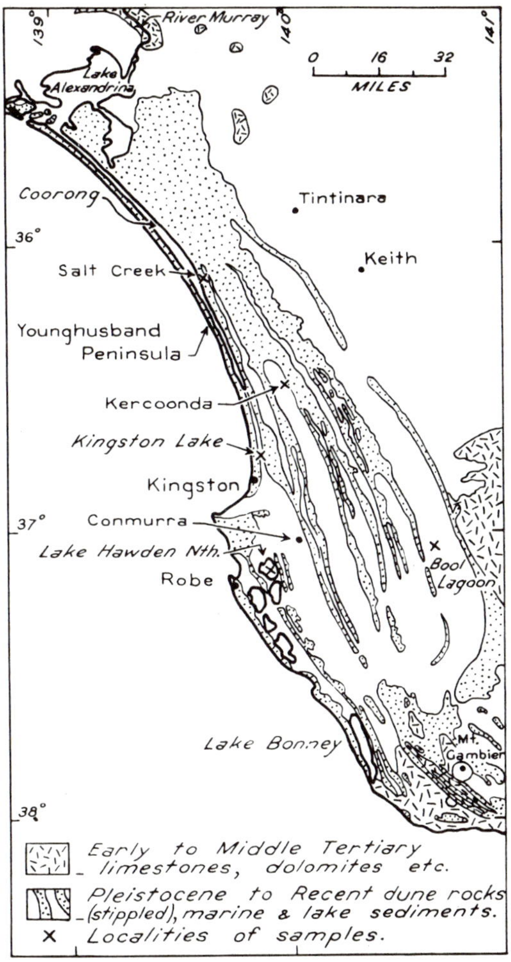

Fig. 1. The South-East of South Australia. Based on S. A. Dept. of Mines geological map, 1953.

grains and many shell fragments. All samples contain minor amounts of clays and halite as well as dolomite and calcite.

Kingston Lake can be taken as typical of many intermittent lakes. Post hole bores put down in this lake and Lake Hawdon North showed 2 to 4 feet of fine carbonate sediment, largely dolomite, above shelly beds and dark sulphur-bearing muds. Fine white sediment occurs on the surface of the Kingston Lake bed at the end of the summer. An annual sedimentation cycle begins when the winter rain accumulates to a depth of 1 to 2 feet in the lake, and plant and animal life begins to reappear in the water. As the weather becomes warmer, water plants growing on the lake floor increase in size and number. It is at this time (November and December) that the lake water may become turbid with fine white sediment which is essentially a mixture of

Table 1
Composition of Carbonate Sediments

	A78.1	A78.17	D.41	D.45	USE 7-4	W 6-4
SiO_2	8.24	6.86	19.42	29.00	17.24	54.34
Al_2O_3	1.11	1.21	1.21	2.28	1.19	13.92
Fe_2O_3	0.81	0.29	0.25	0.40	0.43	4.44
CaO	28.61	34.77	28.61	31.78	30.70	3.11
TiO_2	—	—	—	—	—	0.56
MgO	10.64	11.22	9.42	1.34	10.83	3.75
CO_2	33.26	39.39	32.50	26.05	35.90	6.43
SO_3	1.46	0.27	0.43	0.63	0.02	0.19
NaCl	7.88	1.01	4.29	5.18	0.07	0.48
Total Water	5.64	2.67	1.84	2.26	2.96	12.06
Acid Insoluble Organic Matter as Carbon	1.28	0.34	0.17	0.57	0.42	0.59
	98.93	98.03	98.14	99.49	99.76	99.87

SAMPLE A78.1	10 miles south of Salt Creek (Coorong). Surface layer. Calcite and dolomite in approximately equal proportions.
SAMPLE A78.17	Lake Hawdon North. 12-18 in. depth. Dolomite dominant carbonate mineral.
SAMPLE D.41	Kingston Lake. 1½-6 in. depth. Dolomite dominant carbonate mineral.
SAMPLE D.45	Kingston Lake. 33-36 in. depth. Calcite dominant carbonate mineral.
SAMPLE USE 7-4	Kercoonda. 4-7 in. depth. Calcite dominant carbonate mineral.
SAMPLE W 6-4	Bool Lagoon. 24-36 in. depth. Calcite and dolomite in approximately equal proportions.

calcite and dolomite. The fine precipitate settles onto the lake floor. During the summer the salinity of the lake increases due to high evaporation and low rainfall. The last remnant of the water deposits a crust of salt which, as it dries, forms a loose layer above the dolomite. Most of the dolomite-forming lakes are merely slightly lower areas in broad flat plains and from these the salt layer is removed by the prevailing westerly wind. Where, much more rarely, the lake bed is protected from the wind by trees and sandhills the salt accumulates. In lakes where dolomitic sediment has been forming for very many years the proportion of salt in the consolidated sediment is quite minor even though, as we will see, sodium and chlorine are overwhelmingly the most abundant constituents of the water in the present lakes.

A very similar cycle must prevail in the shallow sea waters at the head of the Coorong. Water level in the Coorong must be mainly controlled by two factors—first, the seasonal rise during the wet winter and the corresponding fall due to evaporation in the summer, and secondly, by the amount of water flowing out of the River Murray. A large volume of water flowing from the Murray mouth will bank up water at the open end of the Coorong and keep the level high. A small flow from the Murray will allow a generally lower level in the Coorong. Variations in these two factors cause differences from year to year but in an average year water-level in the Coorong is high in the winter and low in the summer. In a normal summer, lowering of water-level exposes the bed of the Coorong for several miles at its southern end. The sedimentation sequence must be very similar to that in Kingston Lake. Any salt crystallizing on the surface of the dolomite is removed either by wind erosion or is redissolved as the water rises during the winter season.

The interdunal areas, which seem to be of the nature of former "Coorongs", may show more than one sequence of muds, shelly beds and the fine precipitated carbonate sediment. A drain near Conmurra shows two such sequences separated by a layer of kunkar limestone, the whole section having a thickness of 8 feet. Such sedimentary cycles must have been frequent in the South-East province during Quaternary time.

Partial analysis of a sample of suspended sediment shows: CaO, 39%; MgO, 31%; CO_2, 30%. X-ray data show calcite and dolomite in approximately equal proportions. The sediment (SW 15) in the sample studied was present to the extent of about 0.05 gms in 100 ml of lake water or about 2 percent of the total saline content. As will be seen later the amount of suspended sediment varies from place to place in the lake water.

The composition of natural waters from the Coorong and from Kingston Lake is given in table 2. The Kingston Lake water is typical of water from many similar sources in the southeast. For comparison we give analyses of what can be taken as a typical groundwater in the province (Drain L) and two analyses of sea water, one from the South Australian coast near Adelaide, the other being Sverdrup's (1942) world average.

These analyses show that the water in Kingston Lake has essentially the composition of sea water in various degrees of dilution and concentration. The Coorong water has its saline constituents in similar proportions. Drain L and tributary drains were constructed to lower the water table over a very wide area in the southeast province so that its water can be taken as typical of the groundwater of the region. It differs markedly both in salinity and composition from the lake and sea waters. The high proportions of Na and Cl in all the waters of the region are undoubtedly due to cyclic salt blown inland from the sea by the prevailing westerly winds. The water which accumulates to form lakes in the interdunal flats is rainwater which has dissolved the cyclic salt dust either in the air or distributed over the land. It is from such water, which in composition is essentially diluted or concentrated sea-water, that dolomite precipitates in Kingston Lake, e.g. analysis 5, table 2.

ORIGIN OF THE SEDIMENTS

Microscopic examination of the suspended sediment (sample SW15) referred to above does not show the presence of any structures which would indicate that the sediment had been formed by an organism, although a small proportion of faecal aggregates suggest ingestion by small crustacea. Electron micrographs of this sediment have been made by Mr. J. Farrant of the Chemical Physics Section, C. S. I. R. O. Division of Industrial Chemistry. These showed the presence of well formed rhombohedra of up to 20 microns cross section, but there was no morphological evidence of organic origin.

TABLE 2

Saline Content of Water Samples

	Coorong 3½m. S.E. Salt Ck.			Kingston Lake			Sea Near Adelaide	Sea Sverdrup	Drain L Near Robe		
Month	June	Nov.	Jan.	June	Nov.	Jan.			June	Nov.	Jan.
No.	1	2	3	4	5	6	7	8	9	10	11
Cl	55.9	55.8	55.8	56.0	55.9	55.3	55.5	55.04	48.5	49.9	50.1
SO_4	7.5	7.6	7.6	6.3	6.2	7.5	7.9	7.68	9.5	8.7	9.6
CO_3	0.1	0.2	0.1	0.3	0.5	0.1	0.2	0.41	6.1	5.2	4.8
Na	31.8	31.9	31.7	34.3	34.1	34.2	32.2	30.61	27.3	29.0	28.3
Ca	1.0	0.9	1.0	0.8	0.6	0.4	1.1	1.15	4.0	2.8	1.3
Mg	3.7	3.6	3.7	2.3	2.5	2.6	3.8	3.69	4.6	4.4	5.8
% Salinity	7.05	4.7	5.4	3.33	2.44	22.6	3.8	3.5	0.27	0.25	0.28

Our observations have shown that carbonate precipitation can take place in waters which have only half the salinity of average sea water. Precipitation of carbonates in such solutions could be brought about by a rise in pH. The presence of vigorous plant life in water can have a very considerable effect on the pH of the solution. Baas-Becking (1934) has shown that the extraction of carbon dioxide from lake water by plants during photosynthesis can raise the pH to 9.3 in strong sunlight. The pH may fall to below 8 at night as the water absorbs CO_2 from the air.

That there is a close relation between plant growth, rise in pH, and precipitation of dolomitic sediment in Kingston Lake seems certain. Much work still remains to be done before more complete details of the processes are known. More or less continuous records of pH changes and correlation with precipitation and such factors as sunlight, temperature and salinity appear to be necessary. However, our observations to date have shown that precipitation of dolomitic sediment takes place during the spring and early summer when plant growth is in its most vigorous phase. It has also been observed that during sedimentation the water is most turbid, i.e. the precipitate is most plentiful, where plant masses are most plentiful. Such water has shown a pH of 9.2, although it has not yet been possible to determine pH maxima. The normal figure for these waters in equilibrium with air is 8.2-8.4. It has been mentioned earlier that the most common plant in the lakes and obviously the one most responsible for pH changes is *Ruppia maritima* Linn., although sedges and small algae are also present.

The association described here of shallow saline water, plant growth, elevated pH and dolomitic sedimentation has undoubtedly, we believe, been effective in an otherwise insignificant salt lake near Kingston. We see no reason to believe that this association is not of widespread occurrence. We know that dolomitic sediments of this nature have been formed over hundreds of square miles in the southeast province. We know that identical sediments are forming in sea water in the Coorong today. Admittedly the Coorong is an unusually attenuated gulf but similar conditions could presumably develop in many places where areas of shallow sea water are relatively undisturbed and the climate is suitable for vigorous plant growth.

Although the associations described in this paper may have wide implications, the limited nature of the results so far achieved is freely admitted by the authors. Much more detailed continuous observations are necessary at Kingston Lake which however provides an excellent subject as it has some of the characters of a "closed system" with a well-defined annual cycle. A study of the sediments and sedimentary processes in the Coorong is also necessary and could keep many investigators employed for a long time. However, it could easily supply information of the utmost importance. We still do not know the precise conditions under which dolomite forms in such precipitates in preference to one of the forms of calcium carbonate. Perhaps these dolomitic sediments will provide a solution to this problem.

ACKNOWLEDGMENTS

Our thanks are due to Dr. B. J. Skinner, Professor J. G. Wood, Dr. H. B. S. Womersley, Professor E. A. Rudd and Sir Douglas Mawson for their interest and help both in field and laboratory; to the Director and the Chief Analyst of the South Australian Department of Mines (Mr. T. R. Frost) for analyses of water samples and sediments; Mr. John Hutton, C.S.I.R.O. Division of Soils for anlysis of suspended sediment; Mr. J. Farrant, C.S.I.R.O. Division of Industrial Chemistry for making electron micrographs; and to Dr. B. G. Forbes who collaborated in the early stages of this investigation.

REFERENCE

Baas-Becking, L. G. M., 1934, Geobiologie: The Hague, W. P. Van Stockum & Son.

Bonython, C. W., 1956, The salt of Lake Eyre—its occurrence in Madigan Gulf and its possible origin: Roy. Soc. S. Aust. Trans., v. 79, p. 66-90.

Hossfeld, P. S., 1950, The late Cainozoic history of the South-East of South Australia: Roy. Soc. S. Aust. Trans., v. 73, no. 2, p. 232-279.

Mawson, D., 1929, South Australian algal limestones in process of formation: Geol. Soc. Lond. Quart. Journ., v. 85, no. 4, p. 613-621.

Rogers, L. E. R., Quirk, J. P., and Norrish, K., 1956, Occurrence of an aluminum-sepiolite in a soil having unusual water relationships: Jour. Soil Sci., v. 7, no. 1, p. 177-184.

Sprigg, R. C., 1952, The Geology of the South-East Province, South Australia: Geol. Surv. S. Aust., Bull. 29.

Sverdrup, H. U., Johnson, M. W., and Fleming, R. H., 1942, The Oceans: New York, Prentice-Hall.

School of Geology
University of Adelaide
South Australia

Editors' Comments on Papers 6 Through 12

Holocene Evaporites: Supratidal Flats

No studies have so improved our understanding of the formation of evaporites as have those of the supratidal flats of Laguna Madre, Baja California, and the Trucial Coast. As a result of those investigations, it is now known that there are occurrences of gypsum and anhydrite that are identical in every detail to some ancient deposits whose origin had hitherto been difficult to explain.

The first study of a supratidal evaporite occurrence was that by Masson (1955), whose work is reproduced in this part. Below the sediment surface of the Laguna Madre mudflats on the southeastern coast of Texas, thin beds and crystals of gypsum have formed from lagoonal water blown intermittently across the supratidal flat and concentrated by evaporation within the vadose zone, a conclusion confirmed in later work by Fisk (1959).

In the second paper in this part, Kerr and Thomson (1963) add their own observations to the earlier work on the Holocene deposits of Laguna Madre, and use them in an attempt to understand some features of ancient evaporites. From the belief that individual crystals and rosettes of gypsum found in Laguna Madre sediments would have their ancient counterpart in nodular anhydrite, Kerr and Thomson conclude that the environment of the arid supratidal flat is one which may have produced many deposits of anhydrite and gypsum.

These conclusions are important and were reinforced by very detailed research into the chemical and sedimentological regime that prevailed during formation of the evaporites in the now-classic sabkhas of the Trucial Coast of the Persian Gulf (now called the "United Arab Emirates"). In April 1962 Wells reported the occurrence of Holocene dolomite in supratidal flats around the Qatar Peninsula and in February of the following year, a short but significant note by Curtis, Evans, Kinsman, and Shearman described the presence of considerable amounts of anhydrite within the supratidal flats of the Persian Gulf. That paper is included here for its historical importance because it stimulated other investigations into the nature of the depositional processes proceeding within sabkhas, and because it showed how the vexed question of the equilibrium relations between gypsum and anhydrite could now be viewed with some reference to actual, as opposed to theoretical, models. In addition, it should be noted that the Persian Gulf evaporites occur within carbonates, thereby providing us with a more realistic analog of the very common ancient carbonate/sulfate evaporites than those provided by the sand- and clay-rich sediments of the supratidal flats of Laguna Madre and Baja California.

The effect these investigations had on the study of ancient evaporites is best demonstrated in a provocative paper by Shearman (1966). He speculated that many evaporite sequences, especially anhydrite and gypsum, but not excluding thick halite and potash deposits, had been formed by sabkha-like diagenetic processes. He showed how the sabkha cycle, from lagoonal sediments to nodular anhydrite and minor erosion surface, could be recognized in some ancient sediments. From his position as Lecturer in Geology at Imperial College, London, he has traveled widely, proselytizing for the merits of the sabkha philosophy.

That the supratidal flat environment can be one in which potash salts may form, was demonstrated by Holser (1966), whose paper is included in this part. He found diagenetically formed polyhalite replacing gypsum on the sabkha east of Laguna Ojo de Liebre on the Pacific coast of Baja California. His work provides information of immediate application to studies of ancient occurrences of this and associated minerals. A similar occurrence of polyhalite has been reported from Sabkha el Melah, Tunisia (Perthuisot, 1971); in the supratidal sediments of Kara-Bogaz-Gol, also, potash minerals have formed by diagenetic processes (D. J. Shearman, personal communication).

Kinsman began studies on the Persian Gulf as a student at Imperial College and he has continued to pursue these investigations for the past decade (Kinsman, 1964, 1966, 1967). He is now Associate Professor at Princeton University. In the paper which follows, he summarizes the geology of sabkhas and includes descriptions of continental sabkhas, the coastal sabkhas of the Arabian coast, as well as those near the Colorado River delta, and at the head of Laguna Ojo de Liebre, Baja California.

The chemistry of the diagenetic processes occurring within sabkhas is well described in the work of Butler (1969), whose conclusions were based on the study of 260 samples of interstitial brine from within the Abu Dhabi sabkha. His data, given as an appendix to the original paper, are not included here. The relationships between dolomite and the calcium sulfate minerals in the supratidal environment has been studied by Illing, Wells, and Taylor (1965), and, although not reproduced here, the paper has led to the classifying of dolomite as an evaporite mineral (Friedman and Sanders, 1967, p. 267).

As a natural consequence of the work by Kinsman, Butler, Shearman, and others, ancient sabhkas have been described (e.g., Holliday, 1968, Wood and Wolfe, 1969). Unfortunately, however, the supratidal model is in some danger of being overextended, for not all investigators have used the care that characterized the efforts of the original students of sabkhas. Not all nodular anhydrite represents ancient sabkhas, for it is not just *one* feature, but the complete cycle of features that should be used in any comparative studies of ancient and modern anhydrite occurrences..

Reprinted from *Jour. Sed. Petrol.*, **25**, 72–77 (1955)

6

AN OCCURRENCE OF GYPSUM IN SOUTHWEST TEXAS

P. H. MASSON
Humble Oil and Refining Company, Houston, Texas

ABSTRACT

Gypsum in the form of crystals and rosettes of selenite has been found in subsurface clays and sands of the Laguna Madre mudflats of southwest Texas. The habit of the crystals is unusual, and the dominant crystallographic forms are ($\bar{1}11$) and ($\bar{1}02$). The manner in which the crystals grow larger at depth and become lens-like in shape also is of interest. It is concluded that the gypsum precipitates from highly saline sea water which periodically moves in wind-blown sheets across the surface of the mudflats and sinks into the underlying sediments.

INTRODUCTION

Gypsum in an uncommon form has been revealed during the course of field work in the Laguna Madre mudflats in southwest Texas, and additional information concerning this interesting occurrence has been provided by dredging operations along the Gulf Coast intracoastal canal where it crosses the flats area. The area in which the gypsum was found is an expanse of algae-encrusted coastal mudflats approximately 15 miles in length and 10 miles in width. On most older maps this feature is outlined as part of Laguna Madre proper and appears as a prominent indentation in the coastline of east central Kenedy County midway between Corpus Christi and Brownsville; on new maps the mudflats are shown as part of the mainland (fig. 1).

Most of the gypsum is in the form of selenite, and one of the most interesting features of the occurrence is the habit of the crystals and the progressive changes in their size and form which take place at successively deeper levels in the sediments beneath the mudflats. It seems probable that the selenite is derived from highly saline sea water which periodically moves in thin sheets across the surface of the flats.

MODE OF OCCURRENCE

In the topmost two or three inches of the laminated and thinly bedded clays and algal material of the mudflats, gypsum is present as euhedral crystals less than 0.5 mm long. The wrinkled, leathery crust of algal fibers, clay, and fine calcite which covers most of the

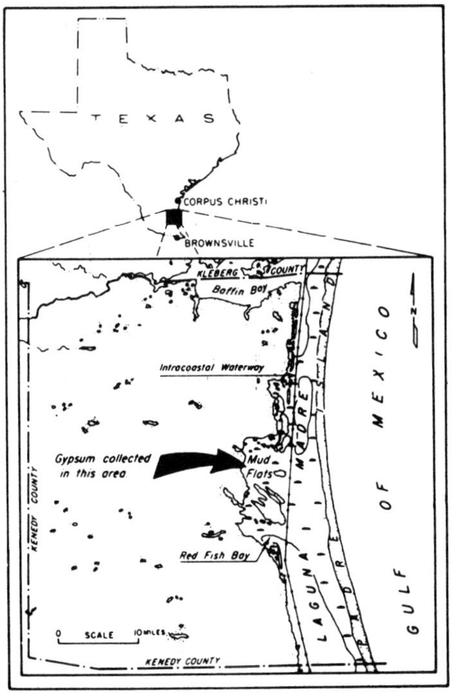

FIG. 1.—Location map showing Kenedy County, Texas, and the Laguna Madre mudflats. Gypsum was obtained in the area labelled Mud Flats and from spoil banks along the east side of the Intracoastal Waterway.

flats area contains euhedral selenite crystals, and in petrographic thin sections these can be seen scattered through the mass of dead algal fibers just below the surface layer of green living algae. Selenite is rare on the surface; euhedral crystals were found enmeshed in the topmost algal fibers, but most of the surface selenite was in the form of ragged windblown grains. In some areas, at depths of several feet, fine gypsum is abundant enough to form thin beds of gypsiferous clay and white layers of nearly pure gypsum; these gypsum layers in most places are less than one inch thick (fig. 2).

FIG. 2.—Interbedded layers of gypsum, clay, and algal material exposed in the wall of a trench. White layers are gypsum; wrinkled layers are buried algal crusts.

At depths below two or three feet the crystals are larger, and some have the form of rough, double-convex discs which bear no trace of rational crystal faces. Single and twinned crystals are present, but more common are delicate clusters of interpenetrating crystals between one and six inches in size. With increasing depth, more of the crystals have the lens-like form and fewer are euhedral or subhedral. Between the surface and depths of four or five feet all gradations from the euhedral form to the discoid form can be seen. The lens-like discoid individuals appear to have grown from euhedral crystals and may well be referred to also as crystals. At greater depths, however, large masses of compactly intergrown subindividuals are found; these aggregates more closely resemble the accretionary structures referred to as rosettes (Pettijohn, 1948). The rosettes and largest crystals were obtained from the sandy spoil bank along the intracoastal canal, having been brought up by suction-type dredges which excavated the canal to a depth of 15 feet. The largest rosette found was 20 inches in diameter (fig. 3).

FIG. 3.—Large selenite rosette dredged from a depth of approximately 15 feet in the Intracoastal Canal where it crosses the Laguna Madre mudflats. Rosette is dark gray in color due to very abundant included sand.

GENERAL CHARACTERISTICS OF THE CRYSTALS AND ROSETTES

Smaller crystals are perfectly euhedral, and all faces are fully and symetrically developed. The habit is that of a flat prism, diamond-shaped in cross-section and terminated by broad faces which cut the prism at a low angle. Unlike crystals with the habit most commonly recorded for gypsum, the Laguna Madre crystals are inclined to the *c* axis. The larger, rough-surfaced, lens-like crystals lie in a plane nearly normal to the *c* axis.

Aggregates of intergrown crystals are common and are composed of distinct crystals which interpenetrate in random manner. Rosettes, on the other hand, are made up of numerous discoid subindividuals closely intergrown in subparallel and radial arrangements. As accretion proceeds, aggregates of crystals apparently grow and become more complex until they assume the characteristics of rosettes. The subindividuals which are added during this later stage, however, appear to grow directly into the discoid form without starting as euhedral crystals. This difference in the manner of growth and mutual relations of the subindividuals is the basis for classifying these structures as rosettes rather than crystal aggregates.

All but the smallest crystals contain included quartz sand. The sand content varies in different crystals, and some have sand-free peripheral zones. In the larger crystals, aggregates, and rosettes, sand is very abundant and in most the selenite is little more than a cement. Crystals and rosette subindividuals, however, have crystallographic continuity so that they can properly be considered as sand crystals and sand rosettes similar to the so-called calcite sand crystals described and figured in some texts (Pettijohn, 1948).

Shells are included in some of the larger crystals, and sedimentary bedding is preserved as included sand layers of contrasting colors. The orientation of these beds is such that the discoid crystals must have grown with the planes of their maximum dimensions vertical or steeply inclined to the bedding.

OTHER SIMILAR OCCURRENCES OF GYPSUM

Gypsum similar to that of the Laguna Madre flats has been described from several other localities. A. Breithaupt (1841) described crystals from Bohemia and Silesia which were similar in habit to those of Laguna Madre. He commented on the discoid or lens-like shape which some of the crystals assumed, and stated further that they occur as nodules and clusters in the clays of tidal flats. C.A. Merritt (1935) described crystals from Alfalfa County, Oklahoma, with a habit somewhat like that of the Laguna Madre crystals; the environment here was a saline marsh. H. R. Bergquist (1942) figured, but did not describe in detail, single and twinned crystals in the Eocene Yazoo clay of Scott County, Mississippi. These crystals attained a maximum length of three inches and appeared to be most abundant along joints and bedding planes. In a recent account by Eardley and Stringham (1952) crystals similar in habit to those of Laguna Madre were reported from subsurface clays at the margin of Great Salt Lake in Utah.

CRYSTALLOGRAPHY

The basic crystal habit from which the rough, double convex crystals and complex intergrowths and aggregates of crystals develop consists of the unit pyramid, *l* ($\bar{1}11$), and broad, rough surfaces approximately parallel to *G* ($\bar{1}02$). The four faces of the monoclinic unit pyramid form a flattened prism; the ($\bar{1}02$) faces truncate this prism at a low angle so that the crystal has broad terminating faces (fig. 4a).

These forms are referred to the crystallographic setting for gypsum described by Palache, Berman, and Frondel (Dana, 1951), and it should be noted that the orientation used here is different from that of other texts and reference books in current use.

It has been reported that faces of the form *e* $(\bar{1}03)$ are usually rough and convex and that, together with $(\bar{1}11)$, they sometimes form rounded, lenticular crystals (Dana, 1951). Angular relations in the Laguna Madre crystals, however, indicate that here the rough terminating surface is parallel to $(\bar{1}02)$ rather than to $(\bar{1}03)$. This surface does not have the appearance of a rational crystal face, but on smaller crystals it is a smooth plane the orientation of which can be measured with results that are reproducible within several degrees from crystal to crystal. Furthermore, the rough, lens-like crystals lie with their longest axes in a plane near $(\bar{1}02)$ (fig. 4d).

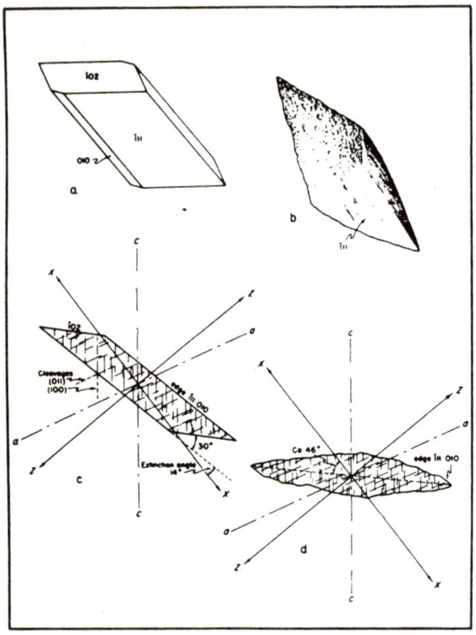

FIG. 4.—Habit and crystallographic orientation of selenite from Laguna Madre mudflats. a. Habit of euhedral crystals showing commonest form. Note that the crystal is viewed from the (−a, +b, +c) octant rather than the view customarily used in crystal drawings. b. Intermediate stage in the development of rough discoid crystals. The irregular surface approximately parallel to $(\bar{1}02)$ has expanded so that $(\bar{1}11)$ faces are partly obliterated and the crystal extends laterally in the direction of the *b* axis. c. Optical and crystallographic orientation of cleavage fragments from euhedral crystals. $\beta = 113°50\frac{1}{2}'$; $Z \wedge c = 51°52'$ (Dana, 1951). d. Optical and crystallographic orientation of cleavage fragments from discoid crystals.

The side pinacoid, *b* (010), which is prominent in crystals of the more common gypsum habit, is here restricted in its development to long, very narrow faces which truncate the acute edges between (111) faces. The unit prism, *m* (110), is present on some crystals as a very small face truncating the acute edge $(\bar{1}11):(\bar{1}02)$, but on most specimens this form is absent.

The crystals cleave readily on (010), and most of the crystallographic and orientation data were obtained on (010) cleavage fragments. Under the microscope, traces of weak cleavages on (011) and (100) can be seen. Treatment of (010) surfaces with dilute HCl accentuates the (100) cleavage traces and produces a rhomboidal etch figure bounded by the traces of (100) and (111). The extinction angle of (010) cleavage fragments is 14° measured from X to the sharply defined edges between $(\bar{1}11)$ crystal faces and the (010) cleavage plane (fig. 4c). In most of the larger rough crystals described below, these $(\bar{1}11)$ edges cannot be found, and the long axes of cleavage fragments are near the plane of $(\bar{1}02)$; the angle between X and these axes of elongation appears to be about 46° (fig. 4d).

Twinning on the front pinacoid is fairly common; the resultant crystal has the shape of an obtuse V.

GROWTH OF CRYSTALS

The changes that take place as crystal enlargement proceeds are of considerable interest. As described earlier, the smallest crystals are terminated by $(\bar{1}02)$; when sizes of one to two inches are attained, the $(\bar{1}02)$ surfaces become irregular and convex and begin to grow at the expense of ($\bar{1}11$) pyramid faces. The crystals extend laterally to points, and a diamond-shaped outline with the longest dimension parallel to the *b*-crystallographic axis is the result (fig. 4b). As the final discoid shape is approached, ($\bar{1}11$) becomes restricted to long, narrow faces fringing the edges of the rough, irregular, convex surfaces, and the outlines of the flattened plate becomes more nearly circular. With the disappearance of all rational crystal faces, a double-convex disc, the maximum dimensions of which lie in a plane approximating ($\bar{1}02$), is formed.

Single discs may have diameters as large as six inches, but most commonly the crystals form an aggregate of interpenetrating discs. The arrangement of discs within the aggregates appears to be random; several small aggregates were examined on the universal stage, and no rational twin relationship between subindividuals could be found.

ORIGIN OF THE GYPSUM

The Laguna Madre flats are two to three feet above sea level and extend across Laguna Madre dividing it into two bodies of water. Padre Island is about 110 miles long and, except for narrow passes at each extremity, isolates Laguna Madre from the Gulf of Mexico. Tidal fluctuations in Laguna Madre are small, but persistent winds from either the southeast or northeast periodically drive lagoonal waters onto the flats. During the summer months, the area frequently is covered by several inches of water. During dry periods the water table may drop as much as one and one-half feet below the surface. It is readily apparent that these thin sheets of windblown water, fed without recirculation from Laguna Madre, will become highly enriched in dissolved salts as a result of evaporation. The trend of this enrichment can be traced from a comparison of three water, analyses quoted in table 1. These analyses represent normal sea water, Laguna Madre water from Red Fish Bay at the northerly end of the south arm of the lagoon, and water from a shallow pond inland on the flats. No complete analysis of water from the moving windblown sheets on the flats surface is available, but the chloride content of one such sample was found to be 79,650 ppm.[1]

On a few occasions, extensive white crusts of precipitated salts including gypsum have been observed, and artificial pits in which water stands and evaporates become filled with coarsely crystalline halite. A small amount of fine gypsum with the crystal habit described above is disseminated through these salt deposits. Usually only minor amounts of salt are to be seen on the surface, and apparently the water recedes from the flats or filters downward through the clay and algal layers without precipitating appreciable amounts of solids. Withdrawal of water by evaporation at the surface during dry periods, however, may bring about precipitation of gypsum beneath the surface algal crust and in the clays near the surface.

Halite is present in the surface layers

TABLE 1.—*Major constituents in ppm of seawater and water from Laguna Madre and environs*

	Sea water[1]	Red Fish Bay[2]	Pond on flats[2]
Chloride	19,443	26,250	66,000
Sulfate	2,711	4,500	10,200
Carbonate	—	24	36
Bicarbonate	143[3]	148	74
Sodium	10,800	16,417	41,189
Calcium	411	671	827
Magnesium	1,303	1,084	2,942

[1] After Sverdrup, Johnson, and Fleming (1942, table 35).
[2] Analysis by Humble Oil & Refining Company Service Laboratory.
[3] All carbon expressed as bicarbonate.

[1] Analysis by M. S. Taggart, Jr., Humble Oil & Refining Company.

and in some places forms a matrix for fine-grained gypsum. The quantitity of halite in proportion to the amount of gypsum, however, is much less than should be expected from the proportions of ions equivalent to these two salts indicated in the water analyses quoted above. It must be assumed, therefore, that water in the sediments and on the surface remains for the most part undersaturated with respect to the more soluble salts; furthermore, there is sufficient fresh water from rainfall on the mudflats and adjacent areas inland to remove most of these salts from the sites of gypsum precipitation.

It is possible that growth of large crystals and rosettes at depth can be attributed to two factors: namely, change in the nature of the clastic sediments with depth, and the greater age of the deeper horizons. Sand beds are thicker and more abundant below the upper topmost few feet of laminated clays. High porosity and permeability in these beds as contrasted with the predominantly argillaceous beds would permit rapid ionic migration to a few large centers of crystallization and thus promote growth of large crystal aggregates and rosettes in the sand beds. The abundance of included sand in the more deeply buried gypsum indicates that it did indeed form in sediments composed almost entirely of sand. This gypsum may have been deposited originally at or near the surface as thin beds of fine crystals and subsequently become concentrated in the form of scattered large crystals and rosettes. Alternatively, it may have been derived entirely from external sources through the medium of moving subsurface brines. Either process would have been operative longer in the older, deeper beds and thus would account for the observed evidence of downward increase in size of the crystals and accretionary masses.

REFERENCES

Bergquist, H. R. and McCutcheon, T. E., 1942, Scott County Geology: Miss. Geol. Surv. Bull. 47, 135 pp.

Breithaupt, A., 1841, Vollständiges Handbuch der Mineralogie: Band 2, Leipzig.

Dana, J. D., 1951, The System of Mineralogy. 7th ed., v. 2, by C. Palache, H. Berman, and C. Frondel; John Wiley and Sons Inc., New York, 1124 pp.

Eardley, A. J. and Stringham, Bronson, 1952, Selenite crystals in clays of the Great Salt Lake: Jour. Sedimentary Petrology, v. 22, pp. 234–338.

Merritt, C. A., 1935, Gypsum Crystals from Alfalfa County, Oklahoma: Am. Mineralogist, v. 20, p. 674.

Pettijohn, F. J., 1949, Sedimentary Rocks. Harper and Brothers, New York, 526 pp.

Sverdrup, H. V., Johnson, M. W., and Fleming, R. H., 1942. The Oceans. Prentice Hall, New York, 1087 pp.

Reprinted from *Amer. Assoc. Petroleum Geologists Bull.*, **47**, 1726–1732 (1963)

ORIGIN OF NODULAR AND BEDDED ANHYDRITE IN PERMIAN SHELF SEDIMENTS, TEXAS AND NEW MEXICO[1]

7

S. DUFF KERR, JR.,[2] AND ALAN THOMSON[2]
Midland, Texas

ABSTRACT

Studies of anhydrite in Permian shelf sediments of Texas and New Mexico indicate similarities to Recent gypsum deposits of Laguna Madre, coastal south Texas.

Permian anhydrite occurs with greatest frequency within cyclical clastic and carbonate shelf sediments. The most common form (other than of pore-filling or replacement origin) is the nodule. Nodules are packed to varying degrees, ranging from widely spaced individuals to dense zones which have been misinterpreted as "beds." Two types of truly bedded anhydrite occur: thin discrete layers, and laminated beds.

In Laguna Madre, rosettes, bladed crystals, and thin beds of gypsum occur below the sediment surface. Thin sheets of lagoon water periodically are blown over mud flat areas where evaporation markedly increases salinity. The concentrated brines percolate downward, supplying hypersaline ground water from which the gypsum is precipitated (Fisk, 1959).

The discovery of Permian anhydrite nodules that are pseudomorphous after euhedral gypsum, and which closely resemble Recent gypsum crystals and crystal aggregates, suggests a similar mode of origin.

INTRODUCTION

General statement.—Evaporite minerals occur in the Permian of Texas and New Mexico in great abundance and wide variety. The evaporite mineral most widely distributed in the subsurface is anhydrite ($CaSO_4$). At the surface or at shallow depths in the subsurface, this anhydrite is converted to gypsum ($CaSO_4 \cdot 2H_2O$), owing to meteoric agencies. Anhydrite is the stable form of calcium sulfate under nearly all conditions in the subsurface.

Purpose of study.—This paper attempts to demonstrate that nodular anhydrite and certain bedded anhydrite of the Permian originated as rosettes, blades, and thin beds of gypsum associated with lagoonal mud flats. The association of evaporites with shelf deposits in the Permian prompted a search for a Recent analogue of such an association. Reports of gypsum occurrences in Laguna Madre by Masson (1955) suggested an area to be investigated. H. A. Bernard (personal communication) and Fisk (1959) presented detailed information on gypsum occurrence in Laguna Madre. Two short trips were made to the area during the fall of 1959 and spring of 1961.

Distribution of anhydrite.—Anhydrite occurs in a very wide variety of conditions in Permian rocks. It is found within or associated with conglomerate, sandstone, siltstone, salt, shale, dolomite, and limestone. It is distributed both over the shelf areas and within basins.

Although anhydrite is found in rocks deposited under a wide variety of environmental conditions, the occurrences may be placed in two groups representing widely differing situations. The first, and probably most prominent, is the anhydrite of the Castile formation (Upper Permian). This thick (average 1,200 feet) sequence is essentially unbroken by non-evaporitic deposits. It typically is composed of varve-like laminations of anhydrite separated by millimeter laminations of organic-rich carbonate (Fig. 5). Its geometry, mineralogy, and paleogeographic position suggest filling of a deep basin. This type of deposit has been widely treated in existing literature, and is not included in the present discussion.

The other situation in which abundant anhydrite occurs is in shelf sediments, both carbonate and clastic.

PERMIAN SHELF ANHYDRITE DEPOSITS

General statement.—Shelf sediments of Permian age in west Texas and New Mexico are strongly cyclical, displaying abrupt changes in vertical sections (with time) but with wide lateral continuity of individual environmental rock types. Rather thin units, on the order of 20–50 feet thick, commonly are traceable over distances on the order of 100 miles. Such units may exhibit only very minor changes in character from place to place. These sediments alternate between clastic and carbonate types, both of which contain anhydrite. As a result, anhydrite is probably the most nearly ubiquitous mineral type of the Permian shelf areas.

Anhydrite occurs in Permian stratigraphic units in a variety of ways: (1) nodules, (2) beds ranging from a few millimeters to several feet in

[1] Manuscript received, October 8, 1962. Published by permission of Shell Oil Company.

[2] Shell Oil Company.

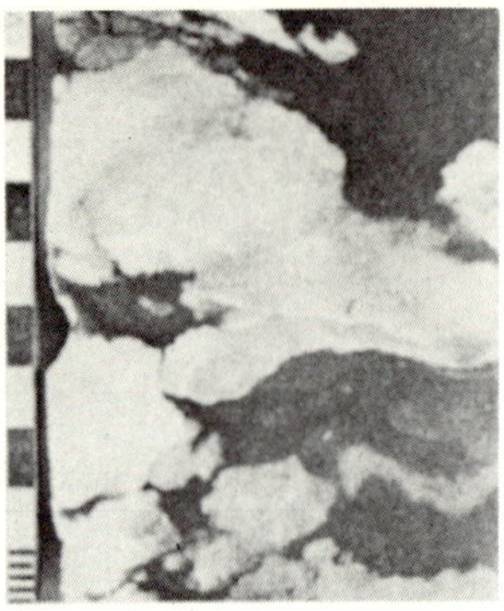

FIG. 1.—Typical irregular ovate anhydrite nodule from subsurface Permian. Matrix is red argillaceous quartz siltstone. Grayburg formation, Eddy County, New Mexico, depth 2,852 feet.

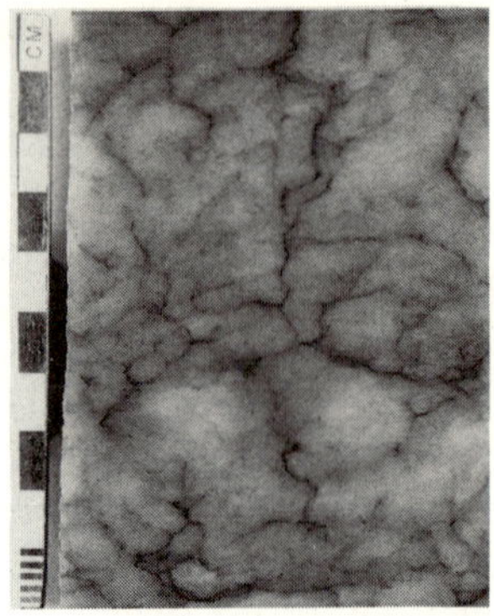

FIG. 3.—Densely packed anhydrite nodules with thin films of primary sediment separating them. San Andres formation, Hockley County, Texas, depth 4,683 feet.

thickness, (3) cement in quartz or carbonate sandstone, (4) cavity fillings, and (5) a replacement of pre-existing carbonate. Replacement and pore-filling anhydrite usually occur as single crystals or simple multiple-crystal masses. These types of anhydrite are not discussed in this paper. Bedded and nodular anhydrite is characteristically a complex multicrystalline aggregate (Fig. 2).

Occurrence of nodules and beds.—The most common form of anhydrite in shelf sediments (other than of pore-filling or replacement origin) is the nodule. The average nodule is ovate, flattened, with the greatest dimension in a horizontal plane (Fig. 1). These nodules are composed of felted masses of tiny rod-like individual crystals (Fig. 2). The masses of crystals commonly show a distinct fabric, usually oriented parallel with bedding. Rarely, anhydrite nodules may have a bladed or otherwise euhedral form.

Observations from cores and outcrops indicate that nodules of anhydrite occur in widely varying degrees of proximity or packing. As density of packing increases, the individual nodules are finally so tightly fitted together that only a thin line separates them (Fig. 3). The pattern of tightly packed nodules is often impossible to distinguish on the outer surface of a core; hence, such zones have erroneously been termed "beds" of anhydrite.

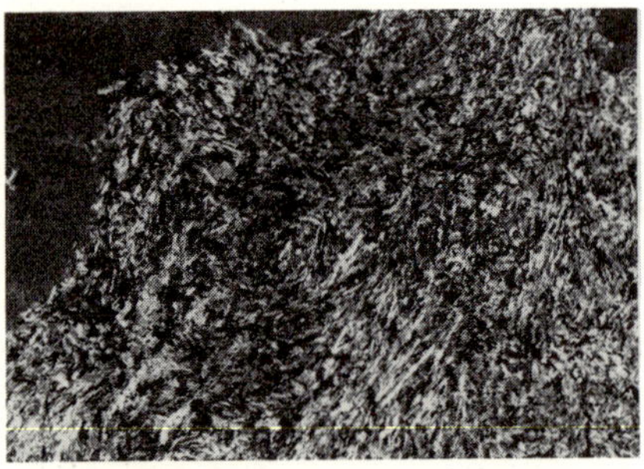

FIG. 2.—Photomicrograph of internal structure of anhydrite nodule in dolomite matrix, showing complex multicrystalline nature. Note fabric. Crossed nicols, 2.65×. San Andres formation, Yoakum County, Texas, depth 5,138 feet.

Two types of truly bedded anhydrite are recognized. The first consists of discrete layers bounded by layers of contrasting rock, generally observed in stromatolitic sequences (Fig. 4). Individual layers of this type are usually not more than 1–2 centimeters in thickness. In contrast to these are layers of laminated anhydrite (Fig. 5) up to several feet in thickness. These are limited in

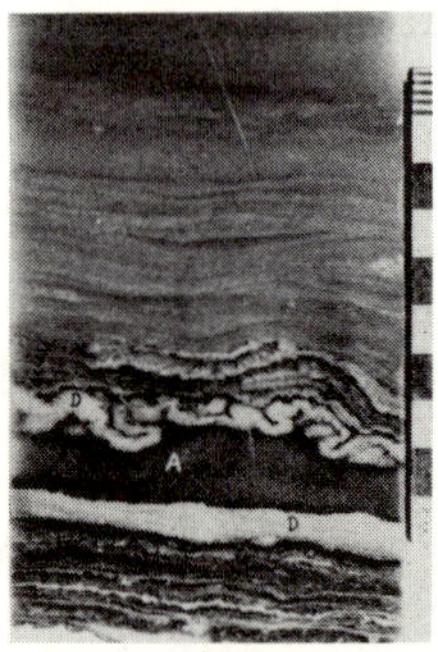

FIG. 4. (left)—Anhydrite layer (A) between dolomite stromatolite layers (D). Note that anhydrite conforms to crinkly underside of overlying stromatolite. Queen formation, Lea County, New Mexico, depth 3,953 feet.

FIG. 5. (right)—Laminated anhydrite (now gypsum) and organic-rich calcite in varve-like sequence. Outcrop of Castile formation, Eddy County, New Mexico.

occurrence, probably as a result of local conditions.

Recent Gypsum Deposits of Laguna Madre, Texas

Regional setting.—Laguna Madre is a linear coastal lagoon in south Texas, separated from the Gulf of Mexico by a practically continuous chain of barrier islands (Fig. 6). Padre Island is the longest of the chain and extends a distance of 110 miles from near the mouth of the Rio Grande to Corpus Christi Bay.

The entire lagoonal depression has an area of 609 square miles, of which nearly 300 square miles is exposed during low water. Depths of water in the lagoon average about 3 feet and reach a maximum of 8 feet in several places. Periodic tides in the region are negligible, but high steady winds can cause a tidal range of 3 or 4 feet, flooding much of the exposed areas (Rusnak, 1960, p. 162).

The exposed areas of the Laguna Madre depression are termed "mud flats" (Fig. 7) and consist principally of fine terrigenous material. Fine sand and silt reach Laguna Madre by eolian transport from Padre Island while clay suspensions enter by way of the passes at either end of Padre Island (Fisk, 1959). Considerable parts of the flats are covered by mats of blue-green algae which act as sediment binders.

Salinities in the lagoon proper range from near normal at the passes at either end to approximately twice that of normal sea water in the areas most distant from the passes (Rusnak, 1960, p. 185). These elevated salinities are due to several factors: (1) no streams enter Laguna Madre and surface runoff is low; (2) evaporation is greatly in excess of rainfall and runoff; and (3) the long, unbroken extent of Padre Island prevents rapid interchange with normal Gulf waters.

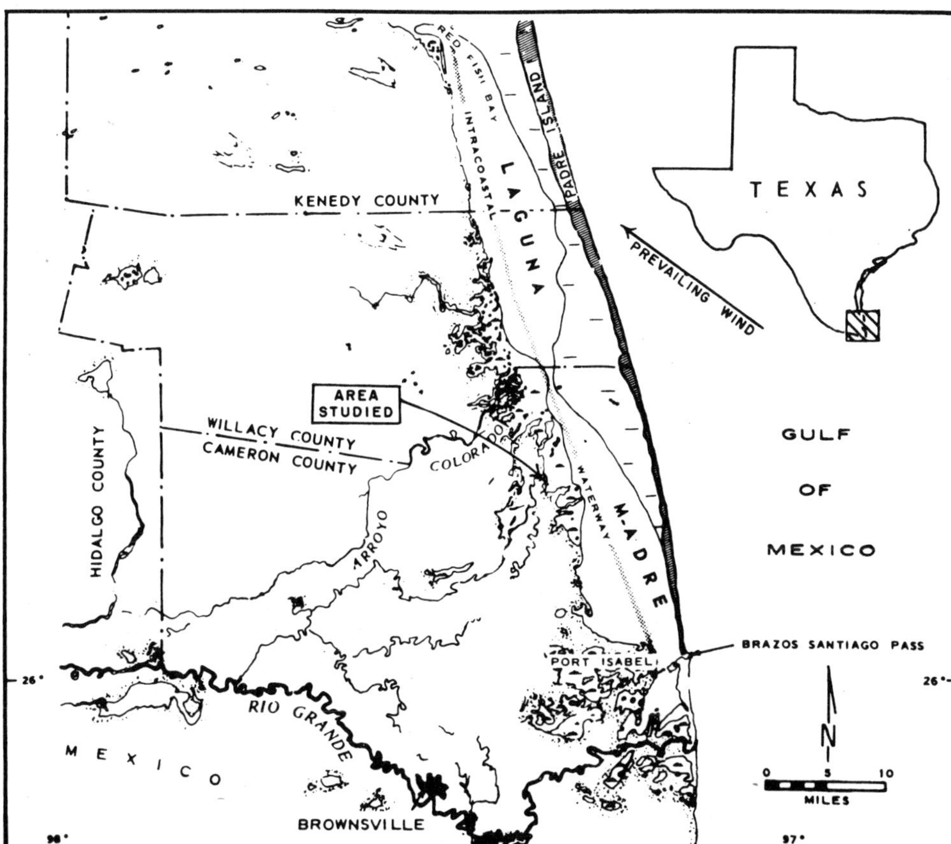

FIG. 6.—Index map of southern half of Laguna Madre, Texas (modified from Fisk, 1959).

FIG. 7.—Partially exposed mud flats of Laguna Madre, viewed from landward side.

FIG. 9.—Bladed gypsum crystals growing *in situ*. Overlying layer of sediment has been removed by wind.

Distribution of gypsum.—Several varieties of gypsum occur in the Laguna Madre region. These are classified as bladed crystals, rosettes, and thin beds.

Bladed crystals are most common near the sediment surface (Fig. 8). In the upper 2 or 3 inches of sediment, the blades are less than 0.5 mm. long but they increase in size with depth so that at a depth of a few feet they may be 3–4 inches in length. These are occasionally exposed on the surface as a result of wind erosion (Fig. 9). The bladed crystals commonly consist of interpenetrating clusters and twins of elongate euhedra (Fig. 10).

Rosettes are compactly intergrown masses of discoid gypsum crystals (Fig. 11). They have been observed in spoil banks along the Intracoastal Waterway from depths below the sediment surface of 10 feet or so. The rosettes increase in size with depth, and a specimen of 20 inches diameter was obtained from a depth of 15 feet by Masson (1955, p. 73).

Thin beds or laminae of gypsum interbedded with clay and algal mat occur in some places. The layers may be as much as 3 inches thick and conform in shape to the overlying material.

Origin of gypsum.—Gypsum precipitates when water salinities are approximately $3\frac{1}{3}$ times that of normal sea water (Posnjak, 1940). Prevailing E-SE winds in the area commonly drive thin sheets of lagoonal water across the flats. This water is subjected to high evaporative rates and consequently its salinity is increased. Surface waters on the flats have salinities averaging 140°/₀₀ and a maximum of 380°/₀₀ has been reported (Fisk, 1959, p. 109).

Salinities of subsurface waters are also abnormally high. Highly saline waters from an advancing sheet seep downward through the porous substratum and mix with ground waters, resulting in ground water salinities averaging nearly 200°/₀₀ (Fisk, 1959, p. 117). Long-lived salt pans do not develop.

Gypsum crystals appear to grow by direct precipitation from these highly charged ground waters, for gypsum does not occur on the surface of the flats but begins to develop within a few inches of the surface. Therefore, it is slightly post-depositional. Blades and rosettes which develop in sandy strata tend to incorporate the

FIG. 8.—Shallow pit in partially exposed mud flats (above) showing bladed gypsum crystals (G) in place. Note pencil for scale.

FIG. 10.—Typical bladed gypsum crystals from Laguna Madre.

FIG. 11. Typical gypsum rosettes from Laguna Madre.

sand and result in "sand crystals." However, those growing in mud are generally relatively clean and appear to have excluded the mud during crystallization, by shouldering aside the soft unconsolidated sediment. In this manner, even very large rosettes can develop.

The gypsum blades and rosettes appear to increase in size downward, but the possibility that blades and rosettes represent stages in a uniform growth spectrum is not as yet supported by field evidence.

Over much of the region the surficial mud flats are underlain by early Recent open lagoonal deposits (Fisk, 1959), and it is from these deposits that the larger rosettes have been extracted. It is unlikely that open lagoons would possess the requisite high salinities for gypsum precipitation. It appears rather that the gypsum rosettes are related to the present surface on which the highly charged ground waters are produced.

Laguna Madre and Permian Shelf Sediments: A Comparison

The crystal aggregates of gypsum in the Laguna Madre area are closely analogous with certain Permian nodular forms of anhydrite (Figs. 12–14). Figure 12 shows monoclinic anhydrite nodules pseudomorphous after (selenite) gypsum. Figures 13 and 14 show anhydrite nodules pseudomorphous after gypsum blades and rosettes, respectively. The common flattened ovate form in the Permian is a result of the influence of overburden pressures, a factor which has not affected the Recent sediments. The differences in internal crystal fabric (larger, discoid aggregates of the Recent gypsum vs. fine, felted blade and fibrous masses of Permian anhydrite) probably result from crystal reorganization upon dehydration of gypsum during conversion to anhydrite after burial. The growth of crystals or rosettes within unconsolidated sediments may result in physical disruption of the layered sediment. This suggests exclusion and shouldering aside of the host sediment.

The variation in density of packing of nodules observed in the Permian is a result of the distribution of crystals in the original sediment; *e.g.*, dense packing of nodules suggests close proximity of original blades or rosettes.

Discrete thin laminae or beds of gypsum develop between algal layers beneath the surface of the Laguna Madre flats. These layers do not exist on the surface, and Fisk (1959, p. 118) ascribes their origin to precipitation from ascending, capillary-transported ground water. A similar feature from the Permian is illustrated in Figure 4.

Layers of fine laminated anhydrite observed in the Permian (Fig. 5) do not have a direct analogue in Laguna Madre, unless they occur in the lagoon proper. Long-lived "lakes" or pans in which salt water is trapped are known from other mud flat environments and, though not present on the Laguna Madre flats, are fairly typical of such environments. Deposition of evaporites in such a pan should be similar to precipitates from other

FIG. 12.—Permian monoclinic anhydrite nodules pseudomorphous after gypsum. Whitehorse gp. (undifferentiated), Crockett County, Texas, depth 2,228 feet.

FIG. 13. (left)—Permian anhydrite nodules pseudomorphous after bladed gypsum crystals. Compare with Recent gypsum blades below. Queen formation, Midland County, Texas, depth 4,136 feet.
FIG. 14. (right)—Permian anhydrite nodules pseudomorphous after gypsum rosettes. Compare with Recent gypsum rosettes in lower left. Queen formation, Midland County, Texas, depth 4,142 feet.

standing bodies of water, and would result in a sequence of fine laminae due to periodic supply and evaporation of water. The thinness and limited occurrence of such laminated layers in the Permian suggest their origin in salt pans.

Riley and Byrne (1961) conclude that nodular anhydrite is formed by early deformation of gelatinous layers of loose anhydrite crystals. No evidence to support this theory has been found in the sediments and rocks studied. The following observations argue against their theory: (1) the presence of widely scattered, discrete nodules in non-contemporaneous host rock with no apparent "layer" source; (2) growth of well defined single crystals or aggregates in the Recent; (3) the presence of Permian anhydrite pseudomorphs after gypsum crystals; and (4) presence of undeformed laminae in both Permian and Recent. Withington (1961) thought that nodular gypsum or anhydrite is formed by precipitation from concentrated interstitial solutions, but rejected the possibility of formation from single-crystal selenite. The present study has demonstrated that multicrystalline anhydrite pseudomorphs after single-crystal selenite occur in the Permian.

Summary and Conclusions

The nature and occurrence of nodular and bedded anhydrite in Permian shelf sediments of west Texas and New Mexico is a close parallel of Recent gypsum deposits in Laguna Madre of coastal south Texas.

Bladed crystals, rosettes, and thin layers of gypsum between layers of contrasting rock type are precipitated beneath the sediment surface by highly charged ground waters derived from the overlying mud flat. Crystals which grow in this manner commonly exclude the host sediment and are preserved as pure mineral masses. Bladed crystals and rosettes of gypsum such as those in Recent sediments are preserved as multicrystalline nodules of anhydrite in ancient sediments. A high density of gypsum crystals growing while excluding soft host sediments will result in densely packed layers of anhydrite nodules which may be mistaken for "beds" of anhydrite.

Sequences of laminated gypsum do not occur in Laguna Madre, and laminated anhydrite is rare in Permian shelf sediments. These probably are deposited in shallow pans from standing bodies of water. Distinction should be made between beds of laminated anhydrite and "beds" of packed nodules, for their mode of origin is different.

The analogy between Recent gypsum crystals deposited from hypersaline ground waters and anhydrite nodules in ancient rocks should be applicable to similar occurrences in rocks of any age in the geologic column.

REFERENCES

Fisk, H. N., 1959, Padre Island and the Laguna Madre Flats, coastal south Texas: NAS-NRC, 2d Coastal Geog. Conf., p. 103–151.

Masson, P. H., 1955, An occurrence of gypsum in southwest Texas: Jour. Sed. Petrology, v. 25, p. 72–79.

Posnjak, E., 1940, Deposition of calcium sulfate from sea water: Am. Jour. Sci. v. 35, p. 539–568.

Riley, C. M., and Byrne, J. V., 1961, Genesis of primary structures in anhydrite: Jour. Sed. Petrology, v. 31, p. 553–559.

Rusnak, G. A., 1960, Sediments of Laguna Madre, Texas, *in* Recent sediments, northwest Gulf of Mexico: Am. Assoc. Petroleum Geologists, p. 153–196.

Withington, C. F., 1961, Origin of mottled structure in bedded calcium sulfate, U. S. Geol. Survey Prof. Paper 424-D, p. D-342–344.

Reprinted from *Nature,* **197**, 679–680 (1963)

8

Association of Dolomite and Anhydrite in the Recent Sediments of the Persian Gulf

A. WELLS[1] has recorded the presence of dolomite associated with aragonite and gypsum in the Recent sediments of the higher parts of the tidal flats around the Qatar peninsula in the Persian Gulf. A similar environment of coastal lagoons and tidal flats extends eastwards from Qatar along the Trucial Coast towards Oman. This communication refers to a short core collected from the higher part of the coastal flats (the sabkha) in the Sheikdom of Abu Dhabi, some 200 miles east of Qatar. The core, 20 cm in length, was taken from the surface of the sabkha on the mainland approach to the Abu Dhabi causeway.

Wells interpreted the dolomite and gypsum around the Qatar peninsula as having been "derived from sea-water left behind in the sediment on the upper part of the tidal flats and concentrated by evaporation during the time between spring or storm high tides, although it has not yet been established to what extent the dolomite is primary and to what extent it is a penecontemporaneous replacement of pre-existing carbonate". The distribution of dolomite in the core from Abu Dhabi, which is described in this communication, indicates that the dolomite is an early diagenetic replacement mineral.

In the Abu Dhabi core, aragonite, calcite, dolomite and halite have been identified. Gypsum, which occurs in the sample described by Wells[1] from Qatar, has not been found, but anhydrite occurs in considerable amounts. The vertical distribution of these components is shown in Fig. 1. The uppermost 8 cm consists essentially of an accumulation of sand- and silt-sized grains of quartz and carbonates, associated with evaporitic halite and anhydrite. This buff-coloured upper part rests on a cream-coloured, very fine-grained carbonate mud, which is well laminated at certain levels. Halite occurs throughout, but anhydrite, which characterizes the superficial parts of the core, has not been identified below 4 cm. At a depth of 9–14 cm from the surface the carbonate is almost exclusively aragonite, but below this level dolomite makes its appearance in detectable amounts. The ratio of dolomite to aragonite increases downwards from less than 1 : 1 at 15 cm to approximately 2 : 1 at 20 cm.

The anhydrite is clearly an evaporitic component, but the vertical distribution and mutual relationships of the dolomite and aragonite suggest that the dolomite is an early diagenetic replacement of aragonite. Although wind-blown grains are scattered throughout the length of

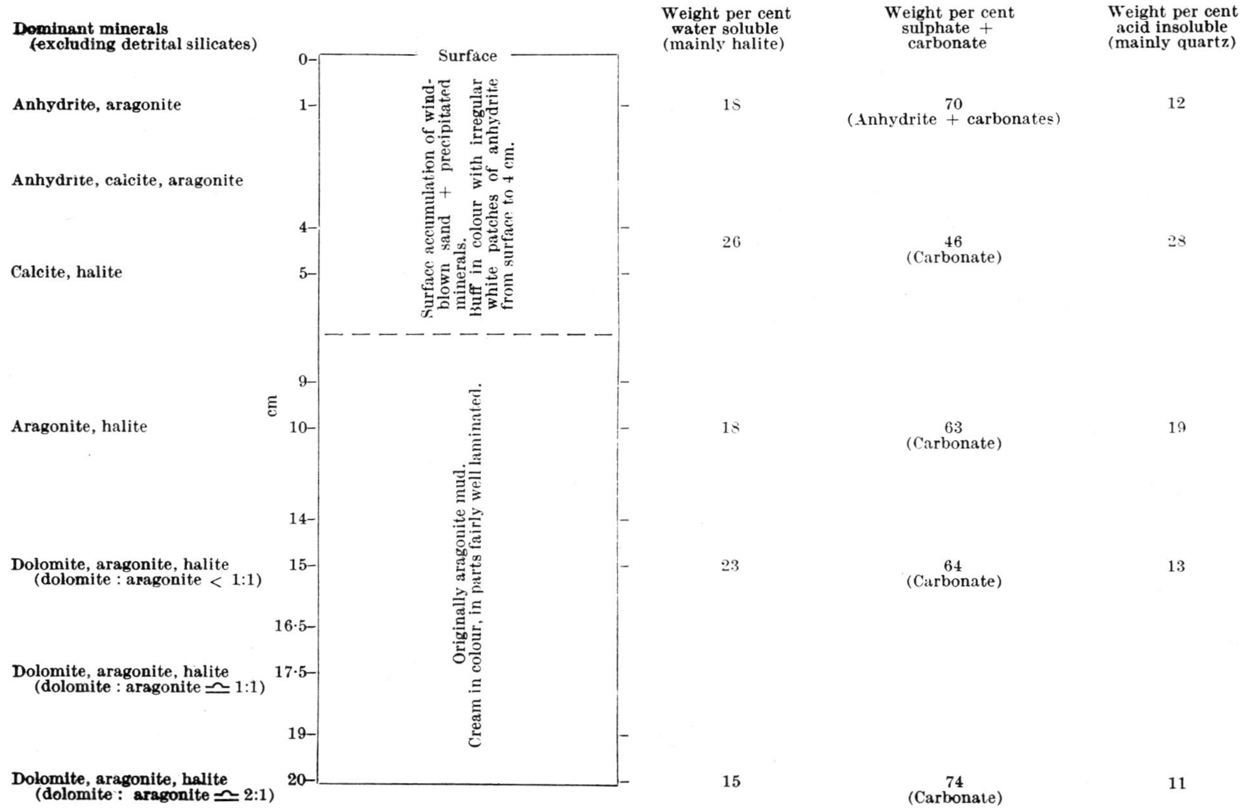
Dominant minerals (excluding detrital silicates)
Anhydrite, aragonite
Anhydrite, calcite, aragonite
Calcite, halite
Aragonite, halite
Dolomite, aragonite, halite (dolomite : aragonite < 1:1)
Dolomite, aragonite, halite (dolomite : aragonite ≏ 1:1)
Dolomite, aragonite, halite (dolomite : aragonite ≏ 2:1)
cm
0
1
4
5
9
10
14
15
16·5
17·5
19
20
Surface
Surface accumulation of wind-blown sand + precipitated minerals. Buff in colour with irregular white patches of anhydrite from surface to 4 cm.
Originally aragonite mud. Cream in colour, in parts fairly well laminated.
Weight per cent water soluble (mainly halite)
18
26
18
23
15
Weight per cent sulphate + carbonate
70 (Anhydrite + carbonates)
46 (Carbonate)
63 (Carbonate)
64 (Carbonate)
74 (Carbonate)
Weight per cent acid insoluble (mainly quartz)
12
28
19
13
11

Fig. 1

the core, there is nothing to suggest that the dolomite is detrital. All the available evidence indicates that it was formed *in situ* and is chemical in origin.

Owing to the extremely low annual rainfall, the pore fluids in the sabkha are supplied by the high salinity sea water of the adjacent lagoons. The waters of the tidal lagoons are alkaline with *p*H values up to 8·0. In the sabkha, which extends inland for 7 miles near Abu Dhabi, the ground-water is acid with the *p*H varying between 6·0 and 6·2. Transitional *p*H values are found in a narrow zone along the seaward margin of the sabkha. During October 1962 the ground water-level in the sabkha stood at approximately 75 cm below the surface, although it varied slightly from place to place. The ground water temperature averaged 35° C. A lateral and vertical migration of these fluids within the sediment is caused by evaporation from the sabkha surface. It is thought that the pore fluids, probably rich in magnesium, bring about the transformation of aragonite to dolomite. However, until a more detailed field and laboratory study has been completed this conclusion can only be a tentative one.

The mineral identifications were made by the X-ray diffraction powder method. The dolomite consistently showed an absence of the superstructure reflexions (221) and $(11\bar{1})$. This feature is characteristic of protodolomites[2]. The dolomite identified by Wells is rich in calcium ($Ca_{54}Mg_{46}$) and has an expanded lattice which suggests that it may also be a protodolomite, although Wells did not record the absence of the superstructure reflexions.

It is of interest at this stage of the investigations, that the dolomite is associated in the Qatar region with gypsum and at Abu Dhabi with anhydrite, in what at first sight appear to be similar environments.

Recent work on a series of short cores taken across the seven mile width of the sabkha near Abu Dhabi has revealed the presence of abundant gypsum. A complex, and as yet not understood, series of relationships exists between the gypsum and the anhydrite. The results of these later investigations will be more fully described elsewhere.

This communication is based on work forming part of a programme of research on the Recent sediments of the Trucial Coast, Persian Gulf, financed by the Department of Scientific and Industrial Research.

R. Curtis
G. Evans
D. J. J. Kinsman
D. J. Shearman

Department of Geology,
Imperial College of Science and Technology,
London, S.W.7.

[1] Wells, A., *Nature*, **194**, 274 (1962).

[2] Graf, D. L., and Goldsmith, J. R., *J. Geol.*, **64**, 173 (1956).

Printed in Great Britain by Fisher, Knight & Co., Ltd., St. Albans.

Reprinted from *Inst. Mining Met., Trans.*, **75**, 208–215 (1966)

Origin of marine evaporites by diagenesis

D. J. Shearman B.Sc.
Department of Geology, Imperial College, London

9

553.068.24:548.535

Synopsis

The possibility is explored that many marine evaporites may be largely diagenetic in origin. It is suggested that sulphate-bearing sequences characterized by nodular anhydrite originated as stratigraphic piles of arid zone supratidal or sabhka facies sediments, similar to those forming along the Trucial Coast at the present day. The soluble salts of sodium and potassium are considered to have been emplaced after burial by a series of replacements and displacements from salts in solution in the connate brines.

Sulphate-bearing horizons in the Purbeckian of Southern England are shown to be composed of a rhythmic repetition of sabhka cycles of sedimentation and diagenesis. Each sabhka cycle consists of the vertical succession near-shore marine, intertidal and supratidal facies sediments, the latter with nodular anhydrite. Evidence for the post-burial origin of potassium and sodium salts in some Permian evaporites is found in the relics of anhydrite nodules, the habits of pseudomorphs after gypsum and the fragments of former algal mats which occur in association with the salts. These vestiges are taken to indicate that the salts were emplaced in sediments of intertidal and sabhka facies.

It has long been firmly held that ancient marine evaporites were formed by the free evaporation of bodies of sea water which had only very restricted exchange with the open oceans. This concept does not appear to have been seriously challenged, despite the fact that many of the deposits show features which are not readily explained by the 'evaporating dish mechanism according to which the normal sequence of evaporites starts with anhydrite–gypsum and ends with potash salts.

A salient fact which has emerged from the petrographic studies of evaporites is that they record a history of mineral replacements more extensive than that encountered in any other group of rocks. In this respect it is useful to refer at the outset to the conclusions of Schaller and Henderson[15] on the potash-bearing evaporites of New Mexico and Texas. They stated that 'the mineral associations now found have undergone many changes since the beginning of the original precipitation' and that the 'saline field, as now existent, in large part therefore does not represent the original minerals first laid down'. This conclusion, although presenting the fundamental fact that much of the profile of salts is diagenetic, begs the question of origin, because it accepts as axiomatic that the deposits originated as precipitates. Stewart[18–20] reached similar conclusions concerning the extent of the replacements in the Permian evaporites of Northeast England, but his discussion on the sequence of changes also appears to have been conditioned by the assumption that the evaporites were initially precipitated within a basin.

Marine evaporite profiles can be divided essentially into two parts: a lower part characterized by calcium sulphate and an upper part which contains the soluble salts of sodium and potassium. The discovery that early diagenetic nodular anhydrite, gypsum and other evaporite minerals are forming at the present day on a regional scale in the Recent sediments of the supratidal flats or coastal *sabhka* of the Sheikdom of Abu Dhabi (Trucial Coast of the Persian Gulf)[5] opens the field for fresh inquiry, because these sediments closely resemble those which make up the sulphate zones of many ancient evaporites.

In view of these facts it was suggested by the present writer[17] that the evidence allows for the possibility that some at least of the marine evaporites may be entirely diagenetic in origin. That is to say they originated as piles of arid zone coast plain sediments in which anhydrite and gypsum were formed during early diagenesis while the soluble salts of sodium and potassium were emplaced subsequently, mainly after burial, as a result of reactions involving the connate brines. The present paper explores that possibility in more detail—most conveniently done by first describing the relevant features of the sediments of the supratidal flats of Abu Dhabi and their diagenetic mineral assemblage, and then by investigating the extent to which similar features can be identified not only in the sulphate zones but also as relics among the soluble salts in the upper parts of marine evaporite profiles. A brief account of the Purbeckian beds encountered in the Warlingham borehole south of London then follows; these provide an example of former supratidal flat evaporites. Finally, the chemistry of connate brines is dealt with, their possible role in the post-depositional emplacement of soluble salts being discussed.

Sediments and diagenetic minerals of the coastal sabhka of Abu Dhabi, Trucial Coast

The shallow sea off the coast of the Sheikdom of Abu Dhabi along the Trucial Coast on the south side of the Persian Gulf is a region of present-day carbonate sedimentation, where virtually the whole spectrum of shelf-sea carbonates is being generated.[6,11] Because of the prevailing onshore winds and wave approach most of the sediment is retained within the area and forced onshore. As a result an extensive coast plain composed of Recent and sub-Recent sediments has been built out, extending for some 200 miles along the coast and in places reaching up to 15 miles in width. In all, the coast plain covers an area of approximately 2000 square miles and its surface, which is an absolute desert, stands only 2 or 3 ft above high spring-tide level. The Arabic word *sabhka*, used to describe flat salt-crusted deserts, is applied to this coastal desert plain. Sabhkas may occur not only along the coast but also inland and far away from the sea. It is important therefore to distinguish

Manuscript first received by the Institution of Mining and Metallurgy on 7 June, 1966; revised manuscript received on 5 July, 1966. Paper published on 10 August, 1966. For discussion at a general meeting of the Institution to be arranged.

between coastal sabhkas, in which the groundwaters are largely marine-derived, and inland sabhkas, where the groundwaters appear to be essentially terrestial. Coastal sabhkas extend along the Mediterranean coast of North Africa, while similar plains are found along arid coastlines elsewhere in the world. As geomorphological entities these coastal sabhkas are the desert zone analogues of the salt marshes of temperate coastlines, and they may be built up of carbonate or non-carbonate sediments, or admixtures of the two.

Along the Trucial Coast, wherever the intertidal zone is a gently shelving feature, as it is in the shelter of islands along much of the eastern part of the area, it is characterized by a spread of rubber-like algal mats which may locally be several inches thick. These mats extend from a little below mid-tide level up towards the high spring-tide line and in places they attain a width of nearly half a mile. Identical algal mats have been described from intertidal zones along shorelines elsewhere, and it therefore seems likely that where similar algal mats are preserved in the fossil state they may be useful indices of former intertidal facies. Over large areas of the sabhka old algal mats may be found buried down to 3 ft below the present surface, thus recording the diachronous advance of the intertidal zone and the build-out of the supratidal flats (Fig. 1).

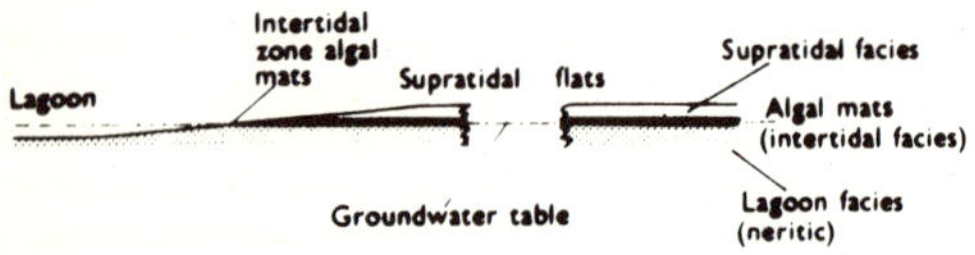

Fig. 1 Schematic cross-section across lagoon and supratidal flats near Abu Dhabi Island showing principal sedimentary facies

Vertical successions exposed in pits dug inland on the sabhka show three distinct facies. The sediments beneath the old algal mats are of marine facies, commonly of lagoonal aragonite muds and pelleted aragonite muds, while the algal mats represent the intertidal facies, and the overlying sediments constitute a supratidal or sabhka facies. The supratidal flat sediments consist in part of lagoonal aragonite muds washed over the sabhka plain when onshore gales coincide with high spring tides, but for the greater part they are comprised of wind-blown sand. Most of this windblown sand is of Recent marine carbonate material consisting of an admixture of skeletal grains, foraminiferal tests, oolites, bahamiths and other aragonite pellets.

On the more exposed parts of the coast algal mats are characteristically absent, and the intertidal zone presents a steep beach face capped by beach ridges.[7] Here the sabhka consists mainly of old beach face sediments.

Although the Abu Dhabi sabhka is strictly a non-marine subaerial environment, the sediments of which it is built are largely marine in aspect and the allogenic constituents of the sabhka are essentially similar to those of the marine sediments offshore. It is the early diagenetic minerals, however, which characterize the sabhka as a distinct facies. Nodular anhydrite, gypsum and dolomite occur there in relative abundance,[5,12,16] while celestite is a minor but common accessory.[8,12] All of these minerals appear to have formed in the sabhka as a result of reactions which involved the marine-derived groundwaters, the groundwaters themselves being concentrated by capillarity and evaporation. So great is the evaporation that the groundwaters of all but the seaward margin of the sabhka plain are virtually saturated with respect to sodium chloride, and halite is locally abundant within the sediments.

The essential features of the Recent nodular anhydrite and the gypsum are now outlined and an inquiry is made into the extent to which similar features can be identified in ancient evaporites.

Anhydrite

Anhydrite is locally abundant in the sediments of the sabhka, characteristically occurring as nodules in the sabhka facies above the old algal mats (Fig. 2). The groundwater table of the sabhka generally lies at about the level of the buried algal mats, and it appears likely therefore that the nodular anhydrite forms in the sediments of the capillary zone. The anhydrite makes its first appearance a short distance inland of normal high spring-tide mark, and in a general way it tends to increase in abundance inland. In places several layers of nodules with an aggregate thickness of up to 18 in have been found within the 30 in or so of sabhka sediment above the algal mats.

The anhydrite nodules have been shown to grow in the sediment by displacement,[4] and among the variety of features resulting from this growth are ptygmatic structures which closely resemble the enterolithic folds displayed by many ancient anhydrites. Most of the Recent anhydrite nodules from the Trucial Coast readily disaggregate when immersed in water and fall apart to produce platy rectangular laths bounded by the three cleavages of anhydrite. Although at first sight they appear to be constituent crystals of the nodules, some of them display stepped interference colours while others have rectangular reentrants—which suggests that they are not discrete crystals but are actually cleavage flakes.

In the field the majority of the anhydrite nodules are moist, putty-like and easily deformed, but those near the surface tend to be dry and cemented by salt. These latter can be collected without disturbing the fabric of the crystals of the nodules. Thin sections reveal that the arrangement of the laths commonly varies from place to place within any one nodule. Locally, they show a divergent radiating fabric; elsewhere, they are decussate; while in other parts they form narrow zones of sub-parallel laths which may suddenly change direction or terminate abruptly. Many of the crystals are cleaved and splayed open, and the cleavage fragments are bent

B209

or broken. It seems likely that this disruption of the crystals was caused by the demand for space of later crystals of anhydrite which grew within the interstices between the earlier-formed crystals. The disruptive effect is cumulative as the nodules increase in size. Because of this 'explosive' growth of the nodules most of them ultimately become aggregates of cleavage fragments. It is possible that the platy habit of the cleavage laths, combined with the moisture films inherent in the capillary zone, facilitates slip between adjacent plates, thus allowing the development of growth structures such as enterolithic folds.

Anhydrite nodules are abundant in the sulphate zones of many ancient evaporites, commonly showing fabrics similar to those of the nodules forming at the present day. In some instances the fabrics are further modified by compaction to form the so-called 'pile of bricks' texture.

The mode of origin of nodular anhydrite has hitherto been a matter of speculation. If, as the evidence now indicates, the nodules record sabhka diagenesis, then it is likely that comparable nodules in the sulphate zones of marine evaporites may serve as criteria for the identification of ancient sabhka facies. Further evidence in support of this possibility is presented later.

It is pertinent at this stage to inquire whether any of the anhydrite associated with halite and other salts in the upper parts of evaporite profiles is also of the sabhka type. The present writer has had the opportunity to examine some short lengths of cores from the halite and potash zones in the Permian of the Fordon No. 1 borehole, Yorkshire. In these rocks anhydrite laths, essentially similar to those released from the present-day nodules, are abundant. Some of them float within the salt crystals, while others occur in abundance in wedges between adjacent salt crystals. Although obvious in thin sections, the detail of the anhydrite laths is best seen by dissolving a small sample of the salt in water and examining the insoluble residue. It can then be seen that many are rectangular in outline, with rectangular inserts controlled by the cleavage. In samples from the halite cores from the Fordon No. 1 borehole small nodules of anhydrite occur virtually intact among the halite crystals. These show the disrupted fabric which characterizes the Recent anhydrite nodules of the Abu Dhabi sabhka. If the identity between the anhydrite laths of the modern nodules and those of the halite- and potash-bearing evaporites is accepted, it is reasonable to suppose that the halite and potash salts grew in a groundmass which contained sabhka-type anhydrite nodules.

It is of particular importance to emphasize that more than one generation of anhydrite occurs in the Permian evaporites, and that not all of it is of sabhka type.

Gypsum

Several generations of gypsum occur in the Trucial Coast sabhka. The specific characteristics of each are now described.

B210

Intertidal zone gypsum

In the intertidal zone small lenticular crystals of gypsum are scattered within the algal mats and the underlying sediments. Locally, these crystals are very abundant and may form a mush several inches thick. The individual crystals range in size from less than a millimetre up to a centimetre or so across, and are flattened in a plane approximately normal to the *c*-crystal axis. No gypsum crystals have been found seaward of the intertidal zone, but similar small crystals occur in the old buried algal mats inland on the sabhka. These small lenticular gypsum crystals appear to be characteristic of intertidal zone diagenesis, and their presence in association with old algal mats in the sediments inland of high-tide level clearly testifies to diachronism in diagenesis. Similar small lenticular gypsum crystals were recorded by Bramkamp and Powers[3] from the Recent sediments elsewhere along the Arabian coast, by Masson[14] from Laguna Madre, Texas, and by Illing *et al.*[9] from the intertidal zone around Qatar.

Typically, these gypsum crystals have a lozenge shape in cross-section, and pseudomorphs after gypsum with this habit are common in many evaporites. Kerr and Thomson,[10] for example, found anhydrite pseudomorphs after gypsum with this and other habits in the Permian of Texas and New Mexico.

Gypsum sand crystals

Large lenticular crystals of gypsum up to 8 in across are locally abundant in the sabhka inland of high-tide level. These are a later generation of gypsum but, like the intertidal zone gypsum, they too are markedly flattened normal to the *c*-crystal axis. In the field they show a preferred orientation in that most grow with the plane of flattening in a vertical or near-vertical position. Where the crystals have grown in carbonate sand they are commonly sand-filled. On (010) cleavage surfaces the enclosed sand grains are often seen to be arranged in a herringbone fashion, owing to displacement of the sand by the advance of crystal faces during growth of the gypsum. In these crystals the axes of the herringbone pattern are parallel to the planes of flattening and approximately normal to the *c*-crystal axes.

Halite pseudomorphs after gypsum showing a similar herringbone pattern were illustrated by Schaller and Henderson from the Permian of Texas and New Mexico[15] and by Stewart from the Permian of Northeast England.[19] In each of these examples the crystals were very large, up to 5 in long, with the length of the pseudomorph and the axis of the 'herringbone' being near vertical. It was suggested[15,19] that these pseudomorphs might represent replacements of large swallow-tail twins of gypsum. The resemblance to the zoning in the gypsum sand crystals is, however, remarkably close.

Gypsum after anhydrite

Over large areas of the inland parts of the Trucial Coast sabhka the anhydrite passes laterally into gypsum—often by transition. That the gypsum is later than the anhydrite is indicated by the presence of corroded anhydrite laths enclosed within the gypsum. These late gypsum crystals are relatively small, usually only a few millimetres in length, and quite different in habit from the intertidal zone gypsums and the gypsum sand crystals described earlier. They are flattened in the *a–c* plane, that is in a plane at right angles to the crystals previously described, and many are twinned on the swallow-tail law. Where they are present in abundance at the surface of the sabhka, they often form open aggregates in which most of the crystals have their *c*-axes nearly vertical.

Field evidence suggests that these gypsum crystals are the result of hydration of anhydrite caused by the influx of less saline groundwaters from the country rocks inland of the sabhka. Illing *et al.*[9] described similar gypsum crystals, flattened in the *a–c* plane, from a sabhka in Qatar. There, too, the gypsum contains corroded relics of anhydrite laths. Although nodular anhydrite has not yet been found in the Recent sediments of the sabhkas of Qatar, these relics may testify to its former presence.

Small pseudomorphs of halite after gypsum, showing the true swallow-tail twin habit, were described by Schaller and Henderson from the cores of the Permian potash field of New Mexico and Texas.[15] It is interesting to observe that where these pseudomorphs are closely packed their shapes indicate that, like the gypsum after anhydrite from Abu Dhabi, the *c*-axes of the former gypsum crystals were near vertical.

Sabhka facies in ancient evaporites

It is evident from the preceding discussion that in their petrography the sulphate-bearing horizons of the ancient evaporites show mineral assemblages and fabrics closely similar to those which have developed at the present day in the Recent sediments of the coastal sabhka of Abu Dhabi. Further, similar assemblages and fabrics can be identified as relics among evaporite profiles where soluble salts are developed. All of this suggests that supratidal flat sedimentation and sabhka diagenesis played a major role not only in the formation of ancient sulphate sequences but were also involved in the generation of the sodium and potassium deposits. It is therefore essential that an attempt should be made to find out to what extent former sabhkas can be identified in ancient evaporites.

Kerr and Thomson[10] found that the various habits of gypsum presently forming in the supratidal flats of Laguna Madre, Texas, could be matched with the shapes of anhydrite pseudomorphs after gypsum which occur in abundance at some levels in the evaporites of the Permian shelf sediments of Texas and New Mexico. They also demonstrated that where the former gypsum crystals had been tightly packed the anhydrite pseudomorphs could produce a structure closely similar to that of nodular anhydrite. It is implicit in this work that supratidal facies were an essential feature of sedimentation on the Permian shelf of Texas and New Mexico.

The present writer was fortunate in being allowed to examine cores from the sulphate-bearing horizons, some 17 ft in thickness, of the Purbeckian in the Warlingham borehole, South London. Before discussing these rocks it is convenient to refer back to the succesion of facies which are developed in the eastern part of the Abu Dhabi sabhka (Fig. 2). This vertical succession from lagoonal sediments up through intertidal zone algal mats into supratidal flat sediments with their assemblage of diagenetic sulphate minerals may be termed the 'sabhka cycle of sedimentation and early diagenesis The whole 17 ft of the Purbeckian evaporites consists of a repetition of these three facies. The most perfectly developed cycle occurs around a depth of 2134 ft (Fig. 3). It consists of lagoonal muds overlain by approximately 1 in of old algal mat sediments—overlain, in turn, by 15 in of nodular anhydrite in a groundmass of carbonate mud. The upper surface of the anhydrite-bearing member is an undulating erosion surface with a thin intraformational conglomerate occupying a depression in the surface. The hiatus is followed upwards by lagoonal muds. Comparison of this Purbeckian sabhka cycle with a typical vertical section through the present-day Trucial Coast sabhka (Figs. 2 and 3) shows them to be essentially similar in all their major features. Another complete sabhka cycle occurs around a depth of 2137 ft, while elsewhere partial cycles are developed consisting of algal mats interbedded with lagoonal muds, and of lagoonal muds followed by the anhydrite member without the intervening algal mat. The cycle between depths of 2142 and 2145 ft is interesting because the upper part of the anhydrite member shows several alternations of nodular anhydrite with thin algal mats—seemingly recording minor oscillations between inter- and supratidal conditions. An enterolithic fold in anhydrite occurs at a depth of 2131 ft, and this is essentially similar to the ptygmatic structures which can be seen in the Recent anhydrite.

Although the Purbeckian sulphate-bearing succession in the Warlingham borehole is only a relatively thin one, it does accord with published illustrations of cores of other ancient nodular anhydrites. To what extent the sabhka rhythm can be identified in other ancient sulphate sequences remains to be seen.

It had formerly been believed that the ancient sulphate sequences were subaqueous precipitates either in shallow lagoons or in partially enclosed deep marine basins with restricted access to the open sea. To the present writer the evidence from the Recent sediments of the Trucial Coast and from the Purbeckian of the Warlingham borehole very strongly indicates that the ancient nodular anhydrites were not precipitates formed in supersaline marine environments but that they were subaerial and diagenetic in origin.

The sabhkas of desert coastlines are the arid zone analogues of the salt marshes of the temperate coasts, for example the salt marshes around the Wash in England and those along the coasts of Holland and North Germany. Also, they are the counterparts of the swamp-

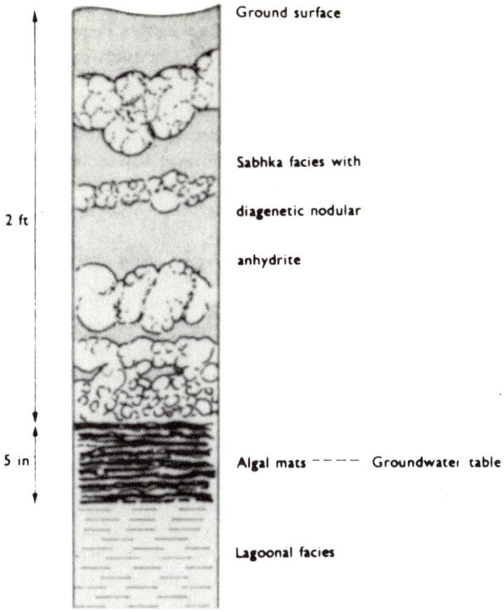

Fig. 2 Schematic vertical section through supratidal flats or sabhka, Trucial Coast, showing sedimentary facies and distribution of early diagenetic nodular anhydrite

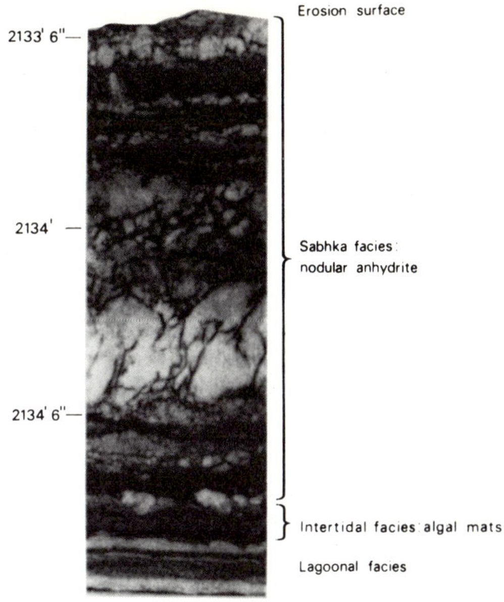

Fig. 3 Complete sabhka cycle (Purbeckian in Warlingham borehole)

forest type of salt marsh which develops in humid tropical regions. It is this latter which the geologist commonly invokes as typifying the environment of coal formation. Thus the intriguing possibility arises that the ancient sulphate sequences are the desert zone analogues of Coal Measures! Indeed, it is reasonable to suppose that if the diastrophic background which led to the accumulation of the 5000 ft of Westphalian coal measures of the Yorkshire–Derbyshire–Nottinghamshire coal field, for example, had operated in a desert region, then the succession would now be represented by 5000 ft of 'sabhka measures' with abundant sulphate.

Connate brines and diagenetic emplacement of soluble salts

When the Recent anhydrite was found in the Abu Dhabi sabhka it was immediately apparent that the anhydrite and dolomite had been produced as a result of reactions which had involved the marine-derived groundwaters.[5] The chemistry of the groundwaters has been investigated by Kinsman and also by Butler, whose results are to be published shortly. For the purposes of the present discussion it is therefore appropriate to deal with the constitution of the groundwaters only in general, if somewhat theoretical, terms.

By virtue of aridity the groundwaters of the Abu Dhabi sabhka are fed from the sea; and these groundwaters are rapidly concentrated by evaporation within the capillary zone above the groundwater table. Over the greater part of the sabhka the groundwaters are saturated brines, and halite is present locally as a salt crust on the surface, as a veneer around the grains in the upper part of the capillary zone and as cubes within the sediment. Thus if the sabhka brines are retained within the sediments the halite and the connate brines contained within a thick sequence of buried sabhka cycles are potentially capable of producing substantial thicknesses of salt as post-depositional emplacements. It is in this respect that the complex series of mineral replacements recorded by Schaller and Henderson,[15] by Stewart,[18-20] and many others may be significant. Are these replacements the results of readjustments within a series of minerals precipitated by the free evaporation of a body of sea water? Or do they record a post-depositional reaction series? The fact that many of the observed replacement series can be traced back to anhydrite, gypsum and dolomite—the relics of which preserve features found in the sabhka-type sulphate zones—suggests that they are actual post-depositional reaction series. Compaction alone would tend to make the brines move upwards through the stratigraphic pile and the expulsion of the brines from the lower part of the pile would leave this little modified mineralogically.

Thus all the inherent features of the former sabhka cycles would be preserved essentially intact in these lower beds. The bulk of the post-depositional reactions would take place in the sediments towards the top of the pile—where a vertical zonation of salts would tend to be produced as the reaction series proceeded. For such a reaction series of replacements to develop it is clearly necessary for the whole sequence to be sealed off above by impermeable strata. In this respect it is perhaps significant to note that many, if not the majority, of the ancient marine evaporites do not carry halite or bitterns but consist solely of sulphate and carbonate. Presumably, in these occurrences the connate brines were not confined and escaped from the formations before reactions could be initiated.

Experimental studies on the sequence of salts resulting from the evaporation of sea water have shown that although the details of the succession and assemblages of salts vary according to physical conditions, the general order can be summarized as follows. Calcium carbonate precipitates first and is followed by calcium sulphate. Next in succession comes sodium chloride, which makes up the bulk of the salts, and this is succeeded by the highly soluble salts of potassium and magnesium—the so-called 'bitterns'. The bitterns should commence with magnesium sulphate, pass up into potassium chloride and finally into bischofite (hydrated magnesium chloride). This generalized succession of salts and their relative proportions is shown in Fig. 4.

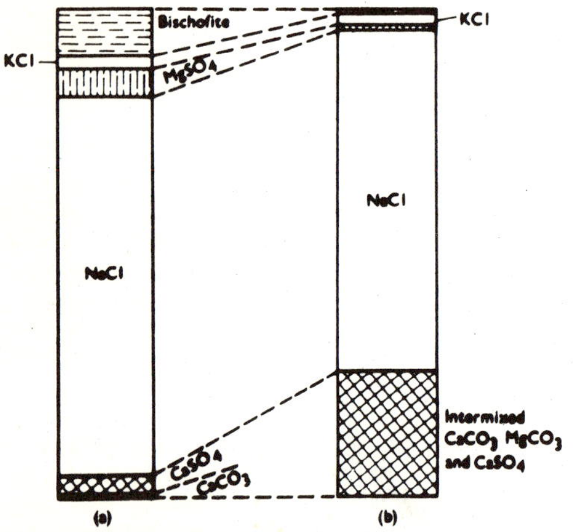

Fig. 4 Comparison of theoretical profile of salts obtained (*a*) by experimental evaporation of sea water; and (*b*) with average profile of various marine evaporites. After Borchert and Muir[2]

Few of the ancient evaporites conform to this arrangement of salts; among these, the evaporites of the Gabon show what must be the nearest approach to the theoretical.[1] The majority of deposits, however, show major discrepancies. Commonly, there is a much greater proportion of carbonate and calcium sulphate than would be expected. Further, the carbonate and sulphate do not follow one another as two discrete zones but are intimately intermixed, while the carbonate tends to be dolomitic rather than pure calcium carbonate. In those where bitterns are developed there is often a major deficiency in magnesium content. Among many workers, Stewart[21] commented on this anomaly, observing that the lack of magnesium in the bitterns is to some extent counterbalanced by the presence of magnesium, as dolomite, at the base of the succession. These anomalies are not easily accounted for on the basis of the evaporating-basin hypothesis, but they do find a reasonable explanation in the mechanism of sabhka diagenesis.

It is important to note that where evaporites show these major discrepancies in their overall chemistry they occur in a background of dominantly carbonate sediments. On the other hand, where the sequence of salts most nearly approaches the theoretical, as in the Gabon evaporites, the host sediments are non-carbonate. This association suggests that the potential bittern assemblage was conditioned by the changes occurring during the sabhka stage of diagenesis. It is therefore necessary to consider the chemical changes which would be expected to occur in brines of, on the one hand, non-carbonate sabhkas, and carbonate sabhkas on the other. This was done by Kinsman[13] and the present writer[17] and is presented schematically in Figs. 5 and 6.

The dissolved substances in normal sea water consist dominantly of sodium (30·61 wt per cent), magnesium (3·69 per cent), calcium (1·16 per cent), potassium (1·10 per cent), chloride (55·04 per cent) and sulphate (7·68 per cent). In a non-carbonate sabhka (Fig. 5), evaporation would lead to the precipitation of calcium sulphate from the marine-derived groundwaters, but since the sulphate content of sea water is much greater than that required to satisfy the calcium, substantial proportions of sulphate ions would remain. Apart from possible readjustments among the clay minerals, no major reactions would be likely to take place between the brines and the minerals of the host sediment. Hence the brines would come to consist essentially of sodium, magnesium, potassium and chloride, with a reduced proportion of sulphate ions. Such brines would be capable of producing the theoretical suite of bitterns carrying magnesium. In a carbonate sabhka, on the other hand (Fig. 6), dolomitization of the host sediment would occur. If this proceeded on the basis of the equation given in Fig. 6 calcium ions would be liberated in equivalent proportion to the magnesium utilized. These calcium ions, together with those inherent in the marine-derived groundwaters, would go a long way towards completely satisfying the sulphate of the sea water as calcium sulphate. Thus diagenesis in a carbonate sabhka tends towards the production of brines carrying sodium, potassium and chloride, but depleted in magnesium and virtually devoid of sulphate ions.[13] It is brines of this composition which could provide magnesium-deficient bitterns. Such early diagenetic evolution of brines by sabhka diagenesis could, of course, be invoked equally in support of the classical 'bar' hypothesis of evaporite formation as it is as an argument in the 'post-burial' theory.

B213

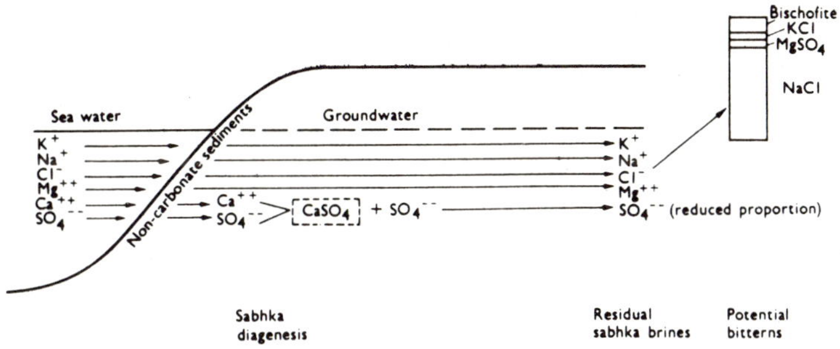

Fig. 5 Schematic diagram showing trend of mineral and brine diagenesis in non-carbonate sabhkas

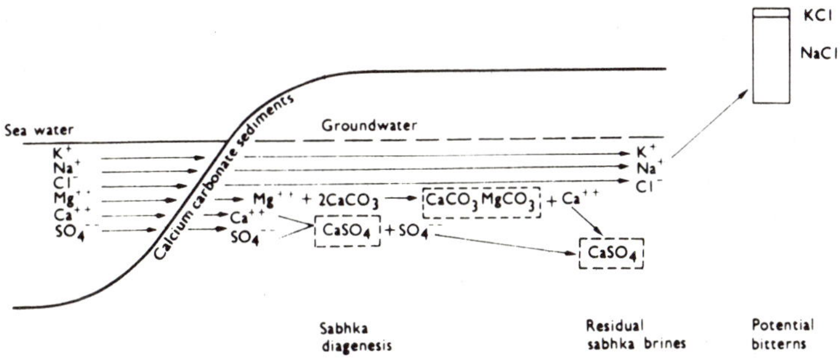

Fig. 6 Schematic diagram showing trend of mineral and brine diagenesis in carbonate sabhkas

In the 'post-burial' theory envisaged here the ultimate emplacement of the soluble salts would be achieved by a series of mineral replacements and displacements. The petrographic studies of Schaller and Henderson,[15] Stewart[18-20] and numerous other workers leave no doubt about the extent of the mineral replacements. Few, if any, of the soluble salts appear to be present as primary precipitates; instead, their fabrics testify to complex series of replacements—all of which tend to lead back to anhydrite and gypsum. On the basis of the available evidence they could equally well be interpreted as series resulting from reactions between connate brines and a host sediment consisting largely of anhydrite–gypsum.

Evidence for the growth of salts by displacement can be found in the arrangement of impurities such as sand grains, clay and anhydrite which occur tightly packed in wedges between the salt crystals. In some cores of apparently pure halite and potassium minerals from the Fordon No. 1 borehole the salts may be dissolved away to leave a delicate self-supporting framework of impurities. Further, in thin sections of these same cores fragments of algal mats essentially similar tho those along the Trucial Coast at the present day are disrupted and displaced by crystals of halite and potassium salts. Organic matter resembling mascerated algal mats is relatively abundant in many of the halite- and potassium-bearing cores from the Fordon No. 1 borehole studied by the present writer. If the diagnosis of this organic matter as algal mat is correct, then it provides corroborative evidence that the original sediment was of tidal flat and supratidal flat facies.

Acknowledgement

Field work along the Trucial Coast was financed by a D.S.I.R. grant. Samples of cores from the Fordon No. 1 borehole were generously supplied by B.P., and permission to study cores from the Warlingham borehole was given by Sir James Stubblefield, F.R.S., Director of the Institute of Geological Sciences, London.

In developing the ideas set out in the present paper the author has drawn heavily on the unpublished observations and researches of his co-workers in the Trucial Coast project: P. Bush, G. P. Butler, Dr. G. Evans, C. Kendall, Dr. D. J. J. Kinsman and Sir Patrick Skipworth. The author would like to thank them for the companionship and enthusiasm both in the field and in the laboratory.

References

1. **Belmonte Y. Hirtz P. and Wenger R.** The salt basins of the Gabon and the Congo (Brazzaville). In *Salt basins around Africa* (London: Institute of Petroleum, 1965), 55–74.

2. **Borchert H. and Muir R. O.** *Salt deposits: The origin, metamorphism and deformation of evaporates* (Princeton, N.J., and London: D. Van Nostrand, 1964), 300 p.
3. **Bramkamp R. A. and Powers R. W.** Two Persian Gulf lagoons (abstract). *J. sedim. Petrol.*, **25,** 1955, 139–40.
4. **Butler G. P.** ***et al.*** Recent evaporite deposits along the Trucial Coast of the Arabian Gulf. *Proc. geol. Soc. Lond.*, no. 1623, July 1965, 91–2.
5. **Curtis R.** ***et al.*** Association of dolomite and anhydrite in the Recent sediments of the Persian Gulf. *Nature, Lond.*, **197,** Feb. 16 1963, 679–80.
6. **Evans G.** ***et al.*** A reconnaissance survey of the environment of Recent carbonate sedimentation along the Trucial Coast, Persian Gulf. *Developments in sedimentology, vol. 1* (Amsterdam: Elsevier, 1964), 129–35.
7. **Evans G.** ***et al.*** Origin of the coastal flats, the Sabkha, of the Trucial Coast, Persian Gulf. *Nature, Lond.*, **202,** May 23 1964, 759–61.
8. **Evans G. and Shearman D. J.** Recent celestine from the sediments of the Trucial Coast of the Persian Gulf. *Nature, Lond.*, **202,** April 25 1964, 385–6.
9. **Illing L. V.** ***et al.*** Penecontemporary dolomite in the Persian Gulf. *Spec. Publs Soc. econ. Palaeont. Miner., Tulsa*, no. 13, 1965, 89–111.
10. **Kerr S. D. and Thomson A.** Origin of nodular and bedded anhydrite in Permian Shelf sediments, Texas and New Mexico. *Bull. Am. Ass. Petrol. Geol.*, **47,** Sept. 1963, 1726–32.
11. **Kinsman D. J. J.** The Recent carbonate sediments near Halat el Bahrani, Trucial Coast, Persian Gulf. *Developments in sedimentology, vol. 1* (Amsterdam: Elsevier, 1964), 185–92.
12. **Kinsman D. J. J.** Dolomitization and evaporite development in lagoonal sediments, Persian Gulf (Abstract) *Program Ann. Meeting, geol. Soc. Am. 1964*, 1964.
13. **Kinsman D. J. J.** Reference 1, 120–2.
14. **Masson P. H.** An occurrence of gypsum in southwest Texas. *J. sedim. Petrol.*, **25,** 1955, 72–7.
15. **Schaller W. T. and Henderson E. P.** Mineralogy of drill cores from the potash field of New Mexico and Texas. *Bull. U.S. geol. Surv.* 833, 1932, 124 p.
16. **Shearman D. J.** Recent anhydrite, gypsum, dolomite and halite from the coastal flats of the Arabian shore of the Persian Gulf. *Proc. geol. Soc. Lond.*, no. 1607, July 1963, 63–5.
17. **Shearman D. J.** Reference 1, 75–6.
18. **Stewart F. H.** The petrology of the evaporites of the Eskdale No. 2 boring, east Yorkshire: pt 1—The Lower Evaporite bed. *Mineralog. Mag.*, **28,** Sept. 1949, 621–75.
19. **Stewart F. H.** The petrology of the evaporites of the Eskdale no. 2 boring, east Yorkshire: pt 2—The Middle Evaporite bed. *Mineralog. Mag.*, **29,** March 1951, 445–75.
20. **Stewart F. H.** The petrology of the evaporites of the Eskdale no. 2 boring, east Yorkshire: pt 3—The Upper Evaporite bed. *Mineralog. Mag.*, **29,** June 1951, 557–72.
21. **Stewart F. H.** Data of geochemistry, 6th edn Chapter Y. Marine evaporites. Prof. Pap. U.S. geol. Surv. 440-Y, 1963, 52 p.

B215

Reprinted from *Amer. Mineralogist,* **51**, 99–109 (1966)

10

DIAGENETIC POLYHALITE IN RECENT SALT FROM BAJA CALIFORNIA

WILLIAM T. HOLSER,
California Research Corporation, La Habra, California.

ABSTRACT

Halite and gypsum now being deposited on extensive tidal flats at Laguna Ojo de Liebre, Baja California, are underlain by up to two meters of bedded evaporites. The porous evaporite sediments are permeated by bitterns whose stage of evaporation increases with distance from the lagoon. Beyond 4 to 8 km from the lagoon, gypsum crystals in the sediments have been altered to fine-grained polyhalite, by a bittern nearly in the potash-magnesia facies (concentration 60× sea water).

Although older polyhalite has occasionally been described as a primary precipitate, this modern occurrence confirms other geological and experimental evidence that polyhalite can replace calcium sulfate and that this can occur at an early diagenetic stage.

INTRODUCTION

Polyhalite, $K_2MgCa_2(SO_4)_4 \cdot 2H_2O$, is a common mineral in evaporite rocks of the potash-magnesia facies, but has not been described in evaporites forming from sea water at the present time. Although some polyhalite is considered to be a primary mineral crystallized directly from brine, most polyhalite is evidently a replacement of other sulfate minerals. The timing of this replacement is generally not known in the rocks, so this first discovery of polyhalite replacing Recent gypsum sediments may be useful in the interpretation of the earlier geological occurrences.

The polyhalite occurs in thin evaporites along the inner shore of Laguna Ojo de Liebre, about half way down the west coast of Baja California, Mexico. Field work was done in cooperation with F. B. Phleger and J. R. Bradshaw of the Scripps Institution of Oceanography. The courtesies extended by Mr. Charles McClaughry, Manager, Exportadora de Sal, Guerrero Negro, B.C.N., and by Mr. Francisco Muñoz, charter pilot at Tijuana, B.C.N., greatly facilitated the field work.

Material was first collected in May 1960, and in August 1961 we returned to core the evaporite sediments. Although study of this material is incomplete, the occurrence of polyhalite seems of sufficient interest to warrant a separate preliminary description.

SALT FLATS OF LAGUNA OJO DE LIEBRE

These salt deposits were first described in detail by Wittich (1916). Recently Phleger and Ewing (1962) described the general sedimentology and oceanography of the coastal lagoons in Baja California, with particular attention to Laguna Ojo de Liebre. Since then, the sediments of the beaches and salt flats have been cored, and we will publish the details when the new material has been thoroughly studied. Here I will only outline those features of the deposits that are pertinent to the occurrence of polyhalite.

Figure 1 shows Laguna Ojo de Liebre (also known as Scammon Lagoon) extending from the Pacific Ocean nearly 40 kilometers into the coastal plain of Baja California. The detailed study by Phleger and Ewing (1962) shows that the lagoon has a U-shaped tidal channel about

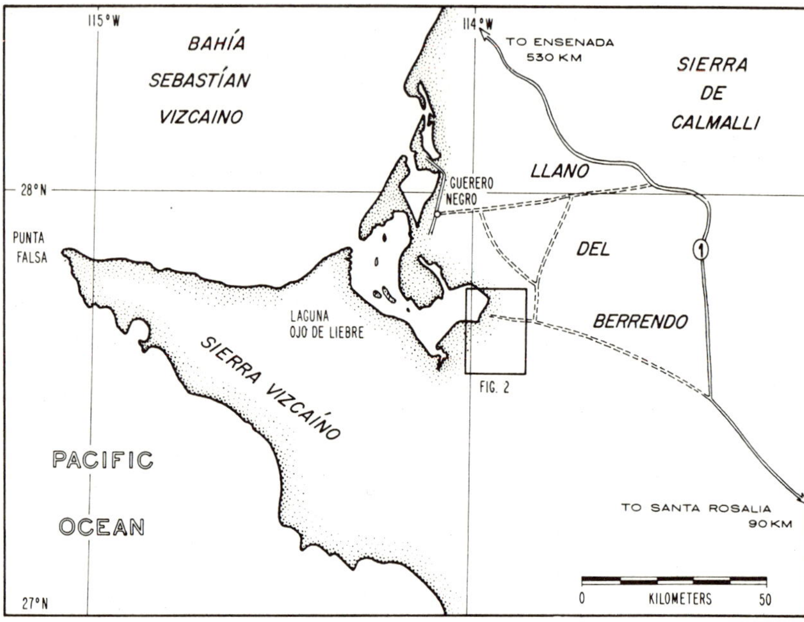

FIG. 1. Index map of part of Baja California.

50 ft deep, bordered by a very shallow shelf. Surface temperatures of up to 25° C. combined with a light but steady northwest wind, lead to an evaporation rate of nearly two meters per year. Consequently, the salinity increases to about 45°/oo (parts per thousand by weight) by the time the water has reached the head of the lagoon (Phleger and Ewing, 1962). However, the lagoon is locally mixed, with no salinity layering. At the upper end of the lagoon, the lagoon shallows are bordered by a low sand beach, cut in a few places where tidal channels flood and drain the flats behind the beach (Figs. 2, 3). The flats are covered intermittently with a few inches of brine with salinity over 70°/oo, driven shoreward by wind as well as tide. It supports a considerable growth of both

algae and salt grass, and in the landward part of this zone gypsum is crystallizing. Beyond this, the flats are white with both salt and gypsum crystals, and are apparently covered with brine only at irregular intervals, probably driven by high winds or storm tides.

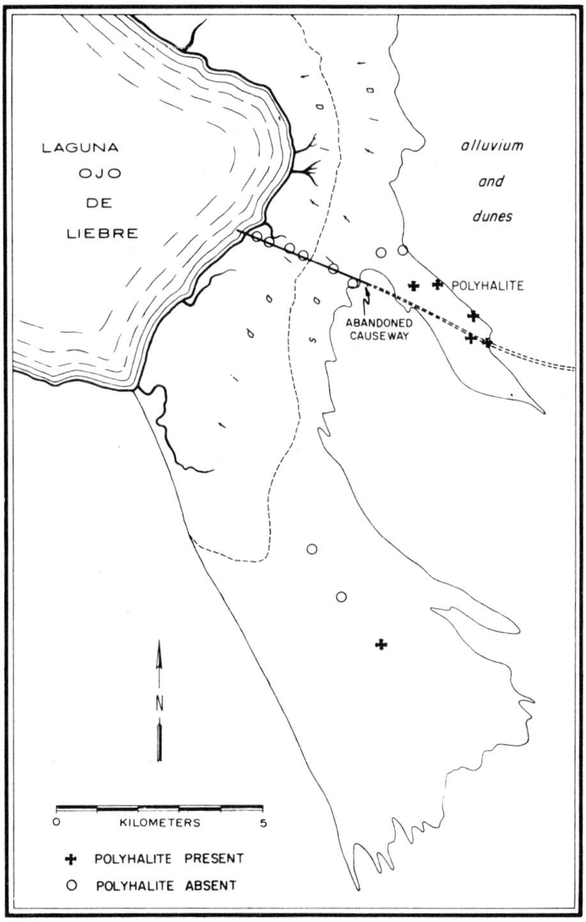

FIG. 2. Polyhalite at Laguna Ojo de Liebre.

The flats are underlain by thinly bedded sediments to a depth of 1–2 meters, mainly gypsum and detrital (quartz and calcite) sands alternating with dark mats of algae and grass. These beds lie on massive and compact gray-green silts and sands that possibly represent an earlier beach deposit. A wide variety of earlier strand lines have been noted along this coast (Wittich, 1912), and 15 km east of the salt flats wells have penetrated 50 m or more of Pleistocene and Recent sands (Mina, 1957). Beneath the salt flats halite is in the upper part of the bedded sediments, only a centimeter or two thick at the outer edge of the flats, increasing to 1.3 m in the centers of the two "bays" that extend eastward and southeastward from the main flat (Figs. 2, 3). Gypsum is widespread: mixed with the halite, as separate layers, as large crystals in the underlying detrital silts, and as windblown surface sand.

Brine lies only a few cm below the surface of the salt flats. It is not only saturated with sodium chloride, but shows the high potassium,

FIG. 3. Salt flats viewed eastward from above the lagoon, with tidal rivers cutting through the beach ridge into the lagoon in the foreground. Tidal flats are darkened by algae and grass. The remains of a causeway and loading dock runs for 3 kilometers across the flats. The thickest deposits of salt, permeated with highly evaporated brines, fill the white "bays" in the background.

magnesium, and bromide indicative of a high degree of evaporation (Table 1). A low rainfall of only 3 cm per year has never resulted in any surface runoff within the memory of local inhabitants. Consequently, the brine has not been diluted by fresh water in recent times. However, Phleger and Ewing (1962) believe that the detritus on the beach and the lagoon had their ultimate origin in floods from the coastal plain in a wet cycle several hundred years ago. The detrital components of the evaporite sediments, principally composed of quartz, hornblende, biotite and plagioclase, probably represents a reworking by tidal currents of these older sediments. The evaporites must be more recent than the wet cycle. Wittich (1916) emphasizes the importance of shoreline emergence in the formation of the evaporites, but he also points out that much of the salt may even have been reformed in the few years since a large but uncertain amount was mined during the last half of the nineteenth century.

Phleger and Ewing (1962) suggest a mechanism for supply of the partially concentrated lagoon water onto the tidal flats, by an asymmetric tidal flow. I should also like to suggest the importance of the on-shore wind, which can be seen to drive a surface layer of brine landward, even as the ebbing tide is flowing outward in the lower part of the brine.

In summary, the present evaporite regine on the tidal flats seems to

TABLE 1. BRINES FROM LAGUNA OJO DE LIEBRE, B.C.S., MEXICO

	1	2	3	4
Na	10.77	13.6	42.5	47.4
K	0.39	0.45	1.05	11.2
Mg	1.33	1.82_6	n.d.	47.3
Ca	0.42	0.52_8	n.d.	0.1
Cl	19.8	26.4_8	61.0	183.1
Br	0.066	0.088	0.188	2.44
SO_4	2.76	n.d.	n.d.	n.d.
Total Solids	35.54	42.97	—	291.5
K/Cl	0.020	0.017	0.017	0.061
Mg/Cl	0.067	0.069	—	0.258
Br/Cl	0.0033	0.0033	0.0031	0.0133
Mg/K	3.4	4.1	—	4.2
Br/K	0.17	0.19	0.18	0.22

1. Normal sea water (Harvey, 1955, p. 4).
2. CRC 17,008-3. From lagoon shallows near beach.[1]
3. CRC 17,008-7. Tidal flat, with gypsum and algae mats, 1 km east of beach.[1]
4. CRC 17,008-20. Brine under surface of salt flat, 4 km east of beach.[1]

[1] Analyses for K and Na by Smith-Emery Company, for all other elements by Gale Baker.

have resulted from a fortuitous combination of low runoff and high evaporation.

POLYHALITE

Figure 2 shows where the polyhalite was found, in two large embayments on the salt flats, 4–8 km from the edge of the lagoon. The polyhalite occurs as a 1 cm thick layer of white hardpan underlying about 5 cm of coarse halite-gypsum sand, interbedded with some organic material (Fig. 4). At the northwesternmost locality (Fig. 2) coring showed scattered polyhalite to a depth 1.2 meters in halite and gypsum-rich silts. This may be the "dünnen Calichebändern" identified by Wittich (1916, p. 26, 28) as calcite.

An *x*-ray diffraction pattern of the white hardpan material showed many lines of polyhalite coincident with the new standard pattern (XRDF 10–355) for polyhalite. A minor amount of polyhalite was also seen in *x*-ray diffraction patterns of an underlying black sand.

A thin section of the hardpan confirmed the presence of polyhalite.

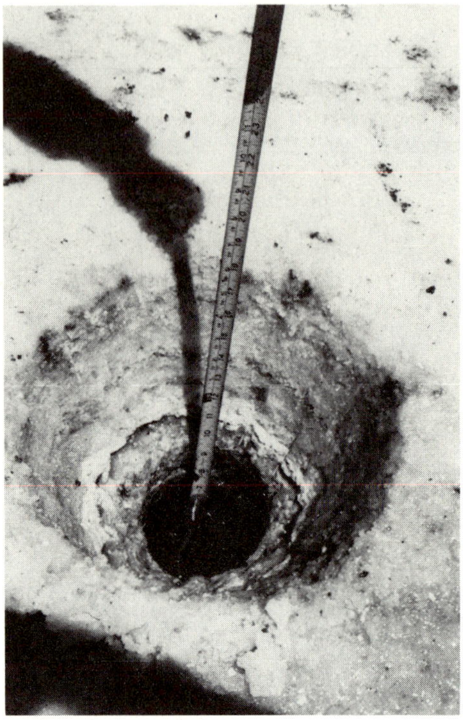

FIG. 4. Evaporite sediments in the polyhalite area. Brine, which normally levels near the top of the pit, was scooped out to display the sedimentary structure. White layers are halite and gypsum, dark layers are mainly decomposed grass, and the dead white layer near the bottom of the pit is polyhalite.

As shown in Fig. 5, the outlines of gypsum crystals are easily seen in plain polarized light. However, examination under crossed polarizers shows that only a small central part of each crystal is really gypsum. This remaining gypsum is surrounded by a narrow band of radially oriented, fibrous polyhalite, and the remainder of the crystal is an aggregate of 0.005 mm crystals of polyhalite. Polyhalite also occurs between evident gypsum pseudomorphs, where it may represent either a cement or replacement of a gypsum cement.

The extremely fine size of the polyhalite crystals precluded detailed optical measurements, but the average refractive index is normal for the mineral. The fibrous polyhalite rims on the gypsum are length-slow. The individual polyhalite crystals have a very irregular extinction, with a suggestion of spherulitic structure. Gypsum crystals in all stages of replacements are found in this layer, although most are nearly completely replaced by polyhalite. The texture is very similar to the mosaic and fibrous polyhalite crystals described from evaporite rocks of various ages and localities (Braitsch, 1961; Celet, 1953; Ladame, 1942; Schaller and Henderson, 1932; Stewart, 1949).

Some crystals are not only replaced by polyhalite, but have rims of an

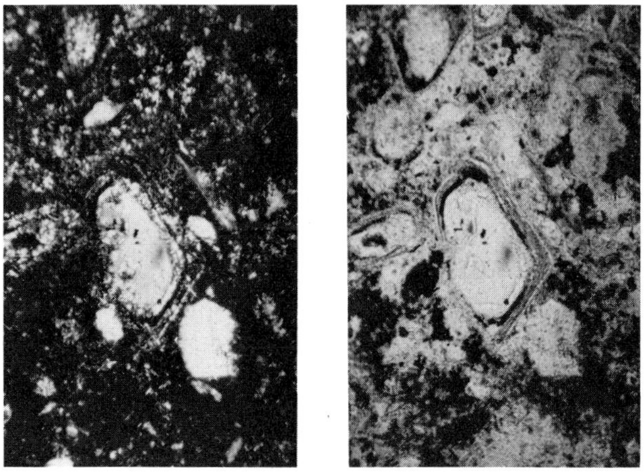

FIG. 5. Photomicrographs of polyhalite replacing euhedral gypsum crystals; interstitial material is also polyhalite. ×40, plane polarized light on left, crossed polarizers on right.

opaque white mineral that also appears as a cement between the grains along with polyhalite. At high magnification this material is found to be composed of very small (less than 0.1μ) equant grains of high refractive index and high birefringence, but it has not been identified.

The occasional detrital grains of plagioclase, hornblende, and biotite are unaltered.

BRINE AND CRYSTAL EQUILIBRIA

The question arises as to why calcium sulfate was deposited first as gypsum and then altered to polyhalite. Can it be explained on the basis of equilibrium with a brine of changing composition, or does one have to assume that the gypsum was crystallized metastably and then converted to a more stable polyhalite?

Considerable information is available on the equilibrium relations of calcium sulfate minerals in the sea water system. Although at 25° C.

gypsum is the stable crystal in equilibirum with sea water concentrated by a factor of up to 5.5×, anhydrite is the stable mineral at higher sea salt concentrations, at least up to 11.5×, where halite starts to precipitate (*e.g.*, D'Ans *et al.*, 1955). On the other hand, laboratory experiments and field observations indicate that the crystal actually growing from these brines is usually metastable gypsum, even along with halite.

The experiments by Van't Hoff and D'Ans (D'Ans, 1933, see also, Autenreith, 1958; Braitsch, 1962) show that polyhalite is the form of calcium sulfate at equilibrium by the time the sea salt is concentrated enough to precipitate potash-magnesia salts at about 63×. In a much-quoted calculation, Janecke (1912) placed the transition anhydrite-polyhalite at about 22× sea salt concentration but there seems to be no good data on which to base such a calculation. However, it is still rather certain that the transition takes place somewhere in the halite facies. After the simple evaporation of sea water to this stage of evaporation, there is not much sulfate available to form polyhalite, it all having been precipitated previously as gypsum or anhydrite. However, if additional sea water or sulfatic meteoric water is added to the system, sulfate should precipitate. Experiments by Lepeshkov and Novikova (1958) indicate that, unlike anhydrite, polyhalite is easily crystallized when such an addition of sulfate is made to the concentrated sea water. There is, therefore, less chance for the metastable crystallization of gypsum from these highly concentrated brines.

This suggests that at Laguna Ojo de Liebre the brine has become much more concentrated since the original crystallization of the gypsum, leading finally to replacement by polyhalite. The conversion of gypsum or other sulfates into polyhalite, within the stability range of the latter mineral, has been shown experimentally by Van't Hoff (1912).

It is interesting to note that no trace of anhydrite was found in any sediments at Laguna Ojo de Liebre, despite a diligent search for it.

Evidence for a change in concentration of the brine since crystallization of the main part of the evaporite is also shown by studies of bromide geochemistry. The pioneering studies of D'Ans and Kuhn (1940) and much subsequent work in Germany, the Soviet Union, and elsewhere, have proved that the bromide in solid solution in halite is a direct measure of the stage of sea salt concentration in the brine from which the halite crystallized. The distribution coefficient for bromide between crystal and brine, based on measurement of both artificial and natural solutions, is such that the first halite crystals to grow from the present-day sea water should have a bromide content of about 70 ppm. As evaporation proceeds, with or without additions of new sea water, the major part of the bromide, rejected from crystallization with the salt, builds up in concentration. Consequently, the bromide content of new salt crystals is higher; and at the point where potash-magnesia minerals begin to crystallize it is about 250 ppm. At Laguna Ojo de Liebre, samples of halite near the polyhalite locality analyzed 51 and 92 ppm.* Brine that flowed from the salt into the hole from which these samples were taken gave the analysis shown in Table 1, column 4. Not only the bromide, but also the magnesium and potassium contents of the brine (relative to chloride) indicate a sea salt concentration of 60×, nearly to the potash-magnesia facies. A brine of this composition should be in equilibrium with halite containing about 270 ppm Br.

In a series of publications, Valiashko (1951; 1959) has emphasized the change of physical conditions that occurs when an evaporite is far enough along in its course of crystallization to form a solid but porous crystalline body. He suggests that the process stabilizes at this point, with little further evaporation, awaiting tectonic or other changes. Valiashko has calculated that such a stage is reached with about 50 per cent porosity, near the beginning of the potash-magnesia facies, and I suggest that loss of brine by reflux (Scruton, 1953) may bring on consolidation of the evaporites at an earlier stage. Also evaporation may still be important from this aggregate, with the assistance of capillary movement of brine upward through the pores. This may result in a higher stage of evaporation of the brine, with the deposition of only a small quantity of late-stage halite around the pores. At Laguna Ojo de Liebre, the occurrence of polyhalite in a band cementing the loose sediments just under the surface of the salt flat suggests such a caliche-type deposit. The high-Br halite that should be associated with the brine might be present in only a small amount that was overlooked in our sampling.

Similar apparent disequilibria of bromide between crystal and brine have been found by analyses of brine inclusions in halite, of Permian age from Kansas and Silurian age from Ontario. These may have a similar origin (Holser, 1962).

Geological Significance

Much of the polyhalite in the geological section has long been recognized as a replacement of gypsum or anhydrite, because of pseudomorphic or other replacement textures. The early experiments of Van't Hoff supported the feasibility of the replacement reaction. The replacement origin has been well documented in many occurrences, from the Permian of New Mexico (Schaller and Henderson, 1932), England (Stewart, 1949), and Germany (Schulze, 1960), and in the Miocene of Ukrainian SSSR (YArzhemskii, 1949). In all of these localities the rocks have been deeply buried since deposition, so that calcium sulfate is also present as

Footnote added 15 June 1972: Later analyses of 25 samples of halite from cores in this area indicate a range of 100–195 ppm Br, which represents, in general, a smaller discrepancy from theoretical values than the earlier analyses (the single earlier analysis at 50 ppm may have been an isolated case of recrystallized halite).

anhydrite rather than gypsum. Pseudomorphs after early gypsum are common, having been replaced by anhydrite, halite or polyhalite. Polyhalite has also been found in lake beds in Central Asia, of Tertiary (Voronova, 1954) and Pleistocene (Norin, 1942) age, but details are lacking.

That the replacement of gypsum or anhydrite by polyhalite may have been diagenetic was suggested by Stewart (1949), Schaller and Henderson (1932), and Schulze (1960), as well as in the reviews by Lotze (1957), and Borchert (1959). However, all of these writers, except Schulze, still have some doubt as to whether the replacement by polyhalite may have taken place at a later time, after consolidation. Polyhalite diagenetically replacing gypsum at Laguna del Ojo de Liebre lends credence to a diagenetic origin for polyhalite in these earlier rocks. It also minimizes the necessity for an intermediate stage of anhydrite or bassanite, which was suggested by Dunham (1949), Schaller and Henderson (1932) and Braitsch (1962), without any direct evidence for either mineral. Such an intermediate stage, for which Stewart has considerable evidence in other transformations, is apparently not always necessary.

The relations between brines and evaporites at Laguna del Ojo de Liebre also suggest that some alterations to polyhalite, and perhaps other replacements, may be caused by a very localized change in brine composition, similar to caliche formation, without requiring the development of an extensive body of concentrated brine. Therefore, an occurrence of polyhalite in evaporite rocks has less significance as a lead to possible near deposits of soluble potash minerals, than might otherwise have been supposed.

References

D'Ans, Jean (1933) *Die Lösungsgleichgewichte der Systeme der Salze ozeanischer Salzablagerungen*. Berlin, Verlagsgesellschaft für Ackerbau M.B.H., 2 vol.

——— and Robert Kuhn (1940) Über den Bromgehalt von Salzgestein der Kalisalzlägerstatten. *Kali*, **34,** 43–46; 59–64; 77–83.

——— *et al.* (1955) Untersuchungen über die Calcium-sulfate. *Kali und Steinsalze*, 1 (9), 17–38.

Autenreith, Hans (1958) Untersuchungen im Sechs-Komponenten-System $K^{\cdot}$, $Na^{\cdot}$, $Mg^{\cdot\cdot}$, $Ca^{\cdot\cdot}$, SO_4'', (Cl'), H_2O mit Schlussfolgerungen für die Verarbeitung der Kalisalze. *Kali und Steinsalz*, 2 (6), 181–200.

Borchert, Hermann (1959) *Ozeane Salzlägerstatten*. Berlin, Gebrüder Borntraeger.

Braitsch, Otto (1961) Zur Kristallographie von Polyhalit. *Beitr. Mineral. Petrog.* **8,** p. 84–91.

——— (1962) *Enstehung und Stoffbestand der Salzlägerstatten*. Springer Verlag, Berlin.

Celet, P. (1953) Etude petrographique de la polyhalite du Keuper salifère du Jura. *Soc. Geol. Nord., Ann.* **73,** 171–181.

Dunham, K. C. (1948) A contribution to the petrology of the Permian evaporite deposits of northeastern England: *Yorkshire Geol. Soc. Proc.* **27,** 217–227.

HARVEY, H. W. (1955) *The Chemistry and Fertility of Sea Waters.* Cambridge, University Press.

HOLSER, W. T. (1963) Chemistry of brine inclusions in Permian salt from Hutchinson, Kansas. *Symposium on Salt, 1st, Cleveland, Ohio, May 1962, Northern Ohio Geol. Soc., Cleveland,* 86–95.

JÄNECKE, E. (1912) Eine graphische Darstellung der Gewichtsverhältnisse bei den ozeanischen Salzablagerungen. *Kali* **6,** 647.

LADAME, GEORGES (1942) Le gisement potassifère de Mamane en Iran. *Schweiz. Mineral. Petrog. Mitt.* **22,** 242–247.

LEPESHKOV, I. N. AND L. V. NOVIDOVA (1958) Formation of primary polyhalite. *Zhur. Neorgan. Khimii.* **3,** 1261–1264 (Calresearch Trans.).

LOTZE, FRANZ (1957) *Steinsalz und Kalisalze.* Berlin, Gebrüder Borntraeger, v. 1, 2nd ed.

NORIN, ERIC (1942) Two salt basins in Central Asia. *Lunds Univ. Geol.-Mineral. Inst., Medd.* **97.**

PHLEGER, F. B. AND G. C. EWING (1962) Sedimentology and oceanography of coastal lagoons in Baja California. *Geol. Soc. Am. Bull.* **73,** 145–182.

MINA U., FREDERICO (1957) Bosquejo geologico del territorio sur de la Baja California. *Assoc. Mex. Geol. Petrol., Bol.* **9,** 139–269.

SCHALLER, W. T. AND E. P. HENDERSON (1932) Mineralogy of drill cores from the potash field of New Mexico and Texas. *U. S. Geol. Survey Bull.* **833.**

SCHULZE, Günter (1960) Beitrag zur Genese des Polyhalites im deutschen Zechsteinsalinar. *Zeit. angew. Geol.* **7,** 310–317.

SCRUTON, P. C. (1953) Deposition of evaporites. *Am. Assoc. Petrol. Geol. Bull.* **37,** 2498–2512.

STEWART, F. H. (1949) Petrology of the evaporites of the Eskdale no. 2 boring, east Yorkshire, Part I: *Mineral. Mag.* **28,** 621–675.

VALIASHKO, M. G. (1951) Volume relations of liquid and solid phases in evaporating sea water. *Akad. Nauk S S S R, Dokl.,* **77,** 1055–1058 (SLA Transl. 61-13083).

——— (1956) Geochemistry of deposits of potassium salts. *In, Voprosy Geologii Agronomicheskikh Rud, Moskva, Izdat. Akad. Nauk S S S R,* 182–207 (Calresearch Transl.).

——— (1959) Die wichtigsten geochemischen Parameter für die Bildung der Kalisalzlägerstatten. *Frieberger Forschungshefte* **A123,** 197–235.

VAN'T HOFF, J. H. (1912) *Untersuchungen über die Bildunsverhältnisse der Ozeanischen Salzablagerungen.* Leipzig.

VORONOVA, M. L. (1954) Discovery of polyhalite and kainite in sulfate salts of Uzun-Su. *Akad. Nauk S S S R, Dokl.* **99,** 449–450, [*Chem. Abs.* **49,** 7453g].

WITTICH, E. (1912) Über Meeresschwankungen am der Küste von Kalifornien. *Deutsche geol. Gesell., Monatsber.* **64,** 505–512.

——— (1912) Die Salzlager am Ojo de Liebre au der Westküste von Nieder-Kalifornien. *Zent. Mineral.* 25–32.

YARZHEMSKII, YA. YA. (1949) Polyhalite in potassium salt deposits: *Akad. Nauk S S S R, Dokl.* **66,** 1157–1160 (*Mineral. Abs.* **11,** 59).

Manuscript received, April 1, 1965; accepted for publication, October 6, 1965.

Reprinted from *Amer. Assoc. Petroleum Geologists Bull.*, **53**, 830–840 (1969)

Modes of Formation, Sedimentary Associations, and Diagnostic Features of Shallow-Water and Supratidal Evaporites[1]

11

DAVID J. J. KINSMAN[2]
Princeton, New Jersey 08540

Abstract Production of brine depends critically on favorable physiographic setting and climate. The physiographic requirements are not common on a large scale, but are common on a small scale in the form of lagoons, pans, and exposed supratidal surfaces. Such environments may be present on the peripheries of large brine basins. The most common evaporite environment is the exposed salt flat or *sabkha*,[3] a fairly level, salt-encrusted surface that only occasionally is inundated. *Coastal sabkhas* are supratidal surfaces formed by depositional offlap of marine sediments, and the associated evaporites are precipitated from seawater-derived basins. *Continental sabkhas* comprise continental or earlier cycle marine sediments, and the associated evaporites are precipitated from evaporated continental waters. Sabkhas are equilibrium geomorphic surfaces, their levels being dictated by the local level of the groundwater. Water-table depths in very few places exceed 1–2 m. Sabkhas are formed where the eolian sand supply is limited and where hard rock lies below the level of the groundwater table. Framework sediments of coastal sabkhas may range from almost pure carbonate to noncarbonate, depending on local supply of terrigenous detritus. Sedimentary evidences of emergence and desiccation are abundant. Evaporite pans may be associated with sabkha environments, but they have distinctive characteristics. Illustrative examples from the Persian Gulf and Baja California are given. Particularly significant are the high rates of shoreline regression (1–2 m/yr), which would allow a sabkha 100–200 km wide to be formed in only 10^5 years. The concept of a *megasabkha*, as being an evaporite environment rather different from a sabkha, probably is erroneous.

Sabkha diagenesis involves interstitial emplacement of evaporite minerals within the host sediments and changes in the host sediments themselves, such as dolomitization. Evaporites are restricted largely to the upper levels, but dolomitization may extend in depth as brines move downward and seaward through the sediment wedge. Sabkha gypsum crystals typically are lensoid and commonly contain inclusions, in contrast to brine-pan gypsum crystals, which are elongate and free from inclusions. Sabkha anhydrite is typically nodular. Gypsum and anhydrite emplaced diagenetically within sabkha sediments in places are crudely layered but not laminated, whereas sedimentary lamination of detrital sediment and evaporite minerals commonly is well developed in brine-pan environments. Brine pans usually accumulate more soluble evaporites than sabkha areas because of the greater evaporation losses from exposed surfaces.

In addition to diachronism of sedimentation, coastal sabkhas show diachronism of diagenesis, the evaporite-mineral belts moving progressively seaward as depositional offlap proceeds. Early stages of diagenesis are masked by later stages and, during this process, the area occupied by the younger, more soluble evaporite minerals increases. Data from the Persian Gulf indicate that with time the coastal sabkha zone dominated by gypsum is shifted progressively seaward but its area remains roughly constant, whereas the area dominated by anhydrite progressively increases. Thus a fossilized sabkha will not necessarily contain all the minerals, or even traces of the minerals, that were developed within it during early diagenesis, even if late diagenetic changes are minor. Many early diagenetic minerals are ephemeral phases only.

Introduction

Ancient evaporite deposits, some of considerable vertical thickness and area, are present throughout the geologic record. Most evaporites show an overall chemical composition and sequence of minerals indicative of probable precipitation from a seawater-derived brine. The significance of brines derived from the evaporation of continental waters, or mixed continental and marine waters, is largely unknown. Evaporite deposition requires situations in which a regime of net evaporation may develop. A large area of the world has an arid climate, but the distribution of Holocene evaporites suggests that, although an arid climate is essential for evaporite formation, it is not of primary significance. The major requirement is that mixing and exchange with the ocean, or other source of diluting waters, be restricted. Even when all climatic factors cause conditions of great evaporation, the annual evaporation loss rarely exceeds 5 m and commonly is less. Therefore, dilution must be severely limited if a net increase in dissolved salt concentration of a solution is to be preserved.

[1] Manuscript received, August 11, 1968; accepted, October 31, 1968.

This paper is a progress report; field and laboratory studies are still under way on many aspects of the work described herein.

[2] Assistant Professor of Geology, Princeton University, on leave at the Sedimentology Research Laboratory, Reading University, England.

The American Association of Petroleum Geologists, American Petroleum Institute, National Science Foundation, and Petroleum Research Fund of the American Chemical Society provided generous support of Persian Gulf studies.

[3] The spelling *sabkhah* which appeared in the AAPG-sponsored research project report, "Early Diagenesis of Carbonate Sediments in a Supratidal Evaporitic Setting" (AAPG *Bulletin*, October, 1968, p. 2071–2072) was in error and should read "sabkha". A discussion of this word is in the present paper.

Physiographic requirements for a barrier sufficient to restrict active ingress or egress of waters in all but very limited quantities are very exacting. The fact that such barriers are uncommon probably explains the absence of large marine-brine basins today and the sporadic development of large brine basins during geologic history. The coincidence of favorable climatic and physiographic factors, although achieved on a large scale in very few places, is more common on a small scale in the form of lagoons, pans, and exposed salt-flat areas during processes of nearshore sedimentation.

In an arid region several different physiographic and hydrologic situations are possible, depending on the degree of isolation of the potential brine from dilution (Table I). The interrelations of the different physiographic environments of evaporite deposition are indicated schematically in Figure 1.

Environment A (Table I) is an unrestricted marine basin with unrestricted normal or near-normal marine peripheral lagoons. Marginal accumulations of normal marine sediments ultimately form exposed supratidal surfaces, and brines develop interstitially within the sediments. Incomplete sedimentation may isolate shallow depressions which may carry semipermanent bodies of brine. Environment B is an unrestricted basin with peripheral restricted lagoons in which standing bodies of brine develop. The marginal supratidal facies also are evaporitic. Environment C is a restricted basin with a large standing body of brine, possibly of considerable depth. All marginal environments are evaporitic.

The evaporite sequences of the lagoons in environments B and C are different; moreover, the supratidal evaporite sequences are different in all three environments. Modern examples of environment C are unknown, although the Dead Sea provides some understanding of it. Environment B also is represented poorly today, although there are a few natural pans and also man-made brine ponds from which

Table I. Environments of Evaporite Deposition in Arid Region

Basin		*Marginal Lagoon-Pan*		*Supratidal Environment (May include Pans)*
Unrestricted	*Restricted*	*Unrestricted*	*Restricted*	
A Marine	—	Marine	—	Brine
B Marine	—	—	Brine	Brine
C —	Brine	—	Brine	Brine

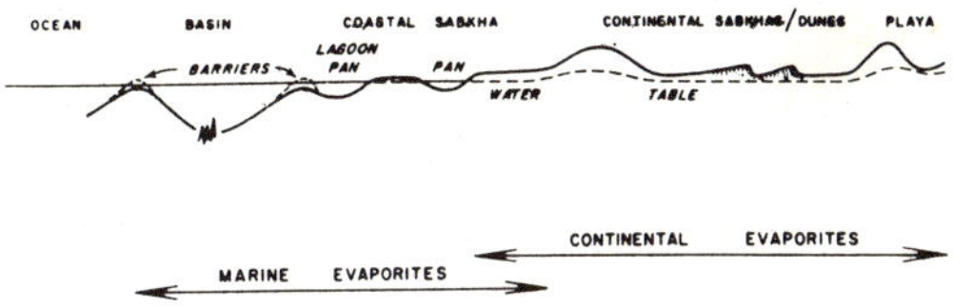

FIG. 1.—Schematic representation of evaporite environments. Evaporites of coastal sabkhas and seaward environments are derived from seawater. Continental sabkha and playa evaporites are derived from continental waters. Mixed zone commonly is developed in inner coastal sabkha areas.

useful information can be derived. Environment A is fairly well represented today, and an example from the Persian Gulf is described and compared briefly with similar areas in Baja California.

The possible geologic significance of the supratidal salt flat or sabkha areas of the Persian Gulf was realized by groups working in Qatar and along the Trucial Coast about 6–7 years ago, and some of this early descriptive work has been published (see references *in* Evans, 1966; Illing *et al.*, 1965; Kinsman, 1964, 1966, 1967). The understanding of this type of environment is incomplete and much study remains to be done. A new program of research, begun in July 1967, is an attempt to acquire a better understanding of sabkha evaporites and sabkha diagenesis.

PERSIAN GULF

Sedimentary Environment

The Persian Gulf and adjacent continental areas are extremely arid. However, dilution by Indian Ocean seawater maintains Persian Gulf seawater salinity at a value only slightly higher than that of average ocean water. The Trucial Coast (Fig. 2) is fringed with islands that partly restrict the coastal lagoons, within which seawater salinity values may be as great as 60–70 parts per thousand. However, no evaporite-depositing, free-standing bodies of brine are known.

Along the Trucial Coast a low hinterland relief, a predominantly onshore wind, and a very low rainfall together reduce terrigenous input into marine areas. Consequently, the marine sediments are almost pure carbonates. A broad shelf on which skeletal sands are accumulating extends offshore. On this shelf are small areas of skeletal-gravel accumulation on local highs and carbonate-mud accumulation in local depressions. Some shoal patches are a reef fa-

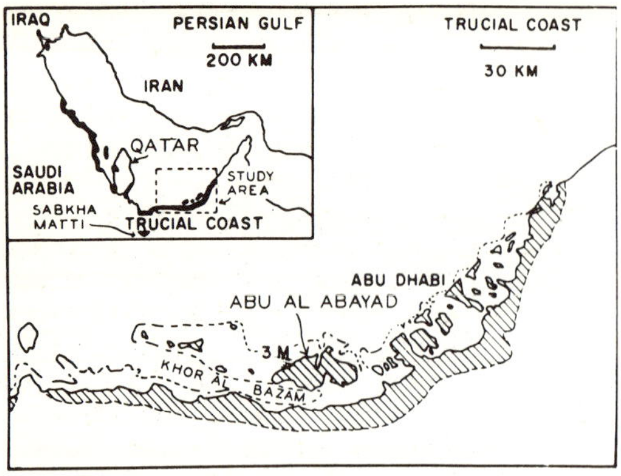

FIG. 2.—Inset map shows Persian Gulf and adjacent countries; solid black indicates locations of major coastal sabkhas; black area below letter *R* of Trucial Coast indicates inland extent of large continental sabkha, Sabkha Matti. Main map shows islands and lagoons of Trucial Coast; hachured areas are predominantly coastal sabkha.

cies. The western part of the island and lagoonal complex comprises a broad shoal area where water depths generally are less than 2 m. This shoal area is dominated by Abu al Abayadh, the largest island along the Trucial Coast (Fig. 2). Landward is a long lagoon, Khor al Bazam, parallel with the coast. On the shoal area coral reefs and skeletal sands are accumulating, and finer grained sediments and small areas of blue-green intertidal algal mats are forming in the lee of some islands. The lagoon itself has a muddy skeletal-sand floor. The landward shore of the lagoon is composed mainly of pellet and skeletal sand and a few areas of intertidal algal mat in sheltered positions. In the general lack of oölite development and paucity of fine-grained carbonate sediments, this western area contrasts markedly with the eastern area.

Barrier islands east of Abu al Abayadh restrict the lagoons (Fig. 2); thus tidal movement of seawater is essentially normal to the coast and largely restricted to tidal channels, rather than parallel with the coast as in Khor al Bazam. The lagoonal barrier sediments consist of oölite sands at lagoon entrances, and coral reef and skeletal sands are accumulating in front of the islands. Coral reefs also are present behind the oölite shoals in some lagoons. The barrier thus consists of an alternating series of oölite and reef facies, and oölite dunes are well developed on some barrier islands. In mid- and inner lagoonal areas, pellet sand and loosely pelleted or agglutinated aragonite mud are accumulating. A spectacular development of intertidal algal mats occurs along many inner lagoon shores. The algal mats trap carbonate sediment and are preserved within the sediment as stromatolitic laminae.

Infilling of the lagoons proceeds by addition of sediment to nearly all shorelines, although rates of shoreline regression differ with orientation of the shore. A shoreline regression rate of 1–2 m/yr has prevailed in many areas during the past 4,000–5,000 years. The infill pattern is complex and, though in supratidal areas the sediment facies are essentially horizontal, they are diachronous. Pits and cores in supratidal areas show a general vertical sucession of subtidal, intertidal, and supratidal facies. The thickness of the intertidal facies in exposed barrier positions is 2 m, whereas in sheltered inner lagoon positions it is only 0.7 m; the difference reflects tidal diminution with increasing distance back into the lagoon.

Lagoonal sediments generally become finer upward, as lagoons have become progressively infilled. Ultimately, behind the oölite and reef facies of the high-energy barrier environment are aragonite muds of the sheltered, low-energy, lagoonal environment. Oölite formation apparently ceases when the tidal exchange of seawater into the lagoons becomes very small. Future development in the barrier area seems to involve formation of a linear, fringing coral reef with associated skeletal-sand facies.

The Holocene carbonate-sediment complex of the inner shelf and lagoonal areas grades landward into broad, salt-encrusted, supratidal surfaces or coastal flats, known in Arabic countries as *sabkha.* The original Arabic word, as translated and rendered phonetically by writers of different nationalities, has taken a variety of forms (sabkha, sabka, sabkhah, sabkhat, sebkha, sebkhat, sebjet). The word usually is applied to salt flats or salt marshes which are inundated occasionally, although in a few places the word is applied to more permanent salt lakes or even very shallow marine lagoons. The writer proposes that the form *sabkha* be used, with the connotation of a salt flat which is inundated only occasionally. In other parts of the world similar features are known by a variety of names, of which *salina* is probably the most common. The form *sebkha* was used by Illing *et al.* (1965) in a study on the Qatar Peninsula, but in later discussion Illing and Taylor (1967) suggested that this spelling be replaced by *sabkha.*

The term "sabkha" normally is applied to both coastal and continental salt flats, although Illing and Taylor (1967) suggested that it be restricted to coastal salt flats because the continental salt flats are "covered by the well established term *playa*." However, more recent field work by the writer and his colleagues has shown that only a few continental salt flats are true playas; most are open hydrologic systems and the most soluble evaporite minerals typical of playas do not accumulate on them. Coastal and continental salt flats share more similarities than differences and it might be inaccurate (in the ancient record it might be impossible) to distinguish between them by calling the former sabkhas and the latter playas.

It may seem unnecessary to refine the definition of sabkha, but the sedimentologic differences between the infrequently inundated surface and the surface which normally is covered by a shallow sheet of standing water, or brine, are significant. The sediment transport and deposition mechanisms are very different and the types, structures, and textures of associated evaporite minerals commonly are distinctive.

There are essentially two types of sabkhas, coastal and continental. Both are equilibrium geomorphic surfaces. Their level is dictated dominantly by the local level of the groundwater table. They are developed where eolian sand supply is limited and where hard rock lies below the groundwater table.

The modern coastal sabkhas are the result, at least in the seaward parts, of depositional offlap of marine sediments of subtidal, intertidal, and supratidal facies. These sabkhas are supratidal surfaces; *i.e.,* they are the normal end product of nearshore marine sedimentation whereby sediments are laid down just above high-water mark and the shoreline gradually moves seaward. Local climate dictates subsequent developments (*i.e.,* whether a salt flat or a freshwater marsh develops). If a salt flat forms, the associated brines normally are derived from the evaporation of seawater.

Continental sabkhas are landward from the coastal sabkhas and comprise continental or earlier cycle marine sediments. Continental sabkhas represent "deflation-sedimentation windows" through to the local water table and their elevation generally increases away from the coast. The water table may be the regional water table or local perched water tables such as those associated with large dunes. Dune-sand areas commonly have hard rock in interdune deflation hollows where consolidated strata are above the water table. Where the water table overlies hard rock, the base level of wind deflation is controlled by the water table or, more correctly, by some level just above the capillary fringe. In the latter example, the equilibrium-deflation surface becomes a salt-encrusted continental sabkha. The water table is 1–2 m from the surface and evaporation brings about brine production and associated interstitial precipitation of evaporite minerals.

The coastal sabkhas of the Trucial Coast consist of Holocene carbonate sediments, at least in the seaward parts, and cover an area of about 2,000 km^2. The continental sabkhas consist of older cycle sediments, commonly noncarbonate, some of which are being redistributed by eolian activity. The coastal sabkhas grade laterally in many places into continental sabkhas without noticeable morphologic discontinuity. Within the belt 100 km from the coast, continental sabkhas cover an area of 6,000–8,000 km^2. This figure includes the innermost limits of the coastal sabkhas and the more elevated southern sabkhas.

Coastal Sabkhas

The normal end product of nearshore sedimentation is a supratidal surface. At that stage, marine sedimentation is essentially complete, and further addition of marine sediment occurs only when storm tides flood the surface. Supratidal surfaces in regions of high net evaporation form salt-encrusted coastal flats or coastal sabkhas. Along the Trucial Coast coastal sabkhas are developed extensively and are locally 8–10 km wide. Most of the barrier and other nearshore islands also are predominantly sabkha surfaces (Fig. 2). The western half of the coastal sabkha strip indicated in Figure 2 is broken by ridges of Miocene rocks into separate coastal sabkhas. Fairly extensive sabkhas also are present along the coasts of Saudi Arabia, Bahrain, and Qatar.

The coastal sabkhas are just above the high-tide level, and rise at the inner margins where they abut older rocks or debris fans. Pits and cores show that the sabkha is a wedge of marine sediments of facies similar to those accumulating in the present marine and intertidal areas, capped by a thin supratidal facies which is overlain by a thin eolian facies. Intertidal stromatolitic algal sediments along a profile near Abu Dhabi Island, and the associated radiocarbon dates, indicate that regressive sedimentation has occurred during the past 4,000–5,000 years.

In addition to their primary marine sediment characteristics, the coastal sabkhas have many other features, particularly early diagenetic minerals. The new minerals are formed in response to the chemical and physical conditions of the sabkha environment. The minerals of the marine carbonate sediments represent a relatively stable marine assemblage, even though aragonite, the dominant mineral, is metastable at earth-surface temperatures and pressures. The fact that aragonite is not replaced by calcite in the marine environment within a span of 10^3–10^4 years indicates its relative stability. However, the temperature and solution conditions of the sabkha environment are very different from those of the marine environment. For example, in the lagoons of the Trucial Coast, seawater salinity values are 1.2–2 times the values of normal ocean water, whereas interstitial pore fluids in the sabkha sediments are seawater-derived brines which, in places, are 10 times as concentrated as normal seawater. In the lagoons the average annual seawater temperature is 29°C, the seasonal range being 15–40°C; in contrast, sabkha brine temperatures at 1-m depth average 34°C, and sabkha surfaces may range from 10 to 60°C.

Evaporation from the sabkha surface causes interstitial pore-fluid concentration. Water loss is made up during storm flooding and perhaps also by lateral flow from a seaward direction. The standing level of the interstitial brine normally is less than 1 m below the sabkha surface. At the inner margins of the coastal sabkhas there is evidence of influx of continental groundwater. Formation of the brine is accompanied by direct precipitation of evaporite minerals and by replacement reactions between the original sediment components and the interstitial brines.

A small amount of early diagenetic aragonite is precipitated within intertidal zone sediments, locally causing lithification of surface sediments into thin crusts. Many of these crusts are broken subsequently and form areas of flat-pebble conglomerate. Gypsum is abundant, first appearing within the upper intertidal zone sediments. Gypsum-crystal mushes up to 1 m thick of flattened discoid or lensoid crystals are common. The crystals are precipitated interstitially within the host sediments and are emplaced by physical or mechanical displacement of host-sediment grains. In places host sediment or even algal laminae are entrapped within the crystals, and in some areas large gypsum sand crystals poikilitically[4] enclose sand grains of the original sediment. Nodular anhydrite, with felted lath texture, is present inland from the gypsum belt, and is the dominant calcium sulfate mineral within the sediments both above and extending to 20 cm below the standing level of the brine across most of the coastal sabkhas.

Early field work showed no evidence of anhydrite pseudomorphs after gypsum. Yet the distribution of gypsum and anhydrite, together with crude mass-balance calculations, indicated that much of the earlier formed gypsum must dissolve and the ions migrate to new sites where they precipitate as anhydrite. More recent field work has shown that gypsum crystals in the upper 4–5 cm of the sabkha are replaced *in situ* by anhydrite, giving rise to well-formed pseudomorphs. The most seaward anhydrite is of this origin. The pseudomorphs begin to lose their shape with time, and primary diagenetic anhydrite (anhydrite with no gypsum precursor) is precipitated upon and between the older pseudomorphs, producing as an end product the larger ovoid and variously shaped anhydrite nodules.

The replacement mechanism is not known —*i.e.*, whether solid-state dehydration or dissolution-reprecipitation has occurred. Where influx of continental waters occurs in inner coastal sabkha areas, anhydrite commonly is hydrated to gypsum. The multicrystalline anhydrite nodules, although many are soft and incompetent, grow within the host sediments by physical displacement. There is little or no evidence that gypsum or anhydrite replaces any of the original carbonate sediment. Some gypsum crystals exposed on the sabkha surface are found to have undergone dehydration and are covered with a thin skin of bassanite. However, that mineral does not extend more than a few millimeters below the surface.

The precipitation of diagenetic aragonite, gypsum, and anhydrite increases the molar magnesium-calcium ratio of the pore-fluid brines. Values in excess of 12 have been recorded. The comparable seawater value is 5. Under these conditions the original carbonate sediments, where fine grained, are dolomitized. However, before appreciable dolomitization

[4] Editor's note: *Dictionary of Geological Terms* defines poikilitic as a "textural term denoting a condition in which small granular crystals are irregularly scattered without common orientation in a larger crystal of another mineral."

takes place, magnesite may be precipitated in the uppermost 10–20 cm of the sediments. Continued dolomitization lowers the molar magnesium-calcium ratio to values of 4 or slightly less. The formation of this dolomite is best described as penecontemporaneous, prelithification dolomitization, and results in formation of a dolomite mud of 1–2 μ crystals. Dolomitization of sand-size carbonate sediments apparently is very slow. Consequently, Holocene dolomite is restricted largely to the eastern coastal sabkhas, where fine-grained carbonate sediments are present.

Celestite is a fairly common, early diagenetic, minor mineral of the coastal sabkhas and is most abundant in areas of intense dolomitization, being largely a byproduct of the replacement of aragonite (7,000–8,000 ppm Sr) by dolomite (600–700 ppm Sr). Huntite, $Mg_3Ca(CO_3)_4$, is an uncommon diagenetic carbonate mineral also recorded from the coastal sabkha. Halite forms a crust on much of the sabkha surface but is not an accumulative phase. Some of the halite is blown away by the onshore winds. Perhaps a more important removal mechanism is the flooding of the sabkha by storm water, some of which subsequently flows seaward, removing the dissolved halite in solution. Other fractions of the storm waters percolate downward into the sabkha and recharge the brine reservoir, or evaporate from the sabka surface to form a new halite crust.

Continental Sabkhas

Coastal sabkhas, with underlying carbonate sediments, may either abut directly against Pleistocene or Miocene rocks or grade laterally without noticeable geomorphic discontinuity into inland continental sabkhas. Alluvial fan deposits may be present at the passage from sabkha to slightly elevated Miocene outcrops. The continental sabkhas are composed mainly of noncarbonate sediments (fine- to medium-grained quartz sand), although carbonate particles from surface outcrops are present in the finer fractions. Continental sabkhas near the coast commonly have a high calcite-grain content, derived from the Pleistocene "Miliolite limestone" by wind action. Continental sabkhas normally are associated with areas of dune sand. In places they are deflation hollows in a dune-dominated landscape, whereas in other areas, such as the Sabkha Matti (Fig. 2) southeast of Qatar, the continental sabkha may dominate and dunes cover only a minor part of the surface. The eolian character of all the sediments is notable in the continental sabkha-dune association. However, sedimentary structures of the sabkha and dune facies are markedly different; the sabkha shows nearly horizontal sheet lamination, in places with minor rippling, whereas the dunes show large-scale cross-bedding. The toes of dune cross-beds may be preserved in sabkha areas if the base level of wind deflation has risen since the dune cross-beds were deposited.

Continental sabkhas are equilibrium deflation-sedimentation surfaces, the continental groundwater table lying at depths of about 1 m. The water table or, more correctly, some level above the capillary fringe marks the base level of wind deflation. The largest continental sabkha near the Arabian shore is the Sabkha Matti (Fig. 2), which extends 80–100 km inland from the coast, has an area of about 6,000 sq km, and has a seaward slope of 1/2,000 (00°02′). At its northern or coastal margin the Sabkha Matti passes into a coastal sabkha of carbonate sediments. Throughout the area of the Sabkha Matti hard rock generally lies beneath the groundwater table. The origin of this large geomorphic feature is not clear (*i.e.*, either structural or erosional processes may have dominated). The depth of the surficial sediment cover in many places exceeds 10 m; hence, if erosional processes produced the Sabkha Matti, the base level of erosion has risen recently more than 10 m. In many parts of the Sabkha Matti, but especially near the margins, there is evidence of the drowning of dune cross-beds; apparently the base level of wind deflation has risen since the older cross-beds were deposited.

Groundwater in peripheral dune areas may have greatly different ionic concentrations. The sabkha surfaces are near the water table and are the sites of large evaporation losses of water vapor, with concomitant concentration of pore fluids. Thus, the sabkhas are the sites of brine formation, irrespective of the total ionic concentration of the peripheral groundwaters feeding into them. Groundwaters from the southern Sabkha Matti are mostly saturated or supersaturated by gypsum and anhydrite, and some samples are nearly saturated by halite. Appreciable halite appears to have precipitated from the most concentrated brine samples, but halite is present only in small amounts at the sabkha surface. Presumably the major fraction of the halite is blown farther inland by the northwesterly shamal winds.

Abundant gypsum and anhydrite are present

in the Sabkha Matti, precipitated from the continental groundwater-derived brines. Discoid or lensoid gypsum crystals, similar to those in coastal sabkhas, are very common. Also common are layers of massive gypsum rock, where the gypsum poikilitically encloses the host sediment to form hard layers, some of which are more than 30 cm thick. Nothing similar has been found in the coastal sabkhas. The anhydrite usually is in the form of nodules with a felted lath texture above the groundwater level. In these characteristics the anhydrite is similar to that in the coastal sabkhas. Abundant evidence has been found in the Sabkha Matti, however, of the *in situ* dehydration of gypsum to form anhydrite. Felted, lath-textured anhydrite forms unusually fine pseudomorphs of former gypsum rosettes. Gypsum at the surface generally is dehydated to bassanite or, more commonly, completely dehydrated to anhydrite. Eolian dispersal of near-surface gypsum, which has been dehydrated to anhydrite, is widespread. As a result, most eolian sands adjacent to, or particularly, downwind from continental sabkhas contain abraded multicrystalline grains of anhydrite, many of which are well rounded and of fine to medium sand size.

Baja California

The writer studied two areas of Holocene evaporite formation in Baja California in association with W. T. Holser. In the northwest Gulf of California, just north of San Felipe (Fig. 3), are extensive supratidal areas which farther north merge into the Colorado River delta complex. The sedimentology of the area has been described by Thompson (1968). The supratidal areas have evolved through the normal processes of subtidal, intertidal, and supratidal sedimentation. The sediments are largely terrigenous sand, silt, and clay, and the tidal range in places exceeds 7 m. Despite the thickness of the Holocene sediment wedge, shoreline regression rates of up to 2 m/yr have occurred during the past few thousand years. The coastal supratidal flats or sabkhas grade laterally inland to large alluvial outwash fans emerging from the mountains. High hinterland relief precludes development of continental sabkhas, except in the far northwestern part of the area.

Late-stage intertidal and supratidal sedimentation has isolated large shallow depressions, which are filled with seawater during storm flood periods. Some intertidal algal mat facies are present and high intertidal and supratidal sediments show widespread desiccation, polygon structures, and other evidences of emergence. Interstitial early diagenetic gypsum crystals are abundant and the main features of their development are similar to those described from the Trucial Coast. Early diagenetic anhydrite is present as a minor mineral phase as small nodules and irregular layers in the uppermost few centimeters of the sediments in the landward parts of the sabkha. Conditions seem to be marginally favorable for its formation and preservation.

The pan evaporites have a central sequence dominated by halite. Episodic flooding of the pan is recorded by crude layering within the halite and also by layers of gypsum crystals. Wind-blown biotite is fairly abundant throughout. The gypsum is in the form of elongate, needle-shaped crystals, free from inclusions. In marginal pan areas is a repetitive vertical sequence of detrital silt and clay, and needle-shaped gypsum-crystal layers, with halite present only as a member of the latest flood episode. Apparently the peripheral halite is dissolved at each recharge of the pan and halite only accumulates in the central pan area. The laminae of detrital sediment and evaporite minerals are true sedimentary layers. Subsequently,

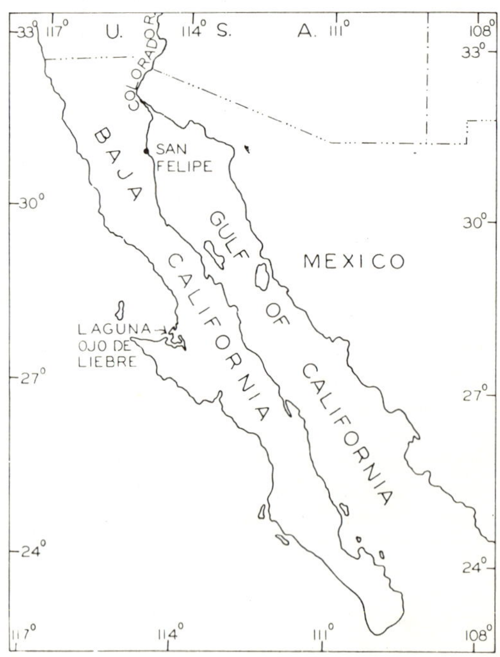

Fig. 3.—Baja California area. Locations of San Felipe and Laguna Ojo de Liebre areas are shown.

subsurface interstitial early diagenetic emplacement formed stubby, lensoid crystals of gypsum, commonly bearing inclusions. These crystals are present in both detrital silt and clay layers and original gypsum-needle layers. The diagenetic gypsum crystals generally are concentrated between the products of two flood episodes at the upper gypsum needle—lower silt layer interface. In places within these high-permeability layers all lensoid gypsum crystals are oriented, with the *c* crystallographic axes being nearly horizontal. The development of elongate crystals in the pan environment probably is caused by the more rapid rate of crystal growth resulting from rapid rates of brine concentration.

On the Pacific Coast of Baja California is a series of lagoons, the largest of which is Laguna Ojo de Liebre. The sedimentology of these lagoons has been studied by Phleger and Ewing (1962). Extensive sabkhas border the lagoons but differ from those of the Trucial Coast and San Felipe in having large areas dominated by halite. The high intertidal and seaward supratidal environments are dominated by early diagenetic gypsum, and anhydrite apparently is absent. The gypsum is in lensoid crystals and locally is fairly well layered, apparently as a result of eolian sheet layering. The prevailing onshore winds blow gypsum crystals and sand grains across the supratidal surface; some leave the sabkha areas entirely and form gypsum dunes, and others are redeposited on the sabkha surface. The layering is caused by differences in the ratio of gypsum crystals to detrital sand grains. Thin layers of fine-grained carbonate minerals (mainly magnesium calcite) which are present within the intertidal sediments commonly are completely dolomitized. Holocene magnesite also has been recorded from this sabkha area and celestite is fairly common in some locations. Early diagenetic gypsum is replaced by polyhalite in landward areas, where interstitial brine concentrations are very high (Holser, 1966). In addition to forming pseudomorphs after gypsum, the polyhalite is present as a primary diagenetic mineral which fills pore spaces between halite crystals. The halite is crudely layered and is several meters thick; it apparently is not a pan deposit, but its occurrence is not understood. Accurate level data are needed to help solve the problem. Possibly the persistent onshore winds force thin sheets of seawater many miles onto the sabkha surface, where they evaporate. The presence many miles inland of marine shells and other flotsam shows that the halite surface is sometimes flooded by seawater.

Discussion and Summary

The local tectonic environment during the past few million years has controlled the shelf and hinterland relief along the southern margins of the Arabian Gulf. Low hinterland relief and arid climate combine to reduce terrigenous input into marine areas, and thus the marine sediments are almost pure carbonates. In contrast, the Laguna Ojo de Liebre area of Baja California apparently derives much of its terrigenous sediment from the coastal dune barrier, the sands of which are blown into the lagoon, and the extensive sabkhas in that area grade landward into a sandy desert. In the San Felipe area of Baja California, clay-rich sediments are derived from the Colorado River, with local alluvial additions, and high hinterland relief has developed an alluvial fan-sabkha association. Where low hinterland relief is present, as in the Trucial Coast, the typical association is coastal sabkha with continental sabkha and/or dune-sand facies, except in areas detached from any landmass, where coastal sabkha facies alone may be developed.

The arid climate and the nature of the primary marine sediments determine the early diagenetic changes which characterize supratidal coastal sabkhas. Once shoreline regression occurs, evaporites are deposited in the upper parts of the marine sediment wedge, and dolomitization of fine-grained carbonate sediments ensues. Dolomitization releases calcium ions which are precipitated as additional gypsum or anhydrite by the interstitial pore fluids. Because the molar ratio of sulfate to calcium in seawater is approximately three, almost three times as much calcium sulfate is precipitated in areas of intense dolomitization as in areas where there is no dolomitization. In areas of dolomitization, brines may be almost depleted of sulfate ion. The dolomitization of the carbonate sediments continues until the molar ratio of magnesium to calcium has decreased to a low value; field data indicate values of four or less. No further dolomitization can take place until pore fluid with a higher magnesium-calcium ratio is added. All the early diagenetic phenomena of the sabkha environment first occur at the surface and extend in depth with time. Evaporite minerals are restricted largely to the upper levels. Dolomitization in depth, with the formation of calcium sulfate minerals as a byproduct, depends on convective movement or reflux of pore fluids. The formation of evaporite minerals and the dolomitization of the sediments depend on the production of interstitial brines. The rate of brine production is related directly to the net evaporation or the balance between losses of water vapor from the sabkha surface and the diluting effects of storm waters, rain water, dew, and lateral seepage from the sea, or from the land in inner sabkha areas where dilute continental groundwater is added to the sabkha reservoir. In the same general environment, water-vapor loss by evaporation from a standing body of brine is greater than the loss from the sabkha surface. Consequently, more soluble salts may accumulate in pans than can accumulate within sabkha sediments.

Continued regressive sedimentation causes diagenetic mineralization to move seaward, and early stages of diagenesis are masked by later ones. In this process some minerals may disappear and be replaced by more stable phases. For example, the first-formed calcium sulfate mineral is gypsum, but it may be dissolved later and replaced by anhydrite. As regressive sedimentation proceeds, the coastal sabkha belt dominated by diagenetic gypsum moves seaward; the area of the gypsum belt remains approximately constant, whereas the area dominated by anhydrite continually increases. The formation of magnesite depends on the slow reaction kinetics of dolomitization, whereby the pore fluids pass temporarily into the magnesite stability field, in the zone of the highest magnesium-calcium ratios. Continued dolomitization lowers this ratio, and magnesite becomes unstable and is dissolved. If dolomitization does not occur, as in noncarbonate sabkhas or carbonate sabkhas composed mainly of coarse skeletal grains, the magnesium-calcium ratio of the interstitial brine remains high and magnesite, once formed, remains as a stable phase. Thus, a fossilized sabkha will not necessarily retain all of the minerals or even traces of the minerals which were developed within it during early diagenesis. Many early diagenetic minerals represent only ephemeral phases. In ancient equivalents, the more soluble minerals such as halite very commonly are leached away completely and even the calcium sulfate may be destroyed.

The development of continental sabkhas depends on low hinterland relief, small eolian sediment supply, and arid climate. If these requirements are met, equilibrium deflation-sedimentation surfaces, controlled by the water-table gradient, may be developed extensively. Depression of hard rock below the present continental groundwater table is required. During the Holocene cycle of sedimentation, the Trucial Coast dune areas have been starved of new sources of sand. Because the prevailing wind is onshore, there presumably has been a net movement of sand toward the interior of the Arabian Peninsula, with concomitant opportunity for development of continental sabkhas in the belt just inland from the coastal sabkhas.

Mean sea level controls the elevation of the seaward extremity of the continental groundwater table. The landward slope normally is related to hinterland relief. If the surface is covered with unconsolidated sediments, as in the Sabkha Matti (Fig. 2), the slope is related to the permeability of the upper sediments. In a continuously subsiding area, mean sea level and the groundwater table rise in relation to older sediment surfaces; thus, extensive continental sediments (including cross-bedded eolian facies, with or without early diagenetic evaporites) may be depressed below the equilibrium deflation level. The nature of the deeper groundwater then determines whether the evaporites are preserved. The rise in the base level of wind deflation in the Sabkha Matti area may have been caused by either tectonic subsidence of the foreland, eustatic rise in sea level since the Wisconsin glacial episode, or both.

Temperature extremes in continental sabkha areas are greater than those of the coastal sabkhas, and minimum relative humidity is generally much lower. The modifying influence of the seawater mass reduces extremes of temperature and relative humidity over coastal sabkhas. The effect of these differences is seen in the contrasting calcium sulfate mineral development in the two sabkha areas. The amount of bassanite in coastal sabkhas is very small and there is little *in situ* dehydration of gypsum to form anhydrite, whereas the dehydration of gypsum to form anhydrite is a characteristic phenomenon in continental sabkhas (the dehydration reactions are functions of temperature and water-vapor pressure). On the basis of combined data from Laguna Ojo de Liebre, San Felipe, and the Trucial Coast, it appears that the formation and preservation of supratidal anhydrite require a mean annual temperature above 22°C. Yet, in many other areas where mean annual temperatures are higher than 22° the only calcium sulfate mineral present is gypsum. Perhaps the initial nucleation of anhydrite takes place at the rather higher summer temperatures in the surficial sabkha sediments. If so, early diagenetic anhydrite may be formed only in arid areas with a continental type of temperature distribution. In oceanic areas, even though mean temperatures are fairly high (for example, much of the Indian Ocean has a mean temperature in excess of 25°C), the high summer temperature is not extreme. The Trucial Coast and San Felipe data suggest that a seasonal temperature in excess of 35°C is an essential requirement in the precipitation of anhydrite.

The high rate of carbonate sediment deposition along the Trucial Coast, very low hinterland slopes, and paucity of eolian sand apparently are the causes of the very extensive coastal and continental sabkha development. The rate of shoreline regression has been 1–2 m/yr during the past 4,000–5,000 years. In 10^5 years this rate of sedimentation would form a sabkha 100–200 km wide and in 10^6 years, 1,000–2,000 km wide, if any original platform of this dimension were present. Continued subsidence during this sedimentation period, however, would cause marine sediment facies to rise across the sediment belt. Continued infilling of the first-formed sabkha area with additional supratidal storm-wash, alluvial, eolian, and evaporite facies probably would occur. A fairly thick evaporite-bearing section could be formed, its thickness dependent on rate of subsidence and duration of sedimentation. The evaporite section conceivably could form a seal over the underlying marine sediments. In tectonically active areas such as San Felipe, coastal sabkhas will develop as linear belts at basin margins, but are unlikely to become very extensive. The possible extent of a sabkha is related largely to the original physiography and geometry of the depositional surface, which in turn is related to tectonic and geomorphic history. Times of the order of 10^5–10^6 years are sufficient for the formation of very extensive sabkha areas.

Recently, the term "megasabkha" has appeared in the evaporite literature. The concept that megasabkhas are fundamentally different from sabkhas seems to be gaining ground. They are conceived as being an order of magnitude larger than sabkhas (*e.g.*, 500 km compared with 50 km). In addition, only a megasabkha can have a full evaporite succession, including potash salts, whereas a sabkha has only calcium sulfate minerals and minor halite facies. The writer believes that these differences are not real, and that there is no sound basis for distinguishing a megasabkha from a sabkha. The examples discussed herein have shown (1) the variability of the sabkha environment, (2) the fact that even late-stage evaporite minerals (polyhalite) may be present, and (3) that rates of sabkha development may be very rapid. There has been a time interval of only 4,000–5,000 years for the formation of the present-day coastal sabkhas, and it is probable that nearshore relief is greater now than it was during much of the geologic past, when less drastic sea-level oscillations prevailed. The writer suggests that use of the term megasabkha is based on mistaken concepts of time and sedimentation rates.

The presence of large continental sabkhas inland from the southern shores of the Persian Gulf surely is no freak coincidence. The required combination of hinterland relief and climate must have been developed along many ancient basin margins. In such locations the near-basin continental sediments (including sabkha-type evaporites)—dominantly gypsum-anhydrite but in places including halite—will be depressed below the base level of erosion as basin downwarp continues. The continental facies nearest the basin margin, particularly in the continental-marine facies overlap zone, are most likely to be preserved; the continental facies farther inland are less likely to be buried and preserved. The evolution of a playalike situation of local internal drainage will be rare in basin-margin positions, because any surface drainage and groundwater underflow will move basinward. It may be expected that playas will be more common in areas distant from the basin margin. However, as the chance of burial and preservation of these distant areas is perhaps remote, continental playa evaporites can be expected to be a rare component of the geologic column.

In the light of the new data presented herein and the speculations advanced, reexamination of many basin-margin and continental sedimentary sequences may yield worthwhile results. Studies of Holocene processes of basin-margin and adjacent continental sedimentation and early evaporitic diagenesis suggest that (1) marine inner-shelf or basin-margin sedimentary successions should contain abundant supratidal surfaces, and evidence for their former presence has been given (desiccation and minor erosion features, stromatolitic algal-banded sediments, supratidal evaporite development, particularly the formation of nodular anhydrite, capping of the supratidal sequence with a thin red-brown eolian sediment, *etc.*); and (2) continental sediments and associated evaporites may have accumulated without the earlier presence of large inland lakes or "seas." The suggested continental sabkha environment seems to offer a more reasonable interpretation than earlier paleogeographic hypotheses calling for the presence of large standing bodies of water in areas of apparently low hinterland relief and arid climate.

Data assembled from the Trucial Coast and from other arid regions such as Baja California provide much information for interpreting ancient shallow-water marine carbonate or noncarbonate sediments and their relations with adjacent continental sediments. In addition, the data available on associated evaporite minerals probably are sufficient to show whether they were precipitated from a standing body of brine or interstitially in a sabkha environment. The Holocene areas provide the geologist with sedimentologic models for interpreting possibly analogous ancient sediments. Although no single model may be satisfactory for interpreting a particular ancient situation, a judicious combination of parts of various sedimentologic models may provide a reasonable interpretation. From the primary sediment characteristics, some reconstruction can be made of the physiography of an ancient area and of the geometry of the land and sea. If supratidal facies sediments are present, the early diagenetic developments will indicate the climatic variables.

References Cited

Evans, G., 1966, The recent sedimentary facies of the Persian Gulf region: Royal Soc. London Philos. Trans., ser. A, v. 259, p. 291–298.

Holser, W. T., 1966, Diagenetic polyhalite in recent salt from Baja California: Am. Mineralogist, v. 51, p. 99–109.

Illing, L. V., and J. C. M. Taylor, 1967, Discussion: London, Inst. Mining and Metallurgy Trans., sec. B, v. 76, p. 883–884.

——— A. J. Wells, and J. C. M. Taylor, 1965, Penecontemporary dolomite in the Persian Gulf, *in* Dolomitization and limestone diagenesis: Soc. Econ. Paleontologists and Mineralogists Spec. Pub. 13, p. 38–111.

Kinsman, D. J. J., 1964, Recent carbonate sedimentation near Abu Dhabi, Trucial Coast, Persian Gulf: Ph.D. thesis, London Univ., 302 p.

——— 1966, Gypsum and anhydrite of recent age, Trucial Coast, Persian Gulf: *in* Second symposium on salt, v. 1: Cleveland, Ohio, Northern Ohio Geol. Soc., p. 302–306.

——— 1967, Huntite from a carbonate-evaporite environment: Am. Mineralogist, v. 52, p. 1332–1340.

Phleger, F. B, and G. C. Ewing, 1962, Sedimentology and oceanography of coastal lagoons in Baja California, Mexico: Geol. Soc. America Bull., v. 73, p. 145–182.

Thompson, R. W., in press, Tidal-flat sedimentation of the Colorado River delta, northwestern Gulf of California: Geol. Soc. America Mem. 107.

Erratum: Page 80, line twelve, column 1 should read: "cipitated from seawater-derived brines. *Continental*"

Reprinted from *Jour. Sed. Petrol.*, **39**, 70–81, 89 (1969)

MODERN EVAPORITE DEPOSITION AND GEOCHEMISTRY OF COEXISTING BRINES, THE SABKHA, TRUCIAL COAST, ARABIAN GULF[1]

12

GODFREY P. BUTLER
University of California, Riverside, California

ABSTRACT

Near Abu Dhabi, Trucial Coast, diagenetic evaporite minerals are developed within an accumulation of detrital Pleistocene sands and Recent carbonate sediments which form a supratidal flat (sabkha) 8 miles wide.

Field studies were carried out by the author and his co-workers of Imperial College in 1964. A further visit to the Trucial Coast was made by the author in 1967 under the auspices of the Dry Lands Research Institute, Riverside. A study of 260 interstitial brine samples showed that the distribution of the evaporite minerals, which include gypsum, anhydrite, dolomite, and halite, is largely controlled by the hydrology of the area. Interstitial brines within the sabkha sediments are replenished with sea water by abnormally high tides and storm surges. These surface waters, which soak down into the sediments, are concentrated both by solution of soluble salts and by evaporation. The subsequent equilibration between brines and host sediments leads to precipitation of aragonite, gypsum, and halite; alteration of gypsum to anhydrite, and dolomitization of carbonate sediment. A direct correlation exists between the frequency and extent of flooding, the mineral phases developed, and the chemical composition of interstitial brines.

Flooding of a 2- to 3-mile wide belt roughly concentric to lagoon margins, occurs at monthly or longer intervals. In this belt, a surface gypsum crystal mush up to 12 inches thick is present which, landwards, is overlain and progressively replaced by anhydrite. Dolomitization is extensive toward the landward margin of this zone. From the lagoon laterally across the 2- to 3-mile wide belt to its landward margin, chlorinity increases from 44‰ to 155‰; sulfate concentration rises from 6 to 17 gm/kg and then falls to less than 3 gm/kg; the Mg^{2+}/Ca^{2+} mole ratio increases from 5.3 to 35 and then decreases to about 4. Concentration of these components decreases with increasing depth below the sediment surface. Gypsum is the stable calcium sulfate mineral in contact with brines of chlorinities less than 145‰ and anhydrite at chlorinities greater than 145‰.

An additional 5-mile wide zone is flooded only at 4- to 5-year intervals. The original gypsum crystal mush in this zone is completely replaced by up to a 12-inch thick layer of chicken-wire mesh structure anhydrite, and the carbonates are highly dolomitized. Interstitial brines are saturated with NaCl at chlorinity 160‰ and specific gravity 1.20. These brines contain less than 1 gm/kg sulfate and have Mg^{2+}/Ca^{2+} mole ratios of about 4. In this zone, constant Mg^{2+}/Ca^{2+} values with increasing magnesium concentration suggest that there has been an approach to mutual equilibrium between brines and evaporite-carbonate minerals.

The remaining 1 to 2 miles at the landward margin of the sabkha are underlain by Pleistocene sands and are not flooded by marine waters. Ground waters in this zone are in part terrestrial in origin. Inland across this zone, chlorinity decreases from 160‰ to 107‰; the sulfate concentration is 2 gm/kg and the Mg^{2+}/Ca^{2+} mole ratio averages 3.5. Locally, anhydrite is hydrated to gypsum where brine chlorinity is less than 145‰.

Processes of gypsum precipitation and dolomitization are essentially restricted to the 2- to 3-mile wide belt close to the lagoon margin. In the mid-sabkha, equilibrium conditions probably prevail between brines, diagenetic minerals, and host sediments. Diagenesis at the landward margin of the sabkha is in part retrogressive. Both a progressive and a retrogressive diagenetic facies can be visualized as moving out with the progradation of the shore line.

The similarity between the stratigraphy and mineralogy of some Ordovician, Silurian, Pennsylvanian, and Jurassic evaporite sequences and the sabkha on the Trucial Coast suggest that evaporites of supratidal origin may be more common in older evaporite rocks than previously realized.

INTRODUCTION

The environments in which modern evaporites form, the factors that control their composition and texture, and the significance of primary structures offer a key to interpreting ancient deposits. Perhaps the best developed evaporite environments occur on the sabkhas of the Arabian Gulf.

Field studies in the Trucial Coast were initiated at Imperial College, London, in 1962 by G. Evans, assisted by D. Kinsman. These workers, who were later joined by D. Shearman, concentrated on the physiography of the Trucial Coast (Evans and others, 1963; Kinsman, 1963; and Evans and Shearman, 1964). Subsequently, Sir Patrick Skipwith (1966) and C. Kendall (1966) studied the lagoon and intertidal areas of the Khor al Bazam (Kendall and Skipwith, 1966) and J. Twyman (unpubl.), the oolite deltas near Abu Dhabi Island. In 1963, Kinsman carried out an extensive program of sediment and water sampling across the inshore areas southwest of Abu Dhabi, and made a reconnaissance study of the Abu Dhabi sabkha (Kins-

[1] Manuscript received March 15, 1968.

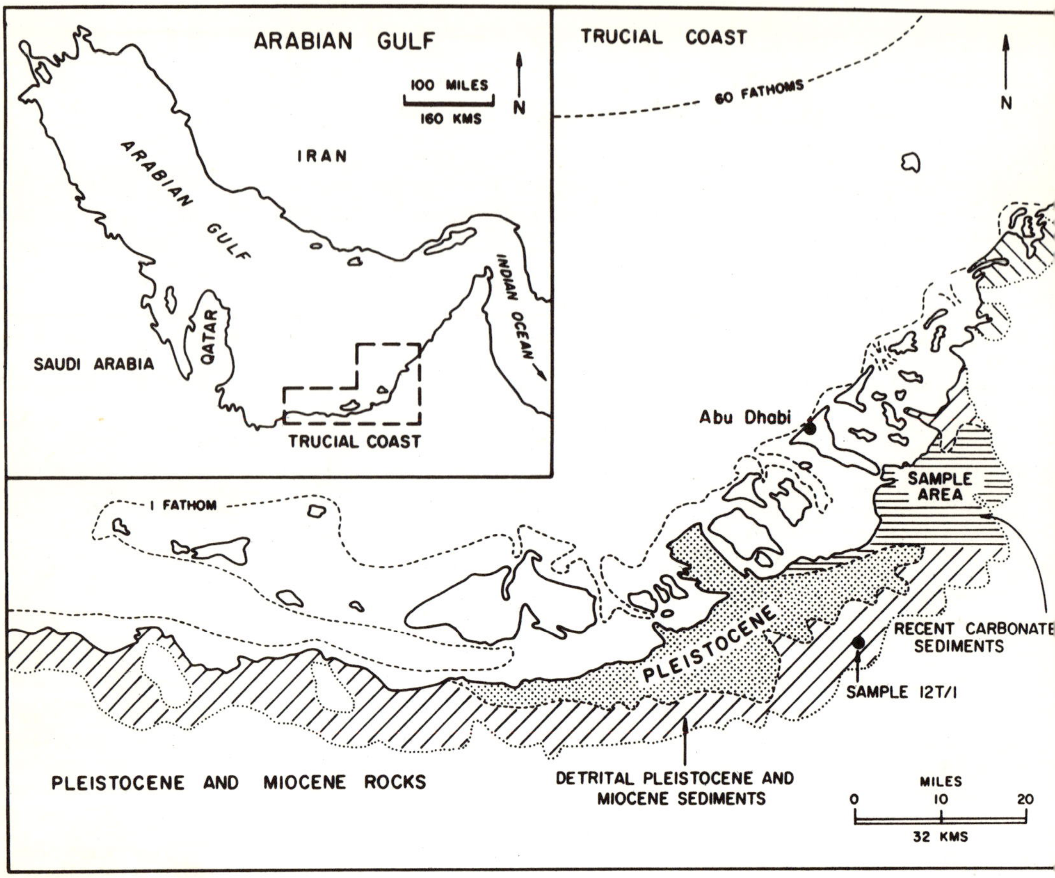

FIG. 1.—Location map of the Arabian Gulf and Trucial Coast. Area of sabkha approximately indicated by ornamentation.

man, 1963, 1964, 1966, and 1967). The author joined the project in 1963 with the purpose of continuing this initial study of the evaporite minerals and the chemistry of the associated brines in the sabkha. This paper essentially presents data collected by the author in 1964 (Butler, 1965). Additional data stem from a second visit to the Trucial Coast in 1967 with Frank Dickson under the auspices of the Dry Lands Research Institute of the University of California, at Riverside.

The sabkhas along the Trucial Coast, Southern Arabian Gulf, contain the classical sequence of evaporite minerals, dolomite, gypsum, anhydrite, and halite. The sabkha environment is considered to be analogous to that of many ancient evaporite forming environments, such as those described by West (1965), Gray (1967), Goldberg (1967), and Roehl (1967).

The purpose of this paper is to (1) describe the environment and stratigraphy of a sabkha on the Trucial Coast and (2) to show how the development of diagenetic evaporite minerals in this sabkha is correlated with the hydrology and the chemistry of interstitial brines, based on a study of 120 cores and 260 samples of brine squeezed from sediments in the field.

PHYSICAL ENVIRONMENT

Sabkhas along the Trucial Coast (fig. 1) have formed by the infilling of lagoons by detrital Pleistocene quartzose sands and Recent carbonate sediments. The evaporite minerals are best developed in a relatively small sabkha on the mainland behind the island of Abu Dhabi. Their occurrence coincides with the largest accumulation of carbonate sediments (fig. 1).

The Abu Dhabi sabkha, the area studied, is about 8 miles wide. The seaward margin is delineated by algal mats, the landward edge feathers out against low hills underlain by sedimentary rocks of Pleistocene and Miocene age.

The stratigraphy of the sabkha sediments is shown in figure 2. Unlithified Recent carbonate sediments and Pleistocene eolian quartzose sands overlie Miocene sedimentary rocks. The

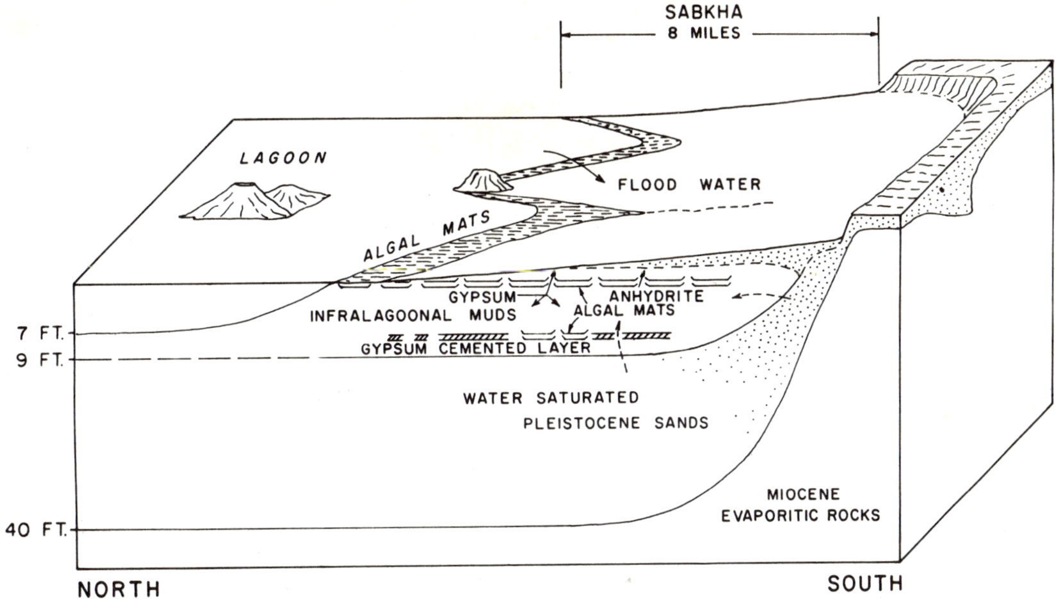

FIG. 2.—Stratigraphy of the Abu Dhabi sabkha. (Schematic cross section.)

quartzose sands form a unit up to 30 feet thick, which crops out at the landward margin of the sabkha to form a zone up to 2 miles wide. The carbonate sediments form a wedge shaped unit up to 6 miles wide and 7 feet thick. Towards the landward margin of the sabkha the unit grades into Pleistocene quartzose sands.

The upper intertidal zone is characterized by a living algal mat. An older algal mat can be traced inland for about 6 miles at the same horizon in the sabkha subsurface, signifying a long period of continued intertidal sedimentation and marine regression of the sea without change in sea level. The two layers of algal mat in the sabkha (fig. 2) show that two cycles of sedimentation have formed the carbonate unit. The second algal mat, which is only locally developed in the subsurface, is laterally continuous with a 3-inch thick gypsum cemented shell layer.

Inland from the intertidal zone, the present algal mat is buried beneath an increasing thickness of detrital quartzose sand. In the mid-sabkha region, these sands are up to 2.5 feet thick. Most of the gypsum and almost all the anhydrite of marine origin occur within these sediments overlying the older algal mat.

A most unusual feature of the Abu Dhabi sabkha is its almost planar surface, which rises inland from the lagoon with a slope of about 1:1000. The surface has formed primarily through reworking of the sediments by flood waters and by the accumulation of windblown dust filling in small surface irregularities. Cavities in salt crusts and perched objects suggest that some local deflation of the surface is taking place. Deflation probably plays an important role locally in maintaining the surface at a fixed level relative to the water table (Evans and others, 1964). The surface is interrupted by elongate depressions, roughly normal to the shore line, which coincide with buried tidal channels in the former lagoon floor. These depressions act as conduits for flood and seepage waters into the sabkha interior.

The movement of water onto the sabkha surface plays a vital role in the development of the chemistry of the interstitial brines. A direct correlation exists between the frequency of flooding, the composition of the interstitial brines, and the nature of the diagenetic minerals developed.

The term *flood recharge* is suggested as being descriptive of the process which rejuvenates older interstitial brines by the delivery at the surface of sea water accompanied by downward soaking. The importance of flooding in supratidal areas has also been recognized by Masson (1955) and Holser (1966).

Flood recharge should not be confused with the reflux process described by Deffeyes and others, (1965) in the salina and supratidal flat of Bonaire, Netherlands Antilles.

The sabkha can be divided into five zones on the combined basis of tides and frequency of flooding, nature of diagenetic minerals devel-

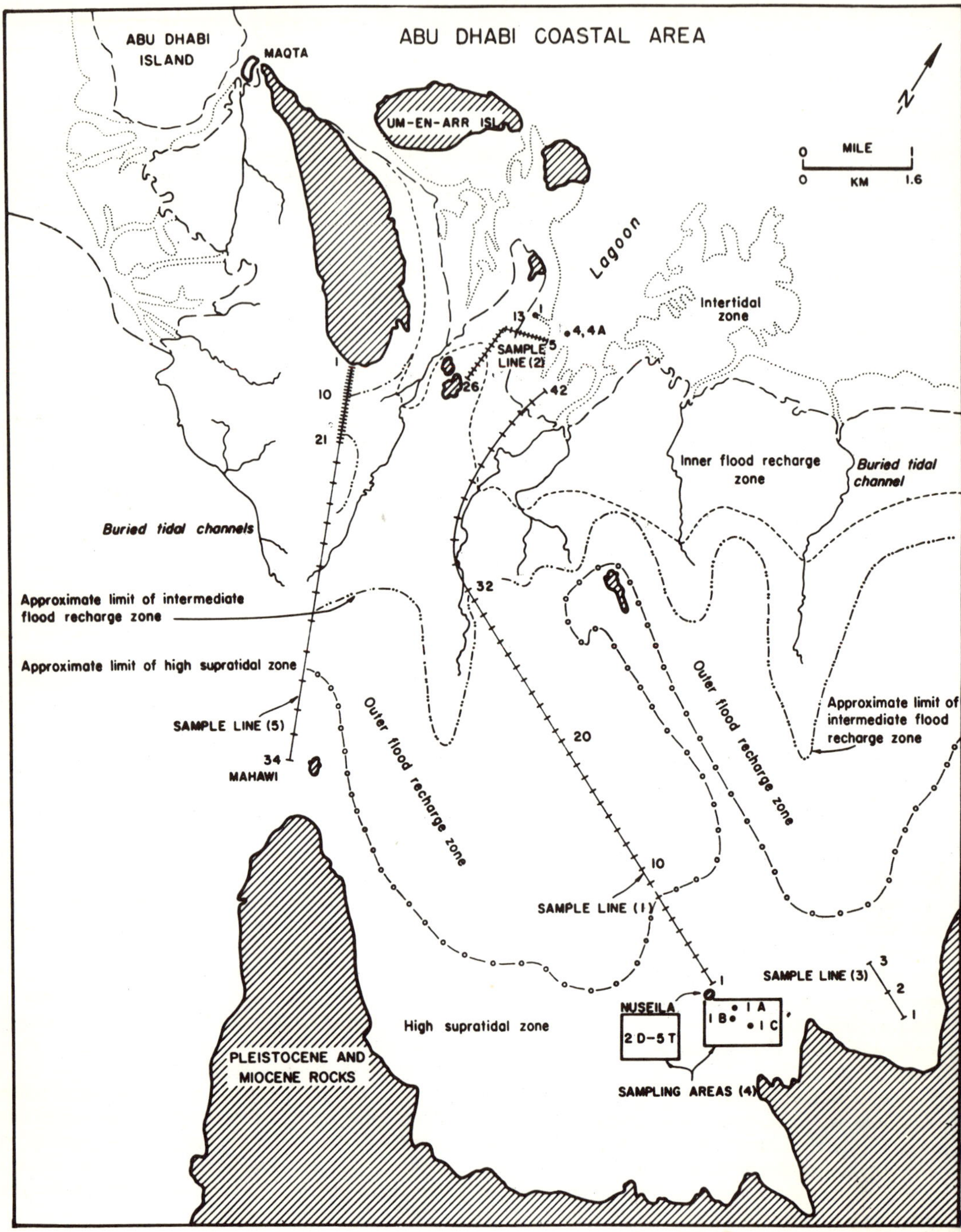

FIG. 3.—Sampling traverses and flood recharge zones across the sabkha.

oped, and nature of the host sediments (fig. 3). These zones are, in order of proximity to the lagoon:

(1) Intertidal zone, 0.25 to 0.5 mile wide, which is regularly flooded by tide waters. The upper intertidal zone is covered by living mats of blue-green algae. The algal mats are interlayered with lime muds and landwards contain pods of small euhedral gypsum crystals, which are less than 0.125 inch long. Below the algal

mats are lagoonal and intertidal carbonate muds.

(2) Inner flood recharge zone, 1 to 1.5 miles wide, which is subjected to monthly or more frequent flooding. The limits of this zone coincide approximately with a surface layer of randomly oriented gypsum crystals (gypsum mush), which is up to 12 inches thick. These gypsum crystals range in length from 0.015 to 0.75 inches with an average lenth of 0.25 inches. Most of the flood waters sink into the underlying sediments while the rest flows back into the lagoon along shallow surface channels. Underlying carbonate muds contain small amounts of dolomite and small aggregates of gypsum crystals.

(3) Intermediate flood recharge zone, 0.25 to 1 mile wide, flooded at intervals longer than one month. The gypsum mush is locally overlain by a skin of reworked detrital quartzose sand. The surface gypsum crystals are replaced by anhydrite nodules. The underlying carbonate muds contain much dispersed fine-grained dolomite and large lenticular gypsum crystals, ranging from 2 to 4 inches long. These crystals are similar to those from the Laguna Madre described by Masson (1955).

(4) Outer flood recharge zone, 0.5 to 5 miles wide, flooded at four- to five-year intervals, with a sequence of sediment types similar to zone 3. The anhydrite progressively replaces the gypsum mush until, at the landward margin of the zone, the anhydrite forms a layer up to 12 inches thick, composed of small ($<$ 0.5 inch diameter) nodules of anhydrite separated by thin films of dolomitic carbonate sediment and quartzose sand (Butler, 1965). The anhydrite displays a rough chickenwire mesh structure of the type commonly observed in ancient anhydrite. Much thin-bedded anhydrite is also present in highly contorted layers within 18 inches of the surface. Underlying carbonate sediments contain fine-grained dolomite and abundant large lenticular gypsum crystals up to 10 inches long.

(5) High supratidal zone, 1 to 2 miles wide, bordered by Pleistocene and Miocene sedimentary rocks. This area is not flooded by sea water and the groundwaters are in part terrestrial (Butler and others, 1965). The underlying sediments are eolian Pleistocene sands, largely uncemented, which contain both gypsum and anhydrite, halite, and some polyhalite and sylvite. The components of these minerals have been supplied in some localities by terrestrial groundwaters. The gypsum occurs largely as an interstitial cement and the anhydrite as nodules concentrated in layers of variable thickness. At a location outside the area considered in this paper (fig. 1, 12T/1) the anhydrite forms a layer 8 feet thick with nodules ranging from less than 1 inch to 6 inches in diameter, separated by films of detrital clays (see analysis of associated brine, sample 12T/1, appendix).

The high supratidal zone is characterized by the local alteration of pre-existing anhydrite to gypsum, as a result of the dilution of interstitial brines with less saline terrestrial groundwaters. The secondary gypsum crystals are zoned with corroded anhydrite laths, and are up to 4 inches long. They are flattened in the a-c plane and differ from those formed in the intertidal and flood recharge zones, which are flattened approximately normal to the c-crystal axis. Illing and others (1965) described similar gypsum crystals, flattened in the a-c plane, from a sabkha on Qatar (fig. 1).

The Lower Purbeckian evaporite beds of Southern England (West, 1965) exhibit almost identical anhydrite textures and structures as those in the Trucial Coast sabkha. West considers that the anhydrite in the Purbeckian also originated from a gypsum precursor. The origin of anhydrite in the sabkha, its texture, and the primary structures developed have already been noted (Butler, 1967) and will be discussed in detail elsewhere (Butler, 1968).

CHEMICAL ENVIRONMENT

Climate and Temperature

The climate of the Trucial Coast is hot and arid. Air temperatures range annually from 16°C to 44°C, and the average rainfall is less than 1.5 inches. Relative humidity varies from an average maximum of 90 percent during the night to an average minimum of 30 percent during the day. The net rate of evaporation is consequently high. Evaporation rates for the Arabian Gulf are estimated by Privitt (1959) to be as much as 50.4 inches per year, compared with a value of 39.6 inches for a similar evaporite-forming environment on Bonaire (Deffeyes and others, 1963).

Temperatures within 0.5 inches of the sabkha surface show a daily range of from 18°C to 53°C (February to April 1964). In general, the highest surface temperatures were recorded in the inner flood recharge zone, where they were about 8°C higher than elsewhere. Diurnal temperature fluctuations were not detected below a depth of 12 inches. Below 12 inches, temperatures remained constant to a depth of about 25 inches and then increased almost linearly with depth to the Pleistocene sands in which temperatures were constant near 31°C. At a depth of 20 inches the average monthly temperature

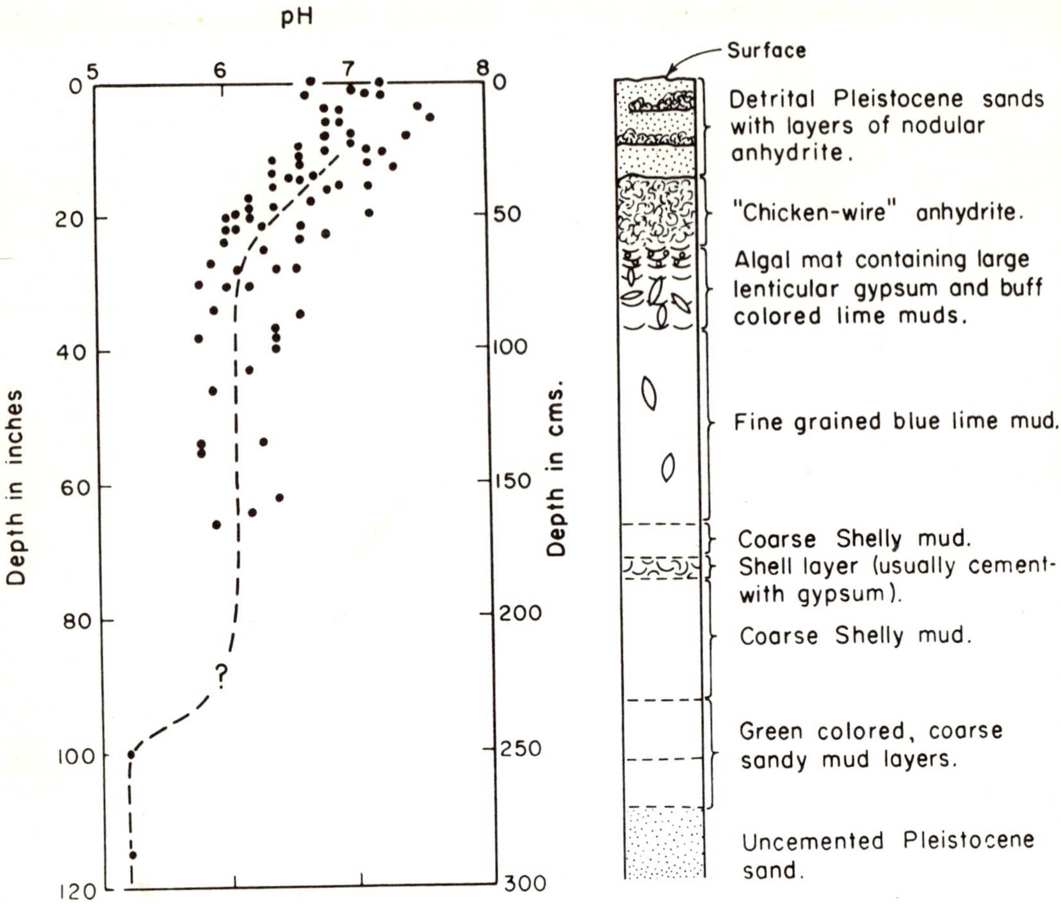

FIG. 4.—Vertical profile of sediment pH and generalized vertical sediment section in the outer flood recharge zone.

variation is calculated to be ±3°C. A probable annual temperature range at 20 inches depth is 23°C to 41°C with an average temperature of 32°C.

pH Measurements

Lagoon waters are recorded as having pH values of about 8.3, interstitial algal flat waters 7.5, which in the mid and inner sabkha fall to 6.0 to 6.4 (Kinsman, 1966). Data on pH collected by the author were not extensive. In general, pH decreased inland from the lagoon and then rose again towards the inland margin of the sabkha. Marginal lagoon waters averaged pH 7.8, which decreased to 6.8 across the inner flood recharge zone, and reached a minimum of 6.0 in the outer flood recharge zone. The pH then rose to about 7.2 across the high supratidal zone. In the outer flood recharge zone, pH decreased from about pH 7 at the surface to a limiting value of between 5.8 and 6.4 at depths of 30 to 40 inches (fig. 4). Values of pH 5.2 were, however, recorded in the Pleistocene sands and immediately overlying sediment. The overall acidity of the brines in the sediments has probably been caused by decomposition of organic matter releasing carbon dioxide and hydrogen sulphide, as indicated by the following analysis of a sample of the usually abundant gas in these sediments. The composition of air is shown for comparison.

	Sabkha gas (mol %)	Normal air (mol %)
N_2	94.99	79.9
O_2	4.76	20
CO_2	0.19	0.03
CH_4	0.06	—
H_2S	>0.01	—

Brine Chemistry

Two hundred and sixty interstitial fluids squeezed from the sediments constituted the bulk of the samples used for analysis. Techniques of sampling and chemical analysis of the fluids are given in the appendix, together with the detailed analytical results, field temperatures, and host sediment type.

The brines in the evaporitic-carbonate sediments, as observed in samples taken along traverses landward from the lagoon margins (fig. 3), are of increasing ionic strengths. They range from lagoon water concentrations to saturation with sodium chloride. The minerals enclosing the brines are dominantly aragonite, calcite, gypsum, anhydrite, dolomite, and halite. Other minerals of minor occurrence are celestite (Evans, and Shearman, 1964) and huntite (Kinsman, 1967). Except for anhydrite and dolomite, the mineral association resembles that group of solid phases which initially separate from sea water subjected to progressive evaporation (Usiglio, *in* Stewart, 1963). The presence of anhydrite and dolomite reflects the combined processes of evaporation and reaction of brines with pre-existing gypsum and calcium carbonate.

Chlorinity.—From the lagoon across the sabkha, brine chlorinity rises to a maximum and decreases again towards the inland margin (fig. 5). The greatest change takes place across the intertidal zone where a three-fold increase in chlorinity was observed. From the intertidal zone inland over the rest of the sabkha, chlorinity rises more slowly to a maximum of 170‰ and then falls to 107‰ or less close to the landward margin of the high supratidal zone. The maximum recorded chlorinity of 170‰ is unusual. The average maximum chlorinity is 160‰ with a corresponding brine specific gravity of 1.20. A brine specific gravity of 1.20 agrees approximately with the data of Usiglio (1963) and Miller (1961) for initial precipitation of sodium chloride from evaporated sea water (fig. 6).

The brines across a smaller sabkha in Qatar (Illing and others, 1965) show similar changes in chlorinity from the lagoon landward, but a maximum of about 125‰ (150 gm/liter) was reported and no decrease in chlorinity towards the inland margin was observed.

Chlorinities across the sabkha, excluding the high supratidal zone, are consistent with the process of flood recharge (fig. 7). During and

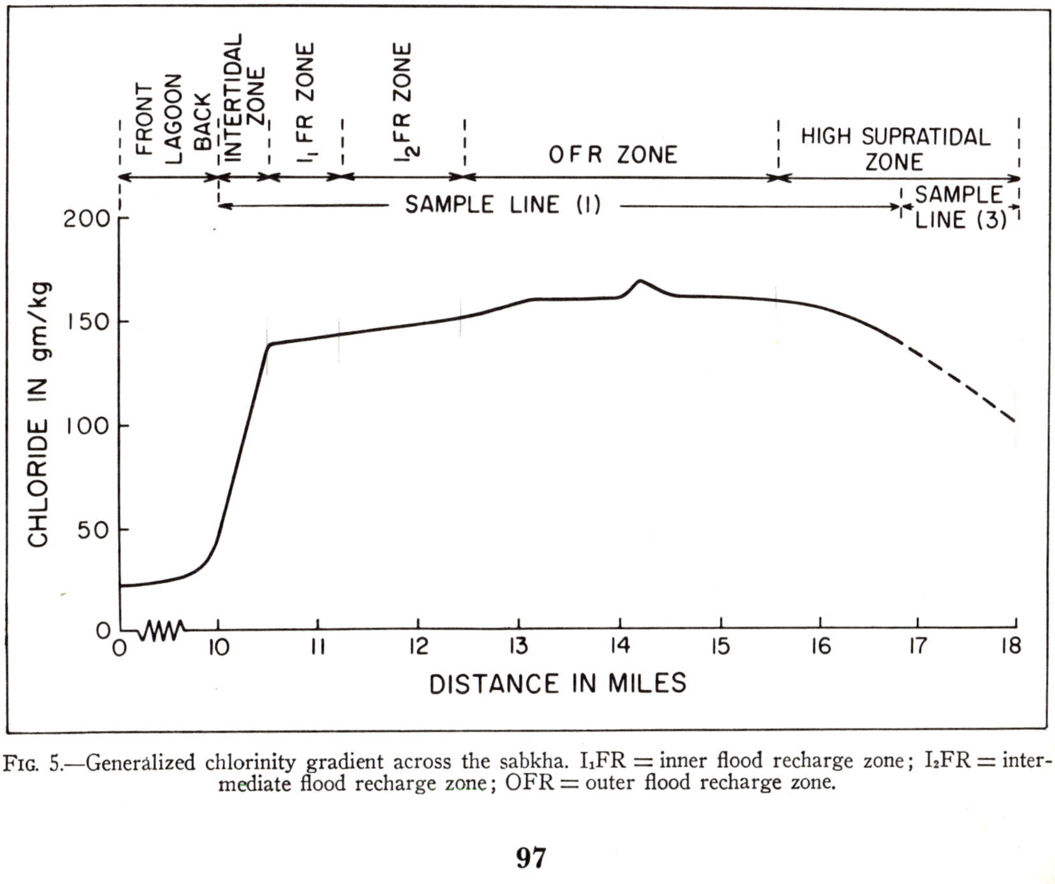

FIG. 5.—Generalized chlorinity gradient across the sabkha. I_1FR = inner flood recharge zone; I_2FR = intermediate flood recharge zone; OFR = outer flood recharge zone.

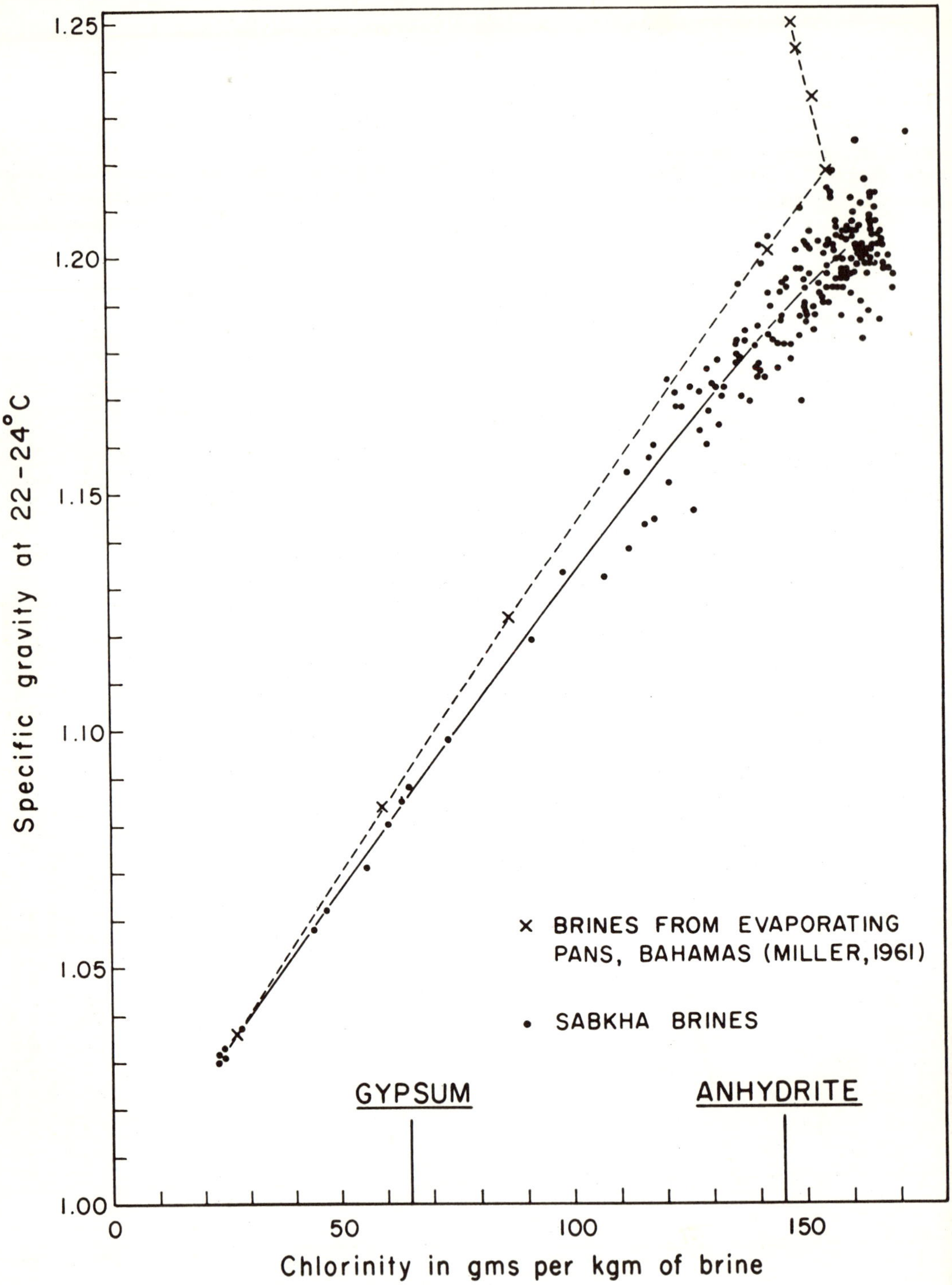

FIG. 6.—Change in brine specific gravity with increasing brine concentration. Gypsum precipitates at chlorinity 65‰, specific gravity 1.085. Gypsum-anhydrite equilibrium at chlorinity 145‰, specific gravity 1.186.

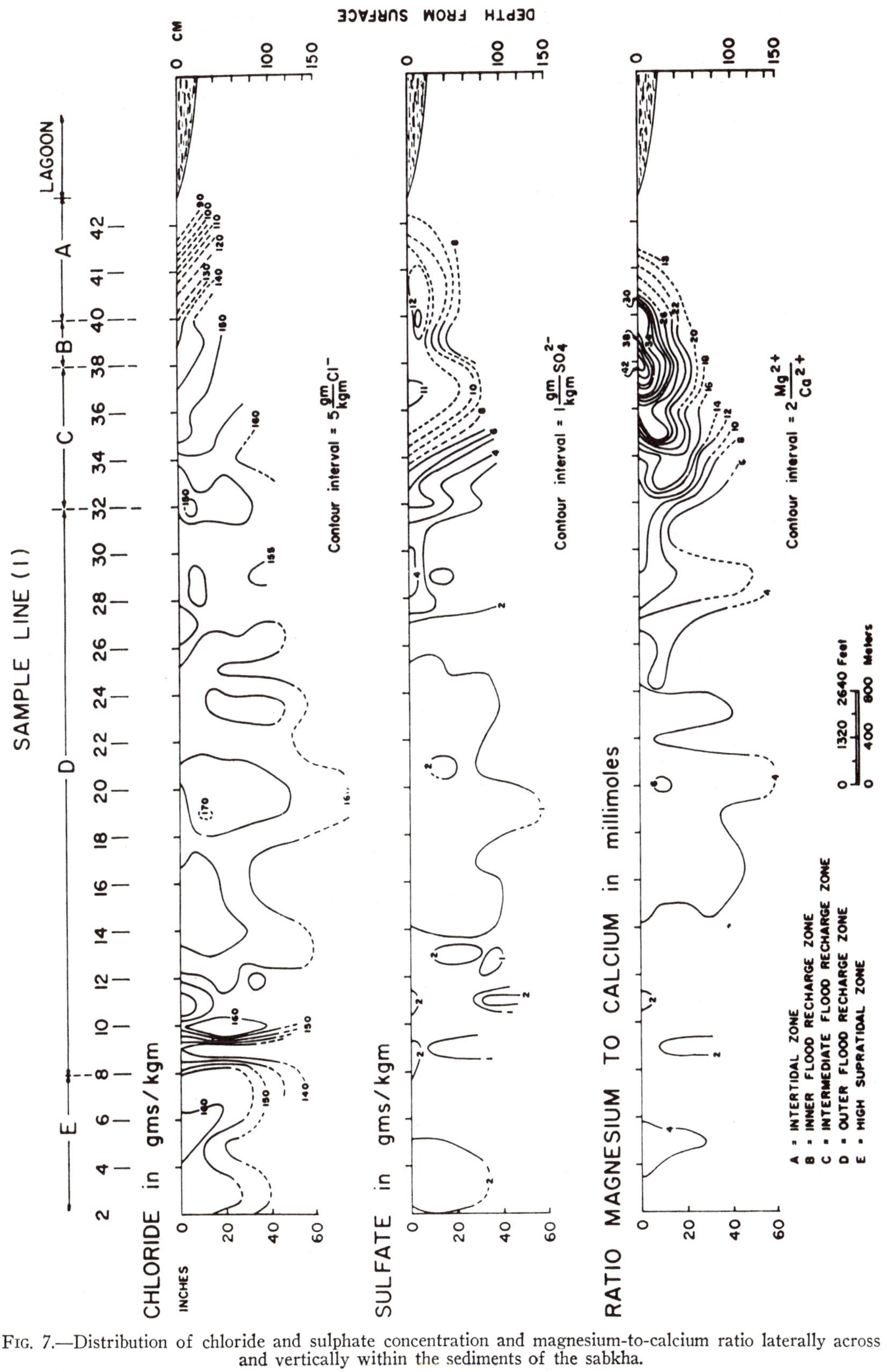

FIG. 7.—Distribution of chloride and sulphate concentration and magnesium-to-calcium ratio laterally across and vertically within the sediments of the sabkha.

immediately following flooding, the sea water concentrates by solution of soluble salts from the surface and within the sediments. Between flooding the brines further concentrate by evaporation.

Chlorinity increases with depth across the 2- to 3-mile wide belt of sabkha just inland from the lagoon margin, which has a high average frequency of flooding. Chlorinity, on the other hand, decreases with depth across the outer flood recharge zone, where flooding is relatively infrequent. Chlorinities in this zone suggest a chemically mature environment which has not been seriously perturbed by the influx of brines at the surface or from depth and where ionic diffusion possibly plays a role of smoothing out the concentration gradient. The brines in the outer flood recharge zone are saturated with respect to sodium chloride. Halite occurs as thin (< 0.5 inch) ephemeral crusts at the surface, as hopper crystals in sediments, and veneers around sand grains within the upper 2 to 4 feet of sediment.

The composition of brines from the high supratidal zone (fig. 7) differs significantly from that of brines to the seaward of the zone. Chlorinity, the Mg^{2+}/Ca^{2+} ratio, and potassium are all relatively low, and sulfate and pH are both relatively high in comparison with brines to seaward. These high supratidal brines are considered initially to have been sea water which has been chemically modified by simple evaporation, and to a greater degree, by influx of terrestrial groundwaters.

Note the distribution of isochlors about station 9 (fig. 7), suggesting the influx from depth of terrestrial brines with chlorinities lower than those of the surroundings. Influx of terrestrial brines may also have occurred in the past at station 19 (fig. 7) where the previously mentioned subsurface shell layer is locally uncemented. This would account for the abnormally high chlorinity of 170‰. Brines derived from concentrated terrestrial waters attain sodium chloride saturation at a chlorinity of about 170‰ and specific gravity between 1.21 and 1.23.

Comparison between the first field occurrence of gypsum in the intertidal zone and anhydrite in the inner flood recharge zone, and the chlorinities of co-existing brines, shows that gypsum and anhydrite first occur in contact with brines of average chlorinity 65‰ and 145‰ respectively (sample line 2, appendix). The chlorinity for gypsum precipitation agrees with that found experimentally by Posnjak (1940). In the high supratidal zone where hydration of anhydrite to gypsum takes place, gypsum occurs in contact with brines with average chlorinity less than 145‰ and anhydrite with average chlorinity greater than 145‰ (sample area 4, appendix). Thus anhydrite is the stable calcium sulphate mineral in the sabkha brines where chlorinity exceeds 145‰.

Sulfate.—Figure 7 shows the adjustment of sulfate concentration to a progressive increase in brine ionic strength. Adjustment occurs by the precipitation of gypsum. The lateral and vertical distribution of sulfate in the brines across the sabkha is similar to the distribution of chlorinity. Sulfate increases from 3.2 to 3.9 gm/kg in the lagoon to a maximum of about 16 gm/kg at the landward margin of the inner flood recharge zone, and then decreases rapidly to less than 1 gm/kg in the outer flood recharge zone. Across a large portion of this zone, sulfate remains fairly constant with values slightly greater than 0.5 gm/kg. Sulfate concentration increases to more than 2 gm/kg across the high supratidal zone and probably reaches values in excess of 4.4 gm/kg (sample line 3, appendix).

The vertical variation of sulfate within the sabkha reflects the vertical variation in ionic strength of the brines (fig. 7).

Magnesium-to-calcium ratio.—The distribution of the Mg^{2+}/Ca^{2+} ratios both laterally across and vertically within the sabkha parallels that of the chlorinity and sulfate (fig. 7). The ratio increases from about 5.3 in the lagoon to a maximum of approximately 35 at the landward margin of the inner flood recharge zone and decreases rapidly to about 10 across the intermediate flood recharge zone. Across the outer flood recharge zone the Mg^{2+}/Ca^{2+} ratio remains nearly constant between 3 and 4. The brines from the high supratidal zone have Mg^{2+}/Ca^{2+} ratios which vary between 1.5 and 4, but average 3.5 (fig. 8). Vertically within the sediments, seaward of the high supratidal zone, the ratio decreases in value downward from the surface (fig. 7).

High Mg^{2+}/Ca^{2+} ratios are attained by the precipitation of calcium ions as carbonate and sulfate minerals. The subsequent decrease in the ratio across the sabkha is attributed to dolomization of pre-existing aragonite and calcite (Curtis and others, 1963; Illing and others, 1965; Kinsman, 1966) by the reaction:

$$\underset{\text{sediment}}{2CaCO_3} + \underset{\text{brine}}{Mg^{2+}} = \underset{\text{dolomite}}{CaMg(CO_3)_2} + \underset{\text{brine}}{Ca^{2+}}$$

In addition, a study of three sabkha cores suggests a quantitative relationship between dolomization and gypsum formation (table 1, data in appendix). All samples contained dolomitized carbonate mud. Sample A at station 32 con-

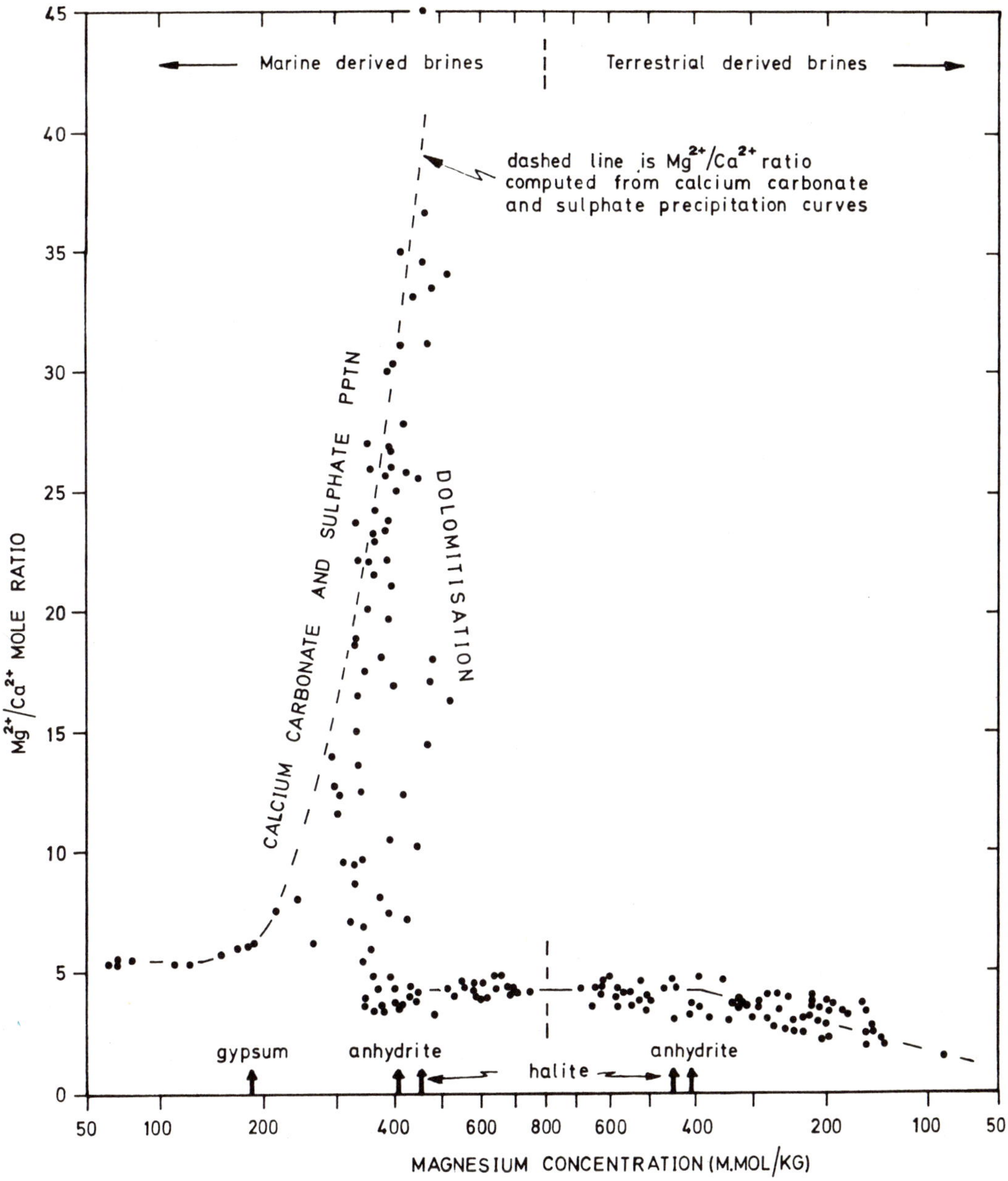

FIG. 8.—Variation in magnesium-to-calcium ratio with magnesium concentration across the sabkha. Brines from outer flood recharge zone have magnesium-to-calcium mole ratios of approximately 4 and magnesium concentration greater than 450 m.mol/kg. See data points near base of figure.

tained anhydrite; the remaining samples, gypsum crystals.

The calculated addition of calcium to the deeper brine samples as a consequence of the carbonate/brine reaction, and the subsequent adjustment of equilibrium with sulfate ions, compared remarkably well with the observed increase of calcium in the deeper brines. Wells (1962) similarly concluded that dolomitization and gypsum precipitation were concurrent processes.

An almost constant Mg^{2+}/Ca^{2+} ratio with increasing ionic strength and magnesium concentration would suggest that near equilibrium conditions existed between brines and associated dolomite, gypsum and (or) anhydrite, and halite. Such a constant ratio was observed in the brines from the outer flood recharge zone (fig. 8).

CONCLUSIONS

(1) The brines and diagenetic evaporite minerals developed in the Abu Dhabi sabkha show zonation based on the frequency of periodic flooding by sea waters and its subsequent soaking down into the sediments (flood recharge).

(2.) The present elevation of evaporite development in the sabkha is determined by the level of flood water above the level of high tide. Subsequent development of evaporites above this elevation in the same area would only come about by down warping of the sabkha and (or) rise in sea level. Continued prograding of the shore line would yield a widespread deposit of evaporites of uniform thickness without change in sea level or down warping.

(3.) Most of the anhydrite in the sabkha is of secondary origin; the component ions being derived from gypsum. There is no conclusive evidence to suggest that any anhydrite has formed by direct precipitation from concentrated sea water.

(4.) The Mg^{2+}/Ca^{2+} balance of evaporite brines is controlled in part by evaporation leading to precipitation of calcium carbonate and gypsum, and in part by concomitant dolomitization causing gypsum precipitation.

(5.) Three major diagenetic facies occur in the sabkha. Gypsum precipitation and dolomitization are essentially limited to a 2- to 3-mile wide belt adjacent to the lagoon margin. This belt is characterized by a high average frequency of flooding. In the mid-sabkha, brine compositions are consistent with an approach to mutual equilibration of brines and evaporitic-carbonate minerals. The landward margin of the sabkha is characterized by retrogressive diagenesis. The location of the progressive and the retrogressive diagenetic facies has moved seaward with the progradation of the shore line.

(6.) The similarity between the stratigraphy and mineralogy of some Ordovician, Silurian, Pennsylvanian, and Jurassic evaporite sequences and the sabkha on the Trucial Coast, suggest that evaporites of supratidal origin may be more common in older evaporite rocks than previously realized.

ACKNOWLEDGEMENTS

This paper forms part of the results of a study initiated at Imperial College, London, and continued under the auspices of the Dry Lands Research Institute at the University of California at Riverside. I should like to thank H. H. the Sheikh of Abu Dhabi, the Iraq Petroleum Company (Abu Dhabi), Frank Dickson, Graham Evans, and Douglas Shearman both for assistance and for making this study possible. I am deeply indebted to Professor Frank W. Dickson for suggesting the continuation of the project at Riverside, and for advice and stimulation during our visit to the Trucial Coast in 1967. I should also like to thank those who helped with advice and editing this manuscript: Christopher Kendall, Werner Raab, Robert Rex, and Seymour Schlanger.

* * * * * * *

REFERENCES

Bather, J. and Riley, J., 1954, The chemistry of the Irish Sea: Part 1. The sulphate-chlorinity ratio: Extrait du journal du Conseil International pour l'Exploration de la mer, v. 20, p. 145–152.

Butler, G. P., 1965, Early diagenesis in the Recent sediments of the Trucial Coast of the Persian Gulf: Unpubl. M.Sc. Thesis, University of London, 162 p.

Butler, G. P., Kendall, G. G. St. C., Kinsman, D. D. J., Shearman, D. J. and Skipwith, Sir Patrick, A. d'E., 1965, Recent evaporite deposits along the Trucial Coast of the Arabian Gulf [abs.]: Proc. Geol. Soc. Lond., v. 1623, p. 246–252.

Butler, G. P., 1967, Dependence of texture and structure on origin of Recent anhydrite, Trucial Coast, Arabian Gulf [abs.]: Geol. Soc. Am. Annual Meetings, p. 30–31.

——— 1968, Origin of Recent anhydrite, its texture and primary structures, the Sabkha, Trucial Coast, Arabian Gulf: (in preparation).

Culkin, Frederick, 1965, Chemical Oceanography. London Academic Press Inc., Ltd., v. 1, 693 p.

Curtis, Raymond, Evans, Graham, Kinsman, D. D. J., and Shearman, D. J., 1963, Association of dolomite and anhydrite in the Recent sediments of the Persian Gulf: Nature, v. 143(3607), p. 679–680.

Deffeyes, K. S., Lucia, F. J., and Weyl, P. K., 1965, Dolomitization of Recent and Plio-Pleistocene sediments by marine evaporite waters on Bonaire, Netherland Antilles, *in* Dolomitization and limestone diagenesis—a symposium: Soc. Econ. Paleontologists and Mineralogists Spec. Publ. No. 13, p. 71–88.

Evans, Graham, Kinsman, D. D. J., and Shearman, D. J., 1963, A reconnaissance survey of the environment of the Recent carbonate sedimentation along the Trucial Coast, Persian Gulf, *in* L.M.J.U. Straatan ed. Developments in sedimentology, v. 1. Deltaic and Shallow Marine Deposits: Amsterdam, Elsevier Publishing Co., p. 129–135.

Evans, Graham, Kendall, C. G. St. C., and Skipwith, Sir Patrick A. d'E., 1964, Origin of the coastal flats, the sabkha of the Trucial Coast, Persian Gulf: Nature, v. 202(4934), p. 749–761.

Evans, Graham, and Shearman, D. J., 1964, Recent celestine from sediments of the Trucial Coast, Persian Gulf: Nature, 202, p. 385–386.

Goldberg, Moshe, 1967, Supratidal dolomitization and dedolomitization in Jurassic rocks of Hamakhtesh Haqatan, Israel: Jour. Sedimentary Petrology, v. 25, p. 760–773.

Gray, R. S., 1967, Cache field—a Pennsylvanian algal reservoir in Southwestern Colorado: Am. Assoc. Petroleum Geologists Bull., v. 51, p. 1959–1978.

Holser, W. T., 1966, Diagenetic polyhalite in Recent salt from Baja California: Am. Mineralogist, v. 51, p. 99–109.

Illing, L. V., Wells, A. J., and Taylor, J. C. M., 1965, Penecontemporary dolomite in the Persian Gulf, *in* Dolomitization and limestone diagenesis—a symposium: Soc. Econ. Paleontologists and Mineralogists Spec. Pub. No. 13, p. 89–111.

Kendall, C. G. St. C., 1966, Recent carbonate sediments of the western Khor al Bazam: Unpub. Ph.D. Thesis, University of London, 436 p.

Kendall, C. G. St. C., and Skipwith, Sir Patrick A. d'E., 1966, Recent algal stromatolites of the Khor al Bazam, Southwestern Persian Gulf [abs.]: Geol. Soc. Am. Special Papers, 108.

Kinsman, D. D. J., 1963, The Recent carbonate sediments near Halat al Bahrani, Trucial Coast, Persian Gulf: *in* Straatan L.M.J.U., ed., Developments in sedimentology, vol. 1. Deltaic and Shallow Marine Deposits: Amsterdam, Elsevier Publishing Co., p. 185–192.

——— 1964, Recent carbonate sedimentation near Abu Dhabi, Trucial Coast, Persian Gulf: Unpubl. Ph.D. Thesis, University of London.

——— 1966, Gypsum and anhydrite of Recent age, Trucial Coast, Persian Gulf, *in* Second Symposium on Salt: Northern Ohio Geol. Soc., v. 1, p. 302–326.

——— 1967, Huntite from a carbonate-evaporite environment: Am. Mineralogist, v. 52, p. 1332–1340.

Masson, P. H., 1955, An occurrence of gypsum in Southwest Texas: Jour. Sedimentary Petrology, v. 25, p. 72–77.

Miller, D. N., 1961, Early diagenetic dolomite associated with salt extraction process, Inagua, Bahamas: Jour. Sedimentary Petrology, v. 31, p. 473–476.

Posnjak, Eugene, 1940, Deposition of calcium sulphate from water: Am. Jour. Sci., v. 238, p. 559–568.

Privett, D. W., 1959, Monthly charts of evaporation from the Northern Indian Ocean (including the Red Sea and the Persian Gulf): Q.J.R. Met. Soc., v. 85, p. 424–478.

Roehl, P. O., 1967, Stony Mountain (Ordovician) and Interlake (Silurian) facies analogs of Recent low energy marine and subaerial carbonates, Bahamas: Am. Assoc. Petroleum Geologists Bull., v. 51, p. 1979–2032.

Siever, Raymond, 1962, A squeezer for extracting interstitial water from modern sediments: Jour. Sedimentary Petrology, v. 32, p. 329–331.

Skipwith, Sir Patrick A. d'E., 1966, Recent carbonate sediments of the Eastern Khor al Bazam, Abu Dhabi, Trucial Coast: Unpubl. Ph.D. Thesis, University of London, 407 p.

Stewart, L. U., 1963, Marine evaporites, *in* Data of Geochemistry: U. S. Geol. Survey Prof. Paper, 440-Y, 52 p.

Wells, A. J., 1962, Recent dolomite in the Persian Gulf: Nature, v. 194, p. 274–275.

West, I. M., 1965, Macrocell structure and enterolithic veins in British Purbeck gypsum and anhydrite: Proc. York. Geol. Soc., v. 35, p. 47–58.

Editors' Comments on Papers 13 Trhough 18

Models of Evaporite Deposition

Long after it became known that seawater provided the source of the salts in many deposits of gypsum, anhydrite, halite, and potash, there still remained two perplexing problems: seawater contains a great variety of ions, but many evaporite deposits consist of only one mineral; and if a volume of seawater 1000 m high were evaporated, only a mere 1 m of gypsum would result, clearly implying that the formation of hundreds of meters of such a rock by static evaporation would require a basin of unreasonable depth. Add to this the knowledge that there are no thick evaporites forming anywhere on earth, and it will be appreciated why speculation on a geologic model to explain thick monomineralic deposits takes up much space in the literature on evaporites.

By the turn of the present century, the Bischof–Ochsenius "bar theory" was generally accepted as a reasonable explanation of thick accumulations, for a continuous supply of salts and persistent subsidence of a basin in an arid region would lead inevitably to such deposits. Branson (1915) realized that the unmodified "bar theory" would allow for no segregation of the major evaporite minerals in marine basins and he, therefore, attempted to explain the monomineralic composition characteristic of many evaporites.

In the first of the papers reproduced in this part, R. H. King proposed a realistic model of evaporite deposition. Although "backflow" of water from barred basins had been mentioned by Ochsenius to explain the absence of potash salts in some deposits, it was King's "theory of reflux" (1947) that opened the door to rational analysis of the origin of monomineralic precipitates. "Sedimentation in Permian Castile Sea," a major advance in evaporite geology, provides a remarkably clearly stated modification of the "bar theory" to account for the Castile Formation, a thick deposit of anhydrite with but minor halite.

The $NaCl:CaSO_4$ ratio in normal seawater is about 30; in the Castile Formation it is about 0.25! Clearly most of the brine that contained only halite and bittern salts escaped from the basin before becoming so concentrated that precipitation of those salts was initiated. In King's model, after anhydrite or gypsum has been precipitated, the remaining brine escaped as a dense countercurrent beneath the inflowing evaporating seawater. This "reflux" theory is accepted as an explanation for the sedimentation of many thick halite or anhydrite deposits, including the Zechstein and Paradox sequences.

P. C. Scruton lent observational support to the theory of reflux by applying oceanography to the stratigraphic reasoning of King. Using modern examples of water circulation, he showed that water movement in restricted basins, including the Mediterranean Sea and Baffin Bay (southeastern Texas), was as described in the theory of reflux, and that friction between the differentially moving water masses, and between the outgoing countercurrent and the basin bottom, would augment the physical barriers to free circulation. This paper, included here, established that students of evaporites had much to learn from oceanographers and from oceanographic principles.

Stratigraphic analysis of several major evaporite sequences showed that the Bischof–Ochsenius–King model of sedimentation was not universally applicable. In particular, the existence of a narrow restricted opening to every basin became a matter of doubt when such a small channelway could not be identified at the margins of many ancient saline basins.

This lead to the development of other models that allowed for sufficient restriction to promote the required concentration levels without the necessity of a narrow channelway through which all inflowing water moved. To Richter-Bernburg (1955) goes credit for the origin of the "saturation shelf" concept, in which marine water is preconcentrated on a broad shelf before entering a partially restricted basin.

The natural extension of the "saturation shelf" concept, included here, is the study of the Lower Cambrian evaporites of the Siberian platform by Arkhangel'skaya and Grigor'yev (1960). They propose that these deposits were formed on a very broad, very shallow shelf, free of any *physical* barrier to circulation; hypersalinity developed because of the great distance the water was driven over this shallow shelf. The rocks display a lateral transition from normal marine sediments to halite, and even potash, deposits. This model, which is described in a theoretical way in the book by Shaw (1964, p. 10–13), has no modern counterpart. The Upper Jurassic Todilto Formation of northern New Mexico formed at the far southern reaches of the Sundance Sea in an environment that may have been similar to that described by Arkhangel'skaya and Grigor'yev.

A unique feature of the article by Arkhangel'skaya and Grigor'yev is the mathematical treatment of the salinity changes that can take place in epeiric seas of the tropics and subtropics. The authors show that at any point within the basin, the salinity is dependent only on the distance from the open ocean, and, assuming sufficient shallowness to prevent reflux, is independent of the average depth and rate of evaporation.

Nataliya Andreevna Arkhangel'skaya graduated from the Moscow Academy of Mines in 1930, and she has been working at the Geological Institute of the Academy of Sciences of the USSR since 1936. Her colleague, Vladimir Nikolaevich Grigor'yev, graduated from the Moscow Oil Institute in 1950 and has been working since then at the Geological Institute.

These Soviet geologists made their study in the late 1950s when the diagenetic formation of sulfates and dolomite in the supratidal environment was unrecognized. The authors, therefore, did not consider the possibility that at least part of the Lower Cambrian sulfate/dolomite sequence may have formed in the supratidal flat environment. One cannot help but wonder if the sabkha cycle is well represented. The authors emphasize the shallowness of the basin and the presence of numerous oncolites and stromatolites and their occasional disturbance by wave action. It seems probable that epeiric sea evaporites and supratidal flat evaporites formed commonly in spatial and temporal proximity.

A. J. Pannekoek of the University of Leiden in the Netherlands, has summarized, neatly and succinctly, all the essential arguments that support the idea that evaporites can form in deep water. First proposed by Richter-Bernburg (1950, 1955), the basic contention is that inasmuch as subsidence is generally slow and halite deposition unusually rapid, basin subsidence must have occurred before halite deposition began in order to accommodate the enormous thicknesses known. Two Soviet geologists, Yanshin (1961) and Ivanov (1967), take opposite sides in this argument; their accounts, not included here, are available in English (see References).

An interesting epilog to the "saturation shelf" and deep-basin models is the analysis of the Paradox evaporites by R. J. Hite, a research geologist with the Branch of Organic Fuels and Chemical Resources of the United States Geological Survey in Denver, Colorado. Hite has spent much of his professional career working on these Pennsylvanian evaporite cycles (Hite and Gere, 1958; Hite, 1960, 1961, 1968; Peterson and Hite, 1969), and in the paper included here, expresses the opinion, which, in the eyes of many who have looked at the Paradox sequence, labels him as an iconoclast, that both "saturation shelf" and deep-basin models are applicable to these cycles. He considers that banks of phylloid algae and other aspects of carbonate deposition developed on a broad shelf adjacent to the main Paradox basin within which the more soluble salts precipitated at the same time and in deeper water.

The efficacy with which oceanographic studies may be applied to the investigation of evaporite deposition was once again demonstrated by the appearance of the paper by Brongersma-Sanders and Groen (1970). Margaretha Brongersma-Sanders, now retired, was lecturer in Oceanography at Leiden University in the Netherlands, and is an earth scientist of some repute, being particularly well known for detailed and voluminous studies of catastrophic mass mortality of marine life. She and Groen demonstrated that under certain circumstances of wind and water circulation, evaporite deposition could be induced within a deep basin, and their analysis was based on observations in existing deep basins such as the Gulf of Cariaco and the Gulf of California.

Reprinted from *Bull. Amer. Assoc. Petroleum Geologists,* **31**, 470–477 (1947)

13

SEDIMENTATION IN PERMIAN CASTILE SEA[1]

RALPH H. KING[2]
Washington, D. C.

ABSTRACT

The Permian Castile formation is a favorable starting point for a study of precipitated marine sediments because of its relatively simple stratigraphy. It is 1,300 to 2,000 feet thick and consists principally of "banded" or laminated anhydrite, but it contains some salt and other materials. It was deposited within the deep steep-walled Delaware Basin of western Texas and southeastern New Mexico. Throughout Castile time communication with the open sea was maintained through a restricted channel near the southwest corner of the basin. Paucity of detrital material indicates that the deposit consists almost exclusively of marine evaporites.

It is suggested that the water within the basin consisted of a body of brine lying below average wave base and a less dense surface layer lying above average wave base. Some of the dense brine continuously returned to the open sea; otherwise the volume of halite deposited would have been about 30 times as great as the volume of anhydrite instead of somewhat less than the volume of anhydrite. The barrier must therefore have been incomplete. Either the top lay partly below wave base or the barrier was permeable. This theory of reflux is presented as a modification of Ochsenius' bar theory.

The calcium sulphate presumably was deposited as anhydrite rather than as gypsum because of the salinity of the brine. The banding is ascribed to seasonal variation in temperature. The average net rate of evaporation in Castile time is calculated as 114 inches a year.

INTRODUCTION

The chief purpose of this paper is to stimulate interest in and discussion of the deposition of precipitated sediments. Mere passive acceptance or prejudiced rejection of the reasoning and conclusions would be disappointing. The subject holds more than academic interest, for any data pertaining to chemical precipitates, and any inferences that may be drawn therefrom, have a bearing on the deposition of potassium salts and other saline minerals.

The Castile formation was deposited within the Delaware Basin near the end of Permian time. The area originally covered by Castile rocks includes all of Loving and Reeves counties and parts of Pecos, Ward, Winkler, and Culberson counties, Texas, and parts of Eddy and Lea counties, New Mexico (Kroenlein, 1939, Fig. 1; King, 1942, Fig. 33). The basin is 165 mile long (north-south) and 96 miles wide (Adams, 1944, p. 1598).

The Castile formation consists principally of "banded" or, more correctly, laminated anhydrite. It contains some small calcite lenses, some massive anhydrite, and some halite (Adams, 1944, pp. 1603–08). A basal layer a few feet thick is calcareous (King, 1942, p. 611). The thickness of laminated and massive anhydrite ranges from 700 feet at the north end of the basin to 1,800 feet at the south end, and averages 1,250 feet. Halite is most prominent in the northern part of the basin, where it locally reaches a maximum thickness of 900 feet. The thickness of the Castile formation as a whole ranges from 1,300 feet at the north end of the basin to 2,000 feet at the south end (Kroenlein, 1939, p. 1686). Adams (1944, p.

[1] Manuscript received, September 7, 1946.

Published by permission of the director of the Geological Survey, United States Department of the Interior.

[2] Geological Survey.

1604) states that the "average" thickness is between 1,500 and 1,850 feet and that the maximum exceeds 2,100 feet.

Two main problems are considered. Perhaps the more controversial is the problem of accounting for the scarcity of salt in the formation, which is present in insignificant quantity in comparison with the great amount of anhydrite, as contrasted to the large ratio of sodium chloride to calcium sulphate in sea water. A modification of Ochsenius' bar theory, termed the theory of reflux, is offered to account for this discrepancy.

The second major problem is the original composition of the calcium sulphate, whether anhydrite or gypsum. Reasons are given for believing that it was anhydrite.

Some inferences on the nature of the barrier and a computation of the average rate of evaporation are also offered.

Discussion is restricted to the Castile formation because the sediments of that age were subjected to the fewest complicating factors during their depositional and post-depositional history. The boundaries of the area are well defined, the stratigraphy of the formation has been revealed by numerous drill holes, and the region has escaped violent tectonic movement, so that the deposits remain unaltered except by weathering.

GEOLOGICAL AND CHEMICAL DATA

Geologic history.—As a result of the slight regional uplift that occurred at the close of Guadalupe time, the Permian sea withdrew from a large area in southeastern New Mexico and western Texas, and during Castile time it occupied only the Delaware Basin in that area. The Delaware Basin had already received non-saline Permian sediments of great thickness prior to Castile time, but it continued to subside, so that throughout Castile time the water was several hundred feet deep (Adams, 1944, pp. 1598, 1617). Within the basin deposition was continuous, although the change in lithologic character was abrupt. The steep basinward face of the Capitan reef, which formed during Guadalupe time and rose 1,200 to 2,000 feet above the basin floor (Adams, 1944, p. 1598), constituted the margin of the basin. Because the sides of the basin were nearly vertical, the area covered by Castile sediments coincided almost exactly with the area of the water surface that was exposed to evaporation. Communication with the open sea was maintained through a shallow channel at the southwest corner of the basin (King, 1942, pp. 758–59).

At the end of Castile time the sea withdrew still farther, so that locally an erosional unconformity separates Castile sediments from the overlying Salado formation (Adams, 1944, p. 1608).

Laboratory data on evaporation of sea water.—Usiglio's data, recomputed by Clarke (1924, pp. 220–21) in more commonly used units, indicate that when a sample of sea water is evaporated, iron oxide and calcium carbonate begin to precipitate before the volume has been reduced by half. When the volume has been reduced to 19 per cent of the original volume, all the calcium carbonate and about one-third of the calcium sulphate are precipitated. When the volume is further reduced to 9½ per cent of the original volume, more than 85 per cent of the calcium sulphate is precipitated, together with slightly more than 10 per cent of the sodium chloride and traces of more soluble compounds. At this point, however, the weight of sodium chloride precipitated already amounts to more than twice the weight of calcium sulphate precipitated. The total volume of sodium chloride precipitated by complete evaporation of sea water is about 30 times as great as the total volume of calcium sulphate calculated as anhydrite.

Solubility of calcium sulphate.—The solubility of anhydrite in water decreases as the temperature increases. Addition of sodium chloride up to about 6 per cent increases the solubility of calcium sulphate in either form at 30°C., but further addition of sodium chloride decreases the solubility of calcium sulphate. In sea water at 30°C., gypsum is less soluble than anhydrite until the volume has been reduced to about 20.8 per cent of the original volume, but thereafter anhydrite is the less soluble form, and further evaporation yields anhydrite (Posnjak, 1940, pp. 564–65). At 42°C. only anhydrite is precipitated at any concentration (Posnjak, 1938, p. 268).

INFERRED DEPOSITIONAL PROCESS

That the anhydrite comprising most of the Castile formation is a chemical precipitate rather than a detrital deposit is a generally accepted belief, but the probable process of deposition has not been fully discussed.

Source of material.—The shallow-water foreland-facies deposits of Guadalupe age, which were exposed to subaerial erosion after the withdrawal of the sea into the Delaware Basin, consist chiefly of detrital materials, dolomite, and gypsum. These beds probably were the source of the minor amount of clastic material contained in the deposits of the Castile formation, and of a comparable amount of chemically precipitated material. That any major portion of the Castile sediments was derived from this source, however, seems unlikely. The ratio of clastic to precipitated sediments in the Castile formation is very small, whereas the reverse would be expected if the mode of transportation were predominantly terrestrial water or wind. Furthermore, the older strata do not show evidence of erosion extensive enough to account for the great mass of anhydrite that constitutes most of the Castile formation. It seems likely, therefore, that the chemical precipitates of the Castile were derived principally from evaporating marine waters, and that the clastic deposits and possibly a minor part of the chemical precipitates were introduced by terrestrial water or wind.

Distribution of water density.—Evaporation takes place only from the surface of a body of water. In that part of a body of water that lies above wave base and which is termed the surface layer, wave action mixes the water so that uniform density is maintained from top to bottom of the surface layer, although not necessarily throughout the horizontal extent of the layer.

In an isolated body of water the density of the surface layer increases as evaporation proceeds, but simultaneously the denser liquid tends to settle by gravity, so that vertical uniformity of density throughout the entire depth is approached if not quite attained. If evaporation from one part of the surface is more rapid, tending to produce a denser column of water beneath that part, such water also moves horizontally, displacing lighter less concentrated solution, so that horizontal uniformity of density is also approached.

In Castile time the Delaware Basin, in which the Castile sediments were deposited, was not completely isolated, but was connected by a channel to the

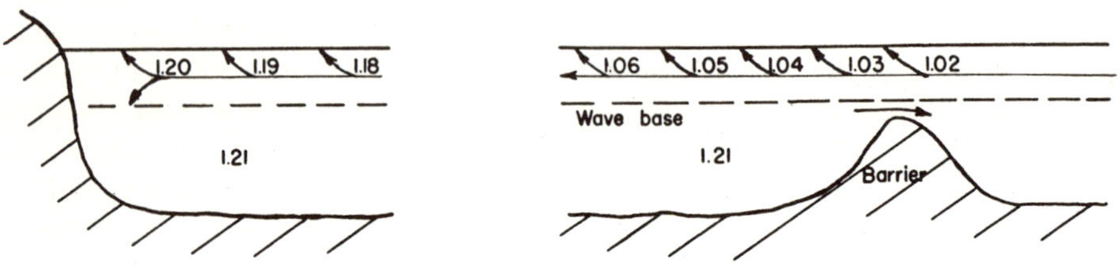

FIG. 1.—Profile through Delaware Basin indicating distribution of water density and direction of movement of water in Castile time.

open sea on the southwest. New sea water could enter the basin to replace the water lost by evaporation. Movement of water from the sea to the basin was not only possible but necessary to maintain the hydrostatic equilibrium between the two bodies of water. The normal sea water entering the basin was obviously less dense than the water already in the basin, which had been concentrated somewhat by evaporation; hence the new water tended to flow over the residual brine. As the new water was subjected to evaporation it became more dense, but as its density reached that of the underlying water the two mingled, so that throughout the basin the new water remained on top of the residual brine until the new water had been concentrated to a density equal to that of the brine.

Only at the entrance of the basin was the new water normal sea water. As it flowed toward the farther edges of the basin it became a progressively more concentrated solution of various salts, its density continually increasing. Inasmuch as the underlying brine was, in effect, an isolated body, presumably its density was virtually uniform throughout. Only at the innermost shores of the basin, therefore, would the water of the surface layer attain sufficient density to sink and mingle with the brine already present. Conversely, the maximum density attained by the surface layer would determine the density of the underlying brine.

In summary, the Delaware Basin probably contained, during Castile time, a body of brine of approximately uniform density, and an overlying surface layer that was less dense than the brine and that increased in density with distance from the entrance.

Theory of reflux.—The volume ratio of halite to anhydrite in the Castile formation as a whole is less than 1:1 (Adams, 1944, p. 1620) as compared with a volume ratio of sodium chloride to calcium sulphate in sea water of about 30:1. The volume of water that must have been evaporated to yield the volume of anhydrite contained in the Castile sediments was approximately 2,500 times the volume of the basin itself, as is discussed subsequently. Either incomplete evaporation of a single body of water or deposition and subsequent removal of the halite in a single volume of water is therefore an inadequate explanation of the scarcity of halite. It is necessary to account either for non-deposition of halite during a protracted period of continuous evaporation and replenishment or for subsequent removal of halite deposited with the anhydrite.

Although locally some halite and some anhydrite were removed from the top of the formation by the pre-Salado erosion, the perfect preservation of the delicate laminae in the anhydrite refutes the suggestion that halite was interbedded with the anhydrite but was subsequently removed.

The alternative to be followed may be stated in another way. It is necessary to account for cessation of evaporation when the surface layer of water reached the concentration at which halite precipitates, that is, when the specific gravity is about 1.21. Inasmuch as evaporation takes place only from the surface layer and inasmuch as the surface layer does not sink until its density equals that of the underlying brine, it follows that the specific gravity of the brine must not exceed 1.21 if deposition of halite is to be avoided. A specific gravity of 1.21 would have been exceeded early in Castile time, however, unless the dense brine could escape from the basin (Adams, 1944, p. 1620). Escape of the brine by percolation into the substrata would require permeable substrata and an outlet at lower altitude, that is, far below sea level. It seems more likely that the brine returned to the sea either by overflowing the barrier or by percolation through it. This theory is offered here as a modification of Ochsenius' bar theory and is termed the theory of *reflux.*[3]

The barrier.—If the theory set forth is correct, the Delaware Basin was connected to the sea by a channel through which new sea water entered the basin to replace the water lost by evaporation and by reflux and through which the reflux returned to the sea. In view of the small amount of clastic material deposited thereby, the volume of terrestrial water can be disregarded, and the volume of sea water entering the basin can be regarded as equal to the volume evaporated plus the volume that returned to the sea as reflux. The previously cited critical specific gravity of 1.21 is attained when about 90 per cent of sea water is evaporated. It follows that the reflux would have been about 10 per cent of the water that entered the basin. The ratio of the mass velocities of the entering sea water and the departing brine would then be 10:1.

The new water entered the basin as the surface layer, which extended down

[3] Adams had the opportunity of examining the first draft of this paper before he published and he courteously refrained from using more of the material than was necessary, and from stating the theory, lest he divert any credit to himself. This accounts for the similarity of treatment in the two papers.

to wave base, whereas the reflux was heavy brine of the underlying layer. If the linear velocity of the influx and reflux were equal the ratio of the cross-section area of the channel above wave base to that below wave base would likewise be 10:1. Depth to wave base is a fluctuating factor and its average value in Castile time is unknown, so the absolute height of the barrier cannot be determined.

Alternatively, if the barrier consisted of loose sand the reflux could pass out of the basin by percolation through the barrier. Such a permeable sand bar would not prevent reflux even though it extended up to wave base. Presumably it could not long persist above wave base.

Rate of evaporation.—The laminae in the anhydrite range from 0.2 mm. to 7.0 mm. in thickness and average 1.63 mm. in Gresham and McAlpine's core test hole Flood No. 1 in Culberson County, Texas (Udden, 1924, p. 353). On the reasonable assumption that the laminae represent annual layers or "varves," and on the basis of the average thickness of the laminae studied and the known thickness of the laminated anhydrite in this hole, Udden calculated the duration of Castile time as 306,000 years.

The area of the Delaware Basin is estimated as 10,000 square miles. The average thickness of the anhydrite in the Castile formation is 1,250 feet. The volume of anhydrite is calculated therefrom as roughly 350×10^{12} cubic feet. A factor of 10.9 cubic feet per ton is used to convert this figure to 32.5×10^{12} tons. Assuming that evaporation of sea water to about 10 per cent of its original volume yields 0.113 per cent anhydrite, deposition of the Castile anhydrite would have required evaporation of 28.75×10^{15} tons of sea water, equivalent to 928.5×10^{15} cubic feet before evaporation. Inasmuch as the area of the basin is only 283×10^{9} square feet, this volume of water would have filled the basin to a depth of 3,280,000 feet if it had all been present at one time. The evaporation of this volume of water to 11.2 per cent of its original volume would have yielded about the amount of halite present in the Castile formation, so it is unnecessary to make any additional allowance for halite. (Elsewhere the figure 10 per cent has been used for simplicity.)

It is assumed that 11.2 per cent of the 3,280,000 feet of sea water, or 370,000 feet, returned to the sea as reflux, and that 2,910,000 feet evaporated. Dividing this last figure by Udden's figure of 306,000 years for the duration of Castile time gives an average net rate of evaporation for the entire basin of $9\frac{1}{2}$ feet, or 114 inches, a year.

It may be suggested that a direct calculation of the amount of evaporation required to deposit an average annual thickness of 1.63 mm. (Udden's figure) of anhydrite should give the desired result. Such a computation, however, would not take into account such factors as the difference in concentration of the water of the surface layer at different points in the basin, difference in rate of evaporation at different points, the presence of nonlaminated anhydrite, and the effect of currents in the basin. If the "varves" are pictured as wedges or lenses the weakness of the method becomes obvious.

Original composition of calcium sulphate sediments.—According to the theories herein presented, at the beginning of Castile time the Delaware Basin contained normal sea water, but as evaporation proceeded while circulation was restricted by the barrier the salinity increased, and the more concentrated solution ponded behind the barrier. Soon the concentration throughout the basin reached the point of saturation with respect to calcium carbonate, and the thin basal calcareous layer of the Castile formation was deposited. Subsequently the concentration reached the point of saturation for calcium sulphate, and that substance was thereafter precipitated along with the additional calcium carbonate that continually entered the basin in the new sea water.

If the temperature exceeded 42°C. (107.6°F.) all the calcium sulphate was precipitated as anhydrite (Posnjak, 1938, p. 268). Such a temperature seems excessive, although not impossible. If the temperature was 30°C. (86°F.) the calcium sulphate was deposited as gypsum until the concentration of the brine was about 4.8 times that of normal sea water and the specific gravity was about 1.11 (Clarke, 1924, p. 220); thereafter it was precipitated as anhydrite (Posnjak, 1940, p. 565). This concentration would have been attained in slightly more than one one-thousandth of Castile time. In either case, virtually all the calcium sulphate was deposited as anhydrite if, as herein suggested, the concentration of the brine was maintained throughout Castile time at or near that at which sodium chloride begins to precipitate.

Any theory that attempts to show that the major part of the calcium sulphate was originally deposited as gypsum rather than anhydrite should include supporting evidence of a cause of dehydration, of a path for the water in leaving the formation, and of an ultimate destination for the water.

Locally and temporarily the concentration reached the point of saturation for sodium chloride, and halite was deposited. Such periods were relatively brief and may have resulted from temporary raising of the barrier, which would reduce the rate of reflux, or from temporary or local increase in the rate of evaporation.

The comparison of the laminae that characterize most of the anhydrite of the Castile formation with those of varved clay deposits has been interpreted as implying an alternation of periods of deposition and periods of non-deposition similar to that producing the varve markings in clay. That may be the correct explanation, but another possibility should be considered. It must be borne in mind that we are not dealing with isothermal evaporation. The temperature of the surface layer of water in the basin responded to seasonal changes in atmospheric temperature, and the solubility ratio of the calcium sulphate and calcium carbonate likewise varied. The temperature of the underlying brine may have remained virtually constant, but the brine was already saturated with respect to both compounds, so that particles of either substance settling through the brine were unaffected by it.

SUMMARY

It is suggested that in a basin such as the Delaware Basin, enclosed on all sides but connected to the open sea by a restricted channel, the water becomes differentiated into an underlying concentrated brine of uniform density, which lies below wave base and is ponded behind the barrier, and a surface layer of somewhat less average density, the water of this layer being normal sea water near the entrance of the basin. In Castile time a portion of the underlying brine flowed continuously from the Delaware Basin to the open sea as reflux, and new sea water entered the basin continuously to offset the loss by reflux and evaporation. The more soluble salts remained in solution and passed out of the basin in the reflux. The less soluble salts were precipitated.

The Castile formation consists principally of anhydrite. The scarcity of clastic material indicates a scarcity of terrestrial water, which is consistent with the computed rate of evaporation of 114 inches a year throughout Castile time. Hence only a small portion of the precipitated sediments can be attributed to a terrestrial source. The preponderance of anhydrite indicates that the ratio of influx to reflux was about 10:1, as a ratio of, say, 12:1 would have resulted in deposition of more halite than anhydrite.

If the barrier at the entrance to the basin had been continuously maintained at wave base and if it had been impermeable there would have been no reflux of the underlying brine as such, so it is suggested that the top of the barrier lay somewhat below average wave base or else that the barrier was permeable.

If, in accordance with the theories just stated, the salinity of the Castile sea within the Delaware Basin was near the point of saturation with respect to sodium chloride throughout Castile time, the calcium sulphate was deposited as anhydrite rather than gypsum.

It is suggested that the characteristic lamination of the anhydrite of the Castile formation may be ascribed to seasonal variations in temperature and accompanying variations in relative solubility of anhydrite and calcite, rather than to cyclical interruptions in deposition.

The theory of reflux is essentially a modification of Ochsenius' bar theory designed to explain the deposition of the Castile sediments and other similar deposits.

REFERENCES

ADAMS, J. E. (1944), "Upper Permian Ochoa Series of Delaware Basin, West Texas and Southeastern New Mexico," *Bull. Amer. Assoc. Petrol. Geol.*, Vol. 28, pp. 1596–1625, Figs. 1–4.

CLARKE, F. W. (1924), "The Data of Geochemistry," *U. S. Geol. Survey Bull. 770*, pp. 1–841.

KING, P. B. (1942), "Permian of West Texas and Southeastern New Mexico," *Bull. Amer. Assoc. Petrol. Geol.*, Vol. 26, pp. 533–763, Pls. 1–2, Figs. 1–34.

KROENLEIN, G. A. (1939), "Salt, Potash, and Anhydrite in Castile Formation of Southeastern New Mexico," *ibid.*, Vol. 23, pp. 1682–93, Figs. 1–3.

POSNJAK, E. (1938), "The System $CaSO_4$—H_2O," *Amer. Jour. Sci.*, 5th Ser., Vol. 35A, pp. 247–72, Figs. 1–4.

——— (1940), "Deposition of Calcium Sulfate from Sea Water," *Amer. Jour. Sci.*, Vol. 238, pp. 559–68, Fig. 1.

SVERDRUP, H. U., JOHNSON, M. W., AND FLEMING, R. H. (1942), *The Oceans, Their Physics, Chemistry, and General Biology*, pp. 1–1087, Figs. 1–265, Charts 1–7. Prentice-Hall, Inc., New York.

UDDEN, J. A. (1924), "Laminated Anhydrite in Texas," *Bull. Geol. Soc. America*, Vol. 35, pp. 347–54, Pls. 7–10.

Reprinted from *Bull. Amer. Assoc. Petroleum Geologists,* **37**, 2498–2512 (1953)

14

DEPOSITION OF EVAPORITES[1]

P. C. SCRUTON[2]
La Jolla, California

ABSTRACT

Modern oceanographic studies in estuaries and other restricted arms of the sea have shown that a characteristic circulation pattern exists. Surface currents flow from regions of low salinity to regions of higher salinity in response to hydrostatic head and are accompanied at depth by oppositely directed currents flowing from high to low salinity regions because of density distribution. Salts are deposited in restricted estuaries where evaporation exceeds precipitation plus runoff. The necessary restrictions of the estuary or basin are in part dynamic and in part static. Dynamic restriction is caused by the hydrostatic head and by frictional stresses between the oppositely directed surface and bottom currents, and between the bottom current and the channel floor. Static restriction is produced by topographic confinement.

When high concentrations are developed a strong horizontal salinity gradient exists which produces lateral segregation of different salts during precipitation. The escaping deep current returns to the sea those salts which have not been precipitated. Fluctuations in equilibrium caused principally by changes in excess of evaporation or in degree of channel closure cause migrations of the horizontal salinity gradient along the longitudinal axis of the basin which produce vertical differentiation of salts. The order in which salts occur in the vertical column can be predicted from oceanographic and chemical data and agrees well with many vertical sequences which have been described.

INTRODUCTION

A problem which has interested geologists for many years is the deposition of thick beds of salts. The ocean is recognized to be the source of the salts which form the more common deposits, and their origin is of oceanographic as well as geologic interest. The purposes of this report are: (1) to describe certain hydrographic conditions in estuaries in both humid and arid regions which have been found by recent oceanographic studies, (2) to utilize this experience to elaborate on known characteristics of salt-depositing evaporating pans, and (3) to infer a depositional sequence of evaporites which suggests explanations for some of the features of known beds.

Ochsenius (1888) was one of the first workers to develop an hypothesis to explain salt deposits, and his now-famous bar theory has become a classic. Later workers, such as Branson (1915) and R. H. King (1947), have proposed modifications to this theory which seemed necessary to explain relationships which they observed.

Many workers have been troubled by the volume ratio of $CaSO_4$:NaCl of about 1:1 which exists in the Permian Castile beds of West Texas and southeast New Mexico. This ratio is not in agreement with the ratio of approximately 1:30 which would exist if normal sea water were completely evaporated. King (1947, pp. 470–71) explained the discrepancy as follows: the deep Permian Delaware

[1] Contribution from the Scripps Institution of Oceanography, New Series, No. 653. Manuscript received, May 18, 1953.

[2] Scripps Institution of Oceanography, University of California. During its preparation the manuscript was critically read by R. S. Arthur, G. S. Bien, M. N. Bramlette, F. B Phleger, R. R. Revelle, and F. P. Shepard, all of the Scripps Institution. Their helpful comments and suggestions are gratefully acknowledged.

basin of Texas and New Mexico, an arid region where evaporation exceeded precipitation plus runoff, was connected to the open sea by a restricted channel on the southwest. The water within this basin below what he inferred to be average wave base consisted of a uniform brine formed by the excess of evaporation. Normal marine water flowing into the basin through the restricted channel, replaced that lost by the excessive evaporation, moved over the uniform brine toward the distal end of the basin, and was concentrated by evaporation until it became similar in density to the brine below. It then mingled with the brine and sank. To compensate for the volume of sinking brine at the basin's upper end, he suggested that a continuous seaward flow of dense hypersaline water, a "reflux" from the basin, took place in the bottom of the channel below average wave base. King's hypothesis was developed to account for geological relationships; he believed a process of this type was necessary to remove the excess sodium chloride in sea water from the basin and establish the observed volume ratio of $CaSO_4$:NaCl.

In recent years much work has been done to establish the flow characteristics and distribution of properties in estuaries, inlets, and other bodies of water partly restricted from the open sea. Studies of this type furnish information which can be utilized to increase understanding of factors leading to formation of evaporites.

OCEANOGRAPHIC DATA

Most recent studies of restricted arms of the sea are in humid areas where fresh water is introduced in quantity to the sea, so that precipitation and runoff exceed evaporation. Stommel and Farmer (1952b) have compiled numerous data from these sources. Among many other notable studies in this type of region, those conducted by Tully (1937, 1949), Redfield (1950), Ketchum (1951), and Pritchard (1950, 1951) are good examples.

These studies have shown that in an estuary or lagoon with a source of fresh water in the upper end, a horizontal surface salinity gradient exists with salinity increasing seaward. In longitudinal section of the estuary, current and salinity measurements show this gradient to be associated with seaward flow of less saline water at the surface accompanied by landward flow of more saline water beneath. Mixing between these opposing currents occurs in the zone of contact and is accompanied by erosion of the deeper layer. Thus volume and salinity of surface water increase seaward and of the deeper water decrease landward. Mixing may occur between fresh and sea water directly, or it may involve the formation of an intermediate water type which is later remixed, so that more complicated relationships are displayed (Redfield, 1950). Figure 1 shows the nature of some of the relationships which have been found.

From their studies Stommel and Farmer (1952b) have classified estuaries in humid regions on the basis of the degree to which these vertical and horizontal salinity and current characteristics are shown and have recognized four general

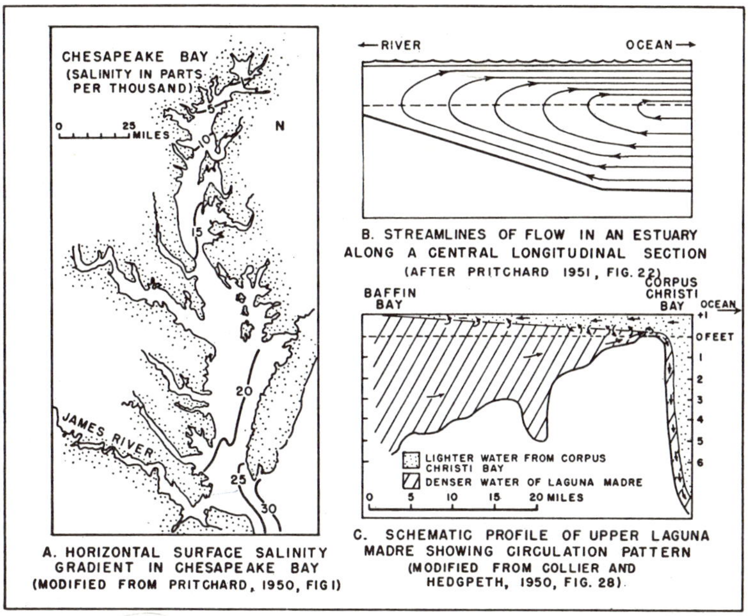

FIG. 1

types. These are as follows. 1. The vertically mixed estuary which shows a horizontal salinity gradient but a minimum of vertical differentiation of water types; the Severn and Raritan rivers are samples. 2. The slightly stratified estuary of which the James River is an example. 3. The highly stratified estuary exemplified by Alberni Inlet. 4. The salt-wedge estuary, a type found at the mouths of strong rivers where the vertical flow and salinity characteristics are best developed and where a salt wedge may extend far upstream. The Mississippi River is an example of this type. In the Mississippi River, Russell (1936), Rhodes (1950), and Simmons (1952) have shown that the distance upstream to which the salt wedge extends is a function of volume of river discharge. During periods of low discharge a small frictional stress is exerted by the superjacent fresh water on the salt wedge, the rate of mixing between the two layers is not great, and the wedge may extend at least 135 miles upstream (Simmons, 1952, p. 70; Russell, 1936, p. 94). Under conditions of high discharge, however, the wedge may be expelled from the river mouth.

There are few studies of estuaries in regions where evaporation exceeds precipitation plus runoff. A group at the Scripps Institution of Oceanography under the direction of G. C. Ewing (Ewing, 1950) investigated the hydrography of San Diego Bay, an evaporation basin during the summer. Collier and Hedgpeth (1950) reported on a series of surface salinities collected in Laguna Madre and Baffin Bay, evaporating pans on the southern Texas coast. The marked salinity gradients in these inlets exist because of the excess of evaporation, so that salinities increase away from the source of normal sea water. Zenkevich (1947) reports a similar salinity distribution in the Sivash (Putrid) Sea, an arm of the Sea of Azov, which he also attributes to excessive evaporation.

Data derived from the San Diego Bay study indicate the existence of a surface drift toward the restricted end of the bay caused by a surface slope produced by excessive evaporation in the bay. Associated with the surface drift is a return, seaward directed, bottom current which is the result of increased density due to concentration of salts in the restricted area of the bay. Collier and Hedgpeth (1950, Fig. 28) deduced a similar system from their observations in Laguna Madre (Figure 1-C). The Mediterranean and Red seas (Sverdrup *et al.*, 1946) provide well known examples of surface currents flowing through restricted channels into evaporating pans with return currents to the open sea of more dense saline water resulting from concentration of salts flowing outward beneath. Sinking of the oxygenated, highly saline, surface waters prevents stagnation of the deep basins (Sverdrup *et al.*, 1946, pp. 649–50).

Although currents and salinity gradients in estuaries in regions of excessive or deficient evaporation are opposite in direction with respect to a fixed point on the estuary, controlling dynamics in the two systems are the same. This can be appreciated if consideration is confined to the relationship between the flow pattern and the salinity and density gradients. In both types of estuaries the surface current flows in the direction of increasing salinity and the bottom current flows toward water of lower salinity. Figures 1 and 2 illustrate this concept. Both Hedgpeth (1953) and Pritchard (personal communication) also have recognized this basic similarity. The vertical pattern of flow in estuaries of either type is a product of surface slope caused by either excessive evaporation or excessive precipitation plus runoff and slope of isobaric surfaces at depth caused by horizontal and vertical density differences. Density differences in these water bodies are caused chiefly by salinity differences (Pritchard, 1951, p. 33), but temperature exerts some effect. Because of the basic similarity of the two types of estuaries, hydrographic characteristics established for one may be applied directly to the elucidation of the characteristics of the other. Figure 2 gives schematically the vertical distributions of salinity, density (A), pressure (B), and currents (C) to be expected in water bodies of either type.

In general, factors which appear to be most important in determining the hydrographic character of estuaries and similar bodies of water are: (1) the balance between amount of fresh water introduced and the amount of evaporation, (2) the depth and configuration of the bottom of the estuary including the sill or bar, and (3) local conditions of wind, temperature, and tides. Conditions of quasi-equilibrium between precipitation, evaporation, and flow to and from the sea tend to be established. Such an equilibrium results in a relatively stable position of the horizontal salinity gradient. Modifications to the equilibrium or steady state conditions are produced, however, by long term as well as seasonal variations in precipitation, temperature, winds, and sea-level changes (Collier and

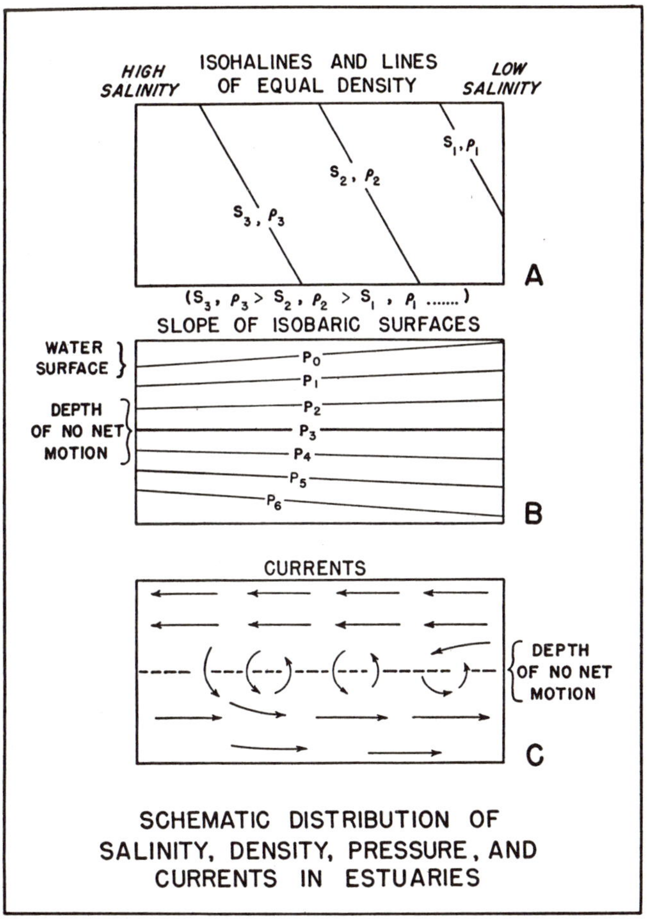

FIG. 2

Hedgpeth, 1950; Pritchard, 1950, 1951), so that the horizontal salinity gradient fluctuates along the longitudinal axis of the basin (on a long term or seasonal basis).

The results of these numerous investigations of estuaries and large evaporation basins amply confirm the basic pattern of circulation suggested in King's general reflux hypothesis. In addition, however, they can be used, together with other data, to extend the hypothesis and to suggest explanations for many of the mineralogic and lithologic variations which are found in evaporite sequences.

CHEMICAL DATA

Usiglio's experiments on evaporation of sea water furnish, so far as is known, the only existing chemical data basic to the entire evaporite problem. Table 1,

TABLE I. APPROXIMATE LIMITING DENSITY AND SALINITY VALUES WITHIN WHICH VARIOUS SALTS PRECIPITATE FROM SOLUTION

(From Clarke, 1924, p. 220)

Salt	*Lower Limit*		*Upper Limit*	
	Density	*Salinity Parts Per Thousand*	*Density*	*Salinity Parts Per Thousand*
Calcium carbonate	1.0500	72	1.1264	199
Calcium sulphate	1.1264	199	1.2015	332
Calcium sulphate and sodium chloride	1.2138	353	1.2570	427
Sodium chloride	1.2778	457	1.3069	523
Magnesium sulphate	1.2138	353	1.3069	523

prepared from these data (Clarke, 1924, p. 220), shows that individual salts tend to be precipitated within a unique salinity and density range. If density values associated with the salinities at which salts are precipitated are substituted in Figure 2-B, isopycnal distribution in a salt-depositing evaporating pan can be reconstructed. Since the density is principally a function of dissolved salts, the isopycnals also can be considered lines of equal salt content. Figures 3 and 4 show how this salinity and density distribution tends to produce deposits of relatively pure salts on the bottom underlying the salinity (density) range within which the particular salt is precipitated (Lang, 1937, pp. 885–88; Hills, 1942; Adams and Frenzel, 1950).

THEORY

Deposition of a salt from sea water requires that the concentration be increased until the solubility product of the salt is exceeded. This can be accomplished on a large scale in nature only by evaporation in areas where restrictions to free circulation exist and normally where evaporation exceeds precipitation plus runoff for a relatively long period of time.

When conditions of excess evaporation over precipitation plus runoff exist, a slope of the water surface develops causing a basinward flow of sea water due to hydrostatic pressure. If conditions remain uniform, an approximate equilibrium between input of fresh and salt water and outflow of bottom water will be established. With the establishment of equilibrium, salinity characteristics do not change, and the horizontal gradient does not shift markedly within the basin.

Restriction to escape of the deep water of the basin is necessary for high salt content to be developed. Such restriction of a static nature is imposed by a sill or similar physical barrier. Dynamic restrictions to escape of deep water also exist, however. Hydrostatic head induces a flow toward the basin head throughout the vertical water column. To counterbalance this flow horizontal density differences (principally the result of salinity differences) are established so that the pressure surfaces at depth slope toward the basin entrance (Fig. 2). The field of pressure produced by hydrostatic head must be exceeded at depth by the

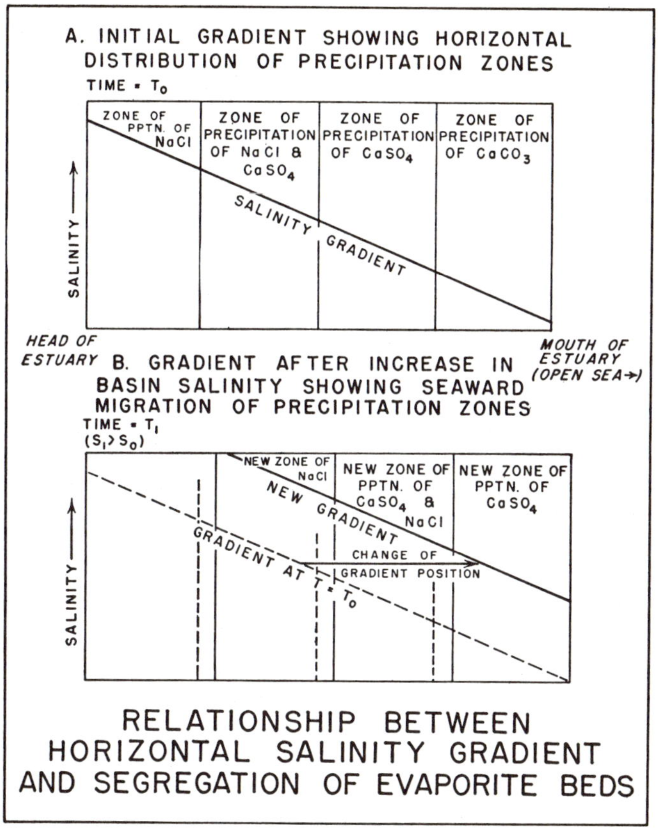

FIG. 3

pressure field due to density. In addition, under conditions of strong stratification in the entrance channel and high rate of surface inflow into a basin caused by a large amount of evaporation, another type of dynamic barrier to brine escape exists. Data from study of the Mississippi River (Rhodes, 1950; Simmons, 1952) have shown that a large flow of fresh water imposes a dynamic barrier to intrusion of salt water because of friction at their interface. Flume experiments on a two layer system at constant volume of surface discharge conducted by Stommel and Farmer (1952a) show the effectiveness of this dynamic barrier to salt water intrusion is increased by constriction of the channel. Similarly, when surface water flows into an evaporation basin, escape of deep brine is inhibited by the establishment of this type of barrier. Existence of a dynamic barrier of this type is not only an observed feature of the flow pattern but may be essential to entrapment of the brine. The effectiveness or completeness of these dynamic barriers is related to both amount of excess of evaporation and degree of restriction of the connecting channel. For this reason, the basin salt, or density, equilibrium must

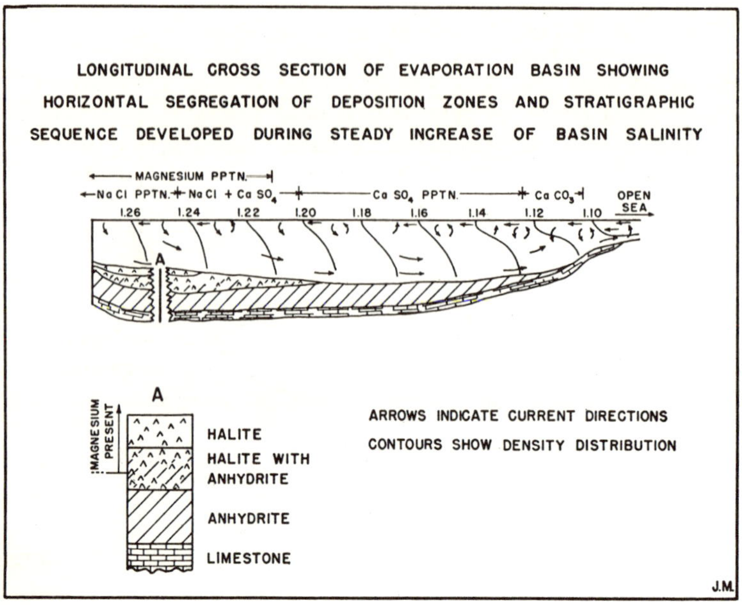

FIG. 4

be considered a dynamic equilibrium which is principally sensitive to fluctuations in these factors. Small scale shifts in equilibrium can be produced by changes in density due to temperature or sea-level variations and by changes in wind stress.

INFERRED PROCESS

The foregoing background information, derived from studies of circulation in estuaries and evaporation basins, chemical experiment, and theoretical consideration, permits a general reconstruction of events and circumstances leading to evaporite deposition.

An excessive rate of evaporation removes water from the restricted basin, causes a slope of the sea surface to develop and a continuous flow of additional water from the ocean to enter the basin in response to this surface slope. At the initiation of the process, basin water has a relatively low and uniform density and therefore concentration of surface water is required to establish strong gradients. A current of dense water flowing to the sea commences when the slope of pressure surfaces at depth (Fig. 2) due to excess density creates a potential for flow which exceeds the hydrostatic head. After this flow is established, the potential also must overcome topographic restraint, bottom friction, and, where strong vertical stratification exists, frictional forces between the oppositely directed currents. Stirring and mixing occur between the surface inflowing water and the outflowing bottom water because of turbulence and friction between the two currents, and takes place at a higher rate in restrictions of the channel. This produces a modification in the character of each of these waters, an increase in the salinity of the surface water and a decrease in the salinity of the bottom water. Thus, soon after the process is initiated, the inflowing current transports water to the basin which is more concentrated than the normal marine water originally flowing from the open sea. This effect may be present at a considerable distance seaward from the basin if the channel is long. The restrictions imposed on the bottom flow are such that free circulation does not develop, and an increase in salinity and density of basin water takes place until conditions of approximate dynamic equilibrium are achieved.

When salinities are high in the inflowing surface water due to mixing, destruction of organisms being swept in will result. Their shells tend to be returned toward the open sea on settling to the bottom, however, because of the seaward directed bottom current.

The hydrographic characteristics of a large basin could be expected to change from a type similar to the salt-wedge river estuary near the entrance to one comparable with the vertically mixed estuary at the basin head as the density differences between surface and deep waters decreased (Fig. 4). The inflowing surface water is concentrated as it flows toward the distal portions of the basin by both evaporation and continued mixing with more saline waters. Because the distal portions of the basin in general contain water of highest salt content, deposition of evaporites is initiated in these regions when the solution is sufficiently concentrated. This seems to be especially true if the distal portions of the basin are large.

Table I shows the approximate order in which the most common evaporites would begin to be formed. Usiglio's order of precipitation indicates that the first material to be deposited would be iron oxide (not shown in Table I). As the water becomes more concentrated, the zone of iron oxide precipitation moves toward the entrance of the basin, and calcium carbonate is precipitated in the distal portion (Figs. 3, 4). Further concentration results in calcium sulphate coming from solution in the distal portion and an entrance-ward movement of the zone of calcium carbonate precipitation. Still greater concentration of the water brings sodium chloride from solution in the distal region. Thus, a vertical section through the resulting deposit would show, at the base, beds containing iron oxide overlain in ascending order by limestone, gypsum or anhydrite, and rock salt, initially with some $CaSO_4$, later containing more magnesium. Deviation from this order can be anticipated, however, since the data of Usiglio give no more than an indication of the order of precipitation. At no place in an evaporation basin could the water be expected to retain "normal marine" ionic ratios existing at the time. Protracted precipitation in the basin and mixing and recycling in the channel and basin tend to concentrate the more soluble ions. Surface drainage in different regions also varies in chemical character. In addition, at high concentrations temperature changes are of great importance in determining character of precipitated material.

Since quasi-equilibrium conditions can be established at any concentration and thus stabilize the position of the horizontal salinity gradient, the thickness of each unit could be a measure of the length of time the material in question was being precipitated over the particular bottom. Furthermore, since the salts are isolated into zones of deposition by differences in solubility products, a longitudinal profile of the basin with the entrance at the origin would show the same sequence. Figure 4 shows horizontal segregation of the deposits and the nature of the inferred vertical sequence.

Seasonal or long-term variations in rainfall such as described by Judson (1952) in the southwestern United States, variations in temperature or sea-level altering the density structure, or variations of wind are capable of altering the equilibrium conditions and causing horizontal movement of the position of the depositing zones. Such movements cause the character of precipitating salt at any point to change.

GEOLOGICAL DATA

It is important to subject the inferred mechanism for producing a migrating salinity gradient to a geological test. This can not be accomplished directly. However, if it is assumed that the channel cross section remained constant and the fluctuations occurred in the excess of evaporation, comparison of the hypothetical sequence of beds which would result with actual sequences described in the literature affords some check.

Stewart (1951a, b) recently has published the results of detailed analyses of cores penetrating the Permian evaporite sequence of Yorkshire, England. Three cycles of deposition are represented, with the middle zone being the most complete. Stewart's generalized subdivision of this zone is as follows (1951a, p. 446).

Top
1. Upper Halite zone
2. Halite-Sylvine zone
3. Halite zone*
4. Halite-Anhydrite zone
5. Anhydrite zone
6. Dolomite zone (upper Magnesian limestone)

* Magnesite is present in the sequence above the middle of this zone, with the lower part containing layers of halite and anhydrite.

This sequence is described in detail and furnishes excellent agreement with the hypothetical sequence. The top Halite zone suggests a migration of the salinity gradient toward the distal portion of the basin following deposition of the Halite-Sylvine zone.

The Permian salt section of Germany (Barton, 1925a, b; Brown, 1934) shown in Table II indicates fluctuating salinity conditions with maxima during deposition of the "Older" potash beds, the "Younger" rock salt, and the "Youngest" rock salt. A considerable break in sedimentation is suggested above the "Older" potash beds. Increase of salinity to a maximum followed by a decrease as indicated by a change in the type of salt also is shown in the German Triassic (Brown, 1934, p. 1245), the Silurian of New York (Phalen, 1919, pp. 55, 57), the

TABLE II. NORMAL PERMIAN EVAPORITE SEQUENCE OF MIDDLE LEINE VALLEY, GERMANY
(After Barton, 1925b, p. 1248)

	Meters	
Upper Zechstein clays	20	Massive redbed clay with anhydrite nodules and basal bed of anhydrite
"Youngest" Rock Salt	50–60	Rock salt
"Pegmatite" Anhydrite	1–2.5	Anhydrite with scattered intergrown salt crystals
Red Salt Clay	20–25	Massive red clay with anhydrite nodules
"Younger" Rock Salt	145–70	Rock salt with some polyhalite and anhydrite
"Main" Anhydrite	20–40	Massive anhydrite, no bedding
"Gray" Salt Clay	4.5	Clay, dolomitic marl, anhydrite, uniform
"Older" Potash beds	8	55% carnallite, 26% rock salt, 17% kieserite
Transition zone	13	Rock salt and kieserite
"Older" Rock Salt	70	Rock salt with anhydrite beds; 94% rock salt, 6% anhydrite
"Lower" Anhydrite (present in normal Stassfurt section)	70–100	Anhydrite with intercalations of bituminous dolomitic beds and 1 or 2 rock salt beds

Silurian of Michigan (Stone *et al.*, 1920, pp. 216–17), and the Permian of Oklahoma (Muir, 1934, pp. 1300–02). Each of these sections can be explained on the basis of fluctuating salt content caused by changes in the evaporation-precipitation budget.

Similar agreement with the hypothetical sequence seems to be indicated by the often-repeated cycle of deposits described by Schaller and Henderson (1932, pp. 8–14) from the potash fields of Texas and New Mexico. This cycle consists of the following.

Top
1. Nearly pure halite
2. Halite (plus some polyhalite)
3. Polyhalite
4. Anhydrite
5. Clay (commonly iron-rich, see Henderson, pp. 24–26)

(Zones 4 and 5 may contain magnesite as well.)

They also attribute genetic significance to the cycle because of its common occurrence (1932, p. 8) and suggest that the polyhalite is formed subsequent to deposition by replacement. However, the possibility of this mineral being primary and resulting from the co-precipitation of magnesium, calcium, and potassium sulphates in concentrated solution with later some solution and redeposition should be considered (see Clarke, 1924, p. 227; Schaller and Henderson, 1932, p. 64, also Pls. 24, 26–33; Hill, 1937, p. 2243; Kroenlein, 1939, p. 1689; Posnjak, 1938, p. 270). From other descriptions of the Castile and overlying Salado formations (Lang, 1937; Kroenlein, 1939; P. B. King, 1942; Adams, 1944) it would appear that the general sequence which is found also corresponds with the inferred section. The Castile beds consist principally of laminated anhydrite in the southern (seaward) end of the basin overlying a thin limestone, but also contain thick halite deposits on the north. Overlying the Castile are the salt beds of the Salado formation. Fluctuations in equilibrium conditions with time producing migrations of the lines of equal salt content are indicated in the vertical column by changes from limestone[3] to anhydrite, to salt, to salt with anhydrite, to anhydrite with salt.

The Pennsylvanian Paradox formation of Colorado and Utah has been described by Baker, Dane, and Reeside (1933). Their most complete section, obtained from driller's records of the J. H. Shafer well No. 1 (1933, p. 969), also suggests general agreement with the hypothesis. Numerous black shales which have special significance occur between evaporite beds of the Paradox formation, however, and suggest extreme fluctuations in the precipitation-evaporation budget, salinity conditions, and the circulation pattern. A change from excess of evaporation to excess of precipitation plus runoff could result in a change from conditions of an oxygenated evaporation pan to those of an euxinic basin, an environment favorable to formation of black shales (Fleming and Revelle, 1939, pp. 95–102), within the same basin framework.

On a much smaller scale the fine lamination of the anhydrites of the Castile formation, consisting of alternations of anhydrite and bituminous calcite (Udden, 1924; Schaller and Henderson, 1932), seem to confirm the hypothesis and to suggest short-term fluctuations in salinity of the water. Udden (1924, pp. 351–52) speculated on the length of time represented by individual laminae and by groups of laminae. He tentatively concluded (1924, p. 352) that the individual laminae represented annual cycles and suggested correlation with sunspot cycles for groups of laminae. These seem to be reasonable interpretations as to time represented, and could be caused by changes in salinity due to variations in temperature, evaporation, sea-level, or wind. Adams (1944, p. 1605) also notes the presence of thin laminae of similar type in many of the limestones associated with evaporites. These laminae perhaps also may be the results of annual fluctuations of salinity.

The validity of the concepts is favored by the good agreement between many evaporite sections described in the literature and the sequence which can be predicted from oceanographic and chemical data and theoretical consideration without recourse to changes in channel restriction. Testing of the inferred process by comparison of hypothetical and actual sequences shows that observed variations can be explained on the basis of changing equilibrium relationships within the basin. In many instances these environmental changes seem to provide the most satisfactory explanation. Such a treatment tacitly excludes from consideration vertical structural movements or periodic closing of the channel. Many variations in described sections, however, have been interpreted to be the results of tectonic movements or channel closures which assume environmental uniformity. These mechanisms also are capable of producing the same types of changes in equilibrium conditions by relaxing or increasing physical restriction to flow in the channel. At any given time deposition of evaporites requires a suitable

[3] It is not implied that all limestones in a sequence containing evaporites are produced by precipitation from a concentrated solution, since freshening, even below normal marine salinities, may occur during fluctuations in excess of evaporation over precipitation plus runoff.

basin within which the processes that have been described can act. Changes of physical character in the basin or its environs, in the evaporation-precipitation budget, or in channel characteristics produce related changes in equilibrium conditions and in the material being deposited. Detailed reconstruction of evaporation and structural history within a particular basin requires equally detailed examination of the rocks in order to detect past fluctuations of salinity conditions and to interpret their causes.

CONCLUSIONS

1. Evaporites are deposited in restricted arms of the sea in areas where evaporation exceeds precipitation and runoff. Large basins, remote from the open sea, are most satisfactory.

2. Circulation in the basin is established similar to that found in modern estuaries, so that a continuous inflowing surface current of partly concentrated solution is counterbalanced in part by a continuous return flow at depth of concentrated brine toward the sea.

3. The salinity achieved by evaporation of surface water represents an equilibrium between evaporation, precipitation plus runoff, and flow from and to the sea. High salinity is developed because of restriction to brine escape. That brine which does escape when equilibrium conditions are established returns to the sea those salts which have not been precipitated.

4. Restrictions to escape of the deep brine are in part static (as physical barriers) and in part dynamic (as relationship between pressures due to hydrostatic head and density distribution, friction between currents flowing in opposite directions, and friction between the deep current and the channel bottom). The basin equilibrium is dynamic and is principally sensitive to fluctuations of excessive evaporation and degree of channel restriction.

5. Horizontal segregations of pure evaporite deposits are produced because of the development of a strong horizontal salinity gradient, salinities increasing from the entrance to the head of the evaporating pan, and by the different salinity ranges at which precipitation of different salts occurs.

6. The vertical sequence of beds which result from salinity changes can be predicted approximately from experiments on evaporation of sea water. The predicted sequence results from lateral migration of the horizontal salinity gradient and shows excellent agreement with natural sequences which have been described in detail. Salinity changes are produced principally by changes in precipitation or channel restriction and to lesser extent by variations of temperature, sea-level, or wind stress.

7. Extreme fluctuations in the precipitation-evaporation budget, so that precipitation and runoff are predominant, can result in a change from conditions of an oxygenated evaporation pan to conditions of an euxinic basin within the same physical framework.

REFERENCES

ADAMS, J. E., 1944, "Upper Permian Ochoa Series of Delaware Basin, West Texas and Southeastern New Mexico," *Bull. Amer. Assoc. Petrol. Geol.*, Vol. 28, No. 11 (November), pp. 1596–1625.

ADAMS, J. E., AND FRENZEL, H. N., 1950, "Capitan Barrier Reef, Texas and New Mexico," *Jour. Geol.*, Vol. 58, pp. 289–312.

BAKER, A. A., DANE, C. H., AND REESIDE, J. B., JR., 1933, "Paradox Formation of Eastern Utah and Western Colorado," *Bull. Amer. Assoc. Petrol. Geol.*, Vol. 17, pp. 963–80.

BARTON, D. C., 1925a, "The Salt Domes of South Texas," *ibid.*, Vol. 9, pp. 536–89.

———, 1925b, "The American Salt Dome Problems in the Light of the Roumanian and German Salt Domes," *ibid.*, pp. 1227–68.

BRANSON, E. B., 1915, "Origin of Thick Gypsum and Salt Deposits," *Bull. Geol. Soc. America*, Vol. 26, pp. 231–42.

BROWN, L. S., 1931, "Cap Rock Petrography," *Bull. Amer. Assoc. Petrol. Geol.*, Vol. 15, pp. 509–30.

———, 1934, "Age of Gulf Border Salt Deposits," *ibid.*, Vol. 18, pp. 1227–96.

CLARKE, F. W., 1924, "The Data of Geochemistry," *U. S. Geol. Survey Bull. 770.* 841 pp.

COLLIER, A., AND HEDGPETH, J. W., 1950, "An Introduction to the Hydrography of Tidal Waters of Texas," *Pub. Inst. Marine Sci.*, Vol. 1, No. 2, pp. 121–94.

EWING, G. C., 1950, "A Survey of San Diego Bay," *U.S.N.R. Research Unit 11-5 Mimeogr. Rept.* (December 31). 21 pp., 9 figs., 3 appendices.

FLEMING, R. H., AND REVELLE, R. R., 1939, "Physical Processes in the Ocean," *Recent Marine Sediments*, Amer. Assoc. Petrol. Geol., pp. 48–141.

HEDGPETH, J. W., 1953, "Introduction to the Biogeography of the Northern Gulf of Mexico," *Pub. Inst. Marine Sci.* (in press).

HILL, A. E., 1937, "The Transition Temperature of Gypsum to Anhydrite," *Jour. Amer. Chem. Soc.*, Vol. 59, Pt. 2, pp. 2242–44.

HILLS, J. M., 1942, "Rhythm in Permian Seas—A Paleogeographic Study," *Bull. Amer. Assoc. Petrol. Geol.*, Vol. 26, pp. 217–55.

JUDSON, S., 1952, "Arroyos," *Sci. American*, Vol. 187, No. 6, pp. 71–76.

KETCHUM, B. H., 1951, "The Exchanges of Fresh and Salt Waters in Tidal Estuaries," *Jour. Marine Res.*, Vol. 10, pp. 18–38.

KING, P. B., 1942, "Permian of West Texas and Southeastern New Mexico," *Bull. Amer. Assoc. Petrol. Geol.*, Vol. 26, pp. 535–763.

KING, R. H., 1947, "Sedimentation in Permian Castile Sea," *ibid.*, Vol. 31, pp. 470–77.

KROENLEIN, G. A., 1939, "Salt, Potash, and Anhydrite in Castile Formation of Southeast New Mexico," *ibid.*, Vol. 23, pp. 1682–93.

KRUMBEIN, W. C., 1951, "Occurrence and Lithologic Associations of Evaporites in the United States," *Jour. Sed. Petrology*, Vol. 21, pp. 63–81.

LANG, W. B., 1937, "The Permian Formations of the Pecos Valley of New Mexico and Texas," *Bull. Amer. Assoc. Petrol. Geol.*, Vol. 21, pp. 833–98.

MUIR, J. L., 1934, "Anhydrite-Gypsum Problem of Blaine Formation, Oklahoma," *ibid.*, Vol. 18, pp. 1297–1312.

OCHSENIUS, C., 1888, "On the Formation of Rock-Salt Beds and Mother Liquor Salts," *Proc. Acad. Nat. Sci. Philadelphia*, Vol. 40, pp. 181–87.

PHALEN, W. C., 1919, "Salt Resources of the United States," *U. S. Geol. Survey Bull. 669.* 284 pp.

POSNJAK, E., 1938, "The System $CaSO_4$—H_2O," *Amer. Jour. Sci.*, Vol. 35A, pp. 247–72.

PRITCHARD, D. W., 1950, "Notes on the Dynamics of Estuarine Waters," *Proc. Colloquium on Flushing of Estuaries*, Woods Hole Oceanographic Inst., Ref. No. 50-37, pp. 49–60.

———, 1951, "The Physical Structure, Circulation, and Mixing in a Coastal Plain Estuary," *unpublished doctor's dissertation*, Univ. of California, Scripps Institution of Oceanography. 60 pp.

REDFIELD, A. C., 1950, "Note on the Circulation of a Deep Estuary—The Juan de Fuca-Georgia Straits," *Proc. Colloquium on Flushing of Estuaries*, Woods Hole Oceanographic Inst., Ref. No. 50-37, pp. 175–77.

RHODES, R. F., 1950, "Effect of Salinity on Current Velocities, Evaluation of Present State of Knowledge of Factors Affecting Tidal Hydraulics and Related Phenomena," *Comm. on Tidal Hydraulics, Corps of Engineers, U. S. Army, Rept. 1.*

ROTH, R. I., 1942, "West Texas Barred Basin," *Bull. Geol. Soc. America*, Vol. 53, pp. 1659–74.

RUSSELL, R. J., 1936, "Physiography of Lower Mississippi River Delta," *Louisiana Geol. Survey Geol. Bull. 8.* 199 pp.

SCHALLER, W. T., AND HENDERSON, E. P., 1932, "Mineralogy of Drill Cores from the Potash Field of New Mexico and Texas," *U. S. Geol. Survey Bull. 833.* 124 pp.

SIMMONS, H. B., 1952, "Salinity Problems," *Proc. Second Conference on Coastal Engineering* (Houston, Texas, November, 1951), pp. 68–85.

STEWART, F. H., 1951a, "The Petrology of the Evaporites of the Eskdale No. 2 Boring, East Yorkshire, Part II, The Middle Evaporite Bed," *Mineralogical Magazine*, Vol. 29, pp. 445–75.
———, 1951b, "The Petrology of the Evaporites of the Eskdale No. 2 Boring, East Yorkshire, Part III, The Upper Evaporite Bed," *ibid.*, pp. 557–72.
STOMMEL, H., AND FARMER, H. G., 1952a, "On the Nature of Estuarine Circulation," *Woods Hole Oceanographic Inst. Tech. Rept. 52-51* (Pt. I, Chapts. I and II).
——— AND ———, 1952b, "On the Nature of Estuarine Circulation," *ibid., Tech. Rept. 52-63* (Pt. III, Chapt. VII).
STONE, R. W., AND OTHERS, 1920, "Gypsum Deposits of the United States," *U. S. Geol. Survey Bull. 697.* 326 pp.
SVERDRUP, H. U., JOHNSON, M. W., AND FLEMING, R. H., 1946, *The Oceans, Their Physics, Chemistry, and General Biology.* 1087 pp. Prentice-Hall, New York.
TULLY, J. P., 1937, "Oceanography of Nootka Sound," *Jour. Biol. Board Canada*, Vol. 3, pp. 43–69.
———, 1949, "Oceanography and Prediction of Pulp Mill Pollution in Alberni Inlet," *Bull. Fish. Res. Board Canada.* 169 pp.
UDDEN, J. A., 1924, "Laminated Anhydrite in Texas," *Bull. Geol. Soc. America*, Vol. 35, pp. 347–54.
ZENKEVICH, L. A., 1947, "Fauna and Biological Productivity of the Sea," *Soviet Science*, Tome II, pp. 393–99.

Translated from *Izvestiya Acad. Sci. USSR, Geol. Ser.*, **4**, 41–54 (1960)

15

FORMATION OF HALOGENIC ZONES IN MARINE BASINS ILLUSTRATED BY THE EXAMPLE OF THE LOWER CAMBRIAN EVAPORITE BASIN OF THE SIBERIAN PLATFORM[1]

by

N. A. ARKHANGEL'SKAYA AND V. N. GRIGOR'YEV

On the basis of material obtained during a study of the Lower Cambrian deposits of the southern and western parts of the Siberian Platform the authors conclude that in an arid climate a halogenic zone may form within an extensive shallow water basin without being isolated by barriers but merging gradually into the areas of normal salinity.

* * * * *

A study of the lithology and facies of the Lower Cambrian deposits on the southern and western borders of the Siberian platform revealed a number of peculiarities of sedimentation in the basin of that age. Of the greatest interest are the conditions of formation of salt deposits which form the subject of this paper.

The Lower Cambrian stratigraphy of the Siberian platform is sufficiently well known now to serve as a basis for paleogeographic and facies reconstructions. As a result of the work of a large group of geologists [3, 5, 9, 13, 16, 20, 25], a unified stratigraphic column has been constructed for the Cambrian period. Using it as a base and taking the lithological data into account, we correlated the Lower Cambrian sections of the entire investigated territory beginning with the Zhura substage (Figure 1). The correlation chart shows that the salt-bearing deposits are restricted to three stratigraphic levels, the Ketema and Tolbachan horizons and the Zhura substage.[2]

The Lower Cambrian strata of the Siberian platform are mainly carbonate rocks, fine-grained and cryptocrystalline dolomites being the most widespread. The uniformity of their texture and composition over long distances leaves no doubt that they are primary dolomites.

Among the limestones, homogeneous micro-granular and fine-grained aphanitic limestones are predominant and they also are of chemical origin.

Very common are carbonate rocks of mixed calcitic and dolomitic composition, which are represented by homogeneous and spotted varieties. In the former, usually microgranular or fine-grained, the grains of calcite and dolomite are uniformly distributed through the rocks. The dolomite grains are usually somewhat larger than the calcite grains, but not infrequently the grains of both minerals are alike in form and size and can be distinguished in thin sections only by staining. The relative amount of calcite and dolomite in different beds varies within broad limits and the rocks form a continuous series from slightly dolomitic limestones to slightly calcitic dolomites. These rocks were evidently formed as a result of simultaneous accumulation and precipitation of calcite and dolomite crystals in the chemical sediment. The spotted textures resulted from diagenetic redistribution of the materials in these sediments.

Phytogenic sediments such as stromatolitic and oncolitic limestones, dolomites and intermediate carbonate rocks are widespread in the Lower Cambrian. They also are essentially chemical deposits but formed by the biological activity of blue-green algae. Zoogenic rocks such as archaeocyathid and fossil detritus limestones are exceptionally rare in the Lower Cambrian.

Side by side with the carbonate rocks an important role in the Lower Cambrian sections is played by such obviously chemical precipitates as gypsum, anhydrite and rock salt, and mixed carbonate-sulfate rocks (anhydrite-dolomite) are of frequent occurrence.

Besides the carbonate and salt deposits, the

[1]Usloviya obrazovaniya solerodnykh zon v morskikh vodoyemakh na primere Nizhnekembriyskogo evaporitovogo basseyna Sibirskoy platformy.

[2]Some investigators [13] refer the lower salt-bearing beds in the Irkutsk amphitheatre (the Usolka formation) to the Sinian age.

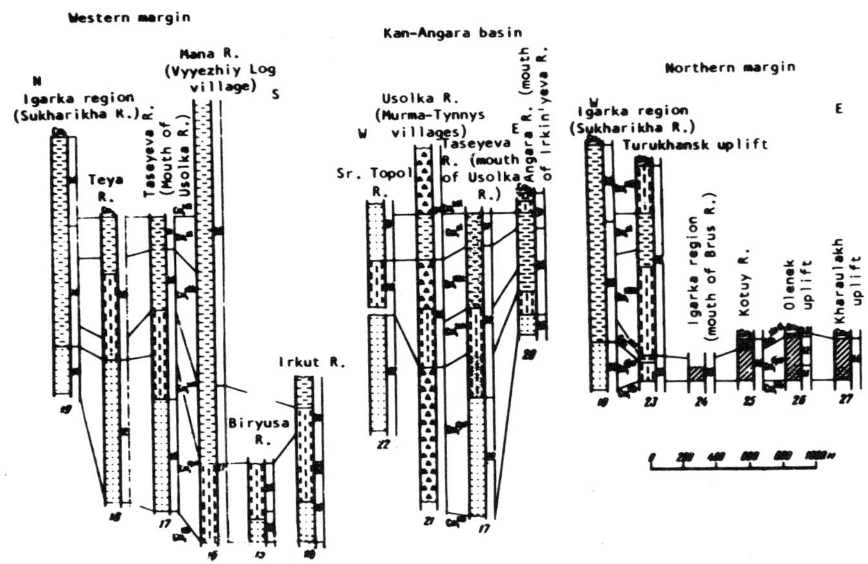

FIGURE 1. Correlation and facies of the Lower Cambrian deposits of the Siberian platform.

Facies: I - facies of normal salinity; II - same facies, thin deposit; III - facies of nearly normal salinity; IV - facies with salinity transitional from normal to high; V - facies of high salinity; VI - salina facies; VII - terrigenous nearshore facies of marine basin; VIII - considerable admixture of terrigenous material in evaporite beds; IX - numbers of exploratory boreholes.

Stratigraphic subdivisions: Aldan stage: Cm_1^{tbl} - Tolbinsk substage, Cm_1^{djur} - Zhurinsk substage. Lena stage: Cm_1^{sin} - Sinskaya zone, Cm_1^{tlbc} - Tolbachan zone, Cm_1^{ol} - Olekma zone, Cm_1^{ket} - Keteminsk zone, Cm_1^{el} - Yelanskoye zone.

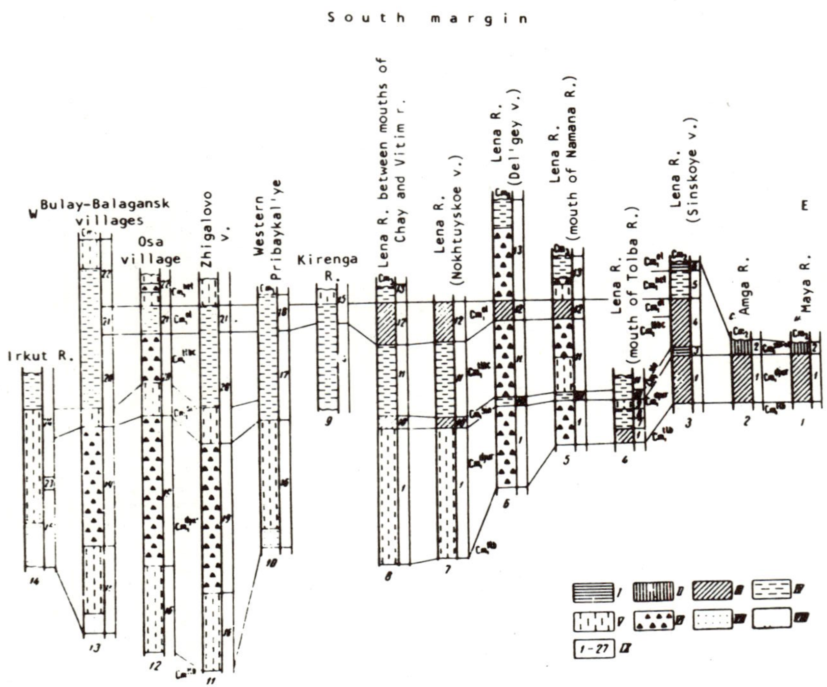

Formations (subformations, members, zones): 1 - Variegated; 2 - Inikan, 3 - Sinskaya, 4 - Kutorgina, 5 - Keteminskaya, 6 - Yelanskaya, 7 - Nokhoroyskaya, 8 - Churanskaya, 9 - Malykanskaya, 10 - El'gyanskaya, 11 - Tolbachansk, 12 - Olekma, 13 - Charsk, 14 - Tukolonskaya, 15 - Munokskaya, 16 - Motskaya, 17 - Ungurinskaya, 18 - Kamenskaya, 19 - Usol'skaya, 20 - Bel'skaya, 21 - Bulayskaya, 22 - Angara, 23 - Perekhodnaya, 24 - Vvedenskaya, 25 - Ust'-Tagul'skaya, 26 - Tal'skaya, 27 - Zhershul'skaya, 28 - Kolbinskaya, 29 - Shalyginskaya, 30 - Klimenskaya, 31 - Agalevskaya, 32 - Nemchanskaya, 33 - Lebyazhinskaya, 34, Izluchinskaya, 35 - Sukharinskaya, 36 - Rechkinskaya, 37 - Platonovskaya, 38 - Kostinskaya, 39 - Ust'-Brusskaya 40 - Ust'-Kotuykanskaya, 41 - Kesyusinskaya, 42 - Yerkeketskaya, 43 - Kuonamskaya, 44 - formation G (zones b - d - e).

43

Lower Cambrian contains great thicknesses of purely continental deposits, but they occur only in some areas of the southwestern border of the Siberian platform in the lower part of the Cambrian section (Zhura substage). They were formed, evidently, as a result of erosion of mountains which existed at that time in these areas.

Inasmuch as most Lower Cambrian sediments are obviously chemical and the determining factor in their deposition was the degree of salinity of the basin of sedimentation, it is precisely this factor that must be taken as the basis for separating the different facies of the Lower Cambrian sediments.

It is necessary to consider the criteria needed for accurate determination of the relative salinity of a basin. First of all, salinity is reflected in the composition of the sediments. For example, a direct proof of the high salinity of a basin is the presence of sulfates and halite, because gypsum begins to preceipitate from sea water only when salinity reaches 13%, and halite when it reaches 27%. The effect of change in salinity on the composition of marine faunas is also well known. An attempt was made to determine the relation between the composition of the Lower Cambrian carbonate rocks and the character of their faunas. As an example, we selected the Lower Cambrian section of the northeast part of the Aldan massif, composed entirely of carbonate rocks and abundantly fossiliferous. It was found that the number of species of trilobites, brachiopods, pteropods and sponges, and the total amount of organic remains are inversely proportional to the average dolomite content in the rocks, so that when this relationship is represented graphically, the curve of content of primary dolomite in the section is a mirror image of the curve representing the total number of the organisms listed above. This indicates that the variation in the dolomite content may be regarded as a clue to the variation in the salinity of a basin equal in value to the faunal change and the content of sulfates and halite.

Using these criteria, obtained by a detailed lithological study of the Lower Cambrian rocks of the Siberian platform, we distinguished a number of facies:

The marine facies characteristic of basins of normal salinity is represented by pure limestones. Here dolomite is exceptional and the sediments contain a rich and varied fauna.

The facies of the zone with nearly normal salinity is also represented by limestones but beds of mixed calcite-dolomite composition are present. The faunal assemblage in these deposits is somewhat less varied than in the preceding facies and the fossils are less abundant.

The facies of the zone of intermediate salinity forms a series illustrating gradually increasing salinity of a basin. This zone is characterized by the presence of pure limestones, mixed carbonate rocks and pure primary dolomites. The proportions of different rocks and the dolomite content vary within broad limits. The intermediata zone is poor in organic remains and the intermediate and extreme members of the salinity series contain only a few species with the individuals being small and obviously stunted.

The facies of the zone of high salinity contains a variety of primary dolomites. Limestones are exceptional and gypsum and anhydrite are occasionally found. No fossils have been found in this zone.

The halogenic zone facies. In this zone we placed all facies containing various salt deposits from dolomites interbedded with gypsum and anhydrite to deposits containing rock salt as the principal constituent.

Besides these facies characteristic of the entire Lower Cambrian, there are others more restricted in time. For example, in the deposits of the Zhura substage there is a group of nearshore marine facies represented by molasse-like deposits of red ripple-marked and cross-bedded polymict sandstones and siltstones with admixed coarse material.

In the Lena stage a special facies must be separated among the facies of the zone of normal salinity occurring only in the eastern border of the Siberian platform and distinguished by a sharply diminished thickness. It is represented by black stongly bituminous (up to 10% organic carbon) rocks, shaly limestones and silty marlstones containing thoroughly silicified beds. These rocks are very similar to the bituminous limestone of the Sinian formation (facies of normal salinity), differing from them mainly in the rather high content of quartzitic silt and intensive silicification [6].

But while the Sinian limestones, about 40 m in thickness, contain a rich fauna correlatable only with the base of the Lena stage (Sinian horizon), the sediments of the facies under discussion, although of the same thickness, contain fossils from almost all horizons of the Lena stage; i. e., they correspond in age to the multifacies sequence of rocks from 350 to 600 m in thickness [15, 24, 25]. At present there is no explanation for this great reduction in thickness.

A characteristic feature of the Lower Cambrian basin as a whole was its shallowness. In the deposits of the normal and nearly normal salinity this is indicated by the presence of bioherms of archaeocyathids and epiphytic algae. The sediments from the zones of higher salinity, including the halogenic zone, contain numerous oncolites and stromatolites. It must be concluded from this that the average depth of the Lower Cambrian basin did not exceed 30-50 m. At such depth wave action could easily affect the bottom of the sea and produce ripple marks and in the molasse-like sands of the nearshore zone and in some carbonate sediments of the other zones. Wave action explains also the occasional destruction of the archaeocyathid and algal bioherms and the formation of interformational breccias. Only the facies with reduced thickness were deposited at a considerably greater depth, as indicated by the very thin unbroken laminations of the rocks and the preservation of a large amount of organic material.

We have studied the distribution of the Lower Cambrian facies on the southern and western borders of the Siberian platform, and constructed schematic facies maps for all stratigraphic horizons. On the basis of these maps and the geological data on the northern regions published by a number of investigators [4, 11, 15], the Zhura substage and Lena stage facies maps were drawn for the entire Siberian platform. The boundaries of the facies zones in the central parts of the platform where the Lower Cambrian deposits are covered by younger rocks are tentative.

At present a considerable amount of material is available indicating that in the Zhura time the Lower Cambrian basin of the Siberian platform was bordered by a sea in the south and west (Figure 2). The southeastern parts of the platform were evidently strongly peneplaned by this time, but high mountainous relief still existed in the west. In the north, northeast and east, the Zhura substage of the platform is characterized by deposits of the zone of nearly normal salinity (the Anabar massif and the eastern part of the Aldan massif). To the south and southwest they are succeeded by deposits of zones of increasing salinity, including the halogenic zone. As the map shows, these zones occupy a vast territory more than 2000 km long and not less that 200 km wide. Farther, in the belt adjacent to the land, salt deposits are followed again by the deposits of the zone of high salinity. To the west of the Anabar massif, salinity also increases but salt deposits are

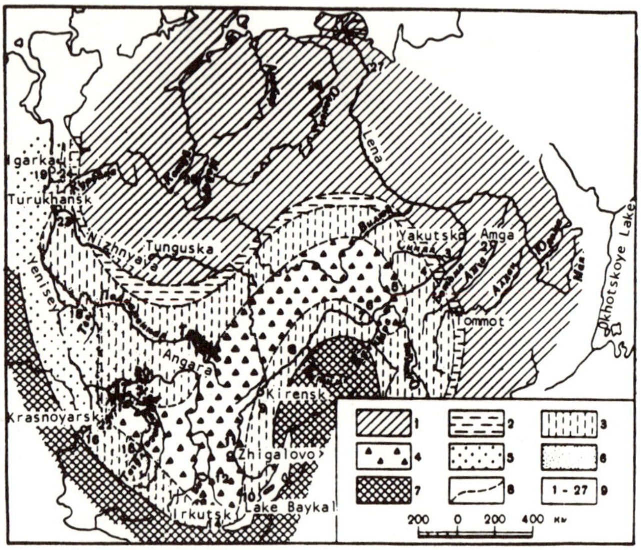

FIGURE 2. Distribution of facies on the Siberian platform in Zhura time.

1 - facies of nearly normal salinity; 2 - facies with salinity transitional between normal and high; 3 - facies of high salinity; 4 - salina facies; 5 - terrigenous nearshore facies of marine basin; 6 - area of evaporites with a considerable admixture of terrigenous material; 7 - strongly dissected provenance areas; 8 - zone boundaries; 9 - numbers of exploratory boreholes.

absent and the sediments of the zone of high salinity pass into the nearshore molasse sediments.

The transitions between the sediments of the different zones are always gradual and consist in the change of composition of the rocks and their faunas. Especially characteristic of the transitional areas are various mixed rocks, dolomitic limestones, calcitic dolomites, anhydrite-dolomites, etc. All this indicates that in the Zhura time a single marine basin existed in the region of the Siberian platform in which the degree of salinity increased gradually from the open northeastern parts of the basin to the northwest in the direction of the nearshore areas and decreased again in the relatively narrow littoral zone.

This regularity in the succession of the zones was retained in a general way by the basin of the Lena time (Figure 3). But the distribution of the zones of different salinity during this time was considerably different as compared with the Zhura time. The zone of sediments deposited in water of high salinity was sharply diminished. Widespread over the entire northeast and east of the Siberian platform are rocks typical of the zone of normal salinity, absent from the deposits of the Zhura substage. In the extreme northeast and east these deposits are unusually thin. To the south and southwest they are replaced by sediments deposited in water of nearly normal and intermediate salinity. These are the most widespread rocks of the Lena stage in the southwestern half of the Siberian platform. Among them in a number of areas are deposits of the halogenic zone bordered by those of the zone of high salinity. As can be seen from Figure 3, these areas lie within the region occupied by the halogenic zone in the Zhura time.

There are no reasons to believe that the geographic setting was different in the Lena time as compared with the Zhura time. The land areas in the south were evidently subjected to more peneplanation. A very noticeable contribution of terrigenous material continued only in the southeast of the Yenisey range.

Individual areas of the peneplaned land were apparently gradually submerged by the continuing marine transgression and a connection

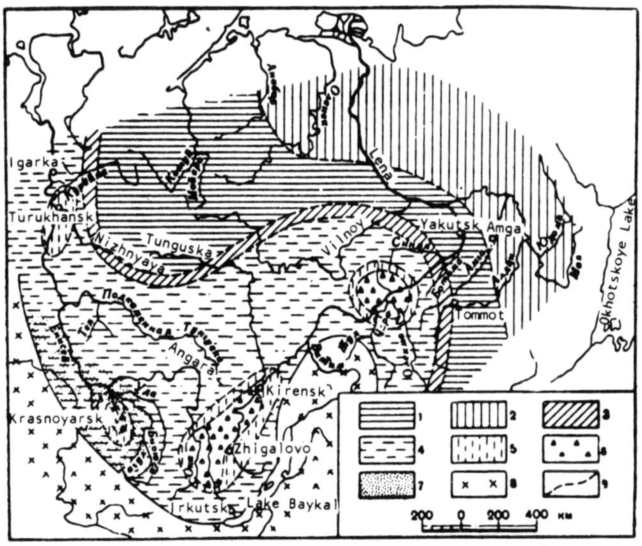

FIGURE 3. Distribution of facies on the Siberian platform in Lena time.

1 - facies of normal salinity; 2 - same facies, thin deposits; 3 - facies of nearly normal salinity; 4 - facies with salinity transitional between normal and high; 5 - facies of high salinity; 6 - salina facies; 7 - areas of evaporites with a considerable admixture of terrigenous material; 8 - peneplaned land areas; 9 - zone boundaries.

formed at this time between the platform sea and the geosynclinal basin lying to the south and southwest. The most complete interchange of water between them began at the very end of the Lower Cambrian as indicated by the presence of similar faunas in the Elanka horizon developed in both basins [14, 23].

Thus, during the entire Lower Cambrian the bast territory of the Siberian platform was occupied by a single very shallow epicontinental sea open to the north, northeast and east. Intensive chemical precipitation occurred everywhere in the basin and its cause must be sought in the increased salinity of the sea water produced by evaporation in an arid climate. This basin may be called an evaporite basin.[3]

It has been mentioned already that the characteristic feature of the Lower Cambrian basin is the gradual change from normal sediments to rock salt, indicating that sedimentation occurred in marine basins of different salinity. The rock salt and the other sediments were deposited in definite areas of the sea, yet there are no abrupt changes suggesting the existence of bars at the boundary between the halogenic zone and the rest of the basin nor any other indications that this zone was an isolated part of the sea during the Lower Cambrian.

To gain a better understanding of the characteristics of sedimentation in the investigated region we must consider the processes causing variation in salinity in shallow basins in general. It is convenient to review these processes one by one and then compare their results.

The fundamental cause of increased salinity of water is evaporation. If, in an evaporite basin freely communicating with the world ocean, evaporation exceeds precipitation, water will flow from the ocean in a slow compensational current and equalize the hydrostatic level. This ocean water will bring new portions of salt into the basin and as a result its area of water of high salinity will increase. Evaporation increases salinity of the surface layer only. But storms mix surface water with the water of the underlying layers and salinity becomes averaged to the depth of wave action, i. e. through the entire depth of a very shallow basin (30 - 50 m).

Let us try by means of a very approximate calculation to determine the character of change in salinity in a shallow evaporite basin receiving water from the ocean. To simplify the problem we shall take a rectangular area of the basin bounded by the shore on one side (section ABCD) and receiving a compensational flow of sea water of normal salinity through section ($A_0B_0C_0D_0$) (Figure 4). Let h be the average depth of the evaporite basin, assumed to be constant, and a the width of the rectangular area of the basin. The distance from the shore to the part of the basin in which salinity is not increasing, i. e. the length of the evaporite basin, will be designated by L. The excess of evaporation over atmospheric precipitation for a given period of time per unit area, q, will be considered constant for the whole evaporite basin.[4]

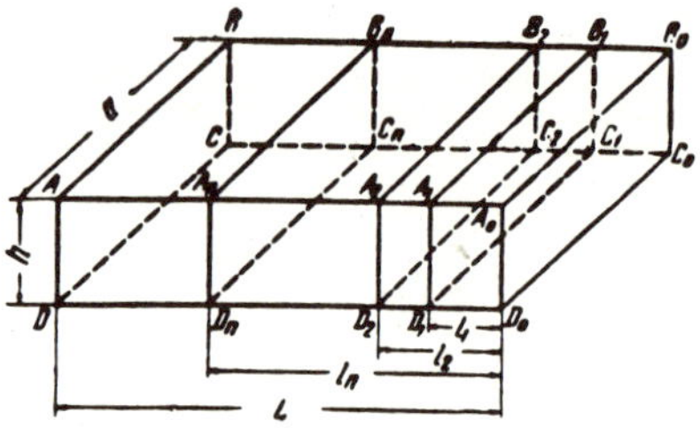

FIGURE 4. Principal parameters of a basin used in calculating the magnitude of compensational flow.

The total amount of water carried by the compensation current through the section $A_0B_0C_0D_0$ during a given period of time must equal the amount of water lost by evaporation during the same time from the entire ABB_0A_0 area. The loss of water is given by product qaL. It is easy to calculate the amount of water (U_0), flowing through a unit area of section $A_0B_0C_0D_0$ in unit time:

$$U_0 = \frac{qaL}{ah} = \frac{qL}{h}. \qquad (1)$$

As the evaporating area decreases, the amount of water flowing through the unit area in every successive section must also decrease.

Thus, in section $A_1B_1C_1D_1$

$$U_1 = \frac{qa(L - l_1)}{ah} = \frac{qL}{h}\left(1 - \frac{l_1}{L}\right) = U_0\left(1 - \frac{l_1}{L}\right). \qquad (1a)$$

In section $A_2B_2C_2D_2$

$$U_2 = \frac{qL}{h}\left(1 - \frac{l_2}{L}\right) = U_0\left(1 - \frac{l_2}{L}\right). \qquad (1b)$$

---

[3]By an evaporite basin we understand an extensive part of a marine basin in which increased evaporation causes chemical precipitation of all salts from carbonates and sulfates to halides. This definition differs from that accepted abroad [27, 28], where the term "evaporite basin" is restricted to areas of precipitation of sulfates and rock salt, i.e. specifically to halogenic basins.

[4]Actually, as the salinity of the water increases, all other conditions being the same, its capacity to evaporate decreases, and within the limits of salinity considered here this decrease may amount to 25%.

In section $A_nB_nC_nD_n$

$$U_n = \frac{qL}{h}\left(1 - \frac{l_n}{L}\right) = U_o\left(1 - \frac{l_n}{L}\right). \quad (1c)$$

Unlike the amount of water, the amount of dissolved salts J_k carried by the compensation current through unit area in unit time remains constant in each successive section ($A_oB_oC_oD_o$ and $A_1B_1C_1D_1$, etc.) provided no salt is precipitated. This may be written:

$$J_k = S_oU_o = S_1U_1 = S_2U_2 = \dots = S_nU_n, \quad (2)$$

where $S_{o,\,1,\,2,\,\dots,\,n}$ is salinity in each successive section.

From expression (2) it follows that:

$$U_n = \frac{S_oU_o}{S_n}. \quad (2a)$$

Substituting the values of U_o and U_n in (2a) and simplifying, we obtain the following expression:

$$l_n = L\left(1 - \frac{S_o}{S_n}\right). \quad (3)$$

Therefore, the distance at which salinity will rise to a given value, measured from the beginning of the evaporite basin, depends on the initial salinity (S_o) and on the entire length of the evaporite basin (L), and is independent of the average depth (h) and the excess of evaporation over precipitation (q) if these values are constant for the whole basin. The magnitude of h and q will affect only the time needed to raise the salinity of the basin to a given level.

Using equation (3) it is easy to calculate l_n for any preassigned salinity (4%, 5%, 6%, etc.). The composition of ocean water of different concentration and the degree of salinity required for precipitation of gypsum, halite, epsomite and potassium salts are known from the work of M. G. Valyashko ([2] and Table 1). The value of l_n can be calculated for each of these salinities. For example, for the beginning of precipitation of gypsum (S = 13.14%), l_n = 0.73L; for that of halite, it is 0.87L; for epsomite, 0.97L; and, finally, for the precipitation of potassium salts (sylvite), l_n = 0.977L.[5]

The values of l_n were used to construct a graph showing variation in the total salinity of water over the area of the evaporite basin in the direction of the compensation current. This graph is given in Figure 5. The values of l_n expressed in fractions of L taken as unity were plotted on the axis of the abscissas and total salinity in precent was plotted on the axis of the ordinates.

The graph shows that the salinity of the evaporite basin does not change uniformly in the direction of the compensation current but increases more and more sharply. Especially sharp is the increase within the halogenic zone. The graph shows also that the precipitation of the salts beginning with calcium sulfate (in the halogenic zone as we define it) is restricted to a zone whose width is small as compared with the area of the evaporite basin, and in the case of a single compensational current it hugs the shore. The smaller the area of the evaporite basin, the narrower is the halogenic zone.

It is easy to show that in the case of two equal and opposite currents the halogenic zone will form in the central part of the basin and the graph of salinity change will be symmetrical. If the two currents are not equal in their effect, for example if the depth of the evaporite basin

Table 1

Composition of Ocean Water at Different Degrees of Concentration in grams per 1000 g of water (after M. G. Valyashko [2])

Degree of concentration of ocean water	$CaCO_3$	$CaSO_4$	$MgSO_4$	$MgCl_2$	Na Cl	K Cl	NaBr	Total salts
Normal ocean water	$Ca(HCO_3)_2$ 0.134	1.276	2.305	3.385	27.667	0.763	$MgBr_2$ 0.092	35.62
Beginning of precipitation:	$MgCa(HCO_3)_2$							
gypsum	0.340	4.90	9.50	14.90	99.10	2.40	0.26	131.40
halite	0.52	0.460	21.0	33.40	214.10	5.20	0.59	275.27
epsomite	2.24	Traces	89.2	158.2	50.50	22.90	2.72	325.70
sylvite	Not det.	"	75.5	169.1	33.80	49.2	Not det.	327.60
carnallite	3.01	"	64.4	218.4	24.2	31.6	3.9	345.51

[5]In calculating l_n and contructing the graph, we considered the change in the primary proportion of salts in ocean water with rise in salinity only from the moment of precipitation of sodium chloride, for it is this salt that constitutes the bulk (77%) of dissolved solids.

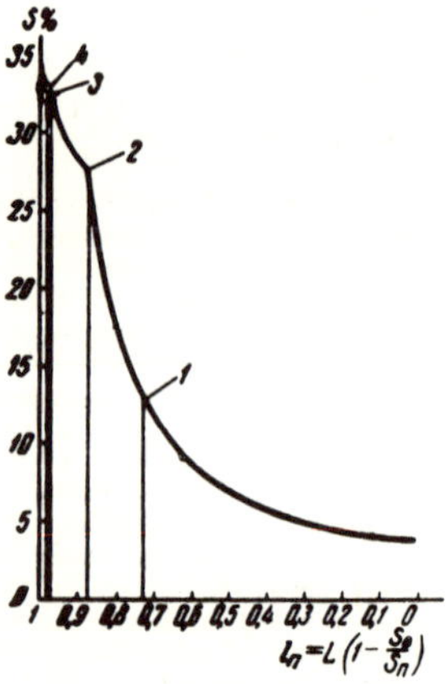

FIGURE 5. Variation in salinity of ocean water within an evaporite basin:

1 - beginning of precipitation of gypsum; 2 - beginning of precipitation of halite; 3 - beginning of precipitation of epsomite; 4 - beginning of precipitation of sylvite.

is not uniform, the halogenic zone will be shifted to one side or the other.

With equal flows from all sides, the halogenic zone in the ideal case will form a circular area in the center of the basin. Here, instead of formula (3), formula

$$l_n = R\left[1 - \left(\frac{S_o}{S_n}\right)^{\frac{1}{2}}\right], \qquad (4)$$

will apply, with R being the radius of the evaporite basin.

In nature if the shape of a shallow evaporite basin is intricate and the compensating currents reaching the halogenic zone from all sides are unequal, the zone may assume various forms and positions. Nevertheless, the areal relationships derived above for an elementary case should hold.

Let us estimate the amount of material carried by the compensation current. The value of J_k (expressed in terms of qL and h) is obtained from equation (2c):

$$J_k = \frac{q\,(L - l_n)\,S_n}{h}\ [\text{cm/sec}]. \qquad (5)$$

For our calculation we shall take q = 0.5 m/year = $\frac{1}{6.10^5}$ cm/sec*, h = 50m = 500 cm, S_n = 27%.

*This considerable excess of evaporation over precipitation is quite usual at present in the high pressure zones to the north and south of the equator.

It is easy to calculate that if L = 10 km, J_k = 0.0012 cm/sec, and that for L = 100 km, J_k is ten times as great (0.012 cm/sec), and for L = 1000 km, J_k is one hundred times as great (0.12 cm/sec). Equation (5) shows that there is an inverse relation between J_k and the depth of the basin h (as h increases by a certain factor, J_k decreases by the same factor).

As zones of different salinity, and therefore of different density, appear in the evaporite basin, processes tending to equalize these differences inevitably arise, i.e. diffusion and outflow of the more dense water. There are two types of diffusion. In a stationary medium only slow molecular diffusion occurs; in a mobile medium such as the water in natural basins the main role is played by the much more active turbulent diffusion always directed towards the zone of lower density.

In turbulent diffusion the transfer of matter through unit area per unit time, J_T, is proportional to the gradient of concentration [8]. For the area of the evaporite basin described above this may be written:

$$J_T = D_T \frac{\Delta S}{\Delta L}\ [\text{cm/sec}], \qquad (6)$$

where D_T is the coefficient of turbulent diffusion and $\frac{\Delta S}{\Delta L}$ is the salinity gradient.

It is very difficult to determine the value of D_T for turbulent diffusion because it depends on many variables and first of all on the magnitude known in hydraulics as the mixing length. When a liquid moves over a horizontal surface the mixing length is the depth of the basin. The larger the value of h, the larger is the coefficient D_T and therefore the larger is the value of J_T.

For a shallow basin, D_T is approximately a million times as great as the coefficient of molecular diffusion; i.e. it amounts to a few square centimeters per second. The experimentally determined coefficient of turbulent diffusion in some modern shallow basins equals 5 cm^2/sec [10]. In large basins it may be a hundred times greater (up to 500 cm^2/sec).

The value of the salinity gradient in different parts of an evaporite basin (i.e. over its length L), or in basins of different sizes, may easily be computed with the aid of the graph of Figure 5. The results are shown in Figure 2. The figures show that only when L is small does the salinity gradient for the cases of high salinity approach thousandths of one percent per centimeter. When values of L are of the order of several hundred kilometers, salinity gradients are quite insignificant.

Substituting D_T = 500 cm^2/sec and $\frac{\Delta S}{\Delta L}$ = 0.0003% cm^{-1} (the value of salinity gradient for high degree of salinity if L = 10 km) in equation (6), we have J_T = 0.15 cm/sec. If L = 100 km,

Table 2

Character of Variation of Salinity Gradient at Different Stages of Concentration with the Size of Evaporite Basin

L km	S %																		
	26	25 24 23	22 21	20 19	18	17	16	15	14	13	12	11	10	9	8	7	6	5	4
	Salinity gradients $\frac{\Delta S}{\Delta L}\left[\frac{\%}{cm}\right] \cdot 10^{-}$[illegible]																		
10	330000	166000	143000	111100	100000	83300	77000	71300	58800	52600	45400	37000	31200	26300	24100	15900	12500	8300	5700
500	6600	3320	2860	2222	2000	1666	1540	1426	1176	1052	908	740	624	526	482	318	250	166	114
1000	3300	1660	1430	1111	1000	833	770	713	588	526	454	370	312	263	241	159	125	83	57
2000	1650	830	715	555	500	416	385	356	294	263	227	185	156	131	120	79	62	41	28

J_T for the case of high salinity will be one tenth of this value (0.015 cm/sec) and for L = 1000 km, one hundredth (0.0015 cm/sec).

Comparison of expressions (5) and (6) and the numerical values obtained for J_k and J_T for different values of L leads to the conclusion that for relatively small values of L (of the order of tens or even hundreds of kilometers) or for a considerable depth of the basin (h = tens and hundreds of meters), $J_T \gg J_k$; and for large values of L (thousands of kilometers) and small h (up to a few tens of meters), $J_T \ll J_k$.

A comparison of expressions (5) and (6) shows also that the transfer of material by the compensation current J_k is directly proportional to the excess of evaporation (q), i. e. depends on climatic conditions, while the amount of material transferred by turbulent diffusion, J_T, is in no way related to q. Hence the hotter the climate and the larger the value of q, the greater the difference between J_T and J_k, other conditions being the same, and the sooner a halogenic zone will form in the basin and the more stable it will be.

Climatic changes disturbing the relation between J_k and J_T will cuase a change in salinity, and in the final analysis this must affect the character of the chemical sediments.

The second process arising with the appearance in the basin of zones of different salinity and density is the outflow of water in the near bottom parts of the basin from the areas of greater density to those of lower density and the reverse flow at the surface (gradient flow). The velocity of the gradient flow must be proportional to the pressure drop, i. e. to the density gradient, which changes as the salinity gradient. It is obvious that the greater the L of the basin the lower is this velocity. Moreover, the velocity of the gradient flow is inversely proportional to the virtual viscosity (friction in turbulent flow [26]). In stratified water with density and salinity changing from layer to layer, this factor is much smaller than in homogeneous water. Therefore in a shallow basin where stratification is lacking because of mixing the velocity of gradient flow must decrease substantially as compared with the velocities of the gradient flow in deeper basins. Thus, the dependence of the gradient flow on the dimensions of the basin is the same as for turbulent diffusion.

A comparison of the character of all three processes leads to the conclusion that in relatively small evaporite basins the concentration of salts cannot reach values characteristic of halogenic zones without additional factors (for example the presence of shallows inhibiting gradient flow). On the other hand, in large basins in which compensation current is overwhelmingly important as compared with other processes the appearance of halogenic zones is quite possible and regular.

It is very difficult to estimate to what extent drift currents inhibited the appearance and preservation of halogenic zones in evaporite basins. Here we cannot use the data on modern seas, for in the present geological period there are almost no very shallow basins of large area. Only some of the Arctic seas of the Soviet Union more or less approach this type of basin, but they have no large permanent currents [12].[6] All modern drift currents are in the oceans or deep seas, for evidently shallow depths prevent their formation. However, even if permanent drifts did exist in an evaporite basin, then, like the modern ocean currents, they would affect only a narrow belt.

According to oceanographic concepts [1, 26] in basins with depths less than 30 - 50 m, temporary variable drifts are not accompanied by opposite near-bottom currents. The wind action in such basins causes only temporary displacements of the entire thickness of water over an area in one direction or another. In an evaporite basin such currents would cause only brief periodic changes of the shape of the halogenic zone but would not remove its brine and freshen it.

In analyzing the unidirectional current in the evaporite basin we concluded that the halogenic zone will hug the shore with its belt of highest salinity against it (Figures 4 - 5). This will be so only if no fresh water is contributed from the land. Near river mouths a freshened zone will form in the littoral belt, its width depending on the amount of water delivered by the rivers.

It is natural that there should exist a transitional belt between the freshened and halogenic zones. Mainly by turbulent diffusion the salts characteristic of each of the two zones will be exchanged. The K, Na, Mg chlorides and Mg sulfates constituting the main part of dissolved salts will diffuse from the halogenic into the freshened zone (Table 1). Modern river waters are rich in calcium bicarbonate and carbonate, even oversaturated with these salts in arid regions [19], and it must be assumed that calcium carbonate and bicarbonate will diffuse from the freshened into the halogenic zone. The dominant precipitate near the halogenic zone enriched in the sulfate radical will be gypsum. Nearer to the shore, with the gradual disappearance of $MgSO_4$ and general freshening, the precipitation of gypsum will cease and dolomite will begin to form. The problem of carbonate parageneses has been fully analyzed in the well known work by N. M. Strakhov [17]. The separation of magnesium sulfate from the salt complex of the halogenic zone as a result of reaction with $CaCO_3$ must change the composition of its water.

After this theoretical discussion of the mechanism of brine formation in an evaporite basin, let us turn to the Lower Cambrian deposits of the Siberian platform and try to use this mechanism to explain some of the regularities in the distribution of their facies. The basic pattern (Figures 2 and 3) is a gradual transition from northeast and east to the south and southwest, from deposits of the zone of low salinity to deposits of the zone of high salinity.[7] As was shown above, during the entire Cambrian time there existed on the Siberian platform a single very shallow evaporite basin of vast area bounded on the south and west by continuous (Zhura time) or nearly continuous (Lena time) land masses. The direction of the compensation current can be perfectly definitely established by the paleogeographic setting. There is no doubt that the main mass of sea water of normal salinity was delivered to the Siberian platform from the northeast and partly from the east.[8] Thus, the conditions which existed during the Lower Cambrian were very similar to those in the basin with the unidirectional compensation current discussed above.

It is quite natural, therefore, that in the nearshore areas of the Lower Cambrian evaporite basin and in the areas far removed from the land and open towards the ocean there existed a halogenic zone attaining its greatest width in the farthest southwestern corner (as follows from equation [3]). As the cited data show, the greatest salt deposits actually lie in the inner part of the Irkutsk amphitheatre.

In the northwest, in the Turukhan region, the length of the compensation current was much less, the change in salinity sharper, and, as shown above, no halogenic zone could form. This explains the wedging out of the salt deposits from south to north along the western boundary of the Siberian platform (Figure 2).

The entry of streams from the land caused the appearance between shore and the halogenic

---

[6]The reason for this is possibly the existence of extensive and long lived ice cover.

[7]The salinity gradients which it is possible to compute for the Lower Cambrian basin are quite commensurable with the theoretical ones (Table 2). Thus, on the northwest slope of the Aldan massif, where sections have been studied in detail, the salinity gradient within the transitional and high salinity zones is 500% $cm^{-1}.10^{-9}$, and in the outer part of the halogenic zone (the belt of deposition of sulfate-dolomite rocks) it is 1500% $cm^{-1}.10^{-9}$.

[8]It is probable that the territory of the modern marginal seas of the Arctic and Pacific oceans, the source of the compensation current, was occupied at that time by basins of other than evaporite type (either deeper basins or basins existing under different climatic conditions).

zone of a belt of water of high salinity and anomalous composition resulting from the mixing of fresh and salt waters. The addition of fresh water was apparently particularly large in the Yenisei region of the Lower Cambrian basin, as indicated by the abundance of terrigenous material in the Zhura and Lena deposits of this area, and for this reason the halogenic zone here lay farther from the shore.

These differences in the distribution of deposits of the zones of different salinity in the Lena stage as compared with the Zhura stage may be explained first of all by the increasing humidity of climate and also, probably, by the appearance of a connection between the Siberian platform basin and the southern geosynclinal sea. Thus the results of theoretical computations agree well with the actual geological facts and confirm our conclusion that in the Lower Cambrian evaporite basin salinity reached the stage of halite precipitation in a halogene zone which was not cut off or separated by bars from the rest of the sea.

This conclusion contradicts the widespread concept that salt deposits of marine origin can form only in lagoons, i. e. in basins barred from the sea or having only a very slight connection with it. This concept is based on modern cases and is actually confirmed by all observations on salt deposition of our time. However, the present geocratic period in the Earth's history substantially differs from many preceding periods, for now there are no broad epicontinental basins in the arid zone.

The idea that salt deposition requires areas cut off from the sea evidently arose from the analogy with modern lagoons and the assumption that the ancient saline basins were also supplied directly with marine water of normal or even low salinity (as, for example, Kara-Bogaz-Gol). The proximity of zones with such sharp difference in salinity that it can be attained only when 80% of the original volume has evaporated is, of course, impossible in a single basin.

In recent years a number of investigators have relinquished the classical concept of salt deposition. For example, M. P. Fiveg [21, 22], considering the well-known lack of correspondence found in many deposits between the composition of the rocks of the salt complex and the composition which should result from evaporation of sea water (for example low content of anhydrite), suggested that thick salt deposits were formed not in lagoons but in peculiar marine halogenic basins. These basins, as a rule, were not directly connected with the sea but communicated with it through intermediate basins..... " [21]. The sea water, flowing through this system of basins with an inhibited exchange of water, was gradually concentrated, and in the more distant, halogenic basins occupying areas of intensive subsidence thick salt beds were deposited from the more concentrated brines. Along the route of flow, in the intermediate basins, the less soluble salts were precipitated, such as carbonates and gypsum, while "the halogenic (halite) basins in which sea water was concentrated occupied only a small part of the total area of concentration" [21].

It is not difficult to see that Fiveg's hypothesis of formation of salt deposits accounts in a general way for the alternation of zones of different salinity observed by us in the Lower Cambrian deposits. However, to explain this process, Fiveg introduces essentially true lagoons, only arranging them in series. The formation of such a series is difficult to visualize, especially as the cause of isolation of the lagoons is explained by the author by the lowering of water level and shallowing of the basins due to intensive evaporation, an impossibility while there is a connection with the open sea.

N. M. Strakhov [18] in discussing the conditions of formation of dolomites arrives at the conclusion that in the extremely shallow Upper Paleozoic sea of the Russian platform lying in the semiarid climatic belt the salinity of the surface water in the pelagic zone was relatively high and hence dolomitic sediments were restricted to it. However, even assuming that salinity in some parts of the open sea was one and a half or two times greater than normal, Strakhov believes that the deposition of anhydrite required the formation of islands and shallows forming a maze in which "true lagoons almost completely isolated from the sea were formed from time to time."

Thus, admitting gradual increase in the concentration of salts in sea water before its arrival in the halogenic zone proper, both authors still consider it necessary to introduce barriers separating this zone from the open sea.

As was shown above (Table 2), the increase in salinity in an extensive evaporite basin must be very gradual, and this may be observed in some very shallow modern basins. For instance, in some of the areas of the Bahamas banks the salinity gradient reaches 102% $cm^{-1} \cdot 10^{-9}$. Within the shallow bays of the Caspian Sea, Mertvyy Kultuk and Kaydak, according to the data of 1934, the mean salinity gradient over a distance of 200 km was 191% $cm^{-1} \cdot 10^{-9}$. The most striking example is that of a branch of the Sivash, a single basin stretching for over 100 km from north to south. Here the average gradient is 1000% $cm^{-1} \cdot 10^{-9}$. These data confirm the possibility of the existence of zones of different salinity within a single basin.

Evidently the inception of evaporite basins with increasing salinity of the type described here was favored by the existence of arid climate over a large territory and by the first stages of a large transgression or the earliest stages of slowly developing regressions, when extensive land areas of strongly peneplaned platforms were covered by shallow epicontinental seas.

It should be pointed out that the process of salt concentration in an evaporite basin suggested here makes precipitation of rock salt in an open marine basin possible but does not always lead to accumulation of large salt deposits. This requires an additional condition, namely, localization of the halogenic zone in an area of the earth's crust undergoing rapid subsidence. As we have seen, this favorable combination of conditions existed on the Siberian platform, where the largest salt deposits are restricted in time to the first half of the extensive Lower Cambrian transgression when the border of the platform adjacent to the mountains of the Baykal folded zone had a strong tendency to subside.

Evidently similar favorable conditions could have existed in the succeeding geological periods and similar regularities can probably be established for salt deposits of different ages. Undoubtedly sulfates and chlorides could have been deposited not only in extensive evaporite basins but also in typical lagoons similar to those existing now and forming most likely during the concluding stages of regressions.

In our opinion, the process of formation of salt deposits in evaporite basins explains certain structural peculiarities of salt deposits which which it is difficult to understand if the deposits are assumed to have formed in lagoons.

For example, it is difficult to imagine typical lagoons corresponding in size to the Verkhnekamsk deposit (225 km in length), the Permian basin of North America (1000 km long) and the Zeichstein basin of Europe extending for 1700 km [7].

Quite typical of evaporite basin deposits is the frequently observed persistence of salt beds along the strike. In the Verkhnekamsk deposit, for example, as pointed out by A. A. Ivanov [7], there existed "an amazing persistence and regional character of the conditions of sedimentation, so that over broad areas of the basin sediments of the same composition, of almost the same thickness and in the same sequence were deposited." Such persistence is hardly possible if it is assumed that deposition took place in numerous partly isolated lagoons lying in a maze of islands and shallows in a dessicating basin.

In all salt deposits there is alternation of rocks of different composition. Thus, halite alternates with anhydrite and potash salts with halite. "It is quite clear," writes A. A. Ivanov [7], "that without a periodic and rather copious addition of sea water to the basin (concentrated and with modified composition) diluting the brine of the basin, saturated with potassium and magnesium chlorides, and bringing in fresh portions of sodium chloride, such alternation in the composition of the crystallizing salts would be impossible." It is difficult to explain such frequent periodical invasions of sea water into lagoons, and it is hard to understand how this sea water became concentrated and changed in composition in advance. If the salts in an evaporite basin accumulated as described above, the intensity of evaporation would vary with climatic changes and thus changes in the composition of the chemical precipitates could occur without disturbing the general course of sedimentation.

REFERENCES

1. Berezkin, V. A., Dinamika morya. [DYNAMICS OF THE SEA]: Gidrometeoizdat, 1947.

2. Valyashko, M. G., Obyemnyye sootnosheniya zhidkikh i tverdykh faz v protsesse ispareniya okeanicheskoy vody kak faktor, opredelyayushchiy obrazovaniye mestorozhdeniy kaliynykh soley. [VOLUME RATIOS OF LIQUID AND SOLID PHASES AS FACTORS DETERMINING FORMATION OF POTASSIUM SALT DEPOSITS]: Doklady, Akademiya Nauk SSSR, t. 77, no. 6, 1951.

3. Geologiya i neftegazonosnost' Vostochnoy Sibiri. [GEOLOGY AND OIL AND GAS OCCURRENCES IN EASTERN SIBERIA]: Gostoptekhizdat, 1959.

4. Demokidov, K. K., O stratigraficheskom raschlenenii kembriyskikh otlozheniy severa Sibirskoy platformy. [ON STRATIGRAPHIC SUBDIVISION OF THE NORTHERN PART OF SIBERIAN PLATFORM]: Tr. N.-i. in-ta geol. Arktiki, t. 67, 1957.

5. Demokidov, K. K., N. Ye. Chernysheva, Ya. K. Pisarchik and O. I. Nekrasova, Stratigrafiya i fatsii kembriya Sibirskoy platformy. [CAMBRIAN STRATIGRAPHY AND FACIES OF THE SIBERIAN PLATFORM]: Tr. N.-i. in-ta geol. Arktiki, t. 80, 1958.

6. Zelenov, K. K., Litologiya nizhnekembriyskikh otlozheniy severnogo sklona Aldanskogo massiva. [LITHOLOGY OF THE LOWER CAMBRIAN ROCKS OF THE NORTHERN SLOPE OF THE ALDAN MASSIF]: Tr. geol. in-ta Akad. Nauk SSSR, vyp. 8, 1957.

7. Ivanov, A. A., Osnovy geologii i metodika poiskov, razvedki i otsenki mestorozhdeniy mineral'nykh soley. [FUNDAMENTALS OF GEOLOGY AND THE METHODS OF EXPLORATION AND EVALUATION OF SALT DEPOSITS]: Gosgeolizdat, 1953.

8. Levich, V.G., Fiziko-khimicheskaya gidrodinamika. [PHYSICOCHEMICAL HYDRODYNAMICS]: Izd. Akad. Nauk SSSR, 1952.

9. Lermontova, Ye. V., Nizhnekembriyskiye trilobity i brakhiopody Vostochnoy Sibiri. [LOWER CAMBRIAN TRILOBITES AND BRACHIOPODS OF EASTERN SIBERIA]: Gosgeolizdat, 1951.

10. Lineykin, P. S., Ob izmenenii solenosti poverkhnostnogo sloya morya pri vypadenii osadkov. [ON THE CHANGE IN SALINITY OF THE UPPER WATERS OF THE SEA DUE TO PRECIPITATION]: Izvestiya, Akad. Nauk SSSR, ser. geofiz., no. 2, 1953.

11. Mezhvilk, A. A., Stratigrafiya Severnogo Kharaulakha. [STRATIGRAPHY OF NORTHERN KHARAULAKH]: Sov. geologiya, no. 7, 1958.

12. Morskoy atlas t. 1. [MARINE ATLAS V. 1]: Izd. Morskogo gen. shtaba, 1950.

13. Pokrovskaya, N. V., Stratigrafiya kembriyskikh otlozheniy yuga Sibirskoy platformy. [CAMBRIAN STRATIGRAPHY OF THE SOUTH OF THE SIBERIAN PLATFORM]: Vopr. geol. Azii, t. 1, 1954.

14. Repina, L. N., Paleontologicheskoye obosnovaniye vozrasta otlozheniy kembriya zapadnoy chasti Vostochnogo Sayana. [PALEONTOLOGICAL EVIDENCE FOR THE CAMBRIAN AGE OF THE DEPOSITS IN THE EASTERN PART OF SAYAN MOUNTAINS]: Doklady, Akad. Nauk SSSR, t. 110, no. 2, 1956.

15. Savitskiy, V. Ye., O stroyenii i ob"yeme Aldanskogo yarusa kembriya na severe Sibirskoy platformy. [ON THE STRUCTURE AND CONTENT OF THE CAMBRIAN ALDAN STAGE IN THE NORTH OF THE SIBERIAN PLATFORM]: Inform. byull. N.-i. in-ta geol. Arktiki, vyp. 1, 1957.

16. Semikhatov, M. A., Nizhniy kembriy Kansko-Angarskoy vpadiny. [LOWER CAMBRIAN OF THE KAN-ANGARA BASIN]: Byul. Mosk. o-va ispyt. prirody, otd. geol., t. 34, vyp. 2, 1959.

17. Strakhov, N. M., Izvestkovo-dolomitovyye fatsii sovremennykh i drevnikh vodoyemov. [LIMESTONE AND DOLOMITE FACIES OF MODERN AND ANCIENT BASINS]: Tr. In-ta geol. nauk, vyp. 124, no. 45, 1951.

18. Strakhov, N. M., O tipakh i genezise dolomitovykh porod. [ON TYPES AND GENESIS OF DOLOMITIC ROCKS]: Tr. geol. in-ta AN SSSR, vyp. 4, 1956.

19. Strakhov, N. M., et al., Obrazovaniye osadkov v sovremennykh vodoyemakh. [DEPOSITION OF SEDIMENTS IN MODERN BASINS]: Izd. Akad. Nauk SSSR, 1954.

20. Suvorova, N. P., O lenskom yaruse nizhnego kembriya Yakutii. [ON THE LOWER CAMBRIAN LENA STAGE OF YAKUTIYA]: Vopr, geol. Azii, t. 1, 1954.

21. Fiveg, M. P., Geologicheskaya obstanovka sedimentatsionnogo etapa obrazovaniya solenosnykh formatsiy. [GEOLOGICAL ENVIRONMENT OF FORMATION OF SALT-BEARING DEPOSITS]: Vopr. mineralogii osad. obrazovaniy, kn. 3-4. L'vov, 1956.

22. Fiveg, M. P., Tipy solerodnykh basseynov. [TYPES OF SALINE BASINS]: Tr. Vses. n.-i. in-ta galurgii, vyp. 32, 1956.

23. Khomentovskiy, V. V., K tektonike Vostochnogo Sayana. [TECTONICS OF EASTERN SAYAN]: Izvestiya, Akad. Nauk SSSR, ser. geol., no. 7, 1957.

24. Chernysheva, N. Ye., Stratigrafiya kembriyskikh otlozheniy yugo-vostochnoy okrainy Sibirskoy platformy. [STRATIGRAPHY OF CAMBRIAN DEPOSITS OF THE EASTERN MARGIN OF SIBERIAN PLATFORM]: Materialy Vses. n.-i. in-ta, vyp. 7, 1955.

25. Chernysheva, N. Ye., K voprosu o raschlenenii kembriyskikh otlozheniy Sibirskoy platformy. [ON SUBDIVISION OF THE CAMBRIAN DEPOSITS OF SIBERIAN PLATFORM]: Sov. geologiya, no. 55, 1957.

26. Shuleykin, V. V., Fizika morya. [PHYSICS OF THE SEA]: Izd. Akad. Nauk SSSR, 1953.

27. Krumbein, W. C., Occurence and lithologic associations of evaporites in the United States. J. Sediment. Petrol., vol. 21, no. 2, 1951.

28. Sloss, L. L., The significance of evaporites. J. Sediment. Petrol., vol. 23, no. 2, 1953.

Geological Institute,
Academy of Sciences, U. S. S. R.
Moscow

Received August 12, 1959

Errata: Page 135, line fifteen, column 1 should read: "basin to the southwest in the direction of the"

Page 136, line nine, column 1 should read: "vast territory of the Siberian platform was"

Page 138, eighth line from the bottom of the page, column 1 should read: "tained from equation (2):"

Page 138, fifth line from the bottom of the page, column 1 should read: "m/year = $\frac{1 \text{ cm/sec*}}{6.10^5}$, h = 50m = 5000 cm,"

Page 138, eleventh line from the bottom of the page, column 2 should read: "The results are shown in Table 2. The figures"

Reprinted from *Amer. Jour. Sci.*, **263**, 284–285 (1965)

16

SHALLOW-WATER AND DEEP-WATER EVAPORITE DEPOSITION

A. J. PANNEKOEK

University of Leyden, The Netherlands

The paper by W. Sugden (1963) on the hydrology of the Persian Gulf is highly interesting and stimulating because of the support it gives to an alternative hypothesis for the origin of evaporite deposits. The author assumes that a further silting up of the Persian Gulf, whose mean depth is only 25 m, would create circumstances similar to those occurring at present in the shallow and highly saline Bahrain and Salwan gulfs and would eventually favor the deposition of evaporites. By taking this mechanism as a model for ancient evaporite deposits, the embarrassing bar hypothesis could be dispensed with.

Similar ideas favoring a shallow origin have recently been brought forward by other investigators. Deposition of evaporites in a tidal mud-flat environment was assumed by Kerr and Thomson (1963) who demonstrated that nodular and perhaps also bedded gypsum originates within the sediment on evaporation of the pore water, as can be observed in the Laguna Madre, Texas. M. G. Rutten (1954) also advocated evaporite deposition from evaporating pore waters, though in a different environment (continental playas). Ricour (1962, p. 304-312), on the other hand, suggested that nodular and disseminated anhydrite may have been precipitated from pore waters at great depth where high concentrations occur owing to the squeezing out of water from clays during compaction.

Though it seems, indeed, quite probable that numerous evaporites in thin but extensive layers, consisting mostly of gypsum, have also been formed in shallow environments in the past, this does not apply to the thick halite accumulations that occurred from time to time in geological history. Thicknesses of several hundreds of meters, unbroken by clastic intercalations, have been found in many of these evaporite basins (Delaware basin of West Texas, Paradox basin, Zechstein basin in Germany, Ural basin, et cetera), and there are various reasons why these deposits cannot have been deposited in shallow water. First, there was no clastic material that could have caused a silting-up of the basin. Second, as soon as the concentration was high enough for halite to precipitate, the deposition occurred extremely rapidly. On the basis of the old figures of Usiglio as recalculated by Clarke (Stewart, 1963, p. 12) and more recently by Briggs (1958), it is easy to see that with a net annual evaporation of 150 cm (as at the Salton Sea, California; Green, 1961) a seasonal increase in surface salinity from, for instance, 27.4 percent to 29.1 percent would cause the precipitation of 192 g of halite per liter of brine, which is equivalent to an annual layer about 13.5 cm thick. The actual figure will be lower because evaporation is reduced by the high concentration of the brine, and the seasonal variation in concentration may have been less than the given figure. The value then approximates that of the mean thickness of the laminations in halite, which many investigators assume to be annual layers (Richter-Bernburg, 1950, 1953, 1960; Dellwig, 1955, critically reviewed by Braitsch, 1962).

With such a high rate of deposition, a shallow basin would have been filled up by halite in a few score years, unless subsidence could keep pace with sedimentation, which would require a rate of basin subsidence some fifty or hundred times more rapid than actually observed. Therefore, basin subsidence amounting to at least the observed thickness of the rock-salt must have taken place *before* the deposition of halite started, as has been pointed out by Richter-Bernburg, Yanshin (1961), and others.

The required basin subsidence gives at the same time a plausible reason for the existence of a submarine bar or shallow sea at the entrance to the basin, enabling the reflux mechanism to work and the sea water to be preconcentrated. This bar or shallow is nothing but a part of the less-subsided basin rim which since it happened to be submerged could thus serve as the channel for the inflow and outflow of preconcentrated sea water.

Though basin subsidence is a very common process, having acted almost continuously throughout the course of geologic history at one place or another, only occasionally did it occur under the climatic, paleogeographical, and tectonic conditions that caused thick evaporites to deposit. It is a coincidence that these conditions are not realized anywhere at present, but it has often been contended (for example, Green, 1961) that the Red Sea would have developed into an evaporite basin had the bar near the outlet happened to be much shallower than it is.

It would be interesting to try to evaluate the volumes of evaporites formed in the course of geologic history by either of the two processes: the shallow and mud-flat mechanism and the deep basin mechanism. It may well be that the first process is responsible for the deposition of most of the gypsum and the second for most of the halite.

REFERENCES

Braitsch, O., 1962, Die Entstehung der Schichtung in rythmisch geschichteten Evaporiten: Geol. Rundschau, v. 52, p. 405-417.

Briggs, L. I., 1958, Evaporite facies: Jour. Sed. Petrology, v. 28, p. 46-56.

Dellwig, L. F., 1955, Origin of the Salina salt of Michigan: Jour. Sed. Petrology, v. 25, p. 83-110.

Green, Robert, 1961, Palaeoclimatic significance of evaporites, *in* Nairn, A. E. M., ed., Descriptive Palaeoclimatology: New York, Intersci. Publishers, p. 61-88.

Kerr, S. D., and Thomson, A., 1963, Origin of nodular and bedded anhydrite in Permian shelf sediments, Texas and New Mexico: Am. Assoc. Petroleum Geol. Bull., v. 47, p. 1726-1732.

Richter-Bernburg, Gerhard, 1950, Zur Frage der absoluten Geschwindigkeit geologischer Vorgänge: Naturwissenschaften, Jahrg. 37, p. 1-8.

——— 1953, Ueber salinare Sedimentation: Deutsche Geol. Gesell. Zeitschr., v. 105, p. 593-645.

——— 1960, Zeitmessung geologischer Vorgänge nach Warven-Korrelationen im Zechstein: Geol. Rundschau, v. 49, p. 132-148.

Ricour, J., 1962, Contribution à une révision du Trias français: Mém. Carte géol. France, 471 p.

Rutten, M. G., 1954, Continental origin of fossil salt layers: Geologie en Mijnb., new ser., v. 16, p. 61-68.

Stewart, F. H., 1963, Marine Evaporites, chapter Y, *in* Data of Geochemistry, 6th ed.: U. S. Geol. Survey Prof. Paper 440 Y, 52 p.

Sugden, W., 1963, The hydrology of the Persian Gulf and its significance in respect to evaporite deposition: Am. Jour. Sci., v. 261, p. 741-755.

Yanshin, A. L., 1961, On the depth of salt-bearing basins and some problems of thick salt series forming [in Russian]: Acad. Sci. U.S.S.R. Siberian Branch, Geology and Geophysics Inst., no. 1, p. 3-15.

Reprinted from *Third Symposium on Salt,* **1**, 48–66 (1970)

17

Shelf Carbonate Sedimentation Controlled by Salinity in the Paradox Basin, Southeast Utah[1]

Robert J. Hite[2]

ABSTRACT

Deposition of shelf carbonates, adjacent to and contemporaneous with a basin evaporite facies, is closely controlled by salinities originating in the evaporite basin. Changes in depositional environment due to rising sea level in these basins is often misinterpreted, especially when the resultant transgression is linked to freshening and expansion of normal marine conditions. Rising sea level may terminate evaporite deposition in the basin, but anaerobic hypersaline conditions are displaced onto the basin shelf and margins during this transgressive period.

Circulation in marine evaporite basins includes some return flow or reflux. Reflux is controlled by change in sea level, evaporation rates, and size of accessways.

Middle Pennsylvanian evaporite cycles in the Paradox basin of southeast Utah and southwest Colorado probably were due to periodic fluctuations in sea level. Each complete evaporite cycle consists of a transgressive and a regressive hemicycle bounded by disconformities. Beds comprising these cycles grade shelfward into predominantly carbonate facies. Salinities of the evaporite basin controlled carbonate sedimentation on the adjacent shelf to the southwest. High sea level and related maximum reflux carried toxic brines onto the shelf and created unfavorable conditions for life and deposition of biogenic carbonates. During maximum reflux, reducing conditions initiated deposition of black sapropelic shale where normal marine life previously thrived. Only during low sea level, with little or no reflux, were shelf conditions optimum for growth of algal bank carbonates.

Using bromine distribution in halite units of evaporite cycles as an index of basin reflux, cycles favorable for the development of algal carbonate facies can be predicted. Because algal carbonate facies in many marine evaporite basins are major sources of petroleum, bromine geochemistry gains still another use in exploration.

INTRODUCTION

All marine evaporites, unless the deposit is the result of solution and redeposition of preexisting evaporites, have an adjacent and contemporaneous carbonate facies. The rock types and distribution of this facies often reflect radical but rhythmic changes in the depositional environment. Similar depositional rhythms are found in the facies of the neighboring evaporite basin where the depositional response has been clearly shown to be controlled by salinity. In the evaporite basin the specific response of the chemical facies to salinity change can be easily demonstrated by the solubility order of the various salts involved. In contrast the effect of salinity changes on the carbonate facies of the adjacent shelf is not well understood, primarily because it involves changes in the biologic community, and our methods of interpreting the ecology of a biofacies are unprecise. In addition, the deeper, more stable environment of the evaporite basin probably preserves a more complete depositional record than the shallower shelf. In the latter the depositional record may be complicated

1. Publication authorized by Director, U.S. Geological Survey.
2. U.S. Geological Survey, Denver, Colorado

by erosion, nondeposition and onlap and offlap relationships. On a shallow, gentle-sloping shelf even minor changes in sea level can cause significant shifts in facies. Thus, the vertical sequence of units in a depositional cycle of shelf rocks may be incomplete or seemingly without order.

Salinity changes in the evaporite basin can be brought about by runoff, differential rates of evaporation, breaching of circulation barriers, and the like; however, most probably are due to variations in sea level. Because evaporite deposition is such a sensitive record of salinity, it is only logical that these chemical sediments are also our best record of sea level change. Therefore, where evaporites are associated with shelf carbonates it may be possible to use the evaporites as the key to deciphering the complex depositional history of the carbonates.

This paper includes a review of certain principles of evaporite deposition and, through model studies, shows how these principles were involved in deposition of evaporites and contemporaneous carbonates in the Paradox basin of southeast Utah and southwest Colorado. Some of the basic principles on which this paper is based, and conclusions drawn from application of these principles, are as follows:

(1) All evaporite basins undergo some degree of reflux.

(2) Rate of reflux is primarily controlled by sea level. If sea level is high, then the rate of reflux will be great; if, however, the sea level is low, reflux is absent or minimal.

(3) Every marine evaporite facies must have a contemporaneous carbonate facies unless the evaporites are the result of solution and redeposition of preexisting deposits.

(4) The contemporaneous carbonate facies will be deposited in the area of seawater ingress, and thus will also be in the path of refluxing brines from the basin.

(5) Contrary to popular opinion, biogenic carbonates should develop during a low sea level or regressive phase when reflux is minimal.

(6) Studies of bromine distribution in the chloride facies of evaporite cycles can establish the relative rate of basin reflux. The reflux index for individual cycles can be used as a tool in evaluating the cycle's favorability of having a biogenic carbonate facies. Because buildups of these carbonates (usually algal) commonly form petroleum reservoirs, bromine geochemistry may serve as a new tool for petroleum exploration.

(7) In undrilled sedimentary basins the presence of algal carbonate buildups in depositional cycles, around the margins of the basin, might indicate a potential target zone for salt and potash exploration deeper in the basin.

I am indebted to H.V.W. Donohoo of the Texas Gulf Sulphur Company for granting permission to publish the photographs taken in the Cane Creek potash mine and certain drill core data. In addition the manuscript was greatly improved by critical reviews of W.B. Cashion and P.T. Hayes.

SOME PRINCIPLES OF EVAPORITE DEPOSITION

In order to understand better how the evaporite environment can influence the deposition of an adjacent carbonate facies, it is first necessary to review some of the principles governing evaporite deposition. Although there are other equally important principles involved in evaporite deposition, there are three that have special significance to the deposition of contemporaneous carbonates. The principles reviewed here are known to most evaporite geologists but have seldom been applied in studies of carbonates associated with evaporites.

Principle I.

The deposition of each individual mineral facies' of an evaporite deposit must first be preceded by the deposition of the other less soluble facies. For a unit volume of halite to be deposited, calcium carbonate and calcium sulfate must first be deposited, and the unit volumes of these salts must at least equal (they usually exceed) the theoretical seawater ratio. A more soluble salt cannot be deposited in excess of the seawater ratio of the less soluble salts. This simple principle is commonly overlooked.

Seawater is a complex system of many ions but the sequence of salts deposited by its evaporation follows a definite order, at least until the bittern stage of concentration is reached. After that the sequence of salts deposited varies according to changes in physicochemical and even biologic factors. The depositional sequence of rock-forming minerals, which was established by Usiglio (1849), starting with the least soluble mineral is (1) calcite, (2) gypsum or anhydrite, and (3) halite. Thus a marine evaporite deposit containing a halite facies must have an associated sulfate and carbonate facies. Possible exceptions would include removal of the sulfate facies by dissolution or bacterial reduction. Where it is assumed that these facies are missing they are either yet to be drilled or their contemporaneity with the halite has not been recognized. One reason that geologists sometimes conclude that a facies is missing is that they have only examined a vertical sequence and have not considered lateral facies changes. In addition salt, due to its greater mobility, may be displaced from its former associate facies in areas affected by salt tectonism.

Principle II.

All marine basins containing thick evaporite deposits have undergone reflux. Without reflux the evaporite deposits of all basins would include the full spectrum of evaporite minerals, and the relative amounts of one mineral to another would be directly proportional to the ratio of the parent elements in seawater.

During the embryonic stage of evaporite geology it was noted that most natural sequences of evaporites do not conform to the sequence of salts produced in the laboratory by evaporation of seawater. In particular, it was observed that the volumetric ratios of various minerals in the deposits differ sharply from laboratory results. An often cited example is the Castile Formation of the Delaware basin which has an anhydrite/halite ratio of about 1:1 whereas the theoretical seawater ratio is about 1:30. Other deposits containing thick halite layers are totally lacking in salts of potassium and magnesium. The solution to the problem came when King (1947) utilized the reflux theory of Ochsenius (1877) to explain the various mineral imbalances observed in evaporite deposits of the Delaware basin. Mineral imbalance, or the complete absence of some of the more soluble salts, was explained by loss of salts back to the sea during reflux.

Principle III.

Reflux is controlled by rise or fall in sea level, change in carrying capacity of the accessway, or change in evaporation rates.

Of these three controls, change in sea level is probably the most significant. When sea level is high the basin will experience a large amount of reflux. Conversely when sea level is low there will be little or no reflux. This last principle is the key to the proper interpretation of carbonate deposition on shelves bordering evaporite basins.

BARRED BASIN EVAPORITE MODEL

The principles of evaporite deposition, outlined in the preceding section, and the strong influence of the evaporite basin on carbonate deposition in the surrounding environs, can be illustrated best with models. Following the barred basin model of Ochsenius (1877), models of various other evaporite depositional sites, involving both physical and dynamic barriers to oceanic circulation, have been proposed; however, in this paper only the barred basin model is discussed because it fits best with the Paradox basin deposits.

In the barred basin evaporite model, circulation with the open ocean is restricted by topographic barriers. Marine access to the basin may consist of a narrow strait or channel analogous to the Strait of Gibralter and the Mediterranean Sea. In the geologic past most oceanic accessways probably extended across broad gently sloping shelves. Seawater moving slowly across such shelves, which might be several hundred miles in width, can become highly concentrated through evaporation even before it reaches the evaporite basins (Fig. 1). Such features have been aptly named "saturation shelves" by Richter-Bernburg (1955, p. 600). These conditions create strong horizontal salinity gradients which extend across the saturation shelves into the evaporite basins. In response to these gradients chemical sedimentation develops an orderly pattern of facies with the most soluble salts, such as potash, located in the most distal reaches of the basins (Scruton, 1953, p. 2510). In addition to controlling the position of each chemical facies, the salinity gradients exert considerable influence on the deposition of biogenic carbonates. Each biofacies is positioned in relation to the salinity tolerance, or preference, of the organism involved.

The mechanics of basin evaporite deposition can be more clearly understood if stated in terms of salt load carried into the basin and salt load carried back to the ocean. At a high sea level stage (Fig. 1a) the evaporite basin will undergo a large amount of reflux. If sea level is high enough the reflux salt load will equal influx salt, preventing a salinity increase and deposition of evaporites. The Mediterranean Sea is a perfect example of a basin in which the incoming salt load is balanced by reflux so that the salinity of the sea remains well below saturation (Peterson and Hite, 1969, p. 893).

The circulatory system of the evaporite basin model is a dynamic system and, therefore, its salt budget is in a continuous state of change. The rate of influx, and therefore the influx salt load, is controlled by evaporation rates in the basin. If these evaporation rates remain the same, the influx salt load can be considered a constant. The greatest variable in the system is the reflux salt load. If we begin to lower sea level in our model the influx salt load will remain constant and for a while so will

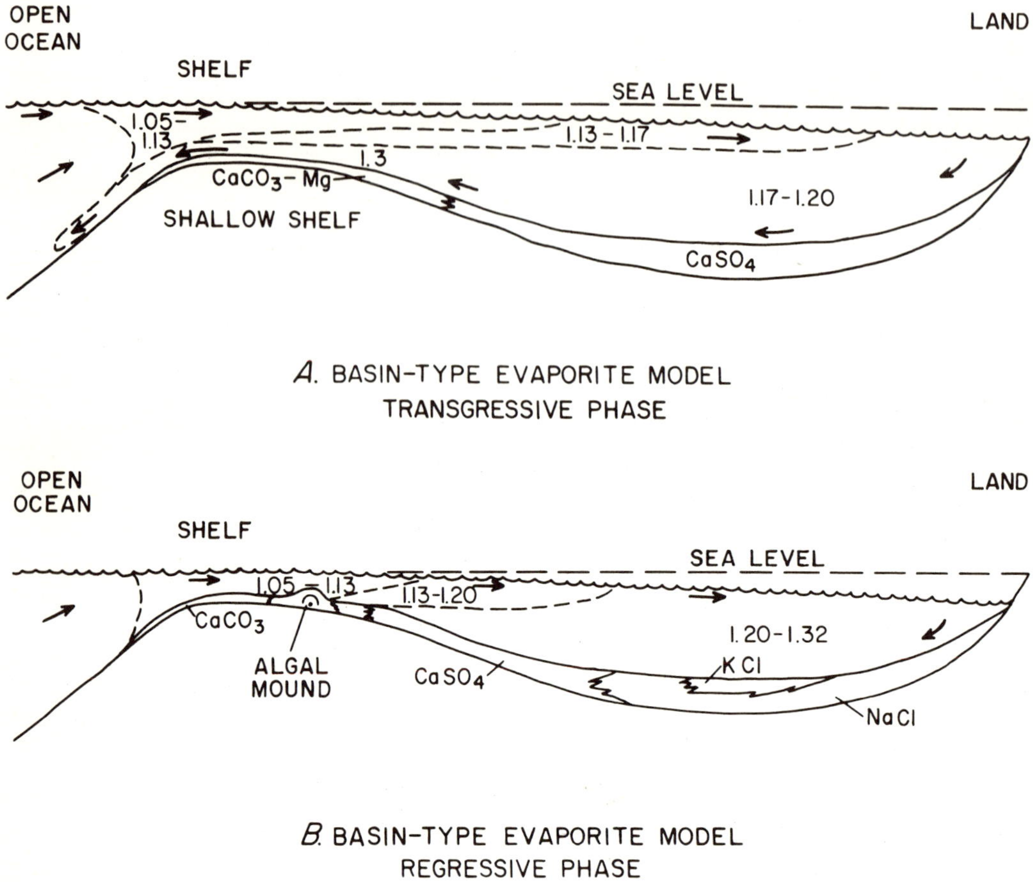

Figure 1. Models of a barred evaporite basin during (A) the transgressive phase (high sea level) and (B) the regressive phase (low sea level). Numbers (representing water densities) are approximate.

the reflux load. However, the velocity of the influx current always dominates the accessway so that any reduction in flow, by constriction of the channel, must take place at the expense of the reflux current. Eventually lowering of sea level will reduce the carrying capacity of the accessway to a point that reflux is inhibited. This creates an imbalance between influx and reflux salt, and the surplus salt retained in the basin raises the salinity. Continued lowering of sea level will increase this imbalance and eventually bring about the deposition of evaporites. Extremely low sea level may so restrict circulation between open ocean and basin that only enough water to replace evaporation losses can enter the basin. At times this inflow may have been completely shut off. Under these conditions very high salinities will build up in the basin and potash salts will be deposited (Fig. 1b).

The models discussed here help to demonstrate the dynamic nature of the evaporite environment which creates conditions that cannot be duplicated by simple evaporation in a laboratory beaker. For example, the salt budget of the evaporite basin must always show a balance between the amount of influx salt and the amount of salt lost to the depositional layer and reflux. If the salt budget of an evaporite basin has reached equilibrium at a point when the brine is saturated for calcium sulfate, only this salt and the less soluble carbonates will be deposited (Fig. 1a). The horizontal salinity gradient will sufficiently separate the solubility fields of these minerals so that an infinite thickness of each facies will result unless there is a sea level change or some other event that alters the basin's salt budget (Zen, 1960). The dynamic system of the evaporite basin also creates another important condition which I refer to as the "funnel effect." Although a basin evaporite deposit may cover a few thousand square miles, the evaporative surface, which affects the salinities in the basin, may cover several hundred thousand square miles. When the nearly saturated brine, generated in this much larger area, is funneled into a relatively small depression very rapid deposition of evaporites can occur.

The principle of a refluxing evaporite basin has received wide acceptance, but the strong influence of reflux brines on sedimentation outside the basin has rarely been considered. Evaporite brines, compared with other natural waters, are strongly active chemical agents. These brines can have a salt content 10 times that of seawater. Compared with seawater, which is weakly alkaline, brines may have pH values of 4.0 and lower. Brines are usually anoxic because of the depletion of oxygen by decay of organic material and the depressed solubility of oxygen in concentrated salt solutions (Fig. 2). Toxic concentrations of hydrogen sulfide are also characteristic of these waters. In addition, high brine temperatures are common owing to the "green house" effect caused by solar radiation passing through the influx layer of fresher seawater into the dense reflux layer where the energy is absorbed. The "green house" effect is being studied as a possible energy source by Israeli scientists who have recorded temperatures of 90°C in dark-bottomed solar brine ponds. Considering all these factors it seems reasonable to assume that an evaporite brine and normal seawater are in chemical disequilibrium. Mixing of these waters should play a prominent role in chemical sedimentation and diagenetic processes on the shelf of the evaporite basin and perhaps on the floor of the open ocean as well.

The concept of the mixing of natural waters as an important process of diagenesis and chemical sedimentation has been proposed and factually supported by Runnells (1969). He showed how the mixing of waters, which differ by one or any combination of the following: (1) salt content,

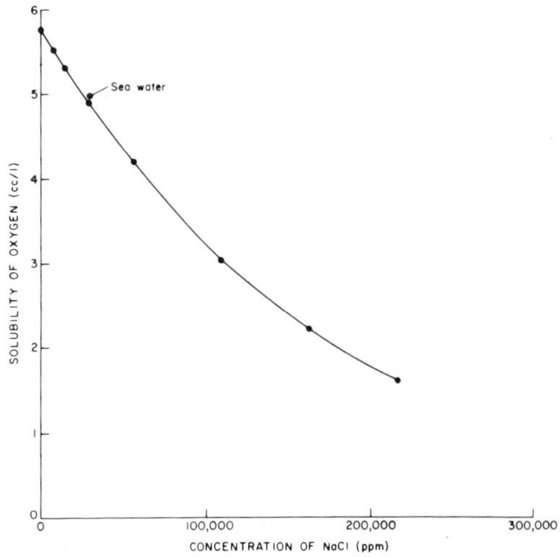

Figure 2. Solubility of oxygen in sodium chloride solutions at 25°C. Data from MacArthur (1916) and Geffcken (1904).

(2) temperature, (3) pressure, (4) pH, (5) content of dissolved organics, and (6) partial pressures of gases, can cause precipitation or dissolution of common minerals such as calcite, gypsum, dolomite, anhydrite, and barite. Direct precipitation of dolomite, calcite, or magnesite might be effected by mixing of the warm hypersaline Mg-Ca-Na-rich brines of the reflux current with colder carbonate-bearing seawater of the overlying influx current. Briggs (1958, p. 50) also proposed the mixing of evaporite brines, of contrasting composition, as a means of precipitating sulfates and carbonates. A similar mechanism has been suggested by Pettijohn (1949, p. 301) who stated "Observations on the calcareous oolites that are forming at the present time indicate that sodium or ammonium-carbonate-bearing waters may be the precipitating agent. Upon introduction of waters containing these ions into waters containing calcium in solution, an exchange takes place, and the less soluble calcium carbonate is thrown down. Under suitable current conditions, the calcium carbonate takes the oolitic form." A solubility differential for silica, between the influx and reflux waters, might also be responsible for the deposition of chert layers which are commonly associated with shelf dolomites.

Refluxing brines may also be responsible for large-scale diagenesis of shelf sediments due to chemical disequilibrium between the brine and the sediment layer. Dolomitization by seepage refluxion has been proposed by Adams and Rhodes (1960); however, dolomitization by magnesium-rich brines moving *over* preexisting carbonates may be an equally important mechanism. Solubilities of calcite, dolomite, anhydrite, and gypsum increase with increasing concentrations of NaCl according to Hem (1959, p. 74, 82) and Runnells (1969). Therefore, dissolution of previously deposited carbonates and sulfates by chloride brines might be expected during periods of substantial reflux.

Added to the important role that refluxing evaporite brines may play in chemical sedimentation and diagenesis is the pronounced influence on the biologic community of the shelf. During reflux the shelf waters are divided into two strongly contrasting environments. The overlying inflow layer of seawater could support the normal nektonic-planktonic biota, but the refluxing layer would bring the anoxic highly reducing environment of the evaporite basin onto the shelf. The survival of benthos under these conditions is doubtful except in areas where a topographic rise might project into the life-supporting inflow layer.

Summarizing, it seems probable that evaporite basins can exert a powerful influence on the chemical and biological sedimentation of the adjacent shelf. Because variations in sea level control basin reflux, and therefore control basin salinities, any given time-rock layer in the chemical facies can be used as a guide to predict the rock type in the contemporaneous shelf facies.

APPLICATION TO A NATURAL SEQUENCE (PARADOX BASIN)

The principles of evaporite deposition described in the preceding section can be applied to problems of contemporaneous carbonate and evaporite sedimentation in the Paradox basin. This basin contains a thick sequence of evaporites of Middle Pennsylvanian age and a thinner complementary facies of shelf carbonates. The evaporite facies, which is predominantly halite, extends over 11,000 square miles in southeastern Utah and southwest Colorado (Fig. 3). The greatest thickness of evaporites occurs in a troughlike depression bordering the ancestral Uncompahgre uplift along the northeast margin of the basin (Fig. 3). This elongate structural depression has been referred to as the Uncompahgre trough (Hite, 1968). The original thickness of evaporites deposited in the trough area is difficult to estimate because of the widespread effects of salt flowage in the numerous diapiric salt

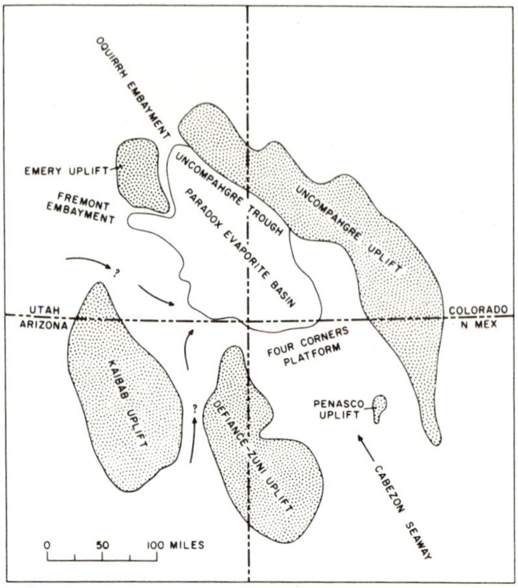

Figure 3. Index and paleogeographic map of Paradox basin during Middle Pennsylvanian time showing land areas (stippled) and possible marine accessways (shown by arrows).

anticlines associated with this region. From observations made in areas where the salt is not so severely disturbed, it is estimated that the original thickness of the evaporite sequence may have been between 5,000 and 6,000 feet.

The evaporite or halite facies forms the middle unit or Paradox Member of the Hermosa Formation. Where the halite facies is present the Hermosa Formation can be divided into an upper member above the halite and a lower member below (Fig. 4). Both of these members are predominantly a carbonate facies, although a few thin anhydrite, black shale, and sandstone layers are present locally. The upper and lower members are respectively the Honaker Trail and Pinkerton Trail Formations of Wengerd and Matheny (1958) and Wengerd and Strickland (1954). The Paradox Member exhibits a remarkable cyclic pattern of chemical and clastic sedimentation. The member can be subdivided into 29 evaporite cycles in the deeper part of the basin and for convenience of correlation each cycle has been given a numerical designation (Hite, 1960). Each evaporite cycle can be correlated across the salt basin into an equivalent cycle within the carbonate facies on the southwest shelf of the basin. Many of the

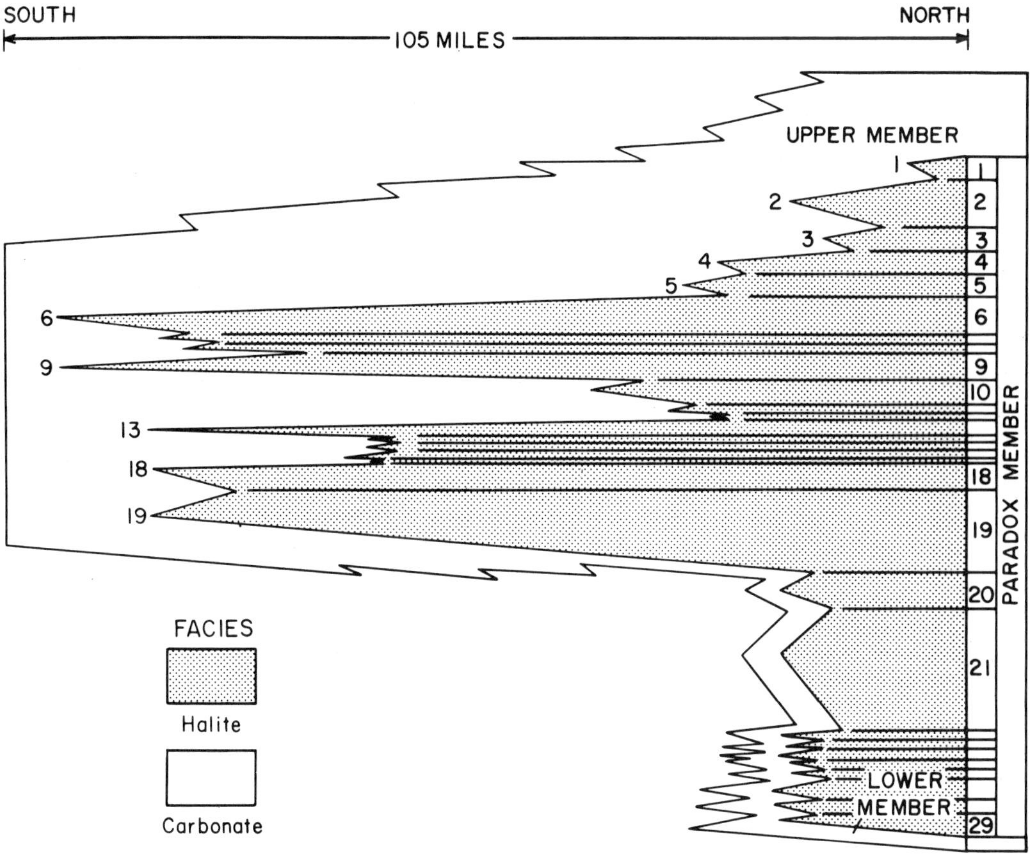

Figure 4. Diagrammatic north-south stratigraphic section across the Paradox basin showing stratigraphy of the Hermosa Formation of Middle Pennsylvanian age. Numbered units are evaporite cycles of the Paradox Member which consist primarily of halite.

carbonate cycles contain petroleum reservoirs, and these producing intervals, which usually span several depositional cycles, have been lumped by petroleum geologists into units known as the Ismay, Desert Creek, Akah, Barker Creek, and Alkali Gulch zones. For a detailed discussion of stratigraphic nomenclature of the shelf carbonates see Baars and others (1967).

The initial deposition of evaporites in the Paradox basin began in the Uncompahgre trough and was restricted to this depression through cycles 29 to 20. It was not until the deposition of cycle 19 that the halite facies began to spread out of the trough toward the western and southern margins of the basin (Fig. 4). This spread of the evaporites probably was due more to filling of the Uncompahgre trough by rapid deposition than to an intensification of the evaporite regimen. During the deposition of cycles 10, 11, and 12 the evaporite facies was again restricted to the trough area suggesting a period of mild downwarping. Maximum expansion of the evaporite environment came during cycles 6 and 9 when halite was deposited far out on the shelf. After the deposition of cycle 6 the evaporite facies retreated rather suddenly back to the trough area and remained there until the basin was finally flushed out by an incursion of normal marine waters.

The evaporite cycles.

The evaporite cycles of the Paradox Member are unequaled in number and completeness by those of other evaporite deposits of the world. The Paradox cycles are also unique in that they afford an excellent opportunity to study the relationship of contemporaneous deposition of carbonates and evaporites. The Paradox evaporite cycles were first described by Herman and Sharps (1956), and later in greater detail, by Herman and Barkell, (1957), Hite and Gere (1958), Hite (1960, 1961, 1968), and Peterson and Hite (1969).

A typical Paradox evaporite cycle consists primarily of chemical rocks; however, a small amount of detrital and organic material forms an integral part of the cycle. The lithofacies of the cycle include halite with or without potash salts, anhydrite, silty dolomite, and black shale (Fig. 5). The halite and anhydrite units of each cycle are entirely a chemical facies; however, the dolomite and black shale are both chemical and clastic, and contain organic material. The dolomite is extremely fine grained and contains abundant uniformly sorted quartz silt. The black shale is the most complex of the facies. About 20 to 30 weight-percent of this rock is carbonate which is divided about equally between dolomite and calcite. The remainder of the black shale is clay-size detritus consisting of quartz, feldspars, various clay minerals, and locally as much as 15 percent organic matter.

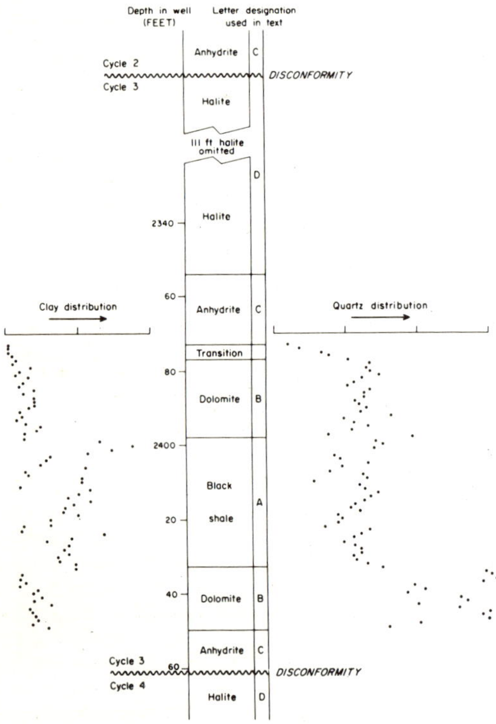

Figure 5. Typical evaporite cycle (cycle 3) of the Paradox Member. Relative distribution of clay and quartz, through the cycle, determined by X-ray diffraction analysis. Data from core hole located in sec. 25, T. 26 S., R. 20 E., Grand County, Utah.

The sequential order of units in a typical Paradox cycle is determined by the degree of solubility of the rock-forming mineral in seawater. If a cycle is defined on the basis of the chemical sequence of rocks deposited, as seawater undergoes a gradual increase and then a decrease in salinity, the sequence is (A) silty, calcareous, dolomitic, argillaceous, organic-rich black shale, (B) silty dolomite, (C) nodular and laminated anhydrite, (D) halite with or without potash salts, (C) anhydrite as before, (B) dolomite as before, and (A) black shale as before. In this sense the cycle can be termed a chemical cycle and, on the basis of the position of lithologic units, shows mirror-image symmetry as A, B, C, D, C, B, A. However, the Paradox chemical cycle is not symmetrical. At the base of the cycle, units A, B, C, and D are deposited as the result of gradually increasing salinity and each unit shows a gradual transition with the overlying unit. The halite (unit D), however, neither shows a gradual decrease in salinity nor grades into the overlying anhydrite unit. Studies by Raup (1966, p. 245) of the bromine content of halite in the Paradox Member have shown that salinities gradually increased all the way to the top, or near the top, of each salt bed. Further evidence of asymmetry can be seen in those cycles that are potash bearing. The potash deposits, which represent the period of highest salinity in the cycle, usually occur at the top of the halite instead of in the normal middle position (Hite, 1968, p. 323). A strongly asymmetrical distribution of detrital quartz through the cycle can also be noted (Fig. 5). The lower dolomite unit (transgressive phase) contains nearly twice as much quartz as the upper dolomite unit (regressive phase). Later it will be shown how this pattern of quartz distribution can be an important factor in correlating facies between evaporite cycles and the carbonate cycles on the shelf.

Boundaries of Paradox cycles have been defined on the basis of rock units, generally the black shales but recently it was recognized that a more workable system of cycle boundaries can be made if disconformities are used (Peterson and Hite, 1969). In the evaporite cycle the base of each halite unit forms a transitional contact with the underlying anhydrite (Fig. 7); however, the upper contact is very abrupt (Fig. 6). This sharp contact, at the top of the halite unit, is a dissolution surface or disconformity along which several feet of halite has been removed. These disconformities, which interrupt the chemical sequence, are the results of major rises in sea level and subsequent freshening in the evaporite basin. Using the disconformities as cycle boundaries, the order of units would be C, B, A, B, C, D. The change from transgressive to regressive conditions occurs somewhere near the midpoint of unit A so that the transgressive hemicycle is C, B, A, and the regressive hemicycle is A, B, C, D. Comparison of the transgressive and regressive hemicycles shows considerable condensation of the sedimentary sequence in the transgressive phase. According to Duff and others (1967, p. 178) this is a normal aspect of sedimentary cycles. A similar observation has been made by Sheldon (1963, p. 147), who concluded that in cycles of the Phosphoria Formation transgression was more rapid than regression.

Figure 6. Photograph of drill core showing sharp dissolution contact between coarsely crystalline halite and finely laminated anhydrite. This is typical of the upper contact of salt beds in the Paradox Member.

The presence of disconformities at the tops of halite units in the Paradox cycles was first suggested by Welsh (1963, p. 489), and has been documented since by drill hole data and by observations from mine exposures. The amount of halite removed along these disconformities is generally unknown; however, it is likely that at least 5 to 10 feet is missing. The disconformity at the top of cycle 5 is readily visible at various locations in the Texas Gulf Sulphur Co. Cane Creek potash mine near Moab, Utah (Fig. 8). Immediately overlying this disconformity is a thin zone of glassy anhydrite (Fig. 9) which probably represents the residuum formed when the underlying halite was taken into solution. Based on the average anhydrite content of the halite immediately underlying the

Figure 7. Photograph of a typical basal contact between a halite bed and underlying anhydrite of a Paradox evaporite cycle. The right-hand end of the upper drill core fits together with the left-hand end of the lower core. Note the gradual transition from massive light-gray anhydrite at the top left of the core into dark-gray halite on the right.

Figure 8. Underground view of abrupt contact (dissolution surface) between light-gray halite and overlying glassy dark-gray anhydrite (Fig. 9); pocket knife is positioned at contact. Contact forms the top of Paradox evaporite cycle no. 5. Photographed in Texas Gulf Sulphur Co. Cane Creek potash mine near Moab, Utah. (Photograph by O.B. Raup.)

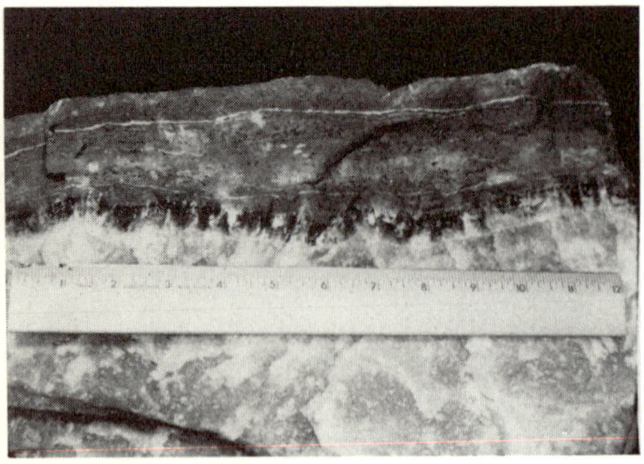

Figure 9. Closeup view of same contact shown in Figure 8. Note 1-inch-thick layer of glassy dark-gray anhydrite immediately above contact with light-gray halite.

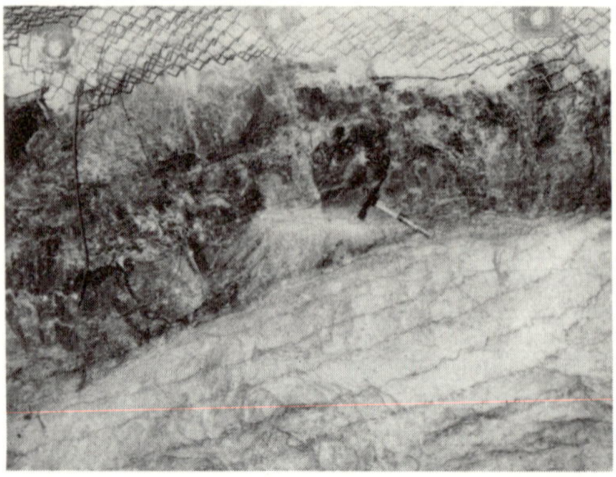

Figure 10. Underground view in Texas Gulf Sulphur Company Cane Creek potash mine of a fold in light-gray halite truncated at dissolution surface by overlying dark-gray anhydrite; end of hammer at contact. Note truncation of thin crenulated anhydrite laminae in the halite underlying the dissolution surface.

disconformity, determined from insoluble residues, the glassy anhydrite layer represents the dissolution of about 10 feet of halite. This thickness compares favorably with local truncation of microstratigraphic units in the upper part of the halite and sylvite of cycle 5 as observed in the mine.

Disconformities at the top of cycles 3 and 4 can also be observed in the Cane Creek mine. In exposures along the incline, connecting the bottom of the mine shaft with the mine workings in the potash deposit in cycle 5, small folds in the halite units of each cycle are truncated by the overlying anhydrite layer of the next cycle (Fig. 10).

In the preceding description of the Paradox evaporite cycles, much emphasis was placed on the disconformity boundaries of the cycles. The importance of the disconformity in the facies relationships between basin evaporites and shelf carbonates is shown in the next section of this paper. The significance of unconformities, or disconformities, in relation to facies has been ably stated by Momper (1966, p. 90d) as follows: "Advocating a search for unconformities does not imply a denial of the existence of facies changes or an attempt to minimize the significance of facies concepts. Quite the contrary is true because correct identification of unconformities is essential to accurate interpretation of facies as well as rock units in general."

The carbonate cycles.

On the southwest shelf of the Paradox basin the carbonate cycles of the Hermosa Formation have been studied in great detail by numerous petroleum geologists. The discovery of petroleum in several of these cycles has resulted in a relatively high concentration of drill holes and therefore detailed subsurface studies of lithofacies have been possible. In addition to subsurface control, excellent exposures of many of the carbonate cycles occur along the San Juan River near Mexican Hat, Utah, (Figs. 11, 12). Most investigations have been concentrated on the two most important zones of petroleum production, the Ismay and Desert Creek.

Figure 11. Photograph of carbonate cycles in the Hermosa Formation in the canyon of the San Juan River near Mexican Hat, Utah. (Photograph by M.O. Glerup.)

Figure 12., Photograph showing algal carbonate buildup facies of Ismay zone along the San Juan River. The algal beds are the dark undulating strata at river level. (Photograph by M.O. Glerup.)

The lithofacies of the shelf carbonate cycles is very diverse, and reflects combinations of chemical, clastic, and biogenic sedimentation. This complex pattern of sedimentation was primarily due to the relatively shallow-water environment of the shelf. Under these conditions only a minor change in sea level might drastically alter the environment at the depositional interface. If local topographic irregularities were present on the sea floor then the depositional environment of a rise could be markedly different from that of an adjacent depression. All of this could lead to abrupt lateral facies changes and dissimilarities in the vertical sequence between contemporaneous rock units. Because of the great variety of lithofacies, described by different authors at different localities for the same cycles, an attempt was made to roughly group the facies of each author into four categories (Table 1). Each facies from the shelf was grouped on the basis of similar lithic character and shown in approximate correlative relationship to the lithofacies of the evaporite cycles. From this comparison it can be noted that the black shale, silty dolomite, and anhydrite of the evaporite cycle have similar facies counterparts in the shelf carbonate cycles. The halite and potash of the evaporite cycle are represented by a totally different facies on the shelf consisting of an algal limestone formed primarily from the leaflike remains of the alga *Ivanovia*. Thus, as shown on Table 1, the boundaries of the facies groups represent time lines; however, several facies probably span these time surfaces. The contemporaneity of the various facies is explained in greater detail in a later section of the report.

All the authors cited in Table 1 have proposed one or more idealized cycles involving the various lithofacies of the shelf. For example, Choquette and Traut (1963, p. 169) recognized (1) an evaporite cycle containing anhydrite and dolomite, (2) a normal-marine cycle with an algal buildup, and (3) a normal-marine cycle without algal buildup. They acknowledged that lateral transition from one type of cycle to another could take place in a few thousand feet. The descriptions of cycles differ also because some are based on intervals in the Ismay and others on intervals in the Desert Creek. In addition, Pray and Wray (1963) made their observations on the basis of surface exposures whereas all the others used subsurface data. Thus it is difficult to diagram the vertical sequence of facies in a "typical" cycle. Much of the emphasis of this paper is placed on explaining why certain carbonate cycles favored the development of an algal carbonate facies. For that reason the idealized cycle described here includes an algal buildup. The following sequence of lithofacies represents a synthesis of descriptions by other authors, and although this version is probably an oversimplification of the problem, it does include elements common to all the cycles.

IDEALIZED SHELF CARBONATE CYCLE

	Next cycle
Cap facies	disconformity
Buildup facies (algal limestone)	
Intermediate facies (silty dolomite and limestone)	
Euxinic facies (black, argillaceous, calcareous shale)	
Intermediate facies (siltstone and sandstone or very silty dolomite)	
Cap facies (pelletal-foraminiferal limestone)	disconformity
Buildup facies	Next cycle

As shown above, the carbonate cycle is bounded by disconformities. Disconformities in the shelf carbonate facies have been recognized and described in the subsurface and in measured sections along the San Juan River by Wengard and Matheny (1958), Welsh (1958, 1963), Elias (1963), and Pray and Wray (1963) but it is probable that some disconformities have been

Table 1. Lithofacies of the Hermosa Formation on the Southwest Shelf of the Paradox Basin, as Described by Various Authors, and the Approximately Contemporaneous Facies of the Inner Evaporite Basin

	Shelf Area					Evaporite Basin
Reference	Choquette and Traut (1963)	Elias (1963)	Gray (1967)	Peterson (1966)	Pray and Wray (1963)	Hite (this paper)
Lithofacies	Laminated carbonate boundstone Shelly calcilutite Black shale Dark carbonate mudstone Sponge-bearing calcilutite	Black shale Sponge	Shelly mudstone Laminated boundstone Black shale Crinoidal limestone Carbonate mudstone	Black shale	Muddy algal Intermediate Sponge	Black shale
	Earthy dolomite Calcareous sandstone and siltstone	Quartz sandstone and siltstone	Earthy dolomite Evaporitic dolomite	Very silty Chalky to dense dolomite	Quartz sandstone and siltstone	Silty dolomite
	Foraminiferal-pelletal limestone Buildup Anhydrite	Fusulinid Pelletal mud Ivanovia Anhydrite	Foraminiferal-pelletal limestone Algal calcirudite breccia Anhydrite	Foraminiferal-pelletal limestone Mound carbonate Anhydrite	Cap Sparry algal (Ivanovia)	Anhydrite Halite and potash

overlooked. In fact, all the carbonate cycles probably are bounded by disconformities. These disconformities may separate facies which are only slightly dissimilar, and the breaks in sedimentation can easily be overlooked. Recognition of some disconformities may come only after detailed mineralogic and geochemical analysis of the vertical sequence. The positions of the disconformities in the idealized cycle described here are different from those as interpreted by Elias (1963, p. 188) who places them at the top of the pelletal-foraminiferal limestones. The same eustatic control, effecting deposition in the evaporite basin, is also in operation on the shelf. Previously it was shown how the initiation of a transgression creates an asymmetrical break in the depositional sequence of the evaporite cycle. If a similar depositional pattern exists on the shelf, as seems reasonable to expect, the disconformity should separate the initial transgressive phase from the underlying regressive phase. Using this concept the logical position of the disconformity would be between the pelletal-foraminiferal limestone (cap facies), which is interpreted here as representing the initial transgressive phase, and the underlying regressive phase of algal limestone (buildup facies).

THE PARADOX BASIN MODEL

The facies relationships in the depositional cycles of the Paradox basin can best be explained through the use of models. However, to model the basin the paleogeographic setting of the region must first be established. Numerous authors, notably Wengerd and Matheny (1958), Fetzner (1960), Ohlen and McIntyre (1965), and Peterson (1966), have shown that the Paradox basin is bordered on the northeast by highlands of the Uncompahgre uplift and on the northwest (Emery), south (Defiance-Zuni), and southwest (Kaibab) by low-lying land masses (Fig. 3). Wengerd and Matheny postulated marine accessways at the northwest (Oquirrh) and southeast (Cabezon) ends and along the west flank (Fremont) of the basin. Elias (1963) suggested that the basin may have been completely enclosed except for two narrow straits which coincide with the Fremont and Oquirrh accessways of Wengerd and Matheny. Facies relationships within the Paradox basin suggest that none of these features, with the possible exception of the Cabezon accessway, were responsible for any significant ingress of seawater. There is some evidence of marine access to the basin through the Cabezon seaway during deposition of Paradox evaporites. However, Fetzner (1960, p. 1408) suggested that a clastic delta from the Uncompahgre uplift provided a sedimentary barrier across the southeast margin of the basin. Additional evidence of circulation barriers between the basin and the Cabezon seaway were found by Peterson and others (1965, p. 2087), who stated that "The Four Corners platform apparently underwent minor growth as a southwest-northeast positive-trending feature during Pennsylvanian time causing shallow-water conditions and shelf carbonate development to extend northeastward along the platform. The combination of the southwest-northeast projection of the shelf carbonate buildup on one side and rapid accumulation of deltaic deposits on the other caused the Paradox basin to be cut off effectively from the main seaway at the southeast." Along the southwest edge of the basin the facies belts within individual cycles tend to broaden, and extend convexly into the basin much like a chemical delta. Facies are distributed in order of solubility, with the least soluble facies in the outer belt and progressively more soluble facies basinward. If seawater were moving into the basin from the postulated Oquirrh seaway, then it is difficult to explain how relatively insoluble calcium carbonate could be carried across 200 miles of saturated chloride brine and finally precipitated at the opposite end of the basin. On the southwest shelf of the basin stratigraphic markers retain their identity in a southwesterly direction through a carbonate facies for greater distances than in any other part of the basin. These relations suggest that the principal marine access was across this broad southwest shelf. Connections between the shelf and the open ocean may have been between the Kaibab and Defiance-Zuni uplifts or around the north end of the Kaibab uplift (Fig. 3).

The barred basin evaporite model, discussed in the first part of this paper, fits well with the lithofacies pattern and geochemical relationships observed in the Paradox depositional cycle. The operation of the Paradox model during a regressive and a transgressive phase of deposition (Fig. 13) can be described as follows:

A. Regressive phase (Fig. 13a). Sea level begins to fall sometime during the deposition of the black shale facies. Lowering of sea level brings about a proportionate decrease in reflux, and salinities begin to increase in the basin. The ratio of calcium ions in the influx layer to magnesium ions in the refluxing brine layer is less, and mixing at the brine interface, combined with mass action effects, causes precipitation of more dolomite than calcite. The increasing salinity within the basin depresses evaporation rates causing a reduction in volume of influx. With less water moving into the basin, there is a proportionate decrease in the amount of organic matter swept in, and the color of the sediment layer deposited is gray rather than black. As the regressive sea retreats across previously winnowed source areas the supply of detritus is reduced, and silt and sand-size particles predominate over clay minerals. Further lowering of sea level and increase in salinities brings about deposition of calcium sulfate and, finally, sodium chloride in the evaporite basin. Extreme lowering of sea level finally restricts the carrying capacity of the accessway to a point where reflux is halted, maximum salinities are achieved, and potash deposits form in the deepest part of the basin. Lack of reflux allows contemporaneous development of algal mounds or banks in the shallows on the shelf. The horizontal salinity gradient from open ocean to the distal end of the basin is strongly developed during this phase. Seawater, moving slowly into the basin across the long and somewhat circuitous accessway, is highly concentrated by the time it reaches the shelf. This warm,

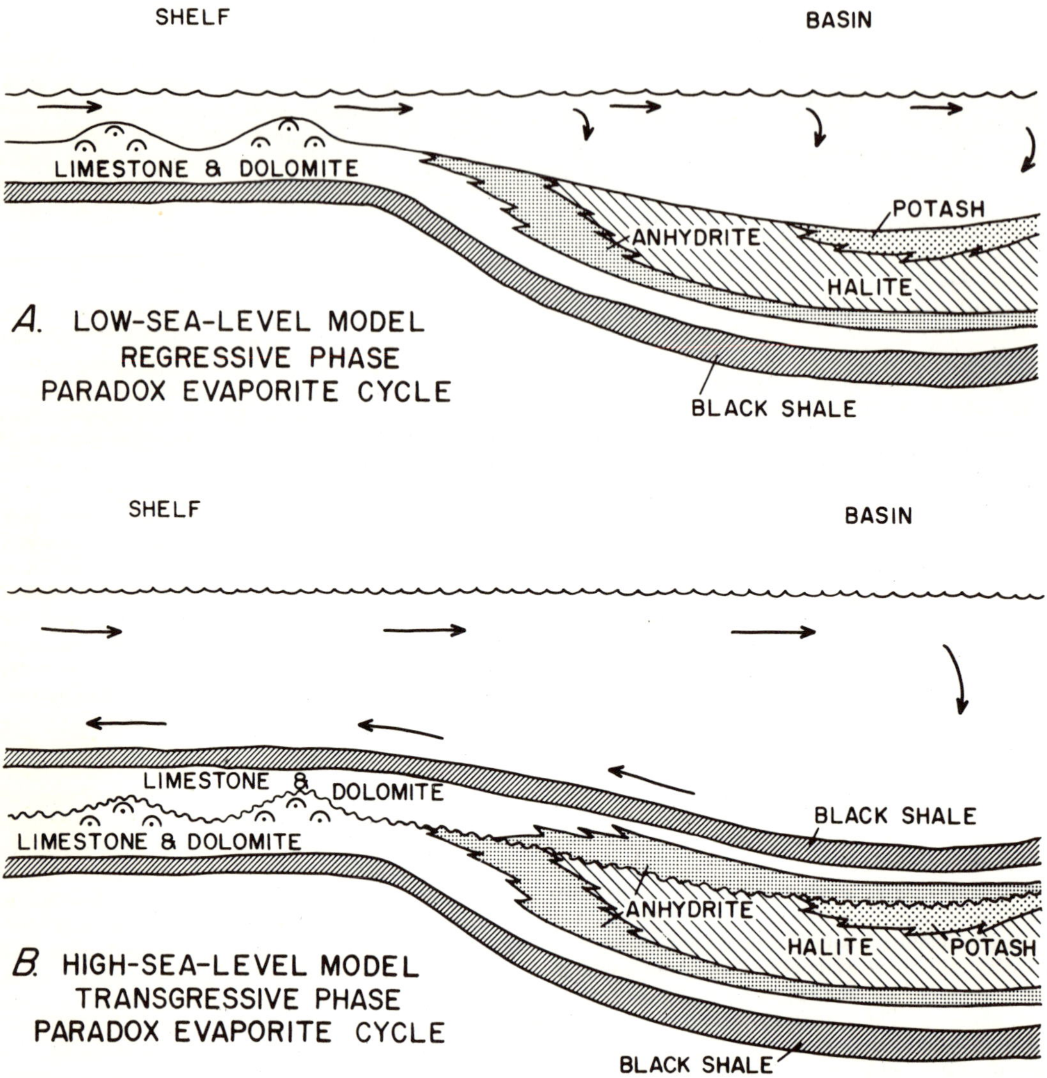

Figure 13. Depositional model of the Paradox basin during (a) the regressive phase and (b) the transgressive phase showing relationships of various basin and shelf facies. Wavy line represents disconformity. Symbol in the shelf limestone and dolomite represents algal mound development.

clear, briny water, carrying a constant supply of nutrients, provides the ideal environment for algal growth while the high salinities eliminate most other competitive forms. Because this current is also vertically stratified, calcium sulfate may deposit in low areas surrounding the algal mounds. In summary, synchronous facies at the close of the regressive phase are (1) algal limestone, (2) dolomite, (3) anhydrite, (4) halite, and (5) potash.

B. Transgressive phase (Fig. 13b). The regressive phase of deposition is terminated by a relatively abrupt rise in sea level. With higher sea level the basin begins to reflux. The reflux layer of toxic brine moves out across the shelf area killing the algal growths. This interruption of sedimentation is represented by a disconformity in shelf area. Reflux creates a deficit in the basin salt budget which reduces salinities. Continued loss of salt by reflux undersaturates bottom brine layers, and corrosion of the previously deposited salt layer begins. The dissolution surface in the basin develops synchronously with the disconformity on the shelf. The first sediments of the transgressive phase on the shelf are foraminiferal-pelletal limestones which grade basinward into "earthy" dolomite, and anhydrite. Limestone pellets form as the result of mixing of the colder carbonate-charged influx layer and the warmer refluxing brine. With further rise in sea level, the corresponding increase in reflux of magnesium-rich brine from the basin results in deposition of most carbonate as dolomite. The strongly transgressing sea begins to churn up sand and silt from formerly emergent source areas. Silt-laden influx water rides across the heavier basin brine depositing extremely well sorted detritus throughout the basin. At maximum high sea level large quantities of marine plankton are swept into the basin along with clays and silts from the higher reaches of the source areas. The accompanying increase in reflux completely covers the shelf with a brine which is anoxic and rich in hydrogen sulfide. At this stage, reflux has extended the euxinic environment of the basin onto the shelf, and only extremely hardy organisms such as sponges survive. A layer of black, organic-rich mud blankets both the basin and the shelf at the culmination of the transgressive phase of deposition.

The model described here is very similar in construction to that of Welsh (1963, Fig. 148); however, it differs greatly in operation. In Welsh's concept of deposition during a Paradox cycle there is no lateral facies change, and the sequence of limestone, dolomite, anhydrite, and salt exist only in a vertical dimension (op. cit., p. 489). Welsh stated, "The writer's interpretation of the salt basin is one of shallow water in which the concentration reached and remained at saturation for the particular salt until the observed thickness of the individual evaporite unit was deposited (op. cit., p. 487)." Apparently, Welsh believed that each evaporite cycle was the result of filling up the evaporite basin, cutting off influx, and allowing complete desiccation of basin water so that first a layer of carbonate, then sulfate, and finally chloride was deposited. Such a static system cannot explain beds of pure halite, several hundred feet in thickness, which are commonplace in the Paradox cycles. Under static conditions it would require a basin nearly 200,000 feet deep to deposit a halite layer 100 feet thick. Accordingly a static evaporite model is untenable with observed relationships in the Paradox basin.

Certain concepts of the Paradox model, as used in this paper, represent a radical departure from previous interpretations by Peterson and Ohlen (1963), Peterson (1966), Peterson and Hite (1969), and Elias (1963). The black shale facies is interpreted here as representing deposition during culmination of the high sea level or transgressive phase of the cycles when basin reflux was greatest. A similar interpretation was made by Pray and Wray (1963, p. 216) for their sponge facies which is probably the shelf equivalent of the black shale facies of the evaporite basin. These authors stated, "We suggest that these rocks were deposited in an anaerobic environment of but slight bottom turbulence. *** We suggest that this facies represents deposition below sill depth at times when the more saline stagnant waters of the Paradox basin to the east and north transgressed highest onto the shallower depositional province represented by the bulk of the Honaker Trail sequence." They also believed that this facies was deposited when water was deepest in the shelf area. The interpretation of depositional conditions during the development of the algal carbonates, as presented here, also differs from that of some of the previously mentioned authors. Evidence suggests that this facies represents conditions of low sea level and high salinities, and that it was deposited contemporaneously with halite and/or potash.

APPLICATION OF BROMINE GEOCHEMISTRY TO SEARCH FOR ALGAL BUILDUPS

Moundlike buildups of algal limestone constitute the major reservoir rock in the petroleum fields of the Paradox basin. Although nearly 30 depositional cycles are known in the area, each of which might be considered a potential host of an algal facies, only a few are known to contain buildups of this facies. Thus, it seems that certain cycles are favorable for the development of this facies and others are not. The models described in preceding sections of this paper suggest that the algal facies developed during the regressive phase of a cycle when most reflux was cut off. The key to exploring for an algal facies, then, might be the determination of which cycles were characterized by a high rate of reflux and which had minimum or no reflux. One simple way of evaluating the favorability of cycles, such as those in the Paradox basin, is on the basis of presence or absence of potash deposits. If a cycle contains a potash deposit then it should have experienced relatively little reflux during that phase of deposition. In the Paradox basin, potash deposits occur in evaporite cycles 5, 6, 7, 9, 13, 16, 18, 19, 20, 21, 24, and 27 (Hite, 1961). Therefore, all these cycles might somewhere contain an algal limestone facies. At present a productive algal facies is known in cycle 5 (lower Desert Creek zone), and possibly in cycles 9 and 19. Production of petroleum has also been established in an algal facies for cycles 2 and 3 (upper and lower Ismay zone) which do not contain potash deposits. The cycles containing the Ismay mounds and perhaps others may have experienced periods of little or no reflux even though potash deposits were not deposited. A more precise means of determining the amount of reflux during the regressive phase of a cycle involves the use of bromine geochemistry. Without going into a detailed discussion of theory it is sufficient to say that normally the bromine content of chloride minerals increases in proportion to concentration of the parent brine. If the distribution of bromine in a salt bed shows no increase from base to top, then it can be assumed that bromine was being lost from the basin by reflux. If the bromine profile shows a rapid increase from base to top of the salt bed then reflux was minimal or absent. One of the best developments of an algal facies in the Paradox basin depositional cycle occurs in the lower Desert Creek zone which is equivalent to evaporite cycle 5. The halite of this cycle, which contains a potash deposit of economic importance, has a flat sloping bromine profile suggesting little or no reflux during the regressive phase of deposition (Fig. 14). In contrast the upper Desert Creek zone (cycle 4) carbonate facies contains a thick buildup of oolitic-pelletal-foraminiferal limestone, much like the initial transgressive facies of other cycles. The bromine profile of cycle 4 halite is almost vertical (Fig. 14). This probably means that the basin was refluxing and there was a close balance between inflow salt and outflow salt plus depositional salt. The refluxing brine during this period created unfavorable conditions on the shelf for algal growth, but did favor chemical precipitation of carbonates by interaction between the refluxing brine and inflowing seawater. At the same time agitation by the overflowing and underflowing currents caused oolitic and pelletal structures to form.

The rate of reflux during the deposition of a halite bed is not always constant. The detailed bromine profile of cycle 2 halite (upper Ismay zone) by Raup (1966, Fig. 3) shows five periods of rapid bromine increase. Each one of these periods represents a low rate of reflux and a favorable time for algal carbonate development. Perhaps this helps explain why at least three separate intervals of algal carbonate have been identified in the upper Ismay zone (Gray, 1967, Fig. 7). Other cycles which show irregular bromine profiles may also contain more than one interval of algal carbonate.

Using the data of Raup (1966); Raup and others (1970), I rated the Paradox evaporite cycles for favorability of algal carbonate development. Each cycle, where data were available, was assigned a reflux index on the basis of the slope angle of the bromine profile for the halite unit of the cycle (Fig. 15). The slope angle was measured from vertical, and a large angle (flat slope) means a low reflux index. Conversely a small slope angle (steep or vertical slope) means a high reflux index. On the basis of reflux index, 14 Paradox cycles may be favorable for development of algal carbonates.

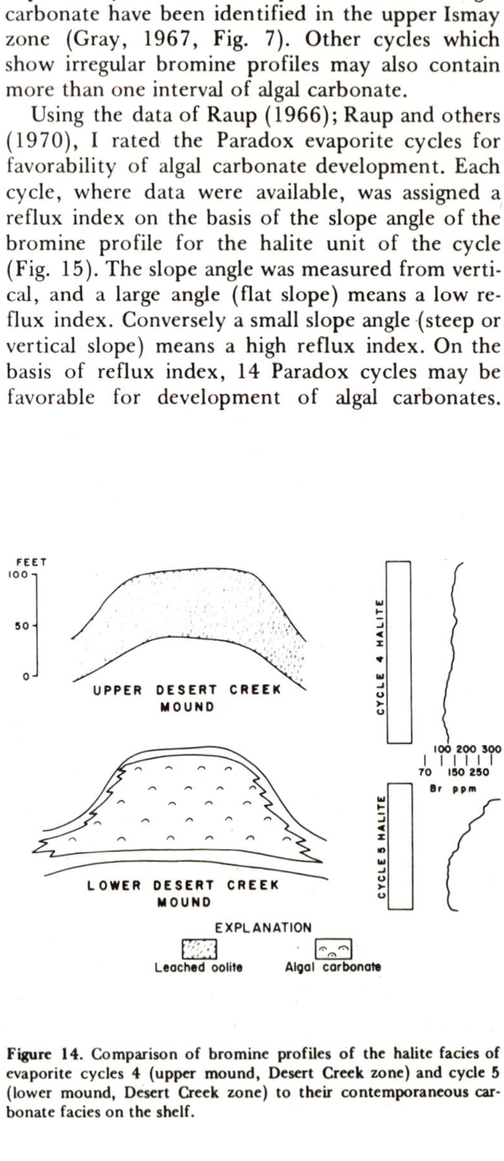

Figure 14. Comparison of bromine profiles of the halite facies of evaporite cycles 4 (upper mound, Desert Creek zone) and cycle 5 (lower mound, Desert Creek zone) to their contemporaneous carbonate facies on the shelf.

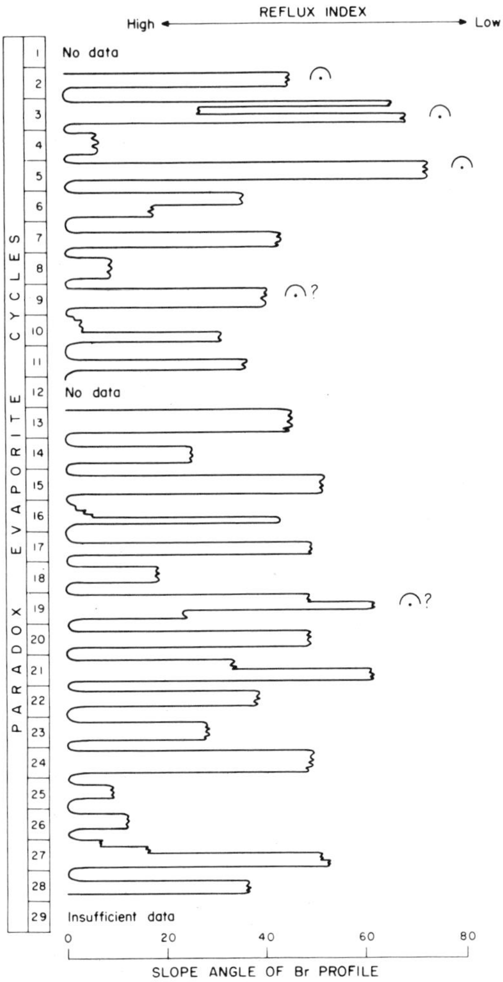

Figure 15. Index or degree of basin reflux during the deposition of the chloride facies of each Paradox evaporite cycle. Reflux index determined by slope angle of the bromine profile for each cycle. Symbol indicates cycles in which the presence of an algal buildup facies has been confirmed; queried where identification is uncertain.

These cycles, ranked in order of favorability starting with the most favorable, are 5, 3, 19, 21, 27, 15, 24, 20, 17, 2, 13, 7, 16, and 9. Cycles 17 and 16 had favorable indexes for a relatively brief period of deposition and if an algal facies is present in these cycles it probably will be thin. At present major developments of algal facies are known in cycle 5 (lower Desert Creek zone) and cycle 3 (lower Ismay zone). A lesser development is known in cycle 2 (upper Ismay zone). Other possible algal developments occur in cycles 9 (upper Akah zone) and 19 (lower Barker Creek zone). The halite units of cycles 19 and 21 are very thick and have very low reflux indexes; therefore, it is suggested that the carbonate facies of these cycles offer potentially important exploration targets for the petroleum industry in the Paradox basin.

The same relationships of an algal carbonate facies to evaporites might also be useful in another sense of exploration. In undrilled sedimentary basins algal carbonate buildups in exposures around the margins of the basin might indicate the presence of halite and potash deposits in the subsurface of the basin.

REFERENCES

Adams, J.E., and Rhodes, M.L., 1960, Dolomitization by seepage refluxion: Am. Assoc. Petroleum Geologists Bull., v. 44, p. 1912-1920.

Baars, D.L., Parker, J.W., and Chronic, J., 1967, Revised stratigraphic nomenclature of Pennsylvanian System, Paradox basin: Am. Assoc. Petroleum Geologists Bull., v. 51, p. 393-403.

Briggs, L.I., Jr., 1958, Evaporite facies: Jour. Sed. Petrology, v. 28, p. 46-56; erratum, p. 384.

Choquette, P.W., and Traut, J.D., 1963, Pennsylvanian carbonate reservoirs, Ismay field, Utah and Colorado, *in* Shelf carbonates of the Paradox basin, a symposium—Four Corners Geol. Soc. 4th Field Conf., 1963: Durango, Colo., Petroleum Inf., p. 157-184.

Duff, P.M.D., Hallan, A., and Walton, E.K., 1967, Developments in sedimentology-v. 10, Cyclic sedimentation: New York, Elsevier Publishing Co., 280 p.

Elias, G.K., 1963, Habitat of Pennsylvanian algal bioherms, Four Corners area, *in* Shelf carbonates of the Paradox basin, a symposium—Four Corners Geol. Soc. 4th Field Conf., 1963: Durango, Colo., Petroleum Inf., p. 185-203.

Fetzner, R.W., 1960, Pennsylvanian paleotectonics of Colorado Plateau: Am. Assoc. Petroleum Geologists Bull., v. 44, p. 1371-1413.

Geffcken, G., 1904, Beiträge zur Kenntnis der Löslichkeitsbeein-flussung: Zeitschr. Phys. Chemie, v. 49, p. 257-302.

Gray, R.S., 1967, Cache field—A Pennsylvanian algal reservoir in southwestern Colorado: Am. Assoc. Petroleum Geologists Bull., v. 51, p. 1959-1978.

Hem, J.D., 1959, Study and interpretation of the chemical characteristics of natural water: U.S. Geol. Survey Water-Supply Paper 1473, 269 p.

Herman, G., and Barkell, C.A., 1957, Pennsylvanian stratigraphy and productive zones, Paradox Salt basin [Colorado Plateau], *in* Am. Assoc. Petroleum Geologists Rocky Mtn. Sec. Symposium on stratigraphic type oil accumulations in the Rocky Mountains: Am. Assoc. Petroleum Geologists Bull., v. 41, p. 861-881.

Herman, G., and Sharps, S.L., 1956, Pennsylvanian and Permian stratigraphy of the Paradox Salt Embayment [Colorado Plateau], *in* Intermountain Assoc. Petroleum Geologists 7th Ann. Field Conf., 1956: p. 77-84.

Hite, R.J., 1960, Stratigraphy of the saline facies of the Paradox Member of the Hermosa Formation of southeastern Utah and southwestern Colorado, *in* Geology of the Paradox basin fold and fault belt—Four Corners Geol. Soc. Guidebook 3rd Field Conf., 1960: p. 86-89.

———, 1961, Potash-bearing evaporite cycles in the salt anticlines of the Paradox basin, Colorado and Utah, *in* Short paper in the geologic and hydrologic sciences: U.S. Geol. Survey Prof. Paper 424-D, p. D135-D138.

———, 1968, Salt deposits of the Paradox basin, southeast Utah and southwest Colorado, *in* Saline deposits: Geol. Soc. America Spec. Paper 88, p. 319-330.

Hite, R.J., and Gere, W.C., 1958, Potash deposits of the Paradox basin [Utah], *in* Intermountain Assoc. Petroleum Geologists Guidebook 9th Ann. Field Conf., 1958: p. 221-225.

King, R.H., 1947, Sedimentation in Permian Castile Sea [U.S.]: Am. Assoc. Petroleum Geologists Bull., v. 31, p. 470-477.

MacArthur, C.G., 1916, Solubility of oxygen in salt solutions and the hydrates of these salts: Jour. Phys. Chemistry v. 20, p. 495-502.

Momper, J.A., 1966, Stratigraphic principles—with some applications to the Permo-Pennsylvanian of the Denver basin, *in* Wyoming Geol. Assoc. 20th Ann. Conf., Symposium on recently developed geologic principles and sedimentation of the Permo-Pennsylvanian of the Rocky Mountains, 1966: p. 90a-90r.

Ochsenius, K., 1877, Die bildung der steinsalzlager und ihrer mutterlaugensalze: Halle, C.E.M. Pfeffer, 172 p.

Ohlen, H.R., and McIntyre, L.B., 1965, Stratigraphy and tectonic features of Paradox basin, Four Corners area: Am. Assoc. Petroleum Geologists Bull., v. 49, p. 2020-2040.

Peterson, J.A., 1966, Stratigraphic vs. structural controls on carbonate-mound hydrocarbon accumulation, Aneth area, Paradox basin: Am. Assoc. Petroleum Geologists Bull., v. 50, p. 2068-2081.

Peterson, J.A., and Hite, R.J., 1969, Pennsylvanian evaporite-carbonate cycles and their relation to petroleum occurrence, southern Rocky Mountains: Am. Assoc. Petroleum Geologists Bull., v. 53, p. 884-908.

Peterson, J.A., Loleit, A.J., Spencer, C.W., and Ullrich, R.A., 1965, Sedimentary history and economic geology of San Juan basin: Am. Assoc. Petroleum Geologists Bull., v. 49, p. 2076-2119.

Peterson, J.A., and Ohlen, H.R., 1963, Pennsylvanian shelf carbonates, Paradox basin, *in* Shelf carbonates of the Paradox basin, a symposium—Four Corners Geol. Soc. 4th Field Conf., 1963: Durango, Colo., Petroleum Inf., p. 65-79.

Pettijohn, F.J., 1949, Sedimentary rocks: New York, Harper & Row, 526 p.

Pray, L.C., and Wray, J.L., 1963, Porous algal facies (Pennsylvanian), Honaker Trail, San Juan Canyon, Utah, *in* Shelf carbonates of the Paradox basin, a symposium—Four Corners Geol. Soc., 4th Field Conf., 1963: Durango, Colo., Petroleum Inf., p. 204-234.

Raup, O.B., 1966, Bromine distribution in some halite rocks of the Paradox Member, Hermosa Formation, in Utah, *in* Second symposium on salt: Northern Ohio Geol. Soc., v. 1, p. 236-247.

Raup, O.B., Hite, R.J., and Groves, L., Jr., 1970, Bromine distribution and paleosalinities from well cuttings, Paradox basin, Utah and Colorado, *in* Third symposium on salt: Northern Ohio Geol. Soc., v. 1, p. 40-47.

Richter-Bernburg, G., 1955, Uber salinare Sedimentation: Deutsch. Geol. Gesell. Zeitschr., v. 105, p. 593-645.

Runnells, D.D., 1969, Diagenesis, chemical sediments, and the mixing of natural waters: Jour. Sed. Petrology, v. 39, p. 1188-1201.

Scruton, P.C., 1953, Deposition of evaporites: Am. Assoc. Petroleum Geologists Bull., v. 37, p. 2498-2512.

Sheldon, R.P., 1963, Physical stratigraphy and mineral resources of Permian rocks in western Wyoming: U.S. Geol. Survey Prof. Paper 313-B, p. 49-273.

Usiglio, J., 1849, Analyse de l'eau de la Mediterranee sur les cotes de France: Annales de Chimie, 3rd ser., v. 27, p. 92-107, 172-191.

Welsh, J.E., 1958, Faunizones of the Pennsylvanian and Permian rocks in the Paradox basin [Colorado Plateau], *in* Intermountain Assoc. Petroleum Geologists Guidebook 9th Ann. Field Conf., 1958: p. 153-162.

———, 1963, Paradox basin salt disconformities (and their relationship to oil production), *in* Oil and gas possibilities of Utah, re-evaluated: Utah Geol. and Mineralog. Survey Bull. 54, p. 487-494.

Wengerd, S.A., and Matheny, M.L., 1958, Pennsylvanian system of Four Corners region: Am. Assoc. Petroleum Geologists Bull., v. 42, p. 2048-2106.

Wengerd, S.A., and Strickland, J.W., 1954, Pennsylvanian stratigraphy of Paradox Salt basin, Four Corners region, Colorado and Utah: Am. Assoc. Petroleum Geologists Bull., v. 38, p. 2157-2199; correction, v.39, p. 259, 1955.

Zen, E-an, 1960, Early stages of evaporite deposition, *in* Short papers in the geological sciences: U.S. Geol. Survey Prof. Paper 400-B, p. B458-B461.

Reprinted from *Third Symposium on Salt,* **1**, 3–7 (1970)

Wind and Water Depth and Their Bearing on the Circulation in Evaporite Basins

18

Margaretha Brongersma-Sanders[1]
Pier Groen[2]

ABSTRACT

Different types of countercurrent systems may occur in subtropical basins. An anti-estuarine circulation develops if there is a horizontal salinity gradient and the wind effect is negligible (1). An estuarine-line circulation develops if the trade wind is blowing the surface water out of the basin, and salinity differences are negligible (2). Where both a horizontal salinity gradient and a counteracting wind are present, an anti-estuarine circulation dominates where the basin is deeper than a certain critical depth, whereas an estuarine-like circulation dominates in the shallow windward end. If the depth above the sill exceeds the critical depth, the subsurface water flows out (3), but if the depth is less than the critical depth an estuarine-like circulation develops above the sill preventing the subsurface water from leaving the basin (4). Type (2) leads to high fertility in the basin and to the development of bituminous deposits in the pre-evaporite phase. Types (1) and (3) lead to very low fertility. Type (4) leads to high salinities in deep basins, and eventually to the precipitation of thick salt deposits. This type (4) does not lead to high fertility in the basin itself, but an area of high fertility will develop above the sill.

In basins in the subtropics there are different types of countercurrent systems. Perhaps best known is the circulation pattern of the Mediterranean Sea with an inflow along the surface and outflow in deeper layers. The primary cause of this circulation is the dry climate involving an excess of evaporation over precipitation plus runoff. This leads to a horizontal increase in both salinity and density towards the head of the basin, the density gradient in turn effecting a downward slope of the sea surface. This slope causes an inward flow of ocean water due to hydrostatic pressure. The pressure field produced by the surface slope is exceeded at depth by the pressure field due to density, and this results in an outflow of deeper layers. In some basins the outflow is mainly restricted to layers above sill depth.

The inflowing water compensates in part for the outflow, and in part for the loss by excess evaporation, the amount required for compensation of loss by excess evaporation being many times smaller than the amount required for the compensation of the outflow (Sverdrup *et al.*, 1946, p. 147). This countercurrent system is called an anti-estuarine circulation since it is more or less the reverse of the circulation in river estuaries.

But one must realize that the anti-estuarine circulation is not the only possibility in the subtropics. In some subtropical basins an estuarine-like circulation is to be found with an outflow along the surface and an inflow in deeper layers. The driving force of this circulation is the wind. The wind effect is only important in the event of a wind of constant direction; the trade winds are such winds. When a semi-enclosed bay or marginal sea is lying with its long axis more or less parallel to the direction of the wind and the opening is

1. Geologisch en Mineralogisch Instituut, Garenmarkt 1, Leiden, The Netherlands (speaker).
2. Instituut voor Aardwetenschappen der Vrije Universiteit, De Boelelaan 1085, Amsterdam 11, The Netherlands.

downwind, the surface water is driven from the inner parts of the basin towards the open sea. This results in a shortage of water (so to say) at the inner end. To compensate for this, subsurface water flows in, and this water ascends towards the surface at the inner end. In this way an estuarine-like countercurrent system develops. An example of a basin with this type of circulation is the Gulf of Cariaco on the north coast of Venezuela. In this area the predominant wind direction is easterly (Gade, 1961a, p. 29); such winds carry the surface water out of the basin, and this outflow is compensated by an inflow of subsurface water over the sill (Richards, 1960; Gade, 1961a, b). Another example is the Gulf of California; during the winter and spring strong winds predominantly from the northwest drive the surface water out of the Gulf, the outflow being compensated below the thermocline by inflow of Pacific water (Van Andel, 1964, p. 263).

The trade winds blow between the subtropical highs (located at about lat. 30°N. and lat. 30°S.) and the equatorial low pressure belt. They are inclined to be dry in their near-subtropic parts, whereas they are more humid on their equatorial margins. In the dry parts of the trade wind belt there are two different forces that may bring about a countercurrent system: (1) the horizontal density (salinity) gradient as a result of the dry climate and (2) the stress of the trade wind. The wind will intensify the anti-estuarine circulation if the opening is upwind, but it will counteract this circulation if the opening is downwind. In the latter case it is a tug of war between the density gradient and the wind. With regard to this tug of war the question arises, when will one or the other predominate?

At the speaker's request the second author of this paper made calculations on the combined effect of these two factors. These calculations will be published elsewhere in detail (Groen, in press). In this paper we shall present only the results, and their bearing on the deposition of salt.

In order to determine the combined effect of the salinity gradient and the wind, a simple two-dimensional model, constructed parallel to the axis of an elongated basin, was evaluated. If no wind were blowing an anti-estuarine circulation would prevail everywhere in the basin, the deeper water layers flowing towards the opening, the upper layers towards the head of the basin (Fig. 1). If on the other hand, there were no salinity differences and with that no horizontal density differences, and a wind were blowing towards the opening of the basin, a pure wind-driven circulation would prevail everywhere in the basin; the upper layers would move with the wind, whereas the lower layers would flow as a compensation current in the opposite direction (Fig. 2). The stress of the wind on the surface of the water, i.e. the force per unit area, is proportional to the square of the wind velocity. Since this force is distributed over the whole water column beneath the surface on which it acts, the force *per unit volume* of the water is inversely proportional to the height of the water column, i.e. to the depth of the water.

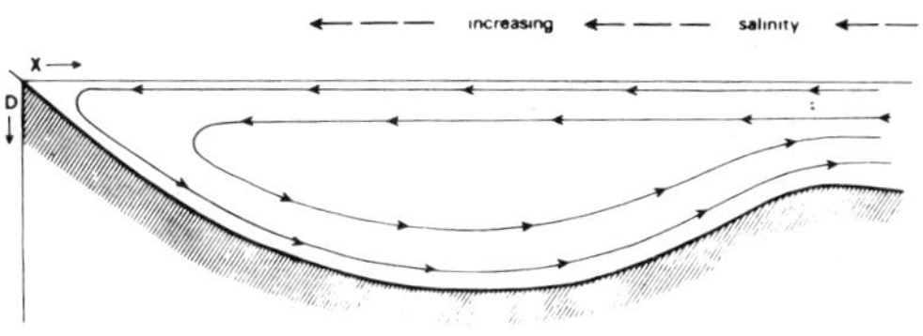

Figure 1. Circulation in basin with horizontal salinity gradient, wind effect negligible; X→ indicates axis and direction of opening of basin.

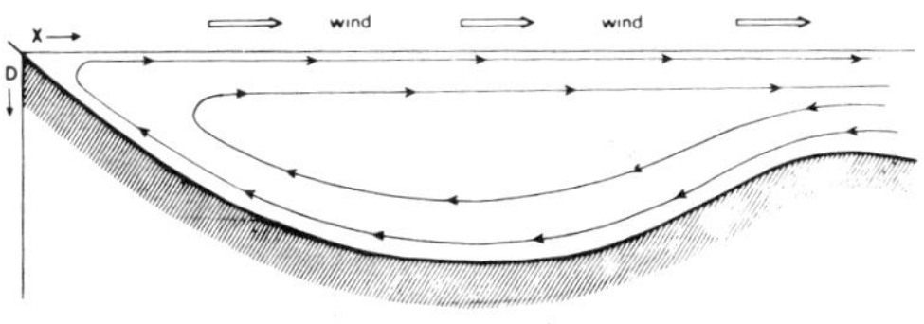

Figure 2. Circulation in basin without horizontal salinity gradient, wind blowing towards opening.

This conclusion is especially important if we now consider the case in which we are interested, namely the case where both a horizontal density gradient and a counteracting wind are present. It now appears that where the depth is small enough the wind-drift will dominate, whereas the anti-estuarine circulation dominates where the depth is great. In order to work this out quantitatively some computations have been made on the basis of the hydrodynamic equations which describe the interplay of forces. For these computations a somewhat simplified model of the hydrodynamic system is adopted, which will, however, give a fair approximation of reality.

It now turns out that, indeed, with a given wind stress and a given density distribution, there is found a critical depth, such that the surface current flows against the wind as long as the depth exceeds this critical depth *(D)*, whereas in places where the depth is less than *D* the wind drift dominates, the surface water flowing with the wind.

The value of the critical depth *(D)* is a function of the surface wind stress (τ_0) and of the vertically averaged horizontal density gradient ($\partial\rho/\partial x$), in such a way that *D* increases with increasing τ_0, but decreases with increasing $\partial\rho/\partial x$. Unfortunately, it is not possible to give a simple formula for *D* as a function of τ_0 and $\partial\rho/\partial x$. It can only be given implicitly, by means of an equation of which it is the root:

$$\frac{4\tau_0}{D^2}\,\text{log. nat.}\frac{D}{z_o} = -g\,\frac{\partial\rho}{\partial x} \quad , \qquad (1)$$

where z_o is the so-called "hydrodynamic roughness" parameter of the water surface and *g* the acceleration of gravity.

If the wind stress (τ_0) and the density gradient ($\partial\rho/\partial x$) are given, one may find the critical depth *D* from equation (1). But one may also put things the other way round, in such a way that, for any given depth *D*, equation (1) gives a critical value of the (vertically averaged) horizontal density gradient, which must be exceeded in the water in order that at the surface it be able to flow against the wind, in an anti-estuarine sense.

As for the deeper layers, it appears from the basic equations of motion that the boundary between the two circulation regimes is tilting down towards the shallower parts of the basin, reaching the bottom in a place where the depth is about one-third times the critical depth.

For a numerical evaluation of equation (1), an empirical relationship is used between the wind stress τ_0 and the wind velocity *W* (τ_0 being proportional to W^2), and for z_o a value 0.5 cm is adopted (valid for moderate and strong winds). In this way one may calculate from (1) for any given wind velocity, pairs of corresponding values of *D* and the horizontal density gradient $\partial\rho/\partial x$, or of *D* and the (vertically averaged) horizontal salinity gradient (the latter follows from $\partial\rho/\partial x$, at least within certain ranges of the salinity, if the temperature is given). A few examples are given in Figure 3 where a water temperature of 25° C is adopted. If the salinity gradients that go with a certain water depth are exceeded, the surface water will flow towards the head of the basin despite the counteracting wind. Lower water temperatures would give smaller values of the salinity gradients with the same *D* and *W*.

D, m ↓ \ Wind, m/sec →	6	8	10	12
20	1,2	2,15	3,4	4,85
50	0,21	0,38	0,59	0,86
80	0,09	0,16	0,24	0,35

Salinity gradients in ‰ per km

Figure 3. Critical values of (vertically averaged) horizontal salinity gradients in % per km. Water temperature 25°C.

It should once more be stressed that the above results are based on a simplified model of the processes involved and therefore should be considered only as reasonable approximations. Nevertheless, they give a picture which is thought to be quantitatively correct.

Now what is the importance of this approach for a better understanding of the geology of evaporites? Assume a deep basin in the subtropics with a horizontal salinity gradient, and a steady counteracting wind. In such a basin either a circulation as shown in Figure 4 or a circulation as shown in Figure 5 will develop depending on whether the

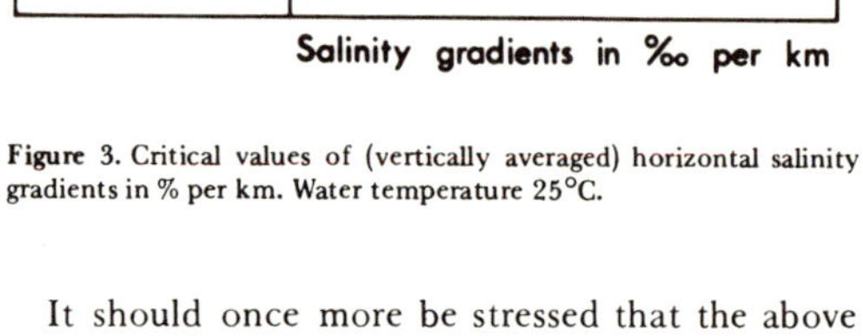

Figure 4. Circulation in basin with horizontal salinity gradient and counteracting wind. Water depth above sill larger than critical depth.

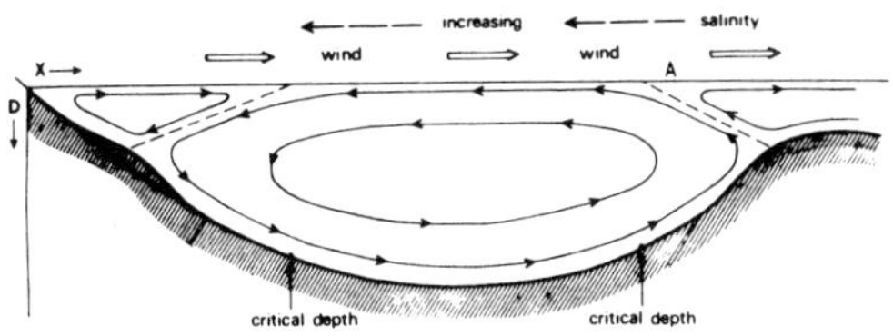

Figure 5. Circulation in basin with horizontal salinity gradient and counteracting wind. Water depth above sill smaller than critical depth. Upwelling at A.

water depth above the sill is larger or smaller than the critical depth corresponding to the locally prevailing density gradients and wind. In the latter case an anti-estuarine circulation dominates in the deeper parts, whereas above the shallow sill the wind is the winning party. This wind-driven estuarine-like circulation is then lying as a watch-dog above the shallow sill, preventing the outflow of the deeper saline layers of the basin, for above the sill a subsurface current is flowing in. By far the greater part of this inflowing subsurface water returns to the ocean before entering into the basin. Only a small percentage (an amount about sufficient to compensate for the loss by evaporation) will enter and mix with the highly saline basin water. The composition of the basin water will hardly be changed by this inflow, even if the latter has the salinity and the composition of normal sea water.

The watch-dog system above the shallow sill combined with strong evaporation and with an anti-estuarine circulation in the deeper parts provides a mechanism that will lead to high salinities in deep basins. In this respect it is noteworthy that in recent years various authors have argued for a deep water origin of thick rock salt deposits. At many times during the geologic past, evaporites, in thin layers, mostly consisting of gypsum, probably have been formed in the shallow environment. It is unlikely, however, that this also applies to the thick halite accumulations which developed from time to time in the geologic past. Richter Bernburg (1960), Yanshin (1961), Pannekoek (1965), and others provided convincing arguments that the basins of sedimentation were deep when the precipitation of halite started, at least as deep as the total thickness of the rock salt deposits. Their arguments for deposition in the deep environment are convincing, but the conditions under which the salinity in deep basins can increase to very high values were not clear. We may suggest that the trade wind played an important part in deposition in such basins. If the opening of the basin was downwind, as long as the sill depth exceeded the critical depth, the highly saline subsurface water could leave the basin; but as soon as it became smaller an estuarine-like circulation would have developed over the sill (at least in the months of strong winds) preventing the outflow of saline water. This must have led to a sharp increase of salinity in the basin.

Another conspicuous problem in the geology of salt is the frequent association of evaporites with bituminous sediments, the bituminosity being particularly characteristic in deposits of the pre-evaporite phase. This problem has already been discussed by the first author at the "Symposium on the geology of saline deposits" in Hannover 1968 (Brongersma in press, further literature in Brongersma, 1968), but an addition must be given in the light of Groen's calculations.

The inorganic nutrients required for the growth of phytoplankton are extracted from the sea in the surface layers by living organisms, and returned to the sea in deeper layers by decaying organisms; this may lead to an increase of the nutrient content with depth. Under certain circumstances the biological process also leads to a shifting of nutrients in a horizontal direction. This occurs if the surface and the subsurface layers flow in opposite directions.

Figure 6 diagrammatically illustrates the process. In the surface water the nutrient content decreases in the direction of flow by the removal of nutrients, whereas in the subsurface water the content increases in the direction of flow by the input of nutrients. If the surface and subsurface layers flow in opposite directions the sum increases in the direction in which the subsurface water is flowing. In basins with an estuarine-like circulation where the subsurface water ascends and overturns in the inner parts (Fig. 2), the sum becomes higher and higher in the course of time until a steady state is

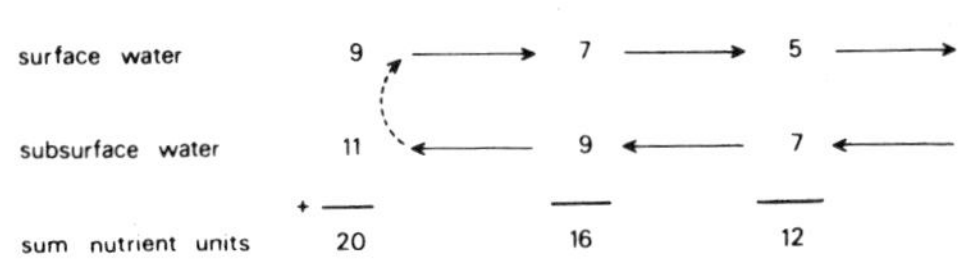

Figure 6. Shifting of nutrients in countercurrents (simplified model, influence of turbulence neglected).
Figures denote content of nutrient units in water, arrows indicate direction of flow.

reached. This produces high productivity in large parts of the basin. On the contrary, an anti-estuarine circulation (Fig. 1) leads to impoverishment, the nutrients being carried out of the basin towards the ocean (cf. Redfield, Ketchum & Richards, 1963, p. 64, Fig. 13 on the nutrient content of the Mediterranean Sea).

High productivity in the surface water leads to high oxygen consumption in the waters below, where large amounts of dead plankton sink down and decay. If high consumption is combined with low oxygen supply anoxic conditions develop. In this way bituminous sediments can be formed in basins with an estuarine-like circulation; this circulation predominates in the pre-evaporite phase when the salinity differences are low.

When speaking in Hannover the first author assumed that with increasing salinity the estuarine-like circulation (Fig. 2) would turn into the anti-estuarine circulation (Fig. 1) involving impoverishment of the basin and disappearance of the bituminous layer. The calculations by Groen have shown, however, that the situation is more complicated, as the circulation of Figure 2 turns either into that of Figure 4 or into that of Figure 5. The circulation of Figure 4 leads to an impoverishment of the basin indeed. In the case of Figure 5 there is hardly any gain from or any loss to the ocean, but there is upwelling above the shallow sill (A in Fig. 5), and this means a region of high fertility at the entrance of the basin. With an occasional change in the force or the direction of the wind, some of the plankton-rich water may enter into the basin leading to the development of a thin bituminous layer between the evaporites. All the same the most appropriate conditions for development of bituminous sediments occur with the estuarine-like circulation of Figure 2 predominating in the pre-evaporite phase.

REFERENCES

Brongersma-Sanders, Margaretha, 1968, On the geographical association of strata-bound ore deposits with evaporites: Mineral. Deposita (Berl.), v. 3, p. 286-291.

———, in press, Hydrological conditions leading to the development of bituminous sediments in the pre-evaporite phase. Paper read at *Symposium on the geology of saline deposits,* 1968 Bundesanstalt f. Bodenforsch. and Nat. Res. Div. of Unesco, Hannover.

Gade, H.G., 1961a, Informe sobre los condiciones hidrograficas en el Golfo de Cariaco, para el periodo que empieza en Mayo y termina en Noviembre de 1960: Bol. Inst. Oceanogr. Cumana, Venezuela, v. 1, p. 27-47.

———, 1961b, Further hydrographic observations in the Gulf of Cariaco, Venezuela. The circulation and water exchange: Bol. Inst. Oceanogr. Cumana, Venezuela, v. 1, p. 359-395.

Groen, Pier, in press, A simplified theory of the combined effect of an anti-estuarine circulation and a superimposed counteracting wind drift: Proc. Kon. Ned. Akad. Wetensch.

Pannekoek, A.J., 1965, Shallow-water and deep-water evaporite deposition: Am. Jour. Sci., v. 263, p. 284-285.

Redfield, A.C., Ketchum, B.H., and Richards, F.A., 1963, The influence of organisms on the composition of sea water: *in* The sea: New York, John Wiley and Sons, pt. 2, p. 26-77.

Richards, F.A., 1960, Some chemical and hydrographical observations along the north coast of South America: Deep-Sea Res., v. 7, p. 163-182.

Richter-Bernburg, G., 1960, Zeitmessung geologischer Vorgange nach Warvenkorrelationen im Zechstein: Geol. Rundsch., v. 49, p. 132-148.

Sverdrup, H.U., Johnson, M.W., and Fleming, R.H., 1946, The oceans: New York, Prentice Hall, 1087 p.

Van Andel, T.H., 1964, Recent marine sediments of the Gulf of California: *in* Marine geology of the Gulf of California: Am. Ass. Petroleum Geologists Mem. 3, p. 216-310.

Yanshin, A.L., 1961, On the depth of salt-bearing basins and some problems concerning the formation of thick salt deposits (in Russian): Geologija i Geophys., Novosibirsk, no. 1, p. 3-15.

Evaporite Petrology

II

Editors' Comments on Papers 19, 20, and 21

19 **Dellwig:** *Origin of the Salina Salt of Michigan*

20 **Wardlaw and Schwerdtner:** *Halite–Anhydrite Seasonal Layers in the Middle Devonian Prairie Evaporite Formation, Saskatchewan, Canada*

21 **Shearman:** *Recent Halite Rock, Baja California, Mexico*

Rhythmic Bedding

A most striking feature of the majority of evaporite deposits is the cyclic repetition of rock types within the vertical sequence. Individual cycles, consisting of two to five lithologic types, range in thickness from less than 1 mm to several hundred meters. The Permian Zechstein evaporites of northern Europe have been studied in detail since the middle of the nineteenth century and have been shown to consist of four major cycles, each of which contains units representing clastic, carbonate, calcium sulfate, halite, and potash deposition. The Pennsylvanian Paradox sequence of Utah and Colorado is perhaps the most remarkable example of cyclic repetition, for within it, twenty-nine major cycles have been identified, and most of them show the complete range of deposits from clastic to potash units (Hite, 1970, Paper 17). Superimposed on the major cycles of the Zechstein, Paradox, and other sequences are smaller cycles that match, in many ways, the repetitions of the larger units.

The cyclic repetitions of rock types, in particular the cycles on a scale of millimeters and centimeters, are referred to as "rhythmic bedding," and it is toward an analysis of these minor cycles that much research has been directed. This section falls naturally into two parts: the descriptive petrography and explanation of the formation of halite–anhydrite cycles, and an analysis of the regional, stratigraphic, and climatological significance of those repetitions of lithology.

Petrology and Formation of Halite–Anhydrite Cycles

Many halite deposits are characterized by regular and repeated alternations of thin beds of halite and fine laminae of anhydrite. These oscillations have been described from such diverse deposits as the Cambrian of Siberia; the Silurian of Michigan,

Ohio, and New York; the Devonian of western Canada; the Mississipian of eastern Canada; the Pennsylvanian of Utah and Colorado; the Permian of New Mexico, Texas, Kansas, and northern Europe; the Triassic of southern Germany; the Jurassic of the Gulf of Mexico; the Cretaceous of Brazil and Gabon; and the Plio-Pleistocene of the Dead Sea region. The origin of these alternations is, therefore, a problem of appreciable significance in the study of the geology of evaporites.

The three papers in this section present the results of three different approaches to this single problem: one a solely petrographic study, the second a geochemical and petrographic investigation, and the third a consideration of lithologic and textural similarities between ancient and modern examples of anhydrite–halite alternations. If the authors reach different conclusions, it serves to demonstrate that perhaps the way to solve the problem is by means of a diverse and varied investigation, not through a single avenue of attack.

L. F. Dellwig, now of the University of Kansas, has been a student of salt deposits for twenty years and has studied them in the United States and, more recently, in Germany. The contribution from which the following excerpts were taken was voted the best paper in the *Journal of Sedimentary Petrology* for 1955, and today still stands as a major contribution to our understanding of halite sedimentation, as much for the petrographic care with which Dellwig approached the problem as for the conclusions that were drawn.

Dellwig examined cores of Silurian Salina salt from two holes drilled in the Michigan Basin. The core from the center of the basin showed alternations of halite and anhydrite–dolomite; the sample from nearer the margin of the basin consisted of halite so recrystallized and deformed that sedimentological inferences were more difficult to draw. In subsequent work in the same basin, Dellwig and Evans (1969) described a variety of sedimentary structures from within the halite in cores or exposed in mines. They were able to show that the differences between halite from the center of the basin and halite from the margin of the basin were the result of different sedimentological regimes.

N. C. Wardlaw and W. M. Schwerdtner, in their study of halite and anhydrite from the Devonian Prairie Evaporite of Canada (1966), supplemented careful petrography with analyses of bromine in halite in an effort to establish precisely the conditions of salinity that produced this major evaporite deposit. They concluded that the halite–anhydrite couplets formed in deep water and that temperature changes influenced precipitation. Seasonal temperature changes were also a feature of Dellwig's explanation of the microstratigraphy of the Salina evaporites.

In the final paper, D. J. Shearman, once again the iconoclast, questions the interpretations of Dellwig, Wardlaw, Schwerdtner, and others who favor the idea that laminated evaporites formed in deep water. In the supratidal flats near the delta of the Colorado River in the Gulf of California, halite–gypsum alternations have formed in shallow depressions as a result of occasional, but not seasonal, or even regular, flooding by marine water. The laminated halite–gypsum rock is similar to the halite–anhydrite rock described by Wardlaw and Schwerdtner, so similar that Shearman uses those authors' own words to describe the rock examined in his investigation!

Reprinted from *Jour. Sed. Petrol.*, **25**, 83–93, 95–102, 107–110 (1955)

19

ORIGIN OF THE SALINA SALT OF MICHIGAN[1]

LOUIS F. DELLWIG
University of Kansas, Lawrence, Kansas

ABSTRACT

The purpose of this study was to determine the manner of deposition of the Salina salt, including the approximate temperature of deposition and the origin of the dolomite-anhydrite laminae.

Specimens were obtained from drill cores from the Michigan basin and from the Detroit mine of the International Salt Company. The bulk of the data was obtained through thin section study and this was supplemented by the examination of polished slabs, water-insoluble residues, and sections of core from which the salt had been dissolved.

Pyramidal shaped hopper crystals of halite, comparable to those crystals produced in the manufacture of grainer salt and those formed in the manufacture of solar salt, were identified in the Salina salt. The pyramidal forms were outlined by liquid inclusions, and from these inclusions the temperatures of formation were determined as being between 32.0 and 48.4° C.

The salt was deposited in layers or bands of three distinct types: (1) cloudy layers of inclusion-rich pyramidal shaped hopper crystals of halite, (2) clear layers of inclusion-free halite, and (3) laminae of anhydrite and dolomite. These units are in part obscured in the recrystallized salt.

The alternation of bands of clear and cloudy halite was the result of temperature changes probably related to the seasons. The salt crystals, initially formed on the surface, grew as pyramidal shaped hoppers. Until the brine became saturated, settling hopper crystals were dissolved. After the saturation point had been reached pyramidal shaped hopper crystals accumulated on the bottom of the basin to form a layer of cloudy salt. With subsequent cooling the brine passed into the supersaturated state and settling hopper crystals provided nuclei for clear overgrowths. These crystals which grew on the bottom of the basin are inclusion-free and form the layers of clear salt. The return of higher temperatures followed again by lower temperatures would cause a repetition of this sequence. This cycle could have been interrupted at any time by the addition of brine to the basin or by periods of unusual climatic conditions.

The deposition of the anhydrite-dolomite laminae may be considered to have been the result of the influx of normal sea water. The addition of brine to the basin as a surface layer would have permitted the maintenance of life in this brine until the concentration reached the point at which mass extermination took place. Subsequent decomposition of the remains caused reducing conditions to develop in the basin and iron precipitated as pyrite. Indicative of this influx is the association with the laminae of carbonaceous material, pyrite, and a petroliferous odor and the lack of a recognizable relationship between the clear or cloudy salt and the anhydrite-dolomite laminae. Iron associated with the halite layers is in the form of hematite, indicating that oxidizing conditions were normal for the basin.

INTRODUCTION

The Salina formation includes the earliest evaporites deposited in the Michigan basin. The Salina formation and the overlying Bass Island dolomite comprise the Cayugan group of the Silurian system. In the Michigan basin the Salina formation has been divided into seven distinct units (Landes, 1945). Three units are of shaly dolomite or dolomitic shale, three are predominantly of salt, and the seventh is a basal dolomite which contains increasing amounts of salt toward the center of the basin.

This study initially centered itself around drill cores filed in the core library of the University of Michigan. A comprehensive examination of cores which cut through the Salina formation showed that the thick salt beds could generally be divided into two contrasting units. The one unit (fig. 1) is even bedded salt with layers of salt three to eight centimeters thick alternating with tissue thin laminae of anhydrite and dolomite. The halite crystals average less than one centimeter in diameter. Generally there

[1] Condensation of Ph.D. Thesis at University of Michigan.

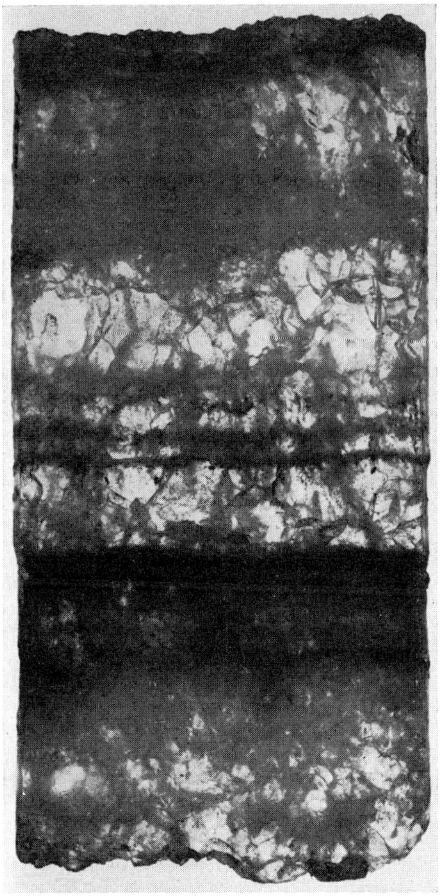

FIG. 1.—A section of core from the Glen Bradley No. 4 well, Newaygo, Co., Michigan; 273 feet above the base of the Salina formation. Core depth—4882 feet. The black areas are predominately anhydrite; the dark grey areas are anhydrite intermixed with halite. Both the clear and the cloudy areas are halite. ×0.7

is a crude layering of clear and cloudy salt. Megascopically the laminae are continuous and flat. In the other unit (fig. 2) the bedding is extremely irregular. The individual halite crystals average several centimeters in diameter, some attaining a diameter in excess of five centimeters. Almost all of the crystals are clear. The laminae are distorted and irregular. Mineralogically, the units show almost no variation. In addition to dolomite and anhydrite, there are contained within the laminae quartz, pyrite, celestite, polyhalite, and carbonaceous material. Complex salts are rare, precipitation apparently being terminated before the development of the polyhalite phase in the theoretical sequence of deposition of salts (table 1).

The first unit (even bedded salt) is now essentially as it was originally deposited; the second unit (irregularly bedded salt) has undergone extensive recrystallization. The disruption or distortion of the laminae indicates that the recrystallization is not a process associated with the original deposition of the halite and the anhydrite and dolomite laminae, and is not involved in the development of a theory of deposition for the Salina salt.

Some phases of Salina salt deposition have previously been described. Near shore conditions of deposition have been studied by Kaufmann and Slawson (1950) and a second paper on deposition in the marginal areas is in preparation by the same authors. The economic importance of the Salina salt has been responsible for several additional studies of a more general nature (Cook, 1914; Grabau and Sherzer, 1910; Sherzer, 1911). The Salina salt of Michigan is also considered in the more general discussions of evaporites by Grabau (1913, 1920), Phalen (1919), and Lotze (1938).

Acknowledgments

The author is most desirous of expressing his thanks to those who helped make this work possible. Dr. L. I. Briggs has offered many hours of consultation and assistance, especially in the development of the physical-chemical aspects of deposition. Professors Landes and Slawson provided specimens for comparative study and freely shared their unpublished findings concerning salt deposition. The carbonaceous material was examined by Professor Chester A. Arnold. Mr. J. C. Buchen of the Leslie Salt Company furnished helpful data regarding the deposition of solar salt. Through the co-

FIG. 2.—A section of core from the Glen Bradley No. 4 well, 17 feet above the base of the Salina formation. Core depth—5138 feet. In this section recrystallization of the salt has caused the enlargement of the halite crystals, the elimination of the banding of clear and cloudy halite, and the deformation of the anhydrite laminae. The apparent thickness of the laminae is due to their wavy character. The black areas are predominately anhydrite; the dark grey areas are anhydrite intermixed with halite. Both the clear and cloudy areas are halite. ×0.7.

operation of Mr. L. E. Read and his staff at the Detroit mine of the International Salt Company, the mine workings were studied and much valuable information was obtained. Special thanks are due to Mr. D. E. Kaufmann of the International Salt Company. His information on manufacturing processes contributed much to the development of certain theoretical aspects of the problem. He also provided numerous samples for study.

Method of Study

The greater part of the data was collected from the core of the Glen Bradley No. 4 well in Newaygo County, Michigan. The samples were taken from "Unit A" (lowermost salt) as classified by Landes (1945). The well is located in the NE$\frac{1}{4}$ SW$\frac{1}{4}$ SE$\frac{1}{4}$ Sec. 11, T. 12 N., R. 13 W. At this point the Salina formation is approximately 1400 feet thick. The base of the Salina formation was intersected at about 5155 feet. The core is about 3.5 inches in diameter. Although the core is not from the center of the basin, the salt represents conditions of deeper water deposition than does the material from the Detroit mine. Additional cores were studied for substantiating or conflicting evidence.

Examination of the workings of the Detroit mine of the International Salt Company yielded much information which was not available from the core study. The mine shaft is located in Oakwood in the NW$\frac{1}{4}$ Sec. 33, T. 2 S., R. 11 E. Mining is being carried on at a depth of about 1140 feet, approximately 666 feet above the base of the Salina formation in "Unit F" as defined by Landes (1945).

Thin sections were supplemented with polished plates and the latter were especially valuable in showing three dimen-

TABLE 1.—*Theoretical profile for the evaporation of sea water at 25°C (Phillips, 1947)*

Kieserite, carnallite, bischofite			Bischofite zone
Kieserite, carnallite	Anhydrite		Carnallite zone
Kieserite, kainite			
Hexahydrite, kainite			Kainite zone
Epsomite, kainite		Halite	
Epsomite			Potash-free magnesium sulphate zone
Blodite	Polyhalite		
Polyhalite			Polyhalite zone
Anhydrite			Anhydrite zone
Gypsum			Gypsum zone
Carbonates			Basal limestone and dolomite

sional relationships. Water insoluble residues were used to study crystal form and to identify some minerals.

In the preparation of the illustrations all photographs and photomicrographs have been oriented with the top of the section or specimen toward the top of the page.

Geologic History of the Michigan Basin and Its Relationship to Salina Deposition

The initial depression of the Michigan basin, probably related to the Keweenawan disturbance (Pirtle, 1932), resulted from the same northeasterly pressure which was responsible for the development of the Lake Superior geosyncline. The basin is approximately 450 miles in diameter and has a slight elongation in a northwest-southeast direction. The average dip of the beds toward the center is 30 to 35 feet per mile. The basin is bounded on the west by the Wisconsin and Kankakee arches and on the east by the Findlay arch. Downwarping of the basin was periodic, and there is a thickening of the sediments in the central portion. Minor structures have developed parallel to the long axis of the downwarp. Coarse clastics are relatively scarce.

The sediments in the basin range in age from Cambrian to Permian(?). Evaporites were deposited during three periods beginning with the Salina formation of Late Silurian age. The problems of deposition of the subsequent evaporites may or may not be related to those of the Salina salt, for, at least in part, the younger evaporites may have been formed through solution and reprecipitation of the Salina salt.

During the Silurian the greater part of North America remained low with the exception of Appalachia. Limestone, dolomite, and subordinate fine clastic sediments are characteristic of lower and middle Silurian formations in the Michigan basin. The first isolation of the basin was during Niagaran time. Salina deposition began with widespread carbonate muds and anhydrite, although in some places salt is found directly above the Niagaran rocks. Salt was deposited above the carbonate muds and anhydrite and this in turn was followed by dolomitic muds. Deposition of dolomite, anyhdrite, halite, and shale continued until the end of the Silurian period.

The base of the Salina in the center of the basin now lies at a depth of slightly more than 9500 feet. The thickening of

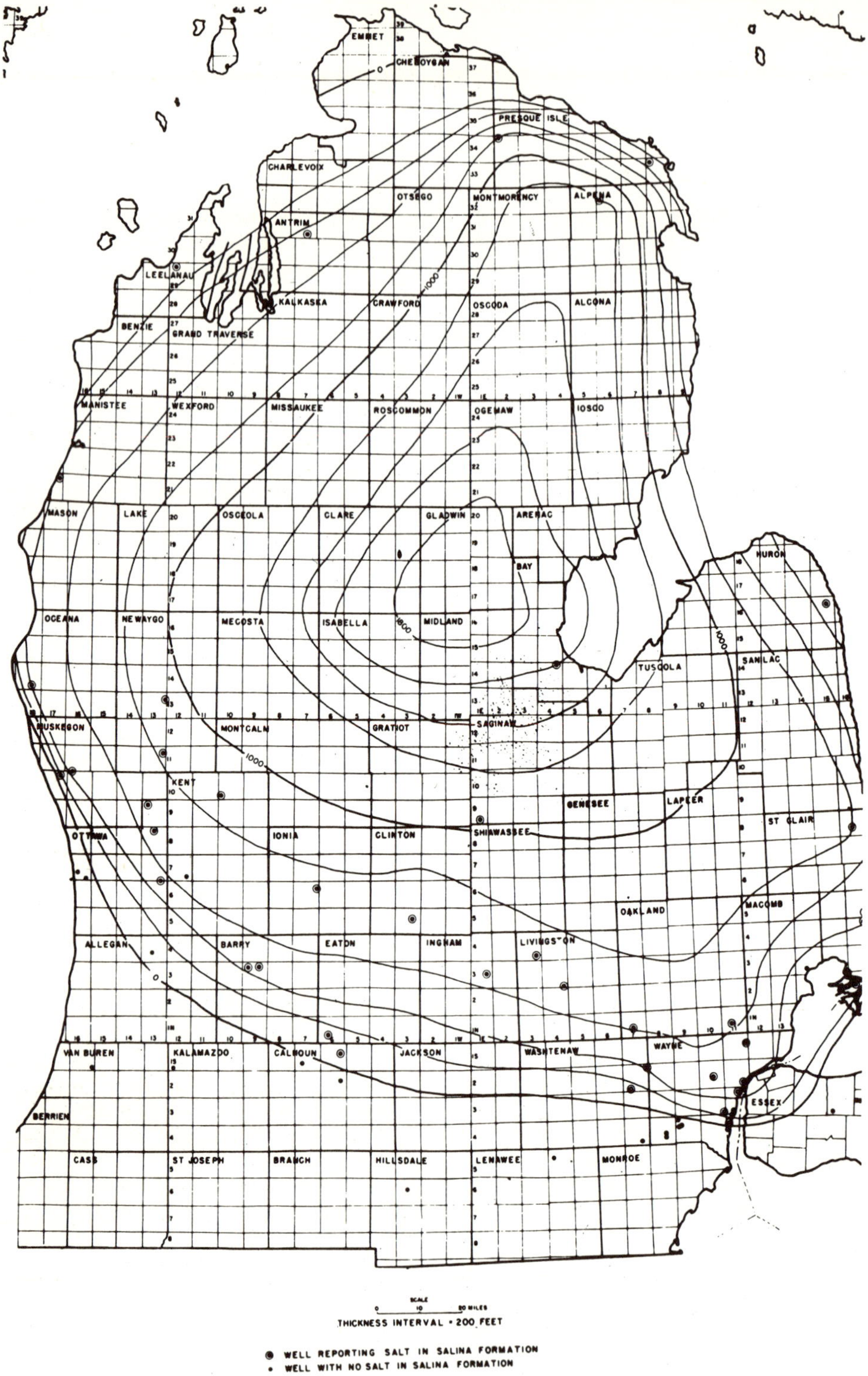

FIG. 3.—Map showing the aggregate thickness of salt in the Salina formation (After Landes, 1945).

the salt toward the center of the basin (fig. 3) combined with the great thicknesses of salt within the basin indicates subsidence of the basin during Salina deposition. It also necessitates the addition of brine during deposition to furnish the great thickness of salt. No seaway through which brine might have been added has yet been identified. Some authors (Dunbar, 1949) place the seaway to the northeast or northwest of the basin although there is no definite evidence of either. Contemporaneous salt and associated gypsum beds are found in eastern Ohio and New York and are probably connected with the Michigan basin beneath Lake Erie between southwest Ontario and Ohio. This presents the possibility of the replenishment of the brine in the Michigan basin through the New York basin.

Most authors agree that the Salina climate was dry (Newcombe, 1932). The postulated conditions of deposition call for a shallow landlocked sea surrounded by a low land mass with bedrock areas of limestone. The replenishment of the water in the basin was apparently intermittent during gradual subsidence of the basin. A delicate balance was maintained between sinking, evaporation, and supply so that at no time did evaporation approach completion.

THE DEPOSITION OF HALITE AND THE BANDED SALT

Introduction

The term, "Jahresringe," has long been used to describe the thin laminae of anhydrite or polyhalite found at relatively close but irregular intervals in salt layers. Early workers (Phillips, 1947) attributed this banding to seasonal deposition. However, some opposition to this theory has developed as a result of the discovery of clay laminae of the same general character which are known to have resulted from an influx of sediment charged waters from an adjacent land mass. This example of a change in the nature of the basin waters as a cause of the deposition of the laminae has led to the postulation that the bands of anhydrite and polyhalite are the result of a similar change—perhaps an influx of water from the land mass or an influx of sea water into the basin.

The banded salt is composed of three distinct units (fig. 1): carbonate and sulfate laminae, clear salt, and cloudy salt. The clear salt and the cloudy salt differ in the manner of crystal growth. The cloudy zones are composed predominately of a peculiar type of halite crystal whose shape is outlined by abundant liquid inclusions (fig. 4). The clear areas are relatively inclusion-free. Normally, in the Glen Bradley core the two types of salt are separated into relatively distinct parallel bands. This relationship is not so pronounced in the Detroit mine. There is no regular sequence or ratio of thicknesses between the two salt units and the anhydrite. In the Glen Bradley core the anhydrite bands are separated by 0.4 to 7.5 centimeters of salt. In the Detroit mine the separation may be as much as 30 centimeters. In general, these bands are bounded on both sides by crystals of cloudy salt.

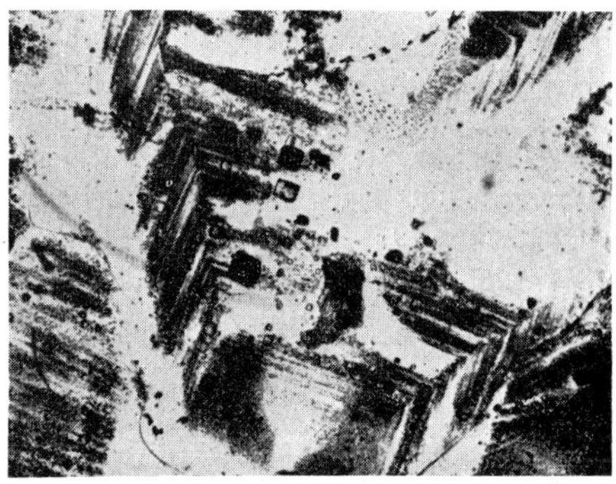

FIG. 4.—Photomicrograph of a thin section showing a pyramidal shaped hopper crystal outlined by brine-filled, negative crystals. The specimen is from the Glen Bradley well core, 264.5 feet above the base of the Salina formation. Core depth—4890.5 feet. Plane-polarized light. ×19.3

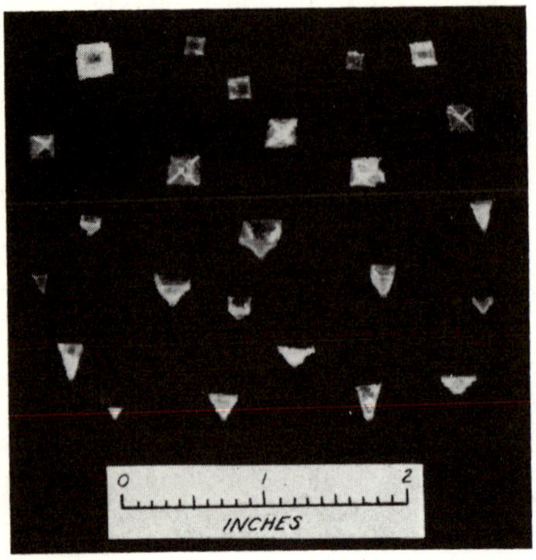

FIG. 5.—Manufactured pyramidal shaped hopper crystals.

Crystal Growth and Deposition of the Salt

That the growth of halite crystals is initiated at the surface of the brine is a fact long recognized. The surface growth of artificial salt crystals in the manufacture of grainer salt is controlled by conditions not unlike those which might exist in a natural basin. Temperatures in this process (49°–60° C) are in excess of those which might be expected in nature but higher temperatures only tend to accelerate crystal growth, and artificially produced hoppers might be expected to be comparable to natural crystals. Industrial usage defines the crystals produced artificially as hopper crystals. The same term, as used by the mineralogist, generally implies a skeletal cube with depressed faces resulting from the greater tendency toward growth along the edges of the embryo cube. As can be seen in figure 5, the industrial crystals differ in form, and result from an entirely different growth process than that responsible for mineralogical hopper crystals.

In the manufacture of grainer salt (Badger and Baker, 1928) crystal growth is brought about by evaporation at temperatures below the boiling point in order to prevent turbulence and to permit the formation of a thin surface film of high-density brine. In this film the halite crystals begin to grow. As growth continues the cube tends to sink under its own weight, although its position at the surface is maintained by surface tension. Because only one face of the cube is in contact with the high density film, growth takes place only along its edges. In this manner, while the crystal sinks, growth continues upward and outward along these edges resulting in the hollow pyramid with the apex pointing downward (fig. 6). When the surface is disturbed, the crystals are broken or swamped and sink. The crystals are extremely delicate and can withstand very little handling. Occasionally a number of crystals in a small area will become attached to one another to form a crust or mat which will eventually sink as a unit. Similar crystals have been observed in the Salina salt. The crystal form is outlined only by the presence of abundant liquid inclusions in the form of oriented negative crystals (fig. 4). For the most part only fragments of crystals

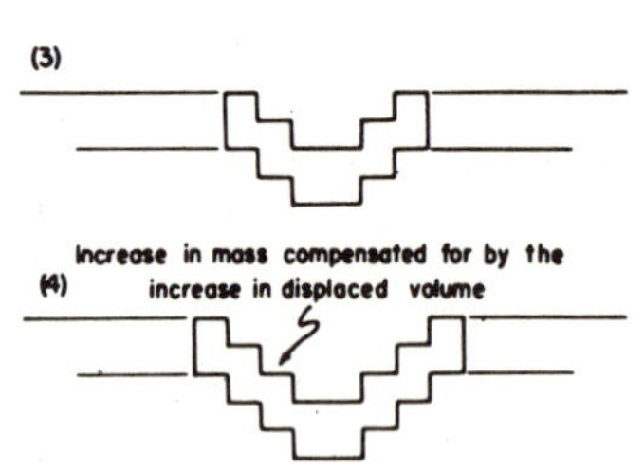

FIG. 6.—Schematic diagram showing in cross section the method of growth of a pyramidal shaped hopper crystal.

FIG. 7.—A pyramidal shaped hopper crystal contained in an oriented overgrowth of clear salt. In the top view the hopper appears in the lower right hand corner with the apex pointing up. In the bottom view the open end of the pyramidal shaped happer is facing out. The specimen is from the Glen Bradley well, 270 feet above the base of the Salina formation. Core depth—4885 feet. ×3.5.

are to be found. The chevron structure (Dellwig and Briggs, 1952) commonly observed in thin section is merely the result of slicing through the corner or edge of a pyramidal shaped hopper.

The maximum measurement across the open end of a pyramidal hopper is 3.6 millimeters, although the size of some of fragmental crystals suggests that the original pyramidal shaped hopper may have been much larger. In some layers it was noticed that there is an orientation of the pyramidal shaped hoppers over a short distance laterally with the vertices of the pyramids pointing downward. This apparently is the result of settling of a portion of a mat or crust which formed through the intergrowth of several individual crystals on the surface. The cloudy layers or masses are the result of lithifaction of an assemblage (generally lacking orientation) of complete and fragmented pyramidal shaped hoppers which grew at the surface of the brine.

The absence of inclusions is not sufficient in itself for concluding that the pyramidal shaped hoppers are not contained in the transparent salt masses. The development of inclusion-free crystals does, however, strongly suggest crystal growth under different conditions for the clear layers as well as for the isolated clear crystals. Isolation of some pyramidal shaped hoppers enclosed in large masses of clear salt yielded cleavage fragments as shown in figure 7. It is apparent that a crystal of this size could not have grown suspended at the surface of the brine. The overgrowth is in crystallographic continuity with the pyramidal shaped hopper. In some cases more than one pyramidal shaped hopper fragment is included in a single large clear crystal. Apparently some fragments acted as a nucleus for overgrowth whereas others have merely been included during growth.

Additional evidence for the development of syngenetic overgrowths on the

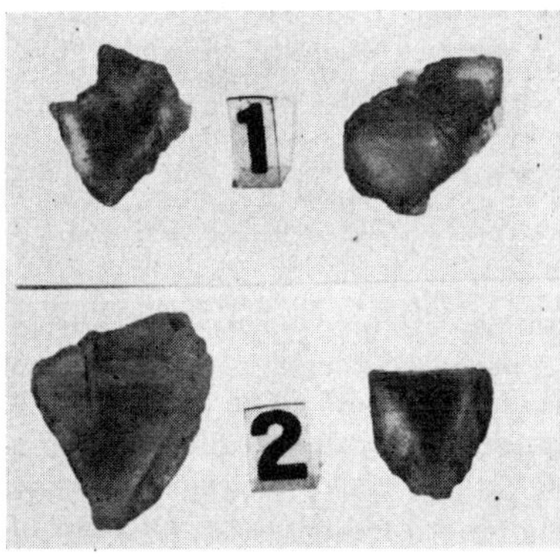

FIG. 8.—Pyramidal shaped hoppers of solar salt. The original hopper which formed at the surface of the brine appears as the dark area of the crystal, and the clear overgrowth appears as the light area. In some of the crystals there is a tendency toward the modification of the hollow pyramid to form a cubic crystal. Natural size.

FIG. 9.—Partly etched section from the Glen Bradley well showing the conformity of the anhydrite in the upper dolomite-anhydrite layer to the underlying halite cubes. The light areas standing out in relief are anhydrite, the dark areas dolomite. The specimen was taken from the Glen Bradley well, 10 feet above the base of the Salina formation, Core depth—5145 feet. Natural size.

pyramidal shaped hoppers can be found in solar salt crystals. Figure 8 shows specimens from two of six sampled localities. In all localities pyramidal shaped hoppers or fragments of pyramidal shaped hoppers are encased in an overgrowth of inclusion-free salt. The crystals are not hollow and it is not possible that surface tension alone could have supported these crystals at the surface of the brine while they grew to this size. The enlargement of the crystals has taken in the bottom of the basin. Not only have the crystals become enlarged, but they have also shown a tendency toward the development of a cube, indicating an entirely different mode of crystal growth.

Figure 9 also attests to syngenetic bottom growth of the salt cubes. The imprint of the cubes of clear halite can be seen in the overlying dolomite and anhydrite, and the undeformed upper surface of the carbonate and sulfate layer attests to the growth of the salt cubes before, or at the latest, during the deposition of this overlying layer. Here again, crystals of this size and form could only be the result of bottom growth.

A further significant feature associated with the pyramidal shaped hopper structures is the absence of crushing or displacement of portions of an individual crystal. At least partial deformation of the pyramidal shaped hopper structure would be expected in the event that this material was subjected to compaction during burial. For the most part the crystals have undergone no deformation. This strongly indicates that little, if any,

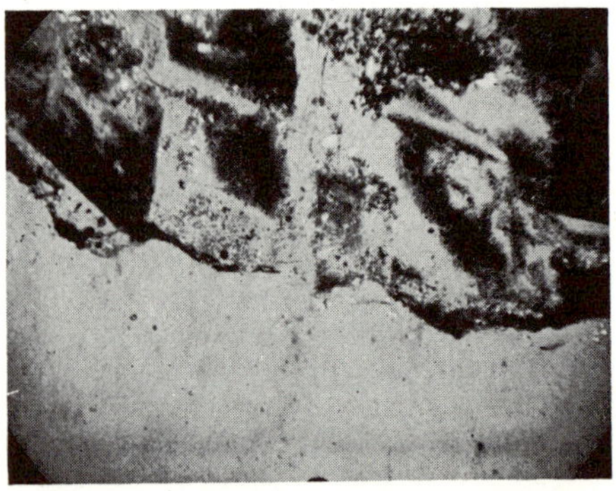

FIG. 10.—Photomicrograph of a thin section showing clear salt separated from pyramidal shaped hopper fragments by an anhydrite lamina. The small grains in the photograph are primarily anhydrite. The large, dark, angular areas are inclusion-rich zones in pyramidal shaped hoppers. The grey background areas are clear halite crystals. The specimen is from the Glen Bradley well, 271.7 feet above the base of the Salina formation. Core depth—4883.3 feet. Nicols partly crossed. ×16.5

pore space existed between adjacent salt grains at the time of burial. Through solution and recrystallization the loose, unconsolidated halite fragments apparently formed a relatively rigid and solid salt mass. Figure 11 shows a typical relationship which is found in thin section study. The angular hopper crystals underlying anhydrite lamina have been beveled, through solution, to form a flat surface on which the anhydrite was deposited. The truncation of the crystals suggests that they were held in a relatively rigid position and that some degree of lithifaction was attained before the development of this surface. Walther (1912), in his description of some present day salt deposits of Mexico, speaks of the salt as being "solid as ice." At the same locality ripple marks were seen developing in this salt crust when the wind moved the overlying water. Ripple marks are also common in the Salina salt in the Detroit area (Kaufmann and Slawson, 1950).

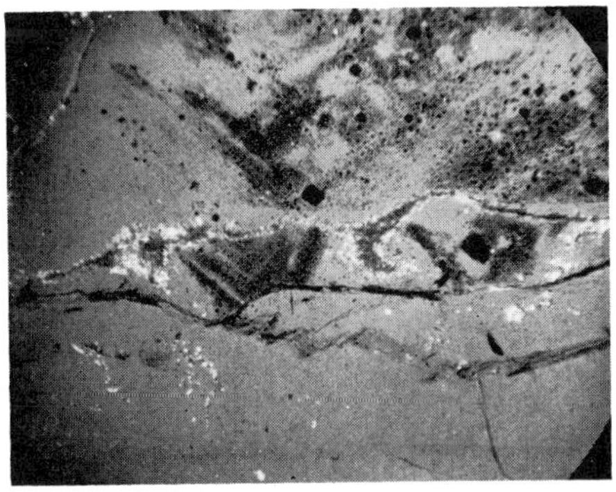

FIG. 11.—Photomicrograph of a thin section showing a sequence of clear salt, pyramidal shaped hopper fragments, an anhydrite lamina, and pyramidal shaped hopper fragments. The dark, chevron shaped areas are inclusion-rich zones in pyramidal shaped hoppers. The linear mass of small, white to grey crystals forms the lamina of anhydrite. The grey background material is clear salt. The specimen is from the Glen Bradley well, 267.4 feet above the base of the Salina formation. Core depth—4887.6 feet. Nicols partly crossed. ×16.5

The data assembled suggest the following process for the deposition of the halite. In a brine-filled basin, growth of the pyramidal shaped hoppers began with evaporation and the formation of a thin film of high density brine. Upon disturbance of the surface the crystals sank. The presence of abundant, flaky crystals soon brought about saturation of the bottom layer of the brine through solution, provided that this condition did not already exist. An increase in temperature changed the saturated brine to an unsaturated condition so that solution of some salt was necessary to reestablish equilibrium. Although the solubility of salt varies but a small amount with temperature, a relatively large amount of salt can be taken into solution with a tempperature change if there is a large enough quantity of water. If saturation took place during a warm period, subsequent cooling brought the brine into the metastable condition and growth began around the existing nuclei (Buckley, 1951). These nuclei were fragments of partly dissolved pyramidal shaped hoppers, or crystals which had dropped subsequent to the saturation of the lower layer of brine. This process was responsible for the growth of large crystals tending toward a cubic habit. With the reestablishment of equilibrium at the reduced temperature, newly precipitated pyramidal hoppers settled onto the crust of clear salt. Repetition of this process was initiated by a subsequent rise in temperature, the amount of solution taking place depending upon the magnitude of the rise and the length of time during which temperature continued to increase. Almost without exception the resulting clear crystals are much larger than the original hoppers.

Although temperature changes can account for solution and reprecipitation of the halite, other factors may aid the process. Some solution may result from pressure at the contact between adjacent grains and recrystallization on noncontact surfaces. Extremely small fragments may go into solution causing temporary supersaturation with equilibrium being reestablished by growth around larger more stable fragments.

* * * * * * *

Deposition of the Banded Salt

The growth processes of the halite crystals provide an important key to the problem of the depositional relationship between the salt and the anhydrite-dolomite layers.

The structural and mineralogical relationships in the Glen Bradley well core are essentially the same as those in the Detroit mine. The material from the Glen Bradley well should be representative of sediments deposited near the central portion of the basin, whereas the material in the Detroit mine should reflect conditions closer to the shore (fig. 3). In the growth of the halite crystals the variation in environmental conditions was unimportant. However, in the deposition of the laminae of dolomite and anhydrite, variations (as a result of the difference in depth environment) between the two localities are quite pronounced.

Central Basin Deposition

The interval between anhydrite laminae in the Glen Bradley well core varies between 1.5 and eight centimeters; the thickness of the individual lamina rarely exceeds 0.5 millimeter. The halite is generally in the form of either clear or cloudy layers, although intermixing of the two types of salt is not uncommon. The individual lamination is relatively flat and uniform in thickness. About 80 percent of the mineral grains in these layers are less than 0.002 millimeter in maximum dimension and only a small percentage of the remaining 20 percent exceeds 0.004 millimeter.

The laminae are composed primarily of stubby tabular crystals of anhydrite but also contain dolomite, pyrite, quartz, celestite, polyhalite, and carbonaceous material. Anhydrite, pyrite, and polyhalite almost always occur as euhedral crystals.

The individual layers show no vertical gradation of grain size nor segregation into mineral bands. The lack of evidence of any consistent relationship between these layers and either type of salt (clear or cloudy) necessitates a cause of deposition entirely independent of the causes of the original growth or the solution and recrystallization of the halite. Although there are significant relationships common to each layer, the salt surfaces of deposition and the layers themselves show considerable variation.

Structure.—Some relationships within the laminae and between the contacts separating the laminae and the enclosing salt are shown in figures 10 through 13.

Figure 10 shows a lamina that attains a maximum thickness of about 0.5 millimeter. The salt below the layer is clear salt and that above is composed of pyramidal shaped hopper crystals separated from one another by contemporaneously crystallizing anhydrite.

A similar relationship is shown in figure 11 except that several pyramidal shaped hoppers have settled onto the clear salt layer prior to the precipitation of the anhydrite. The truncation of the upper portion of the crystals and the filling of the interstices with anhydrite suggest that solution took place prior to the precipitation of the anhydrite.

Extreme irregularity of the underside of a lamination due to control of the depositional surface by the angularity of the pyramidal shaped hoppers is shown in figure 12. In this specimen the pyramidal shaped hoppers have not been subjected to solution prior to the deposition of the anhydrite so that the bottom surface of the layer has a saw-tooth structure. A similar type of irregular surface of deposition is shown in figure 9 where the irregularity is caused by well developed cubes of clear salt. Whether this irregularity is caused by pyramidal shaped hoppers or cubes of clear salt, it is indicative of deposition of a particular lamina without prior solution of the bottom crystals.

In present day natural salt deposits and also in solar salt pans some degree of consolidation takes place even before burial under subsequent layers. Some evidence of similar lithifaction in the Salina salt is shown by the manner in which rigidly supported crystals are beveled to form a flat surface of deposition for the layer above. Figure 13 shows immediately above the beveled surface about 0.1 millimeter of clear salt whose growth apparently preceded and in part was contemporaneous with the deposition of the overlying pyramidal shaped hoppers.

The microfaulting is best attributed to compaction because deformation of salt under pressure would result in flowage rather than in faulting of the salt. The faulting might also be accounted for by recrystallization but there is no evidence of this.

For the most part, although the over-all appearance is suggestive of uniformity, many of the laminations are irregular and discontinuous.

Mineralogy.—The most significant feature of the banded salt is the association of pyrite and carbonaceous material with one another and their localization along the anhydrite-dolomite laminae. The pyrite commonly occurs as euhedral

FIG. 12.—Photomicrograph of a thin section showing the control of the deposition of an anhydrite lamina by angular pyramidal shaped hoppers. The grey and black chevron shaped areas are the hopper crystals of halite above which has been deposited the anhydrite (fine-grained, grey to black) intermixed with and overlain by halite (light grey). The specimen is from the Glen Bradley well, 263.6 feet above the base of the Saline formation. Core depth—4891.4 feet. Partly crossed nicols. ×16.5

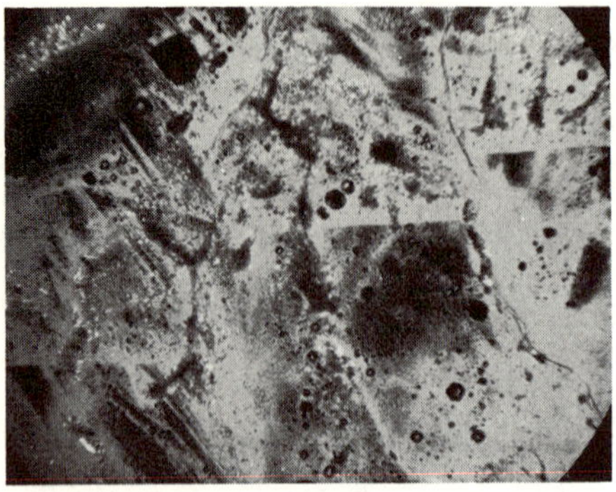

FIG. 13.—Photomicrograph of a thin section showing a faulted contact. Immediately above the flat solution surface (arrows) is approximately 0.1 millimeter of clear salt. Above this have been deposited pyramidal shaped hopper crystals contemporaneous with the anhydrite. The anhydrite lamina is composed of a series of discontinuous masses. The light grey areas are clear salt and the black chevron shaped areas are bands of liquid inclusions. The fine-grained, white to dark grey crystals are anhydrite. The specimen is from the Glen Bradley well, 264 feet above the base of the Salina formation. Core depth—4891 feet. Partly crossed nicols. ×16.5

to subhedral pyritohedrons. Irregular masses are often found occasionally covered with radiating needles of pyrite. Although most of the pyrite is associated with anhydrite-dolomite laminae, a few crystals are contained within the salt layers. It is apparent that the pyrite is genetically associated with the laminae. Those crystals contained within the salt probably settled out of the newly deposited anhydrite-dolomite lamina into the underlying salt. On the other hand, iron in the form of hematite is confined to the salt layers, occurring as either dust-like inclusions or as extremely fine dendritic masses.

The explanation for the iron in the form of pyrite can be found in the association of carbonaceous material with the laminae which would indicate a reducing environment. Carbonaceous material has been reported in the closely related Salina salt of New York (Alling, 1928). However, Alling indicated no specific mineralogical association and the irregular shape of the fragments prevented the determination of their origin.

Fragments of the carbonaceous material are for the most part in the form of hollow tubes. The tubes have a diameter of approximately 0.0195 millimeter and range in length up to 0.18 millimeter depending upon how the fragments broke during separation. The walls of the tubes are relatively thin. The fragments were examined by Professor Chester Arnold of the University of Michigan who stated that there can be little doubt that they are plant material. Although positive identification could not be made, it was suggested that they closely resemble the large tubes of *Prototaxites*—a Devonian and Silurian plant of questionable terrestrial or marine origin (Arnold, 1952).

Substantiating evidence of the association of organic material with these layers is the petroliferous odor detected in a fresh break through one of the laminae, a feature not uncommon in salt deposits. The odor is not found, however, in the salt layers. Several factors suggest that the inclusion of the petroliferous hydrocarbons took place at the time of deposition. Organic carbon is found in these layers. The migration of the hydrocarbons along laminae less than a fraction of a millimeter in thickness, and often discontinuous, is highly improbable.

Anhydrite occurs primarily as stubby, rectangular, euhedral to subhedral crystals generally less than 0.002 millimeter in length. For the most part, the crystals are oriented with their long axes parallel to the bedding.

Dolomite crystals are sparsely scattered through the anhydrite laminae. The crystals show a wide range in shape, varying from well developed rhombohedrons to rounded crystals with a circular cross section. The crystals are pale red-brown in color and often contain a nucleus of a black, dusty material.

Polyhalite is found as isolated individuals in the salt layers and along the marginal areas of laminae and isolated pods. It occurs either as acicular prisms or as rectangular crystals with the latter pseudomorphic after anhydrite. It has not been found with the relict crystal form of the gypsum. The size of the crystal varies little from that of anhydrite (for the most part less than 0.002 millimeter).

Carnallite occurs as rounded or pitted grains or as well developed euhedra. It is found either scattered through, or concentrated in zones in the halite crystals (Lucas, 1954).

Gypsum is usually found concentrated in small circular shaped areas in the salt bands and less commonly in small vein-like zones. Both the circular areas and the vein-like zones transect the halite crystal boundaries. The localization and concentration of gypsum in such restricted zones which intersect the halite crystal boundaries indicate that the gypsum is secondary.

Some isolated grains of gypsum are found in the salt bands and, although the temperature range and salinity at the time of deposition may have been favorable for the deposition of the hydrous sulfate, it does not seem possible that it would survive as such in the dense bottom brine. These crystals too are most likely secondary and may have formed as late as at the time of drilling.

Whether or not any of the calcium sulfate was originally deposited as gypsum cannot be determined for no pseudomorphs were found. However, Hollingworth (1948) points out that slow deposition and burial of primary gypsum (given sufficient time to react with the denser bottom solutions) gives rise to secondary anhydrite that is difficult to distinguish from primary anhydrite. As a result all evidence of primary gypsum may have been removed.

Quartz is common, both as doubly terminated euhedral crystals and as aggregates of anhedral crystals commonly in the form of rosettes. Most of the quartz is associated with the anhydrite-dolomite laminae, but a genetic relationship to these bands is not clear. The anhedral crystals are generally between 0.004 and 0.008 millimeter in length and the aggregates may be several times larger.

Summary.—Several generalizations regarding the deposition of the banded salt are summarized as follows: (1) There is no uniform or cyclic relationship between anhydrite-dolomite laminae and the cloudy or clear salt. The depositional surfaces at the base of the laminae vary from a smooth surface developed through solution to a highly irregular surface resulting from random orientation of either angular pyramidal shaped hopper crystals or recrystallized cubes. The depositional surface is underlain by either pyramidal-shaped hoppers or clear, inclusion-free halite. (2) The confinement of carbonaceous material and an associated petroliferous odor to the laminae, plus the concentration of pyrite along these layers (as opposed to the association of hematite with the salt bands), point to an influx of water into the basin as the cause of deposition of the laminae. As a result of the influx reducing conditions are established in the basin with a change in the Eh value from +0.1 for the deposition of the halite to −0.3 for the deposition of the anhydrite-dolomite laminae (Krumbein and Garrels, 1952).

Replacement has been limited and the sequence is as follows: (1) Gypsum was replaced by anhydrite. (2) Anhydrite was replaced by polyhalite. (3) Halite was replaced by gypsum.

Marginal Basin Deposition

Structure.—Although there is a general similarity of structure in the banded salt in the Glen Bradley well and the Detroit mine, megascopic examination reveals several pronounced differences between the salt deposited in the central portion of the basin and that deposited nearer shore.

One of the most pronounced differences is the absence or poor development of banding of clear and cloudy salt at the Detroit mine. In most instances banding is completely absent and the entire salt mass shows a relatively uniform degree of translucency. In several specimens the banding is poorly developed; in no specimen is banding as well developed as in the Glen Bradley core.

Another pronounced difference is that the anhydrite-dolmite layers at the Detroit mine generally are thicker and more widely spaced than in the Glen Bradley core, and intervals between adjacent bands of eight to 10 centimeters can be found throughout the mine workings. Dolomite and anhydrite appear to be more abundant and this is substantiated by analyses of representative samples. Pyrite associated with these laminations is also more abundant and occurs in larger masses than in the Glen Bradley core. It is also found in a lesser quantity scattered throughout the salt bands.

The anhydrite-dolomite laminae are not only more prominent near the basin margin, but the bedding is undulating, although it is apparent in a wall or face that the average distance between adjacent laminae does not vary (figs. 1 and 14). In some areas the undulation in the individual layer may be attributed partly to ripple marks. Ripple marks are almost continuous along several laminae and are found intermittently associated with a number of other laminae.

Microscopic examination reveals that the salt from the basin margin contains fewer pyramidal shaped hopper crystals and more clear salt than does the salt from the central part of the basin. Fragmentation of the pyramidal shaped hoppers has been extensive and complete hoppers are not found. The increased percentage of clear salt suggests a greater amount of bottom solution and recrystallization. As in the Glen Bradley core, there is no apparent genetic relationship between the clear or cloudy salt and the anhydrite-dolomite laminae.

The environmental differences of deposition are further expressed in the laminae by variations in grain size and crystal form, by more pronounced replacement and recrystallization, and by layering within the individual laminae. Specimens from the mine workings are layered in two distinct ways. Dolomite crystals contained within layers of anhydrite are segregated into poorly developed bands, or the anhydrite layers may be separated into several thin laminae by dark bands of iron oxide (?). The anhydrite adjacent to these bands is commonly oriented with its long axis parallel to the bedding. The layering of the dolomite may suggest fluctuation in conditions of deposition or, along with the orientation of the anhydrite crystals and the layering by films of iron oxide, it may indicate a degree of sorting and orientation of the crystals by bottom currents. Some weight is given to this conclu-

FIG. 14.—Typical banded Salina salt from the Wabash area of the Detroit mine of the International Salt Company. The black areas are predominately anhydrite and the white and grey areas are halite. The specimen comes from approximately 666 feet above the base of the salina formation. Mine depth—about 1140 feet. ×0.5

sion by the persistent association of ripple marks with the laminae. Two thin sections through ripple marks show some degree of orientation of the anhydrite grains and the development of a crude layering by the films of iron oxide.

In the thick anhydrite-dolomite layers in the mine it is often observed that a layer is composed of distinctly separated bands of dolomite and anhydrite. In some specimens the relationship between a pair or series of bands cannot be explained reasonably by any postulated set of physical-chemical conditions of deposition. However, fragments of halite scattered throughout the thick anhydrite-dolomite layer can be identified as pyramidal shaped hopper crystals and appear to represent remnants of a layer of halite which has been removed through solution. Complete solution of thin salt layers could in this way give rise to anomalous relationships which are not the result of changing physical-chemical conditions.

More commonly, however, a layer of anhydrite and dolomite acts as a cap for the underlying salt and retards or prevents its solution. Retardation of solution in this manner has been observed by Grossman (1949) in the sodium sulfate deposits in western North Dakota.

Mineralogy.—In contrast to the stubby, rectangular anhydrite crystals from the Glen Bradley core, the anhydrite from the Detroit mine is usually found in the form of needle-like crystals and to a lesser extent as crystals with the crystal form of gypsum. About 60 percent of the needles are between 0.008 and 0.016 millimeter in length. The long axes of those crystals with the crystal form of gypsum generally measure between 0.004 and 0.008 millimeter, although some of the crystals are considerably larger.

The needle-like crystals occur in interlocking mats or laminae of primary origin, as unoriented individuals replacing halite, in radiating groups of crystals in the salt adjacent to the anhydrite-dolomite laminae, or as bursts of large crystals radiating from a central mass of fine-grained anhydrite. Large crystals radiating from a fine-grained nucleus have been attributed by Dunham (1948) to replacement of anhydrite by halite and concomitant recrystallization of the anhydrite. The needle-like form may, therefore, be of either primary or secondary origin.

Anhydrite pseudomorphs after gypsum are generally found either in the marginal areas of the laminae or as isolated crystals contained within the salt. The replacement is usually complete, although in some crystals remnants of gypsum can be found. The crystals are rarely entire and deeply etched boundaries show replacement by halite. The replacement is almost entirely confined to the gypsum pseudomorphs; in some cases all that remains of the crystal is a group of several fragments roughly outlining the original gypsum crystal form and having the same optical orientation.

It should be pointed out again that the presence of pseudomorphs after gypsum in the Detroit mine, in contrast to their absence in the Glen Bradley core, need not indicate a difference in the temperature or salinity between near shore and off shore environments of deposition. It may rather be merely a reflection of the rate of precipitation and speed of burial of the gypsum. Rapid precipitation and burial of gypsum will result in the replacing anhydrite retaining the crystal form of the gypsum.

Pyrite is found in increased amounts in the mine specimens and primarily as large, irregular masses rather than as euhedral crystals. Some of these masses are in excess of 0.5 millimeter in diameter. The pyrite is not confined as closely to the anhydrite layers as in the specimens from the Glen Bradley well, but the concentration of a large percentage of the grains adjacent to the layers again indicates this genetic relationship. On the other hand, little carbonaceous material is found and the fragments are irregular in shape. It is confined primarily to the anhydrite-dolomite laminae.

Dolomite is more abundant in the Detroit mine than in the Glen Bradley well and is largely concentrated in crude, discontinuous layers. The crystals are anhedral to subhedral, brown in color, and some contain a nucleus of black, dusty material. The character of individual grains and their relationship to anhydrite strongly resembles the material described from the Permian of England (Dunham, 1948; Stewart 1949, 1951a, 1951b, 1953). The occurrence of the dolomite in nearly spherical grains is attributed by Dunham (1948) to a change in composition of the brine before completion of the crystallization of the carbonate. The dark centers found in many of the grains possibly acted as seeds, initiating the crystallization of the dolomite.

An alternate explanation is that the dolomite, at least in part, might be clastic in origin. The low, barren land mass adjacent to the basin exposed carbonate rock and the climate was arid. It seems more than reasonable that some carbonate would be carried into the basin by the wind or by intermittent land drainage. The sphericity of the grains could easily be the result of transportation in this manner.

Unfortunately, the differentiation between clastic and precipitated carbonate becomes difficult if dolomitization (Pettijohn, 1949) and lithifaction (Hatch and Rastall, 1938) have occurred. The existence of almost ideal conditions for the formation and transportation of clastic carbonate, the increase in the percentage of carbonate in the salt nearer the margin of the basin, and the sphericity of the grains suggest in part a clastic origin. On the other hand chemical precipitation is favored by the individual grains being composed of a single crystal, the presence of dusty inclusions in the center of the grains, and the confinement of the grains to the vicinity of the dolomite-anhydrite laminae.

A small number of dolomite grains have been found which contain cores of calcite, this indicating that at least some of the dolomite has resulted from the replacement of calcite. Not uncommon are ring shaped fragments of dolomite, in which the central portion has been replaced by halite. This suggests a differential replacement of the carbonate although there is no indication of the cause or manner of the replacement.

Although calcium sulfate was deposited in the form of both anhydrite and gypsum, little gypsum has survived the replacement by anhydrite. The gypsum is found only as remnants in anhydrite crystals. However, as in the Glen Bradley core, a later generation of gypsum has replaced halite. The crystals are generally euhedral, and are only a fraction of the size of those of the first generation. They are also confined to restricted and generally circular areas in which there is a higher concentration of sulfate than in the surrounding salt.

Polyhalite is relatively rare. The individual grains have retained the original crystal form of the gypsum. The replacement sequence is gypsum-anhydrite-polyhalite.

Summary.—The poor development of bedding or its non-existence in the clear and cloudy salt, fragmentation of the pyramidal shaped hopper crystals, and abundant ripple marks are all suggestive of some degree of turbulence in the marginal areas of the basin. With this turbulence would come repeated vertical mixing of the brine, with subsequent changes in the composition of the bottom layer of the brine causing destruction of the conditions of equilibrium at the depositional interface. In deep water rain and light winds do not suffice to intermix the upper and lower density zones (Harvey, 1925). However, in shallow water meteoric conditions might frequently destroy density zoning.

The halite crystals initially grow as pyramidal shaped hoppers, but, through turbulence and intermixing of the upper and lower density zones, they undergo fragmentation and solution to a greater extent than do those crystals deposited in the deeper waters of the basin. However, the persistent occurrence of the hopper fragments in the salt column indicates that the brine maintained a relatively high concentration even with vertical mixing.

Gypsum and anhydrite were both deposited as primary minerals although gypsum was extensively replaced by anhydrite and to a lesser degree by halite.

It is probable that the carbonate was deposited both as a clastic and as a chemical precipitate.

The complex replacement sequence in these marginal basin evaporites appears to have been as follows: (1) Calcite was replaced by dolomite. (2) Gypsum was replaced by halite. (3) Gypsum was replaced by anhydrite. (4) Anhydrite was replaced by polyhalite. (5) Anhydrite was replaced by halite with attending recrystallization of anhydrite. (6) Dolomite was replaced by halite. (7) Halite was replaced by gypsum.

These observations point to more, and more rapidly developed, differences in the composition of the solution near the margin of the basin. No difference, however, is indicated for the basic principles of deposition.

* * * * * * *

SUMMARY AND CONCLUSIONS

The deposition of the Salina salt has resulted from the evaporation of brine in a semi-closed basin and in an arid or semi-arid region. The addition of brine was apparently intermittent, and there is no evidence of cyclic repetition. Chemically the solution had a pH value of about 8.0 with an Eh value shifting from $+0.1$ to -0.3 with the influx of water into the basin.

Thick beds of anhydrite and/or gypsum are absent below the salt beds. In the lower portion of the salt layers thick bands of anhydrite and dolomite alternate with layers of salt of about the same thickness. These thick bands represent not only periods of influx of brine into the basin, but possibly periods of subsidence during which large quantities of added brine are nearly depleted of calcium carbonate and calcium sulfate. Thus a large volume of concentrated brine accumulated. Occasionally the brine reached the saturation point for sodium chloride before more brine was added, and some salt was deposited. Within these thick layers the alternation of laminations of dolomite and anhydrite suggests seasonal banding. The deposition of the dolomite was retarded during the colder seasons and accelerated during the warmer seasons when the solubility was lowered.

Once the basin became more or less tectonically stable and accumulated a large volume of saturated brine, continued evaporation caused the deposition of alternating bands of clear and cloudy salt. This deposition was occasionally interrupted by the addition of new brine to the basin resulting in deposition of anhydrite and dolomite laminae.

The alternation of bands of clear and cloudy salt is a seasonal function. During the warm periods halite, which is precipitated from the surface as pyramidal shaped hopper crystals, may go into solution so that a saturated, dense brine is developed on the bottom of the basin. Deposition of halite, after the establishment of equilibrium between halite crystals and the solution, then form a layer of cloudy, pyramidal hopper crystals. On cooling the brine becomes supersaturated and crystal growth takes place around hopper nuclei. Thus layers of pyramidal shaped hopper crystals represent deposition at which time the bottom layer of brine is saturated with sodium chloride; and layers of clear salt represent crystallization from a supersaturated brine due to a drop in temperature. This cycle may be interrupted by the addition of brine to the basin or by periods of unusual climatic conditions.

The deposition of the anhydrite-dolomite laminae is completely independent of the deposition of the salt, and it may be considered to be the result of the influx of brine to the basin. Indicative of this influx are carbonaceous material, pyrite, a petroliferous odor and the lack of uniform relationships between the banded salt and the anhydrite-dolomite laminae. Although the only carbon to be identified is of plant origin, some may be of animal origin. The addition of brine to the basin as a surface layer would permit the maintenance of life in this brine until the concentration reaches the point when mass extermination takes place, and the subsequent decomposition of the remains would cause reducing conditions to develop in the bottom of the basin. Iron associated with the halite layers is in the form of hematite, indicating that oxidizing conditions were normal for the basin. The association of the laminae with an influx of brine into the basin does not preclude that there may also be seasonal relationship although such a relationship has not been identified.

The effect of rain water falling into the basin is negligible for the only effect is to dilute temporarily the surface layer and retard precipitation. The effect of runoff would only be a local change in the composition of the marginal waters. There can be little doubt that some wind blown material was added to the basin.

In the thick (several centimeters) layers of dolomite and anhydrite, pyrite and carbonaceous material are associated with both the upper and lower contacts with the salt. This may represent extermination of some organisms due to the intermingling of the lower dense brine and the added sea water, and final mass extermination with the continued concentration of the brine.

The range of temperature of deposition as determined by a study of inclusions in the halite is between 32 to 48.4° C. The presence of langbeinite offers some substantiating evidence for this mineral is deposited at a minimum temperature of 37° C. This temperature range is undoubtedly a reflection of climatic conditions to a limited extent, although with increased concentration of the brine there is a decrease in the rate of evaporation as an increasing part of the solar energy used to increase the temperature of the brine.

Distinct from the deposits near the center of the basin, the marginal deposits lack the layering of clear and cloudy salt and the laminae are highly irregular, both owing to increased bottom turbulence.

REFERENCES

ALLING, H. L., 1928, The geology and origin of the Silurian salt of New York state: New York State Mus. Bull., no. 275.

ARNOLD, C. A., 1952, A specimen of Prototaxities from the Kettle Point black shale of Ontario: Palaeont., v. 93 (B), pp. 45–46.

ATKINS, W. R. G., 1924, On the vertical mixing of sea water and its importance for the algal plankton: Marine Biol. Assoc. U. K. Jour., n. s., v. 13, pp. 319–324.

BADGER, W. L. and BAKER, E. M., 1928, Inorganic chemical technology. McGraw-Hill Book Company, Inc., New York, pp. 15–22.

BIRGE, E. A., 1910, An unregarded factor in lake temperature: Wisconsin Acad. Sci. Trans., v. 16, pt. 2, pp. 889–1004.

———, and JUDAY, CHANCEY, 1922, The inland lakes of Wisconsin—the plankton. I, Its quantity and chemical composition: Wis. Geol. and Nat. History Survey, Sci. ser., Bull. 64, no. 13.

BRANSON, E. B., 1915, Origin of thick gypsum and salt deposits: Geol. Soc. America Bull., v. 26, pp. 231–242.

BUCKLEY, H. E., 1951, Crystal growth. John Wiley and Sons, Inc. New York, 571 pp.

BURY, C. R., 1948, The chemistry of saline deposits: Proc. Yorks, Geol. Soc., v. 27, pp. 205–209.

CLARKE, F. W., 1924, The data of geochemistry: U. S. Geol. Survey Bull. 770, 841 pp.
COOK, C. W., 1914, The brine and salt deposits of Michigan: Michigan Geol. and Biol. Survey Pub. 15, Geol. ser. 12, 118 pp.
DELLWIG, L. F., 1953, Hopper crystals of halite in the Salina of Michigan: Am. Mineralogist, v. 38, pp. 730, 731.
DELLWIG, L. F. and BRIGGS, L. I., 1952, Textural relationships in the Salina salt of Michigan (abs.): Geol. Soc. America Bull., v. 63, p. 1242.
DREYER, R. M., GARRELS, R. M., and HOWLAND, A. I., 1949, Liquid inclusions in halite as a guide to geologic thermometry: Am. Mineralogist, v. 34, pp. 26–34.
DUNBAR, C. O., 1949, Historical geology. John Wiley and Sons, Inc., New York. pp. 573
DUNHAM, K. C., 1948, A contribution to the petrology of the Permian evaporite deposits of northeastern England: Proc. Yorks. Geol. Soc., v. 27, pp. 217–227.
FOWLER, A., 1944, A deep bore in the Cleveland Hills: Geol. Mag., v. 81, pp. 193–206, 254–265.
FULDA, Ernst, 1924, Studie über die Entstehung der Kalelagerstattën des deutschen Zechsteins: Zs. deut. geol. Ges., v. 76, Montasber., pp. 7–30.
GRABAU, A. W., 1913, The origin of salt deposits with special reference to the Siluric salt deposits of North America (with discussion): Min. Met. Soc. America Bull. 57, pp. 33–44.
———, 1920, Geology of the nonmetallic deposits other than silicates: v. I, Principles of salt deposition. McGraw Hill Book Company, New York.
GRABAU, A. W. and SHERZER, W. H., 1910, The Monroe formation of southern Michigan and adjoining regions: Michigan Geol. and Biol. Survey Pub. 2, Geol. ser. 1, 435 pp.
GROSSMAN, I. G., 1949, The sodium sulphate deposits of western North Dakota: North Dakota Geol. Survey Rept. Inv., no. 1, 66 pp.
HARVEY, H. W., 1923, Hydrographic features of the water in the neighborhood of Plymouth during the years 1921 and 1922; Marine Biol. Assoc. U. K. Jour., n. s., v. 13, pp. 225–235.
———, 1925, Evaporation and temperature changes in the English Channel: Marine Biol. Assoc. U. K. Jour,. n. s., v. 13, pp. 678–692.
HATCH, F. H., and RASTALL, R. H., 1938, The petrology of the sedimentary rocks, 3rd edition, revised by Maurice Black, Thomas Murby and Co., London, 383 pp.
HILL, A. E., 1937, The transition temperature of gypsum to anhydrite: Am. Chem. Soc. Jour., v. 59, pp. 2242–2244.
HOLLINGWORTH, S. E., 1948, Evaporites: Proc. Yorks. Geol. Soc., v. 27, pp. 192–198.
INGERSON, EARL, 1947, Liquid inclusions in geologic thermometry; Am. Mineralogist, v. 32, pp. 375–388.
JOHNSTON, JOHN and WILLIAMSON, E. D., 1916, The role of inorganic agencies in the deposition of calcium carbonate: Jour. Geology, v. 24, pp. 729–750.
KAUFMANN, D. W. and SLAWSON, C. B., 1950, Ripple marks in rock salt of the Salina formation: Jour. Geol., v. 58, pp. 24–29.
KINDLE, E. M., 1918, Separation of salt from saline water and mud: Geol. Soc. America Bull., v. 29, pp. 471–487.
KRUMBEIN, W. C., 1951, Occurrence and lithologic associations of evaporites in the United States: Jour. Sedimentary Petrology, v. 21, pp. 62–81.
KRUMBEIN, W. C. and GARRELS, R. M., 1952, Origin and classification of chemical sediments in terms of pH and oxidation-reduction potentials: Jour. Geology, v. 60, pp. 1–33.
LANDES, K. K., 1945, Preliminary map no. 40, The Salina and Bass Island rocks in the Michigan basin: U. S. Geol. Survey, Oil and Gas Inv.
LINCK, G., 1937, Bildung des Dolomites und Dolomitisierung: Chemie d. Erde, v. II, pp. 278–286.
LOEWENGART, STEFAN, 1928, Das Tote Meer, seine Entstehung und die Möglichkeiten seiner Ausbeutung: Zs. prakt. Geol., v. 36, pp. 86–89.
LOTZE, FRANZ, 1938, Steinsalz und Kalisalze. Verlag von Gebrüder Borntraeger, Berlin, 936 pp.
LUCAS, P. T., 1954, Environment of Salina salt Deposition: Master's thesis, University of Michigan, Ann Arbor.
MITCHELL, A. E., 1923, Studies on the dolomite system: Chem. Soc. Jour., v. 123, pp. 1055–1065, 1887–1904.
MITCHELL, P. H. and SOLINGER, J. P., 1934, The effect of land drainage upon the excess bases of sea water: Biol. Bull., v. 66, pp. 97–101.
NEWCOMBE, R. B., 1932, Oil and gas fields of Michigan; a discussion of depositional and structural features of the Michigan Basin: Michigan Geol. Survey Div., Pub. 38, Geol. ser. 32, 293 pp.
PEACH, P. A., 1949, Liquid inclusions in geothermometry: Am. Mineralogist, v. 34, pp. 460–461.
PETTIJOHN, F. J., 1949, Sedimentary rocks. Harper and Brothers, New York, 306 pp.

Phalen, W. C., 1919, Salt resources of the United States: U. S. Geol. Survey Bull. 669, 284 pp.
Phillips, F. C., 1947, Oceanic salt deposits: Quart. Rev., v. 1, pp. 91–111.
Pirtle, G. W., 1932, Michigan structural basin and its relationship to surrounding areas: Am. Assoc. Petroleum Geologists Bull., v. 16, pp. 145–152.
Posnjak, E., 1938, The system, $CaSO_4$—H_2O: Am. Jour. Sci., 5th ser., v. 35-A, pp. 247–272.
———, 1940, Deposition of calcium sulfate from sea water: Am. Jour. Sci., vol. 238, pp. 559–568.
Rankama, Kalervo and Sahama, Th. G., 1950, Geochemistry. University of Chicago Press, Chicago, 911 pp.
Revelle, Roger, 1934, Physical-chemical factors affecting the solubility of calcium carbonate in sea water: Jour. Sedimentary Petrology, v. 4, pp. 103–110.
Rubey, W. W., 1951, Geologic history of sea water: Geol. Soc. America Bull., v. 62, pp. 1111–1148.
Sherzer, W. H., 1911, Geological report on Wayne County: Michigan Geol. and Biol. Survey, Pub. 12, Geol. ser. 9, 338 pp.
Shaw, Sir Napier, 1928, Manual of meteorology; v. II, Comparative meteorology. Cambridge University Press, London, 445 pp.
Sorby, H. C., 1858, On the microscopical structure of crystals indicating the origin of minerals and rocks: Geol. Soc. London Quart. Jour., v. 14, pp. 450–500.
Stewart, F. H., 1949, The petrology of the evaporites of the Eskdale no. 2. boring, east Yorkshire. Part I. The lower evaporite bed: Mineralog. Mag., v. 28, pp. 621–675.
———, 1951a, The petrology of the evaporites of the Eskdale no. 2 boring,east Yorkshire. Part II. The middle evaporite bed: Mineralog. Mag., v. 29, pp. 445–475.
———, 1951b, The petrology of the evaporites of the Eskdale no. 2 boring, east Yorkshire. Part III. The upper evaporite bed: Mineralog. Mag., v. 29, pp. 557–572.
———, 1953, Early gypsum in the Permian evaporites of northeastern England: Proc. Geol. Assoc., v. 64, pp. 33–39.
Trask, P. D., 1937, Relation of salinity to the calcium carbonate content of marine sediments: U. S. Geol. Survey Prof. Paper 186-N, pp. 273–299.
———, (Editor), 1939, Recent marine sediments, a symposium, Am. Assoc. Petrol. Geol. Tulsa. 736 pp.
Van Tuyl, F. M., 1916, New points on the origin of dolomite: Am. Jour. Sci., 4th ser., v. 192, pp. 249–259.
Van't Hoff, J. H., 1903, Physical chemistry in the service of the sciences: Univ. Dec. Pub., 2nd. Ser., v. 18, Chicago, pp. 97–124.
Walther, Johannes, 1912, Das Gesetz der Wustenbildung. 2d ed., D. Reimer, Berlin.
Wells, R. C., 1915, The solubility of calcite in water in contact with the atmosphere, and its variation with temperature: Washington Acad. Sci. Jour., v. 5, pp. 617–622.
Wood, F. W., 1949, Recent information concerning the evaporites and the pre-Permian floor of southeast Durham: Geol. Soc. London Quart. Jour., v. 105, pp. 327–341.

Reprinted from *Geol. Soc. Amer. Bull.*, **77**, 331–342 (1966)

N. C. WARDLAW
W. M. SCHWERDTNER } *University of Saskatchewan, Canada*

Halite-Anhydrite Seasonal Layers in the Middle Devonian Prairie Evaporite Formation, Saskatchewan, Canada

20

Abstract: In the lower half of the Devonian Prairie Evaporite Formation, halite units 2–10 cm thick alternate with anhydrite units, most of which are less than 1 cm thick. Such halite-anhydrite alternations are thought to result from seasonal changes in the basin waters during deposition and are called seasonal layers.

Halite occurs as opaque "chevron" grains with abundant liquid inclusions and as transparent grains without liquid inclusions. The chevron grains are conspicuously elongated perpendicular to bedding and contain a distinctive chevron structure caused by zones of liquid inclusions. They started growth as cubes, which were oriented with (110) or (111) approximately parallel to bedding; subsequently growth took place on only two or three upward-directed faces. These grains are considered to be primary and grown on the Devonian sea floor. Similar chevron grains were produced artificially during competitive growth into a salt solution.

Inclusion-free halite grains tend to have irregular shape and no elongation perpendicular to bedding. They are larger than associated chevron grains and locally appear to have replaced them. Their bromide content is higher than that of associated chevron grains.

Bromide analyses of successive chevron grains above an anhydrite layer show a rapid increase to a maximum value and then a steady decrease as the overlying anhydrite layer is approached. This variation of bromide is most satisfactorily explained by influxes of sea water or brine. Following the deposition of anhydrite, the influxes increased markedly in salinity and then decreased before deposition of the overlying anhydrite layer.

CONTENTS

INTRODUCTION

Thin laminae of anhydrite alternate with halite in the lower half of the Devonian Prairie Evaporite Formation of Saskatchewan. Halite units range from about 2 to 10 cm thick, whereas most associated anhydrite units are less than 1 cm thick. Similar halite-anhydrite alternations, well developed in the Zechstein of Germany, have been called "*Jahresringe*" (Richter-Bernburg, 1950, p. 2), and are referred to here as seasonal layers.

Geological Society of America Bulletin, v. 77, p. 331–342, 6 figs., 3 pls., April 1966

The compositional changes from halite to anhydrite in the Prairie Evaporite Formation are thought to be a reflection of seasonal changes in the basin waters at the time of deposition, although the duration of the "seasons" is unknown. Such changes have been interpreted to result from either changes in temperature or composition of the basin waters.

The significance of the distribution of bromide ions in solid phases of chlorides in relation to the solutions from which they crystallize has long been known (Valyashko, 1956). Principles underlying the distribution of bromide in chlorides and the method of making bromide determinations are treated here briefly.

Bromine does not form bromides during the crystallization of salts from sea water, but occurs in solid solution in chlorides. The quantity of bromide in a chloride bears a definite relationship to the bromide content of the brine from which that chloride crystallized. This definite relationship is expressed by a distribution coefficient which is constant for a particular chloride at the low bromide concentrations encountered in natural brines. The distribution coefficient can be defined as:

$$\frac{\text{weight per cent bromide in the solid phase of a chloride}}{\text{weight per cent bromide in the liquid phase from which the chloride crystallized}}.$$

Different chlorides have different, but definite, distribution coefficients, and where two or more chlorides crystallize simultaneously from a brine, the bromide enters each chloride in the same ratio as if only that one chloride were present.

The effect of temperature on the distribution coefficient of halite is taken to be negligible (Braitsch and Herrman, 1963, p. 372).

Experiments have shown that the distribution coefficients for all chlorides are invariably less than one. Consequently, during the progressive evaporation of a brine, although bromide is removed from solution into the chloride crystals, the concentration of bromide in the brine progressively increases in relation to chloride. Thus, a definite bromide content in the chloride, as well as in the remaining liquid, will correspond to each stage of concentration during evaporation, and a particular bromide content for chloride crystals can be related to a particular concentration of mother brine.

Detailed bromide profiles made throughout successive seasonal layers in the Prairie Evaporite Formation are presented in this paper and their significance in terms of the genesis of the salt and concentration of the brine from which the salt crystallized is discussed. By this method, it is possible to estimate the seasonal changes in salinity of the brines from which the halite was precipitated and to interpret the possible control on deposition. Figure 1 shows the location of the cores analyzed.

ACKNOWLEDGMENTS

Grants in aid of research from the National Research Council, Canada, and the Saskatchewan Research Council are gratefully acknowledged. The writers take pleasure in thanking Mr. J. Sandbrand for his assistance in making bromide analyses and would also like to thank Dr. Robert Kühn (Hannover), Professor Richter-Bernburg (Hannover), and Dr. A. G. Hermann (Göttingen) for helpful suggestions and their time spent in discussion. Drs. W. O. Kupsch and W. G. E. Caldwell, University of Saskatchewan, critically read the manuscript.

METHODS OF ANALYSIS

The method of analysis used is that described by D'Ans and Höfer (1934) (*see* Schwerdtner, 1963, for English summary). Determinations were made using not less than 4 g of halite sample. Blank analyses were necessary to determine the bromide impurities in the reagents, and whenever new reagents were used these were analyzed for bromide. A sample with known bromide content (prepared with potassium bromide solution) was analyzed with each group of approximately eight unknowns, in order to keep a check on the accuracy of results.

The presence of insoluble material in the samples lowers the value of the bromide expressed as a weight percentage of sodium chloride, and where such insoluble material was observed in a sample during solution, a total chloride analysis was made in order to eliminate this error, which in all samples used was small. Six analyses for potassium were made in the lower 50 feet of halite from the Winsal Osler core, and 0.016 weight per cent potassium was the largest value recorded. This potassium content is considered insignificant in its effect on the bromide data quoted. The total error involved in the analyses quoted rarely exceeds ± 0.0005 weight per cent bromide and is insufficient to affect the trends described.

If the bromide weight percentage in halite is to be used as a reliable indication of the salinity of the solution from which the chloride crystallized, one must know that the bromide determined is in solid solution in the chloride lattice, and not bromide which may be present in brines included within crystals. In order to determine the effect of the bromide in brine inclusions on the total weight percentage of bromide in halite samples, samples rich in inclusions were analyzed for total bromide, and subsequently similar samples were analyzed after removal of most of the contained liquid inclusions. Liquid inclusions were eliminated by grinding the samples in absolute alcohol in a ball mill for about 10

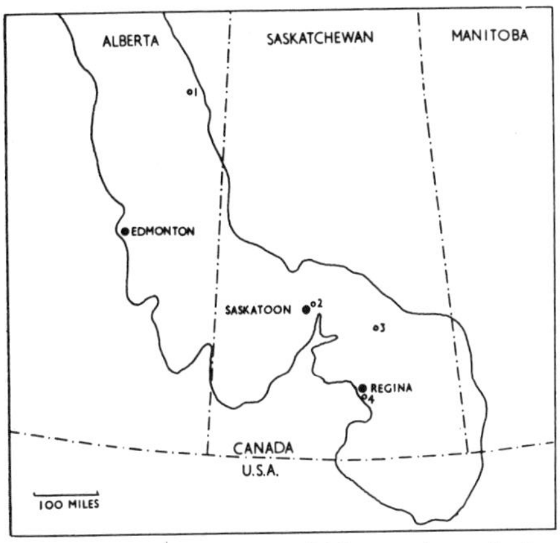

Figure 1. Index map. Solid is zero isopach for salt in the Elk Point Group (*after* Pearson, 1963, p. 233). 1. Industrial Minerals No. 3 Salt Well, McMurray; 2. Winsal Osler (Lsd. 3, Sec. 28, T. 39, R. 4 W3); 3. Krasne Crown No. 1 (Lsd. 16, Sec. 14, T. 30, R. 16 W2); 4. Saskatchewan Power Corporation (Lsd. 9, Sec. 27, T. 16, R. 20 W2)

minutes, then washing in alcohol, and finally drying. Results showed that brine could account for as much as 0.002 weight per cent bromide in crystals with a total bromide content of 0.007 weight per cent.

The relatively large amount of bromide in the brine is understandable in terms of the distribution coefficient for halite, which, in the initial phases of halite precipitation from sea water, is 0.14 (Braitsch and Herrmann, 1963, p. 379).

Using the coefficient 0.14 for halite, it was calculated that a crystal with 0.006 weight per cent bromide in solid solution, but containing 10 per cent by volume of brine inclusions, would have a total bromide content of approximately 0.008 weight per cent, assuming the bromide of the brine is in equilibrium with the bromide in solid solution in halite. This value corresponds closely with that determined experimentally on actual crystals. Inclusions in chevron halite grains are generally small (less than 50 μ across) and more or less uniformly spread throughout the grains. It is assumed that the error due to brine inclusions is approximately equal in different chevron grains, and no corrections for brine inclusions were made.

PETROGRAPHY OF HALITE LAYERS

Two types of halite are observed in cores from the Prairie Evaporite Formation: white, opaque grains with abundant liquid inclusions, and transparent grains without liquid inclusions (Pl. 1, fig. 3; Pl. 2, fig. 4). Both types commonly occur together.

The opaque grains are described as chevron halite because of their distinctive structure. In zones parallel to cube faces, each grain contains layers with a high concentration of fluid inclusions alternating with layers that contain few, if any, fluid inclusions (Pl. 2, fig. 5). The inclusions are almost perfectly cubical, and in some grains, there is a grading in the size of the inclusions (Pl. 2, fig. 5), as observed by Sorby (1858, p. 460) in artifically grown crystals. Sorby stated, "In general the size of the cavities varies inversely as their number, for when the crystals are slowly formed, they are larger, though less numerous."

From Figures 2 and 3 of Plate 1, it can be seen that the halite grains are conspicuously elongated perpendicular to bedding, and that the chevron structure is invariably arranged with the point of the chevron directed upward. This structure provides a right-way-up criterion and is a useful check on the orientation of the core, which sometimes is misoriented in the core boxes.

These grains must have started growth as cubes, which were oriented with (110) or (111) approximately parallel to bedding. Subsequently, growth was unidirectional and took place on only two or three upward-directed faces (Fig. 2).

In artificially grown halite (Pl. 2, figs. 1, 2), it was found that, with competitive growth into a solution, the grains which survived were mainly those oriented with (110) parallel to bedding. Grains with other orientations tend to be eliminated during competitive growth. A possible explanation is that the growth velocity of halite is greater in a [[110]] direction than in a [[100]] direction.

The elongation of grains perpendicular to bedding, the chevron structure indicating unidirectional upward growth, and the regular alternation of inclusion-rich and inclusion-poor zones within grains all suggest that these grains are primary and grew in upward competition with each other on the Devonian sea floor.

Similar structures were recorded from the German Zechstein deposits by Gottesmann (1963), who also concluded that the halite grains grew upward from the sea floor.

Present-day evaporite basins as large as the Devonian basin of Saskatchewan or the Zechstein basin of Germany are unknown, and knowledge of the formation of primary halite is limited to observations from smaller present-day salt lakes and to theoretical or experimental work combined with petrographic observation.

The crystallization of primary halite in salt lakes has been discussed by Lotze (1957, p. 101) who stressed that, through evaporation, surface brines were the first to become saturated with sodium chloride and that halite growth was initiated on the brine surface. Subsequently, the surface-formed crystals sank to the basin floor, where growth may have continued as long as contact with a saturated brine was maintained.

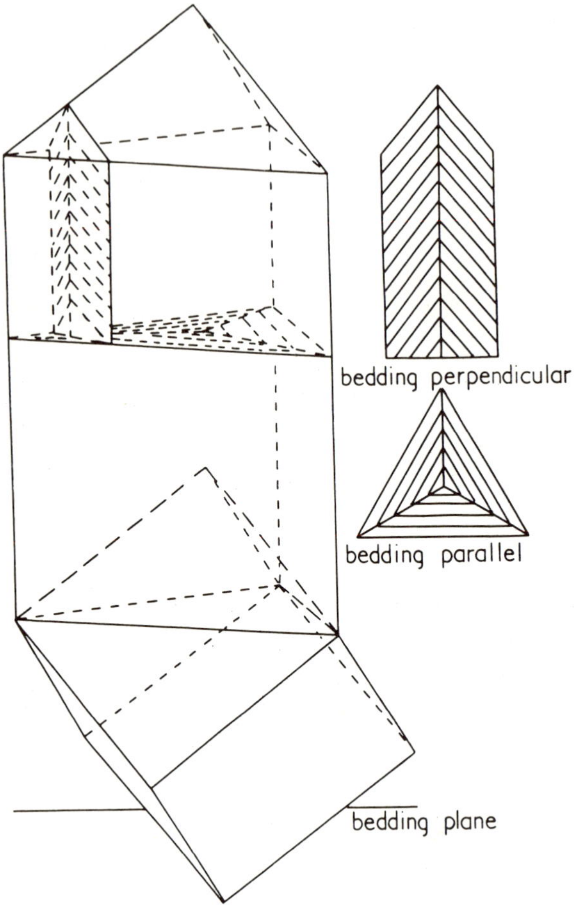

Figure 2. Diagrammatic halite chevron crystal with (111) approximately parallel to bedding. Growth has occurred by extension of 3 upward-directed faces. Bedding perpendicular and bedding parallel cuts through the crystal are illustrated. Compare with real crystals of Plate 1, figures 1 and 2.

Dellwig (1953; 1955) and Dellwig and Briggs (1953) recorded inclusion-rich "hopper crystals" in the Salina Salt of Michigan which they interpreted as having formed in this way. These "hopper" crystals are not described as being conspicuously elongate perpendicular to bedding, nor as having a consistently upward-directed chevron structure, and are not to be confused with the chevron crystals described in this paper.

The crystallization of major amounts of halite on basin floors, as is postulated here for the chevron crystals in Saskatchewan, is less easy to visualize but may have resulted from the cooling of warmer surface layers as they descended to depth along the basin floor.

Textural observations suggest that perfectly transparent grains of halite associated with the chevron grains (Pl. 2, fig. 4) locally replace the chevron grains, and therefore are younger and secondary to the rock. In Figure 4 of Plate 2 it can be seen that two portions of a chevron grain (A and B), of a common orientation, are separated by clear halite, suggesting that they are remantns of a once continuous crystal. The boundaries between clear and chevron grains are abrupt and consertal.

Clear grains tend to be of irregular shape and without elongation perpendicular to bedding. They commonly occur immediately above, rather than immediately below, anhydrite layers. Neither the clear nor the chevron halite shows any consistent grading of grain size within the seasonal layers.

PETROGRAPHY OF ANHYDRITE-CARBONATE LAYERS

Anhydrite is mainly in the form of more or less spherical masses, commonly about 1 mm in diameter, although the masses illustrated in Figure 1 of Plate 3 are smaller, consisting of anhydrite grains about 100 μ long and 10 μ thick. Some acicular grains appear to be arranged in a vaguely concentric manner. Larger, euhedral anhydrite grains occur at the margins of the spherical anhydrite masses, particularly where they are in contact with halite, and tend to be concentrated at halite grain boundaries, suggesting that they may be younger than these boundaries (Pl. 3, fig. 3).

Both the lower and upper contacts of the anhydrite layers with halite are irregular but sharp, and at the lower contacts there is no conspicuous truncation of halite crystals, in contrast some clay layers interbedded in halite in the upper part of the Prairie Evaporite Formation obviously truncate underlying halite crystals (Pl. 3, fig. 4).

The spherical form of anhydrite masses is commonly accentuated by dolomite rhombohedra which occur marginally to them (Pl. 3, fig. 1). Dolomite also occurs along grain boundaries in halite.

Associated with the anhydrite layers, and also occurring in the halite between anhydrite layers, are spherulites of calcite of approximately 1 mm diameter, which are composed of very small, radially arranged fibers (Pl. 3, fig. 2). The origin of these spherulites is not known.

COMPARATIVE BROMIDE PROFILES FOR TOTAL PRAIRIE EVAPORITE

It is considered significant to present three typical bromide profiles for the entire Prairie Evaporite Formation before presenting and discussing small-scale bromide profiles in relation to individual seasonal layers.

The Prairie Evaporite Formation at MacMurray, Alberta, consists of 180 feet of halite without potassium minerals. The halite is mainly in the form of chevron grains, and the bromide content increases from the base of the formation upward (Fig. 3). This is the relationship to be expected in a basin where crystallization occurred from brines which were becoming progressively more concentrated.

The bromide profile for the Winsal Osler core, Saskatchewan, is less simple (Fig. 3). In the lower 150 feet of section, chevron grains are associated with clear halite, the clear halite apparently replacing, and therefore postdating, the chevron grains. The bromide content of clear halite is consistently greater than that of the associated chevron grains, but both types of halite have bromide contents increasing upward in the lower 150 feet of Prairie Evaporite core. The bromide profile in the overlying 250 feet of salt section is highly irregular, and bromide decreases upward. This observation was discussed by Schwerdtner (1964) and suggests that much of the halite in the upper part of the section is not primary but has been subjected to resolution and crystallization since deposition. Schwerdtner suggested that alteration by downward-percolating solutions may have resulted in the anomalous bromide trends in the upper part of the salt sequence. Vertical stringers of clay and carnallite in the core support this explanation. It seems likely that much of the halite in the upper part of the section is secondary and therefore unsuitable for use in determining the salinity of primary brines.

The entire Saskatchewan Power Corporation core of the Prairie Evaporite Formation consists of halite texturally similar to that in the upper part of the Winsal Osler core. The halite is of the clear variety, and no chevron grains were observed. A remarkable feature is that the total bromide profile (Fig. 3), although highly irregular, shows no progressive trend either for an increase or decrease in bromide from base to top of the formation.

Since there is no assurance that any of the halite in this core is primary, it is unsuitable for the purposes of this study; it is typical of other cores examined from the southern portion of the basin.

BROMIDE PROFILES THROUGH SEASONAL LAYERS

Bromide profiles within the halite portions of seasonal layers from two cores are illustrated in Figure 4. Samples from the Krasne Crown 1 well are from within 20 feet of the base of a 600-foot chloride section, whereas the Winsal Osler samples are from within 60 feet of the base of a 400-foot chloride section. Data presented in Figures 4 and 5 are the total results obtained to date.

The bromide values presented in Figure 4, all of which were obtained from chevron grains, presumably grown on the Devonian sea floor, are remarkably regular and, with few exceptions, appear to bear a simple relationship to the seasonal layers. Above an anhydrite layer, the bromide content of halite increases to a maximum value and then decreases to the overlying anhydrite layer. The maximum bromide content occurs about half way between the anhydrite layers.

The bromide values from chevron grains presented in Figure 4 are, with two exceptions, less than 0.007 weight per cent, the bromide content of halite at the beginning of halite precipitation from present-day sea water (Braitsch, 1962, p. 106). Analyses previously published (Schwerdtner and Wardlaw, 1963, p. 243) were of halite samples where no separation was made between chevron and clear halite grains, and it is apparent now, from the data in Figure 4, that primary halite formed from Devonian sea water can have bromide values as low as 0.003 weight per cent.

The significance of this low bromide content has been discussed elsewhere (Schwerdtner, 1964; Schwerdtner and Wardlaw, 1963; Wardlaw, 1964). Assuming that the grains involved are primary growths, either the Devonian seas of Saskatchewan did not have the same bromide-to-chloride ratio as present-day sea water, or else the halites with a bromide content lower than 0.0075 are precipitates from brines which reached saturation partly by re-solution of salt. Such re-solution of halite would be favored

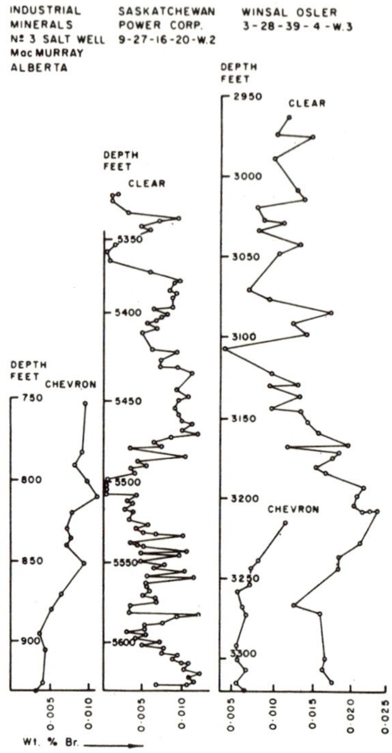

Figure 3. Bromide profiles from three cores through the Prairie Evaporite Formation. Types of halite analyzed are indicated by the words "chevron" and "clear." Formation tops and bottoms, measured in feet from the surface, for the three wells are, from left to right: 730 and 934, 5314 and 5638, 2923 and 3319.

particularly at a stage of evaporation in which the concentration of the brines oscillated about the point of saturation of sodium chloride. Such oscillations are indicated by the numerous layers of anhydrite within the lower part of the halite sequence.

If the chevron grains are accepted as primary, it is unlikely that the trends recorded are of diagenetic origin, for example the product of local solution and deposition by water lost from gypsum in a change to anhydrite.

Figure 1. Bedding parallel-cut through chevron halite grains of Figure 3 of this Plate (4½ ×)

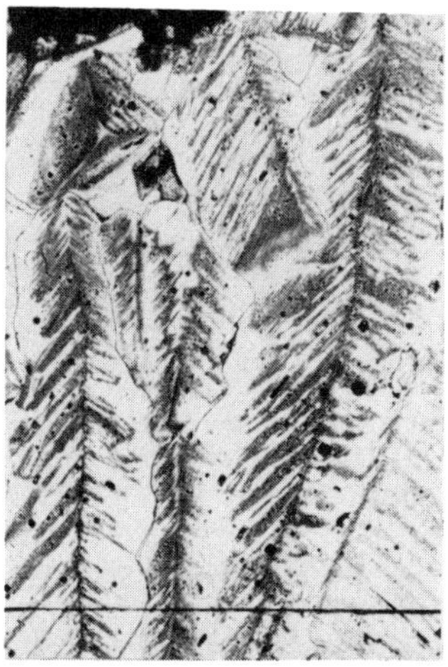

Figure 2. Bedding perpendicular-cut through chevron halite grains of Figure 3 of this Plate (4½ ×)

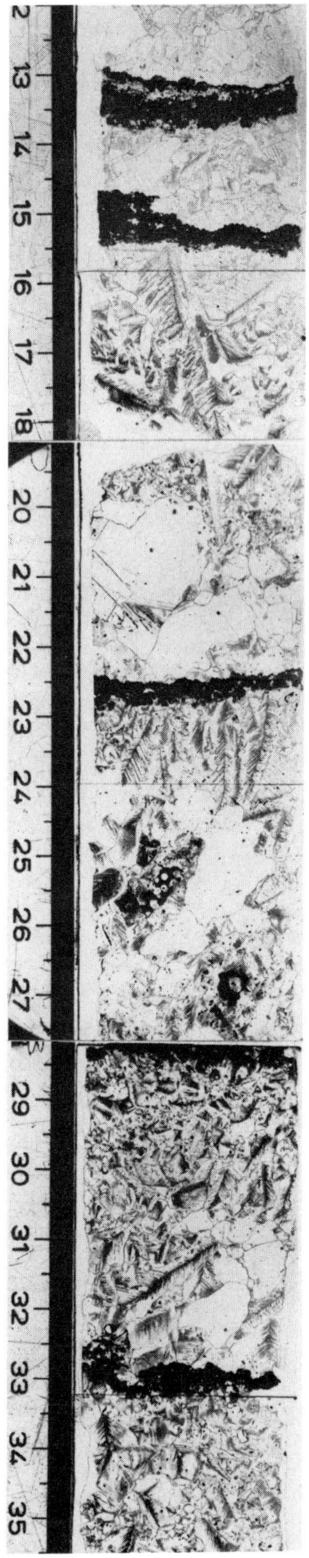

Figure 3. Series of continuous thin sections through five seasonal layers. Clear, inclusion-free grains of halite associated with chevron halite. Horizontal layers of anhydrite and carbonate appear dark. From lower portion of Winsal Osler core, 3275 feet, scale in centimeters

THIN SECTIONS THROUGH HALITE-ANHYDRITE SEASONAL LAYERS IN PRAIRIE EVAPORITE FORMATION, SASKATCHEWAN

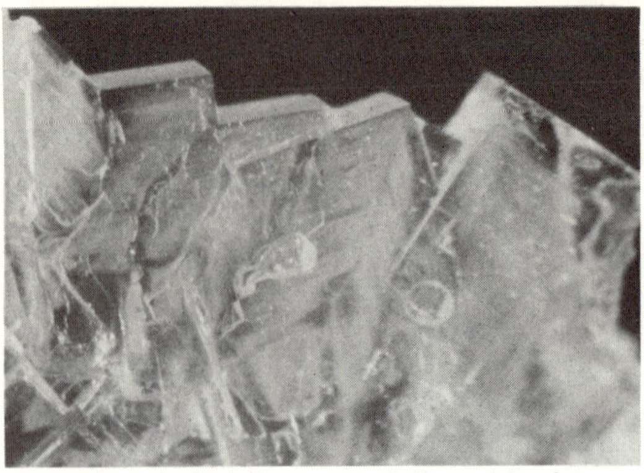

Figure 1. Artificially grown halite with chevron structure. Grains elongated with [110] axes parallel to growth direction. Enlarged portions of Figure 2 of this Plate (18 ×)

Figure 2. Artificially grown halite with chevron structure (3 ×)

Figure 3. A pair of chevron crystals extracted from Winsal Osler core (left). Artificially grown halite showing high inclusion density at corners compared with face centers (right). (3 ×)

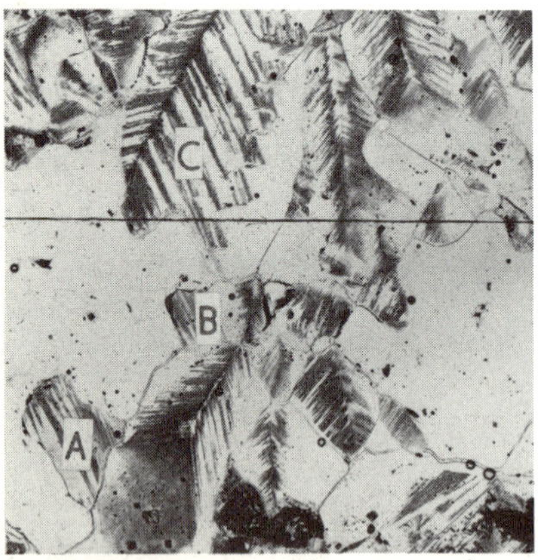

Figure 4. Chevron halite apparently replaced by clear halite. It is suggested that A, B, and C are parts of a once continuous chevron grain. Enlarged portion of Figure 3 of Plate 1 (3 ×)

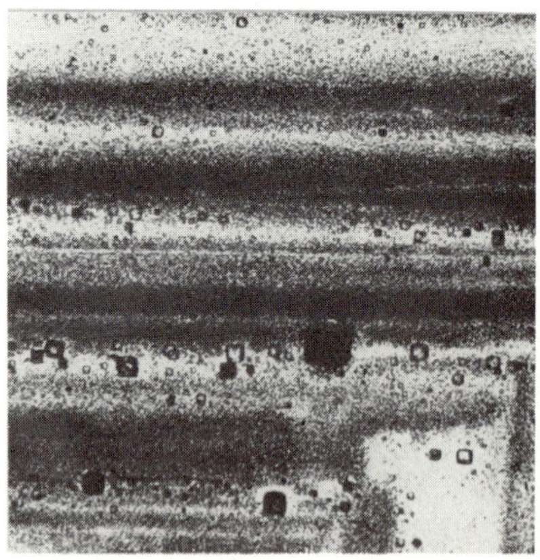

Figure 5. Halite grain with alternating inclusion-rich (dark) and inclusion-free (light) zones. Chevron grain from Figure 3 of Plate 1 (30 ×)

ARTIFICIALLY GROWN AND NATURAL CHEVRON HALITE AND CLEAR HALITE FROM PRAIRIE EVAPORITE FORMATION, SASKATCHEWAN

WARDLAW AND SCHWERDTNER, PLATE 2

Geological Society of America Bulletin, volume 77

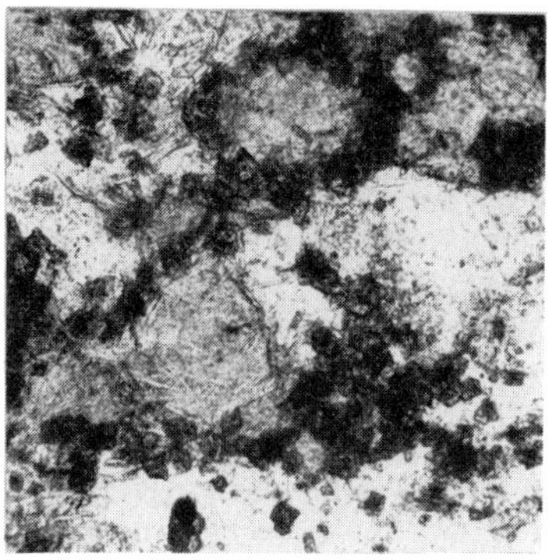

Figure 1. Spherical masses of acicular anhydrite with rhombohedral dolomite. Clear areas are halite. From one of anhydrite layers in Figure 3 of Plate 1 (50 ×)

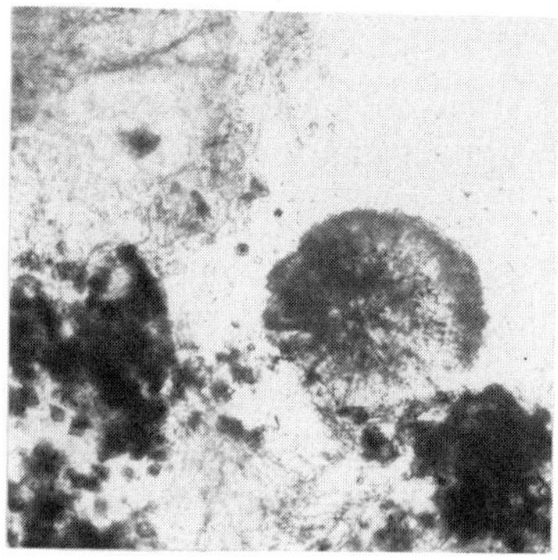

Figure 2. Calcite spherulite composed of small radially arranged fibers. Krasne Crown 1, 4125 feet (20 ×)

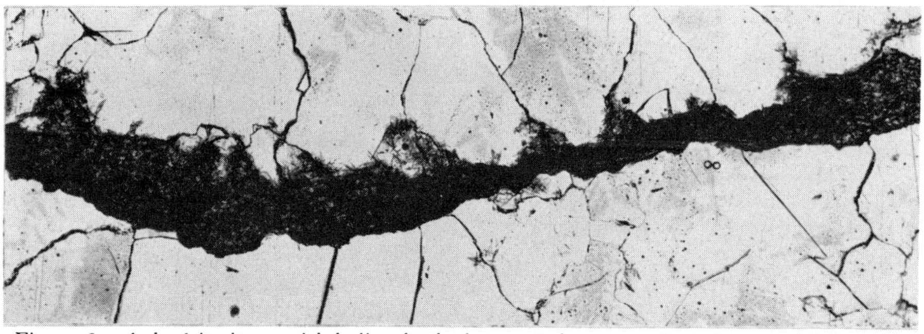

Figure 3. Anhydrite layer with halite, both clear and chevron grains, above and below. At upper contact anhydrite is concentrated at halite grain boundaries. Winsal Osler, 3003 feet (8 ×)

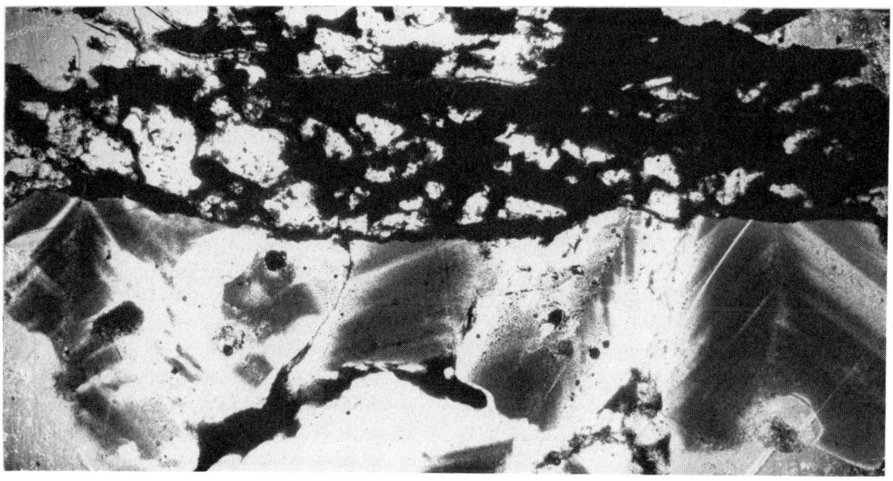

Figure 4. Clay interlayer in upper portion of Prairie Evaporite Formation illustrating truncation of underlying chevron halite. Saskatoon 12–19–37–4–W3, 3312 feet (20 ×)

ANHYDRITE AND CLAY LAYERS FROM PRAIRIE EVAPORITE FORMATION, SASKATCHEWAN

WARDLAW AND SCHWERDTNER, PLATE 3

Geological Society of America Bulletin, volume 77

The upward increase in bromide above an anhydrite layer could be explained simply on the basis of progressive evaporation and consequent crystallization of halite. Under such conditions, progressive enrichment of bromide in the solution would occur and there would be

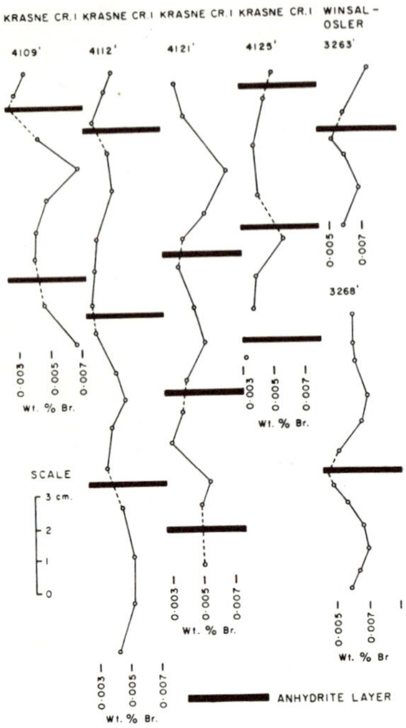

Figure 4. Detailed bromide profiles for the halite of seasonal layers. Thick horizontal lines represent anhydrite layers. All analyses are of chevron grains. Samples are from the lower half of the Prairie Evaporite Formation.

a corresponding enrichment of bromide in the halite crystallizing from the solution.

Certain deductions concerning brine depth can be made from the gradient of the bromide profiles by applying an equation used by Kühn (1955, p. 658):

$$T = \frac{b_2 m}{b_2 - b_1} \cdot f,$$

in which T is the brine depth, and b_1 and b_2 are the bromide contents of two samples which are separated by a thickness m, measured in vertical profile, of halite (b_1 is less than, and underlies, b_2). The term f is defined as:

$$f = \frac{d \,.\, 100}{s \,.\, c},$$

where d is the density of halite, s is the density of brine, and c is the weight percentage of sodium chloride in the brine.

Two important observations are made from the data presented in Figure 4:

(1) Above an anhydrite layer, the bromide content increases to a maximum value over an average thickness of halite of 2.6 cm;

(2) The average increase of bromide in this halite is 0.002 weight per cent (average minimum and maximum 0.004 and 0.006).

Since,

$$b_1 = 0.004 \text{ wt. \% Br},$$
$$b_2 = 0.006 \text{ wt. \% Br},$$
$$m = 2.6 \text{ cm},$$
$$f = 8,$$

if one applies Kühn's equation, it follows that $T = 62$ cm.

This means that 2.6 cm thickness of halite, with initial and final bromide contents of 0.004 and 0.006 weight per cent respectively, would be precipitated from 62 cm initial depth of brine.

The foregoing equation allows a comparison between the rate of increase of bromide in halite in vertical profile and the depth of brine from which that halite crystallized. The more rapid the upward increase in bromide, the less the depth of brine involved. The bromide profiles of Figure 4 have a remarkably rapid upward increase of bromide in halite which, under static-isothermal conditions of evaporation, would correspond with an initial brine depth of less than a meter.

The brine depth calculated is unrealistically small because in nature the conditions were more complex than in the simple model so far presented. Various other factors must be considered. Temperature changes may have affected the precipitation of halite. The basin was not a closed system, but was subject to influxes and perhaps refluxes of sea water or brine, and deposition probably occurred from moving brine streams of variable salinity. The situation probably was complicated by layering of brines of different densities within the basin.

A specific example will illustrate the effect of temperature change on the precipitation of halite. A temperature change from 30° C to 20° C would produce a halite layer 2.6 cm thick, which is as observed in the increasing portions of the bromide profiles in Figure 4, from a depth of brine of 113 m. However, this temperature change would not adequately explain the facts, because the deposition of 2.6 cm of halite from this volume of water would produce a negligible change in the bromide-to-chloride ratio of the solution and would not explain the observed rapid increase of bromide in vertical profile.

It is possible that the rapid increase of bromide in halite, in the halite seasonal layers, is in the upper portions of halite layers. During the deposition of these portions, there must have been continuing decreases of bromide, relative to chloride, in the solution. These could have been due to a reduction in the flow of preconcentrated influxes and introduction of more normal sea water, or dissolving of some previously precipitated halite. Solution of halite

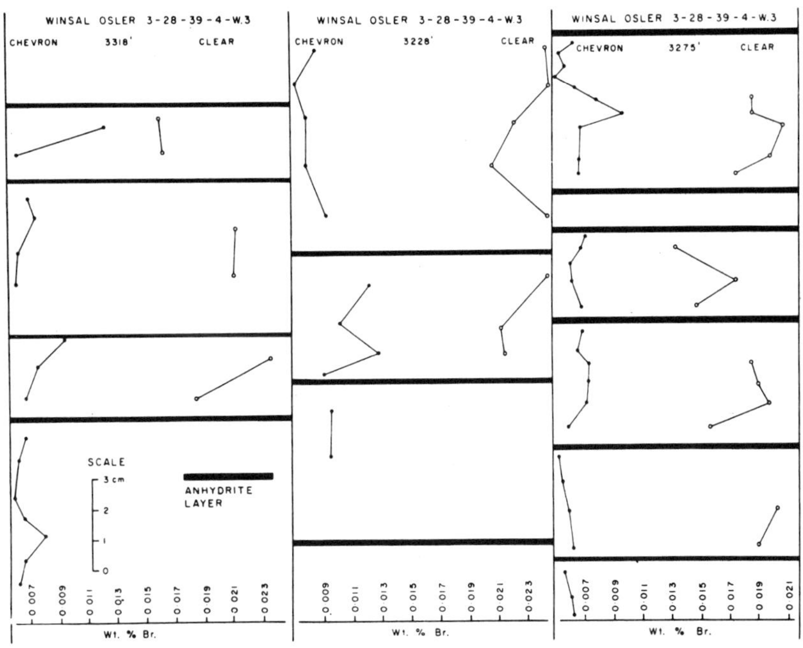

Figure 5. Detailed bromide profiles for the halite of seasonal layers. Separate analyses made of associated clear and chevron halite. Clear halite consistently has a higher bromide content. Thick horizontal lines represent anhydrite layers.

due to influx of preconcentrated sea water which was bromide-enriched, through the precipitation of halite, before the Saskatchewan basin was reached. Such a suggestion is consistent with the paleogeographic evidence, since the open ocean is believed to have lain to the northwest, and very extensive salt deposits of equivalent age occur in Alberta (Pearson, 1963, p. 230).

The bromide profiles of Figure 4 are for chevron halite grains, which grew on the basin floor, and it is likely that the brines from which crystallization occurred were not typical of the entire basin. The effects of brine layering on deposition within the basin are not known.

A further feature of the seasonal-layer bromide profiles is the decreasing bromide content could have taken place in one part of the basin, and, because brines were moving, halite could then have been redeposited elsewhere. Solution of halite could have resulted from influxes of fresh water from the surrounding land which diluted brines below the point of saturation with respect to sodium chloride, or from influxes of sea water undersaturated with respect to sodium chloride.

It might be expected that fresh water draining off the surrounding land into the basin would be accompanied by clastic sediments, but none are found in the lower portion of the evaporite section. The second possibility, resolution of salt by sea water undersaturated with respect to sodium chloride, seems a more likely alternative, since, for other reasons, it appears likely that there were continuing influxes of brines during deposition.

The rapid increase and decrease of bromide in the halite of seasonal layers is most satisfactorily explained, on the basis of the present data, by influxes of sea water or brine which periodically varied in salinity. Evaporation rates may have had an important bearing on the whole process. Following the deposition of anhydrite, increased evaporation rates could have preconcentrated brines beyond the point of saturation with sodium chloride before the Saskatchewan basin was reached. Subsequently, a decreased rate of evaporation may have led to the introduction of less concentrated brines or sea water, from which calcium sulfate was precipitated after the deposition of halite with decreasing amounts of bromide. The bromide profile of a seasonal layer could be explained by a period of high evaporation, followed by a period of lower rates of evaporation.

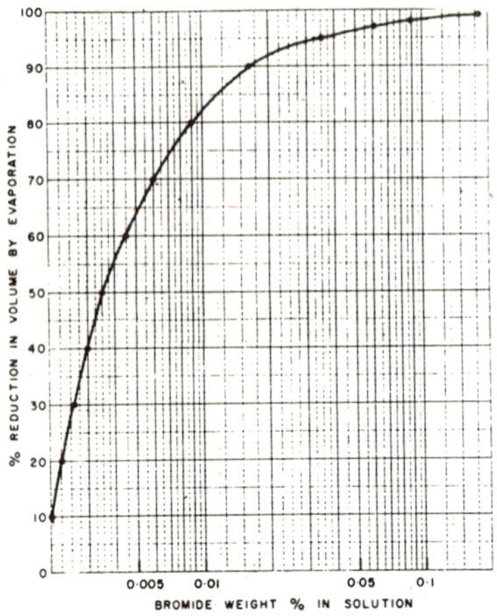

Figure 6. Graph illustrates change in bromide content of a sodium chloride solution with progressive evaporation and crystallization of halite. The solution is an imaginary one made from halite, with a bromide content of 0.007 weight per cent, dissolved in fresh water to the point of saturation. A 96 per cent reduction in volume is necessary to produce a precipitate with the original bromide content of 0.007 weight per cent.

Kühn (1955, Fig. 5) illustrated detailed bromide profiles through seasonal layers (*Jahresringe*) from the Zechstein of Thüringen, Germany. The Thüringen profiles differ from the Saskatchewan profiles in their relationship to anhydrite layers. Kühn concluded that the bromide content of halite reaches a peak value just below an anhydrite layer and then drops sharply in the overlying salt before increasing again to the next overlying anhydrite layer. The gradients of the Thüringen profiles are smaller than those from Saskatchewan, since the range of bromide fluctuation is only about 0.0005 weight per cent, compared with 0.002 weight per cent in the Saskatchewan profiles.

Kühn suggested that evaporation during a dry period caused salt precipitation and progressive bromide enrichment. A rainy period then followed, producing a sufficient dilution that the brines became unsaturated with sodium chloride, and salt deposition was interrupted. Subsequently, saturation was achieved partly by solution of salt, which decreased the bromide-to-chloride ratio of the brine and caused the lowered bromide content in halite immediately overlying anhydrite.

BROMIDE CONTENT OF CHEVRON HALITE AND CLEAR HALITE

Careful separation was made between associated chevron and clear grains, and samples of both types were systematically taken through halite "annual rings." Figures 3 and 5 show that the clear grains consistently have a higher bromide content than the chevrons. In the Winsal Osler core (Fig. 3), both types of halite have an upward increase of bromide in the lower hundred feet of Prairie Evaporite core.

It may be considered that the clear halite formed from residual brines trapped in pore space, and that halite subsequently crystallizing from this brine had a higher bromide content than the associated chevron crystals. In this event, traces of potassium minerals might be expected, and these are not found (Schwerdtner and Wardlaw, 1963). Also, this interpretation would not explain the replacement of chevron grains by clear grains.

Secondary halite, formed by solution of preexisting grains in fresh water and subsequent crystallization, would have a lower bromide content than the original grains (Fig. 6). For example, if halite, with a bromide content of 0.007 weight per cent, were dissolved in fresh water to the point of saturation and then crystallized through evaporation, the new crystals would have a progressively increasing bromide content as evaporation proceeded. This situation is illustrated in Figure 6, which was constructed by use of the Boecke equation

(Braitsch, 1962, p. 102). It can be seen that a 96 per cent reduction in volume would be necessary to produce the original bromide content of 0.007 weight per cent. This makes it very improbable that the clear grains resulted from simple solution of chevron grains in fresh water and subsequent crystallization.

Experiments by Kühn (personal communication) indicate that bromide-chloride ion exchange is negligible between halite and a bromide-enriched solution saturated with sodium chloride. It appears unlikely that such an exchange could significantly alter the bromide content of halite.

Little is known about the effects which increased pressure, or recrystallization in the presence of limited amounts of liquid, may have on the distribution coefficient, and it is not known whether the clear grains formed during recrystallization in the dry state or formed as a result of replacement in the presence of limited amounts of solution, or as a result of replacement with large amounts of brine in pore space.

Whatever the process, clear halite could not have formed from the chevron grains unless extra bromide was introduced to the system. It is possible to visualize bromide-enriched late-stage brines, perhaps diluted by fresh water, percolating through the halite section and depositing bromide-enriched halite, but in this case it is surprising that potassium minerals are not found. Also it would be difficult to explain the stratigraphically upward increase of bromide in both types of halite (Fig. 3).

The texture of the rock suggests that the chevron grains are primary and that the clear grains are younger, and in part replacing them. It has not proved possible to reconcile satisfactorily the geochemical data with these observations.

BRINE DEPTH DURING DEPOSITION

Approximately 35000 feet of sea water would have to be evaporated to obtain a salt sequence 600 feet thick, such as is found in the Prairie Evaporite Formation. It is highly unlikely that such a deep basin existed during Devonian time in Saskatchewan, and the great thickness of salt more probably formed as a result of a continuing influx of sea water during evaporation from a basin with much shallower waters.

It is possible that, during the deposition of halite, reflux of brines out of the basin may have been of limited importance, since the ratio of halite to potassium salts in the central portion of the basin is close to that obtained from present-day sea water evaporated under static-isothermal conditions (Schwerdtner and Wardlaw, 1964). This ratio would not be expected if influx was accompanied by reflux, but in a system with influx of sea water only, the ratio of salts precipitated would be the same as that obtained with evaporation under closed conditions. These considerations could be substantiated if quantitative estimates were available of the amounts of various salts within the entire basin, instead of estimates for the basin center as given above.

The seasonal layers of the Prairie Evaporite Formation consist of an alternation of halite units with an average thickness of about 5 cm and anhydrite-carbonate units, which are commonly less than 0.5 cm. Dellwig (1955, p. 104) recorded that in the production of solar salt in California, about 11.5 to 12.5 cm of poorly consolidated salt are deposited from concentrated brines at a rate of evaporation of about 150 cm per year. Lotze (1957, p. 105) stated that some salinas of Mexico and Egypt are known to have an annual precipitation of 7–14 cm of salt, while in the Elton sea of the Soviet Union, annual precipitation of 2.5–8.8 cm is known. Borchert (1959, p. 14) recorded that recent annual salt layers average 1–5 cm.

The halite in the seasonal layers of the Prairie Evaporite Formation may be the equivalent of 1 year's deposition, since the halite units correspond in thickness with the halite known to be precipitated in 1 year under present-day conditions in salt lakes. However, it should be stated that in these lakes the initial crystallization is mainly on the brine surface, and not on the basin floor as postulated for the Saskatchewan chevron halite. If the annual nature of the Saskatchewan layers is accepted, with an average rate of deposition of 5 cm per year, the 600-foot chloride section in the center of the basin could have formed in less than 4000 years. There are breaks in the record, and it is likely that the time involved was longer than this, but even so, it seems likely that the formation accumulated in a very short time.

Beloussov (1962, p. 210–216) summarized what is known of present-day crustal movements, which only in exceptional circumstances exceed a few millimeters per year, and, if comparison can be made with the past, it seems likely that subsidence was negligible during the short period of Prairie Evaporite deposition.

This has an important bearing on an interpretation of brine depth, because it means that there was no significant subsidence of the basin during deposition of the salts. In the final phases of desiccation, when potassium salts were forming, the brine surface must have been almost coincident with the upper surface of the Prairie Evaporite Formation at that time. If there was no great subsidence of the basin during salt deposition, as suggested heretofore, and if there were no drastic changes of world-wide sea level during this period, it can be stated that the initial water depth must have been equal to, or in excess of, the total thickness of salt. It follows that, as evaporation proceeded and the salt layers were built up, the brine depth decreased until, in the final event, complete desiccation resulted in the deposition of potassium salts.

SUMMARY

In the lower half of the Prairie Evaporite Formation, anhydrite layers are interbedded with halite, and as a result of comparison with salt deposits forming in brine lakes at present, these alternations are considered to be of a seasonal nature and are called seasonal layers. It is not known whether these seasons were longer or shorter than a year. The halite units range in thickness from 2 to 10 cm, whereas most of the anhydrite units are less than 1 cm.

Approximately 35,000 feet of sea water would have to be evaporated to obtain a salt sequence 600 feet thick such as is found in the Prairie Evaporite Formation. It is highly unlikely that such a deep basin existed during Devonian time in Saskatchewan, and the great thickness of salt more probably formed as a result of evaporation from a much shallower basin with a continuing influx of brine.

At an average rate of deposition of 5 cm of halite per year, even allowing for breaks in the sedimentary record, the entire formation would have accumulated in a very short time geologically–on the order of 4000 years. Subsidence was probably negligible during this interval, and, assuming that there were no drastic changes of world-wide sea level, the initial depth must have been equal to, or in excess of, the 600-foot total thickness of salt. It follows that, as evaporation proceeded and the salt layers were built up, the brine depth decreased until complete desiccation resulted in the deposition of potassium salts. These considerations are generally valid even if the seasonal layers were formed over longer periods, of say 10–15 years.

The Elk Point Basin during Prairie Evaporite deposition can be visualized as a shallow inland sea, probably not much more than 600 feet deep in the center of the basin, into which there was a continuing influx of sea water, but out of which there was little reflux to the open sea, at any rate after halite deposition was advanced.

Observations of texture enable the recognition of primary grains of halite, named chevron grains from their distinctive structure, which grew upward in competition from theDevonian sea floor, and younger grains of clear halite, which have partially replaced the chevron grains.

Fluctuations in the bromide content of chevron halite within these annual rings provides a measure of seasonal salinity changes, and a comparison can be made between the rate of increase of bromide in halite crystallized from the Devonian sea and the depth of brine from which this halite crystallized. The bromide profiles for chevron halite between anhydrite layers have remarkably steep gradients (Fig. 4). Under static-isothermal conditions of evaporation these bromide profiles would correspond to an initial brine depth of only half a meter. However, static-isothermal conditions probably did not exist in the Saskatchewan basin, and such depths are inconsistent with the proposed conditions of tectonic stability.

The rapid increase and decrease of bromide in the halite seasonal layers is most satisfactorily explained by influxes of sea water or brine that varied periodically in salinity. Following the deposition of anhydrite, the influx increased markedly in salinity and then decreased again before deposition of the overlying anhydrite layer.

REFERENCES CITED

Beloussov, V. V., 1962, Basic problems in geotectonics: New York, McGraw-Hill Book Co., Inc., 809 p.

Borchert, Hermann, 1959, Ozeane Salzlagerstätten: Berlin-Nikolassee, Gebrüder Borntraeger, 237 p.

Braitsch, Otto, 1962, Entstehung und Stoffbestand der Salzlagerstätten: Berlin, Mineralogie und Petrographie in Einzeldarstellungen, Springer-Verlag, 232 p.

Braitsch, Otto, and Herrmann, A. G., 1963, Zur Geochemie des Broms in salinaren Sedimenten: Geochim. Cosmochim. Acta, v. 27, p. 361–391

Briggs, L. I., 1956, Quantitative aspects of evaporite deposition: Mich. Acad. Science, Arts and Letters, v. XLII, p. 115–123

D'Ans, J., and Höfer, P., 1934, Untersuchungen an Brom: Angew. Chemie, Jahrg. 47, no. 5, p. 71–74

Dellwig, L. F., 1953, Hopper crystals of halite in the Salina of Michigan: Am. Mineralogist, v. 38, p. 730–731

—— 1955, Origin of the Salina Salt of Michigan: Jour. Sed. Petrology, v. 25, p. 83–110

Dellwig, L. F., and Briggs, L. I., 1953, Textural relationships in the Salina Salt of Michigan (Abstract): Geol. Soc. America Bull., v. 63, p. 1242

Gottesmann, W., 1963, Eine häufig auftretende Struktur des Halits im Kaliflöz Stassfurt: Berlin, Geologie, Akademie-Verlag, v. 12, pt. 5, p. 576–581

Kühn, R., 1955, Tiefenberechnung des Zechsteinmeeres nach dem Bromgehalt der Salze: Zeits. deutsch. geol. Ges., v. 105, pt. 4, p. 646–663

Lotze, Franz, 1957, Steinsalz und Kalisalze: Berlin-Nikolassee, Gebrüder Borntraeger, 465 p.

Pearson, W. J., 1963, Salt deposits of Canada, p. 197–239 *in* Symposium on Salt: Cleveland, Ohio, Northern Ohio Geol. Soc., Inc., 661 p.

Richter-Bernburg, Gerhard, 1950, Zur Frage der absoluten Geschwindigkeit geologischer Vorgange: Die Naturwissenschaften, v. 37, no. 1, p. 1–8

Schwerdtner, W. M., 1963, Analysis of small amounts of bromides in the presence of large amounts of chlorides, p. 247–248 *in* Symposium on Salt: Cleveland, Ohio, Northern Ohio Geol. Soc., Inc., 661 p.

—— 1964, Genesis of the potash rocks in the Middle Devonian Prairie Evaporite Formation of Saskatchewan: Am. Assoc. Petroleum Geologists Bull., v. 48, p. 1108–1115

Schwerdtner, W. M., and Wardlaw, N. C., 1963, Geochemistry of bromine in some salt rocks of the Prairie Evaporite Formation of Saskatchewan, p. 240–246 *in* Symposium on Salt: Cleveland, Ohio, Northern Ohio Geol. Soc., Inc., 661 p.

—— 1964, Comparison of a Middle Devonian evaporite profile in Saskatchewan with hypothetical model, p. 266–269 *in* Third International Williston Basin Symposium: Saskatchewan Geol. Soc., 287 p.

Sorby, H. C., 1858, On the microscopical structure of crystals indicating the origin of minerals and rocks: Geol. Soc. London Quart. Jour., v. 14, p. 450–500

Valyashko, M. G., 1956, Geochemistry of bromine in the processes of salt deposition and the use of the bromine content as a genetic and prospecting criterion: Geochemistry, v. 1, p. 570–589

Van der Meulen, H., 1931, Bromo-Jodometrische Onderzoekingen: Chemisch Weekblad, v. 28, no. 82, p. 239

Wardlaw, N. C., 1964, Bromide in some Middle Devonian salt rocks of Alberta and Saskatchewan, p. 270–273 *in* Third International Williston Basin Symposium: Saskatchewan Geol. Soc., 287 p.

Manuscript Received by the Society December 7, 1964

Erratum: Page 199, line twenty-five, column 1 should read: "remnants of a once continuous crystal. The"

Reprinted from *Inst. Mining Met., Trans.*, **79**, 155–162 (1970)

Recent halite rock, Baja California, Mexico

D. J. Shearman B.Sc.
Department of Geology, Imperial College, London

21

553.068.22 : 553.291(722)

Synopsis
Layered halite rock is presently forming in depressions on supratidal flats off the mouth of the Colorado River, Mexico. The halite crystals display a chevron structure essentially similar to that which occurs in many ancient layered salt deposits. It is apparent in the field that the layers are not the records of annual events but that each one is the product of a long period of 'reworking' of brine-pan halite and groundwater-brine diagenesis.

In 1966 Holser[1] published an account of the discovery of Recent polyhalite in association with gypsum and halite on supratidal flats around Laguna Ojo de Liebre on the Pacific Coast of Baja California, Mexico. Polyhalite is a common potassium-bearing mineral in many of the ancient marine evaporites, and it is not without significance that the first occurrence in present-day marine sediments should have been found in a supratidal setting. The halite itself was not without interest, however: it forms a 'bedded' deposit, approximately 1 m in thickness, in which layers of halite alternate with laminae of gypsum and detrital sand. The bedding of the deposit is further accentuated by layers of organic matter, composed mainly of marine plants which appear to have been washed on to the halite flats during storms at spring tides. The general setting of the evaporites in relation to the geomorphology and sedimentology of the complex of lagoons and tidal flats was discussed by Phleger.[2] He commented on the fact that the halite was reminiscent of ancient layered salt deposits, and suggested the possibility that they too might have been supratidal in origin.

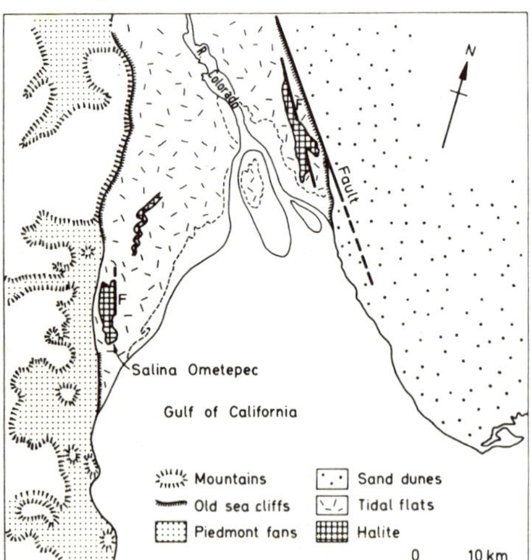

Fig. 1 Tracing from space photograph showing location of halite flats, Baja California, Mexico

Manuscript received by the Institution of Mining and Metallurgy on 16 June 1970. Paper published on 10 August, 1970.

On the basis of the observations of Holser[1] and Phleger[2] the present author visited other coastal halite flats in Baja California in order to establish the extent to which these Recent supratidal halite deposits resembled, or differed from, those of the stratigraphic record. Attention was directed mainly to an area of salt accumulation, known as Salina Ometepec, on the supratidal flats which extend down the east coast of Baja California south of the mouth of the Colorado River (Fig. 1). The sedimentation history of these intertidal and supratidal flats was dealt with by Thompson,[3] and his work provides a useful background for study of the halite deposition. The halite of Salina Ometepec occupies a slight topographic depression near the landward margin of the supratidal flats. Salt is widespread on the surface and its distribution is clearly revealed on the space photograph of the head of the Gulf of California.[4] Over part of the area the surface is underlain by halite rock, which locally reaches 25 cm in thickness. It is with this halite rock that the present paper is primarily concerned, for it has many of the essential petrographic features which characterize ancient layered salt deposits.

Halite flats of Salina Ometepec: general features

When the writer first visited Salina Ometepec at the beginning of March, 1970, part of the halite flat was covered by shallow natural brine pools which were separated from one another by a polygonal network of low ridges composed largely of gypsum sand with accessory detrital silicate grains. By the beginning of April the brine pools had dried out completely, but residual brine remained as groundwater only a centimetre or so beneath the floors of the pans. Professor D. Gorsline (University of Southern California, Los Angeles) had visited the area early in February, and at that time the halite flats were a standing body of water. It seems likely therefore that the flats were flooded by the sea during the January spring tide cycle. If so, then the sea water which was left stranded in the depression had evaporated to completion in the course of a period of approximately two months, leaving thicknesses of up to 8 cm of halite in some of the pans. There can have been little significant loss of water by drainage, because the flats as a whole are built mainly of clayey sediments. The sea did not encroach on the supratidal flats at any time during the spring tide cycle of March, and it appears therefore that the depression is only inundated either in the course of the maximum spring tide cycle of year (January was such an occasion) or when storms coincide with spring tides. Thompson[3] recorded several inundations during his survey of the flats, all of which coincided with maximum spring tide cycles.

Two phases of halite formation occurred in the course of the drying-out of the salt flats during March, 1970. These will be referred to as (1) the brine-pool phase and (2) the dry salt pan phase. Halite precipitation during the brine-pool phase was by crystals nucleated at the surface of the brine. These rapidly formed small pyramidal hopper crystals which hung suspended from the surface. The hoppers aggregated together to form rafts, some of which sank to the bottom, whereas others were blown by the wind across the pools to bank up on the lee-side. Such rafts of pyramidal hopper crystals are of common occurrence in natural and artificial brine pans, and have been described, for example, by Dellwig[5] and by Müller and Irion.[6] The halite rafts are so light that on windy days some of them were blown out from the brine pools and carried across the intervening dry ground. Those rafts which sank to the bottom commonly developed into aggregates of cubes by enlargement of the original pyramidal hopper crystals. By the time the pools had dried out, a loose aggregate of up to 5 cm of these halite rafts had accumulated on the floors of some of the pans.

Once the pools had dried out, small protuberances of halite began to appear as an efflorescence on the surface of the earlier-formed brine-pan halite. This marked the beginning of the dry salt pan phase. Each of these protuberances consisted of an aggregate of minute cubes of halite saturated by brine. The moisture content of the aggregates was so high that they were virtually fluid. It was evident that these efflorescent masses were produced as a result of groundwater brine being drawn up to the surface by capillarity. In the course of only a few days the efflorescent halite increased in abundance, and fast became a dominant feature of the surface. At this time too, the halite rafts which had formed during the brine-pool phase, and were now exposed dry at the surface, began to lose their definition, and the impression was gained that the apparently dry halite was being redistributed in some way. It is possible that this was effected by halite going into solution in dew at night and then reprecipitating the next day. It was at this stage that the writer had to leave the area.

There can be little doubt that part at least of the salt is ephemeral, and that it will dissolve away the next time that the flat is flooded by the sea. The water which is left standing in the depression will thus become enriched in sodium chloride. It can readily be appreciated that successive episodes of marine flooding and evaporation must lead to a progressive increase in the amount of salt. That this has, in fact, happened over the course of time is revealed by the presence of a layer of halite rock beneath the salt pans. In some places the present salt pans rest directly on the halite rock; but, elsewhere, they are separated by a layer of gypsum sand up to 2 or 3 cm in thickness. In hand specimen the halite rock is quite different in appearance from the halite of the overlying salt pans. It is massive, in the sense that it requires use of a pickaxe and much effort to break through it, as it reaches a thickness of up to 25 cm. It is distinctly bedded, with layers of halite from 1 to 8 cm in thickness which alternate with thin layers of gypsum sand (Fig. 2).

B156

Fig. 2 Specimen of halite rock, Salina Ometepec

The rock is highly cavernous both on a macro- and a micro-scale (Fig. 2), and a vertical grain can be seen in the halite layers. The latter results, in part, from the verticality of the smaller dissolution pipes and, in part, from the elongate shape of some of the halite crystals. The layering of the rock is similar to that which is so abundantly developed in ancient evaporites, the difference between them being in the mineralogy of the sulphate laminae, which are gypsum in the Recent and anhydrite in the ancient. However, diagenetic alteration of gypsum to anhydrite is well substantiated, so this point of difference is not significant.

Salt working has been carried out in part of the area over a number of years, and the possibility has to be considered that the subsurface halite may be an artifact resulting from modification of the ground in the course of these operations. At the present time work is restricted to manual scraping of the superficial layer of salt. No machinery appears to have been used apart from trucks. The subsurface salt is present not only in the area of salt working but extends into ground which shows no sign of ever having been disturbed. If it were not for the layer of subsurface salt, which acts as a pavement, it would not be possible to get trucks on to the working area, which is otherwise clay and could not safely support the weight. It appears fairly certain therefore that, even in the area of salt exploitation, the subsurface halite rock is an inherent feature.

The gypsum of the gypsum sand is a precipitate resulting from the evaporation of sea water. Two crystal habits are present, and these often occur intermixed. The most abundantly developed variety is prismatic to short acicular, elongate in the *c*-direction. Many of these crystals show swallow-tail twin re-entrants. The other habit, which appears to be less common, is tabular in the *ac* plane, and these too are often present as swallow-tail twins. The two habits show the same assemblages of crystal faces, but differ in the relative development of these faces. Transitional forms between the two habits are rare and it appears likely that they were formed in response to different physico-chemical conditions at various stages in the desiccation history. Anhydrite is present as an accessory mineral in the gypsum sand.[7] It is not nodular like that in the Recent sediments of the supratidal sabkhas along the Trucial Coast of Arabia but pervades the sand in a manner which resembles an advanced stage of dry rot in wood. However, Kinsman[8] found small nodules and irregular layers of anhydrite in the sediments of the supratidal facies on the same system of flats in the area north of San Felipe.

Gypsum crystals which are totally different in habit from those of the gypsum sand are found scattered in the brown clay. They are lens-like and flat in a plane approximately normal to the *c*-crystal axis and resemble those which grow interstitially in the present-day intertidal zone algal mats of the Trucial Coast.

Petrography of the Recent halite rock

In thin section under the microscope a remarkable similarity is immediately apparent between the Recent halite rock and the halite in the 'seasonal layers' of the Middle Devonian Prairie Evaporite Formation of Saskatchewan, as described by Wardlaw and Schwerdtner[9] (compare Fig. 3 with plate 1, Fig. 2, of reference 9). The Recent halite is best described by quoting from their description: 'it can be seen that the halite grains are conspicuously elongated perpendicular to bedding, and that the chevron structure is invariably arranged with the point of the chevron directed upward'.

'The ... grains are described as chevron halite because of their distinctive structure. In zones parallel to cube faces, each grain contains layers with a high concentration of fluid inclusions alternating with layers that contain few, if any, inclusions. . . . The inclusions are almost perfectly cubical, and in some grains, there is a grading in the size of the inclusions . . .' (compare Fig. 4 with plate 2, Fig. 4, of reference 9).

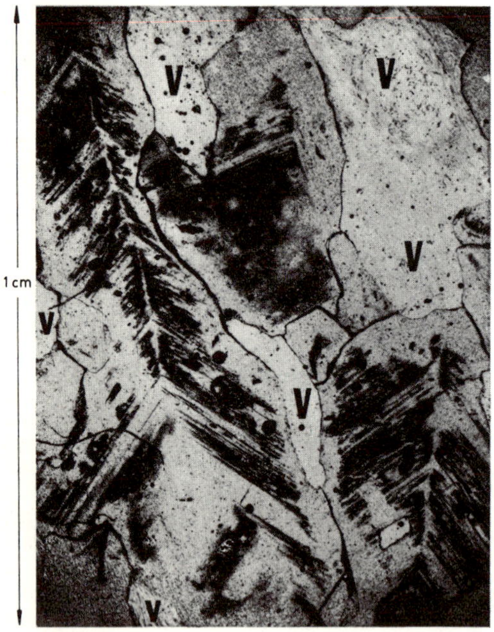

Fig. 3 Photomicrograph of halite rock, Salina Ometepec. Elongate zoned crystals of halite with apices of chevrons directed upwards. Dissolution cavities indicated by V

Fig. 4 Photomicrograph showing detail of brine inclusions in zoned halite. High power magnification

The salt in the Prairie Evaporite Formation consists of two types of halite—the zoned halite previously referred to and clear halite without inclusions. In the Recent halite rock cavities take the place of much of the clear halite of the ancient. If these cavities were to be infilled by clear halite, there would be little if any difference between the two. Halite rocks which display essentially similar petrographic features appear to be a common rock type in many other salt deposits, and have been described from the late Silurian Salina Salt of Michigan, U.S.A., by Dellwig[5] and from the early Permian salt of the Wellington Formation, Kansas, by Jones.[10]

Wardlaw and Schwerdtner[9] argued that the elongate shape of the grains of zoned halite, their near-vertical orientation and upward-pointing chevrons were the results of competitive growth between halite crystals which grew upwards from a surface at one and the same time. The fabrics of the Recent halite support this suggestion. It is apparent from the geometry of a cube that if a halite crystal is growing uniformly, the coigns will advance more rapidly than the edges and the edges more rapidly than the cube faces. Thus crystals which are initiated with a coign or an edge directed away from the surface will rapidly override those which have a cube face parallel to the surface (Fig. 5(*a*)). The

B157

more favourably orientated cubes continue to grow upwards and produce near-vertical elongate crystals with compromise boundaries between them. The zones within the halite crystals are parallel to the cube faces, and, in consequence, the zoning in the elongate crystals appears as chevrons which have their apices directed upwards. Thus chevron halites are not a special variety of halite but are simply halite crystals which are zoned parallel to the cube face. The elongate shape and upward-directed chevrons are merely the expression of a particular manner of growth.

In the Recent halite rock many of the crystals consist in part of zoned halite and in part of clear halite (Figs. 3 and 6). In individual crystals whole groups of zones are abruptly truncated and replaced laterally by clear halite. The boundaries between the zoned and clear areas of these crystals are commonly arcuate, and clear areas often cut deep embayments into the zones. Reference was made earlier to the dissolution cavities which have locally removed parts of the halite crystals. These cavities embay the zoned crystals, truncate the zones and exhibit precisely the same relationships as do the areas of clear halite. It is immediately apparent that if the cavities were to be infilled with clear halite in lattice continuity with that of the embayed grains, they would become indistinguishable from the other areas of clear halite in relation to the zoned parts of the crystals. Fig. 6 illustrates a former dissolution cavity which has been partly infilled with clear halite. There can be little doubt therefore that the areas of clear halite record

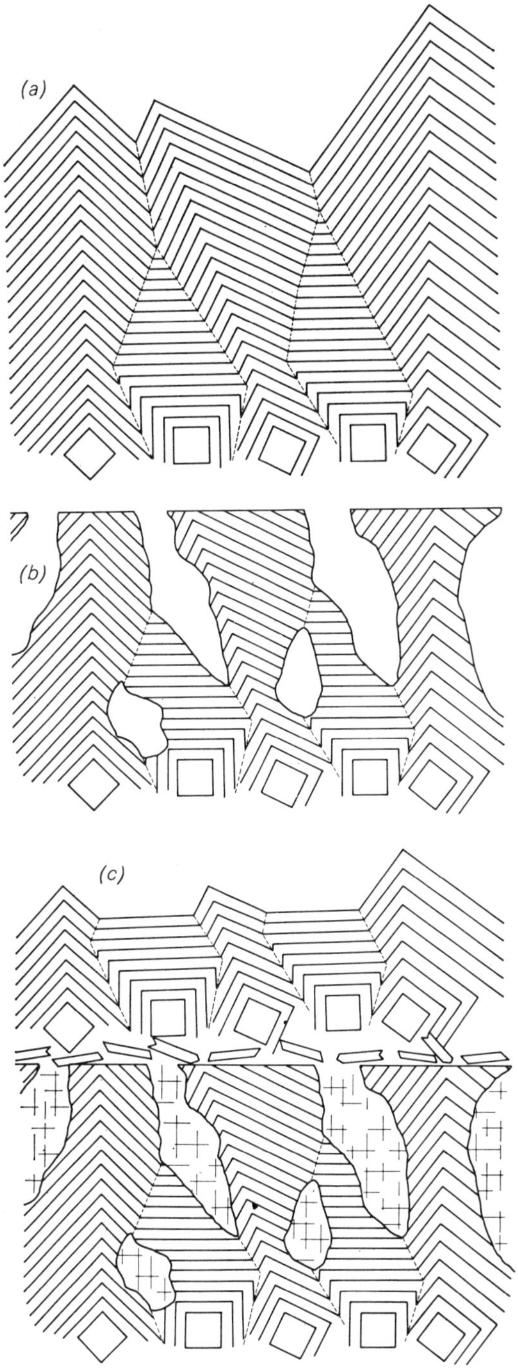

Fig. 5 Schematic diagrams showing mode of development of layered halite rock

B158

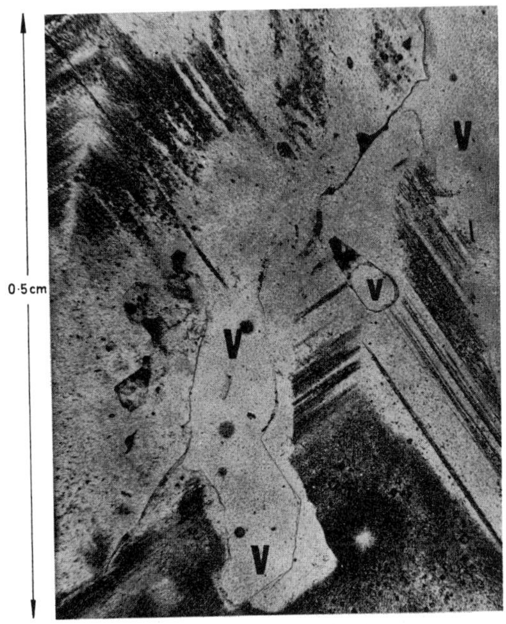

Fig. 6 Photomicrograph of halite rock, Salina Ometepec. Former dissolution cavity in zoned halite partly infilled with clear halite (V = cavity)

former dissolution hollows. Many of the present cavities cut across not only zoned areas but also across clear parts of the same crystals. This whole arrangement of cavities, areas of clear halite and areas of zoned halite within one and the same crystal testifies to a history of repeated episodes of dissolution and precipitation. Such processes would account for the consertal outlines of the boundaries which characteristically occur between adjacent clear grains and between clear grains and zoned ones. The clear halite is diagenetic and the zoned areas within the crystals are all that now remain of the original halite. This conclusion assumes, however, that the original crystals were zoned throughout—an assumption which may not be absolutely true, but which is probably a valid generalization. It can be readily appreciated that all transitions could be generated between a rock which consists essentially of chevron halite on the one hand and clear halite on the other, depending on the extent of the penecontemporaneous reconstitution.

Such an explanation for the relationships between zoned and clear halite would not be at variance with the petrographic observations of Wardlaw and Schwerdtner[9] on the halite of the seasonal layers in the Prairie Evaporite Formation of Saskatchewan.

Analogous rock types can be found in other salt deposits. The layered salt in the late Silurian Salina Salt of Michigan, U.S.A., exhibits a superficially similar relationship. In this deposit Dellwig[5] argued that, where zoned and clear halite are present in the one crystal, the zoned areas mark the outlines of former pyramidal hoppers and the clear areas are later overgrowths in lattice continuity. In the Recent halite rock the geometry of the zones in halite crystals is not consistent with such an interpretation.

There is a strong tendency for many of the vertical dissolution cavities to develop along the boundaries between adjacent halite crystals. This is shown not only by the distribution of the present generation of cavities but is also implied by the presence of a narrow zone of clear halite which commonly intervenes between the zoned parts and the edges of the crystals (Fig. 3). Gottesmann[11] described analogous anhydrite–halite rocks, kieserite–halite rocks and carnallite–halite rocks from the Zechstein of Germany. In these the anhydrite, kieserite and carnallite display a textural relationship to the zoned crystals of halite similar to that shown by the cavities and the zoned halite in the Recent sample.

The upper surface of the Recent halite rock is a dissolution surface, and many of the present generation of cavities pipe down from this surface (Fig. 5(*b*)). This much is evident in the field. It is also revealed under the microscope by the fact that the zones within the uppermost halite crystals are truncated at the surface. Older dissolution surfaces can also be identified immediately beneath some of the gypsum layers (Fig. 5 (*c*)). In some places small lobes of gypsum sand cemented by clear halite hang down from the surfaces in a manner which indicates that gypsum sand had lodged in the upper parts of former dissolution pipes. Similar dissolution surfaces were described by Wardlaw and Schwerdtner[9] in the Prairie Evaporite and by Dellwig[5] from the Salina Salt.

On each occasion that the area is flooded by the sea the water will rapidly approach saturation by dissolving halite from the salt pans of the previous evaporation cycle. Some of this new brine will percolate down into the underlying salt rock, where it will cause a measure of dissolution. Downward movement of the water is evidenced by the vertical trend of many of the cavities and by the fact that the upper surface of the salt rock is a dissolution surface. Solar evaporation of the overlying body of brine will bring about the precipitation of a new generation of brine-pan halite in the manner described earlier. After completion of the salt-pan stage, the brines which are now subsurface become further concentrated by capillarity and evaporation. It is probably during this late stage in the evaporation cycle that the diagenetic, clear halite is precipitated in the cavities of the underlying halite rock.

Problem of origin of the halite rock

The origin of the halite rock poses problems, as it is apparently unlike the various forms of halite which were found to be forming in the brine pools above. Three possibilities come to mind: (1) that it was produced by modification of former brine-pan halite; (2) that it formed in brine pools, but under conditions slightly different from those presently prevailing; or (3) that it was generated *de novo* in the sediments beneath the floors of the surface pans. The fact that the early generation of crystals in the halite rock are zoned with inclusions, many of them displaying the elongate chevron habit, suggests that they were formed at the surface. All the samples of brine-pool halite that have been examined so far are characterized by an abundance of brine-filled inclusions, and in this respect they contrast markedly with the later generation of inclusion-free clear halite which precipitated from the groundwater brines. Inclusions of brine suggest that the crystals grew rapidly, and the close spacing of the zones would therefore appear to record repeated episodes of rapid precipitation. By virtue of exposure to the atmosphere the brine-pool environment is one of abrupt changes in physical conditions. The writer has seen, for example, a whole crop of pyramidal hopper crystals initiated at the surface of the brine apparently as a direct result of a wind squall blowing over the flats. The crystal fabrics of the elongate crystals and the upward-pointing chevrons testify to growth from a surface, as was argued by Wardlaw and Schwerdtner[9] and also by Gottesmann.[11] Such a fabric is unlikely to develop where crystals grew by displacement of a host. Although, on balance of the evidence, the halite appears to have originated at the surface, the problem of its precise mode of origin must remain open until field relationships have been more fully investigated.

B159

Possible tectonic setting of the halite rock at Salina Ometepec

The halite at Salina Ometepec occupies a depression, albeit a very slight one, near the landward margin of the supratidal flats. Such a depression is an anomalous feature not readily explicable in terms of tidal-flat and supratidal-flat evolution. The Gemini IV spacecraft photograph of the head of the Gulf of California[4] shows that the eastern boundary of the salt is virtually a straight line trending north–south (see Fig. 1). There is also a large halite-filled depression on the supratidal flats of the east side of the Colorado Delta. This too has a strong linear trend along part of its eastern boundary. The feature lies parallel to, and only a few kilometres from, the fault which forms the eastern boundary of the Sonoran Mesa. Such linear features are alien in a normal supratidal environment, and the probability that they are fault-controlled has to be considered. If this is so, then the halite depressions may be the direct results of present-day tectonic subsidence. Other lines of evidence support this possibility. To the north and south of Salina Ometepec the supratidal flats are backed by low cliffs cut in the toes of the piedmont fans which spread down from the mountain backbone of Baja California. However, cliffs are absent in the vicinity of the halite accumulation, where one can walk directly without impediment from terrestrial piedmont fan on to marine supratidal flat in a matter of yards. There is no apparent reason why cliffs should not have been cut along this section, and one is tempted to enquire whether they do not now lie buried beneath the inner part of the halite flat. It was in this area that Thompson[3] recorded marked differences from the normal stratigraphic sequence of the tidal-flat sediments. He found, for example, that the former lower intertidal and subtidal facies which can be traced in borings under the whole of the supratidal flats are displaced downwards by several metres in the region of the halite flats. In places the former intertidal sediments are overlain by gypsiferous muds which extend down as much as 2 m below present mean sea level. Thompson explained this local anomaly as the 'product of upward growth of the mud flats and depositional regression accompanying the final stages of late Wisconsin sea level rise . The alternative—a present-day tectonic depression—merits consideration.

Discussion on concept of annual layering in salt deposits

In an earlier paper[12] a general supratidal origin was argued for nodular anhydrite rocks on the basis of similarities between these rocks and the nodular anhydrite which is presently being formed in the Recent sediments of the supratidal flats or coastal sabkhas along the Trucial Coast of the Persian Gulf. Where such an origin is accepted for nodular anhydrite rocks, it poses the problem of environment of deposition of the halite with which they are often associated. Such an association casts doubt on the validity of the commonly held belief that the halite accumulated beneath a deep body of brine. The crystal fabrics of many halite rocks show evidence of widespread mineral replacements, and in some instances they also display evidence of crystal growth by displacement. This led the writer to explore the possibility that they might be post-burial in origin, and the products of reactions involving connate brines and 'sabkha measures'. Subsequently, however, the writer had the opportunity to examine a number of borehole cores from thick halite successions, and it became evident that although some halite rocks clearly originated in the manner suggested, the mechanism was far from being of universal application. Holliday[13] expressed the contention that layered salt with chevron-type halite could develop in a supratidal setting. The finding of these features in the supratidal halite at Salina Ometepec has confirmed his conviction.

The virtual complete petrographic identity between the Recent supratidal halite rock and ancient layered salt deposits impels enquiry into the validity of the arguments which have been used in support of a deep-water origin for these sediments.

Support for the deep basin theory appears to rest in large part on the assumption that the layers of salt record annual precipitation cycles. Because of the thicknesses of individual layers and total thickness of the deposit, it would have required excessively rapid rates of relative subsidence to accommodate the salt had it all been deposited in a shallow-water or emergent environment. Two examples will suffice to illustrate the argument. Richter-Bernburg[14] made careful counts of halite 'Jahresringe' over wide areas of the Zechstein of Germany, and obtained rates of accumulation in the order of 500 m in 8000 years. Wardlaw and Schwerdtner[9] estimated, on the basis of the average thickness of the 'seasonal layers', that the 600-ft salt succession of the Middle Devonian Prairie Evaporite Formation of Saskatchewan was deposited in approximately 4000 years. To postulate a shallow-water environment for such rates of sedimentation would have required a continuous background of subsidence of between 4 and 6 cm per year. If, however, the layers were not annual, then one of the prime objections to a shallow-water origin of the salt is removed. The evidence from the Recent salt of Baja California challenges the Jahresringe concept, because it is clear that each of the layers of the halite rock is the product of a long period of repeated reworking of brine-pan halite and groundwater-brine diagenesis. At Ojo de Liebre Phleger[2] also concluded that the layers in the Recent salt do not record yearly events.

Richter-Bernburg[14] estimated that the entire column of Zechstein sediments in Germany represents a period of 500 000 years of sedimentation. This is but a small part of the 5–10 m.y. believed to be the span of the Zechstein. He used this discrepancy as an illustration of the general principle that more of geological time is represented by breaks in the stratigraphic record than is preserved in the rocks themselves. However, Borchert and Muir[15] pointed out that such an argument is not readily applicable to what are supposed to have been chemically precipitated sediments in a deep brine-filled basin If the sediments represent only one-tenth of the time available, one is entitled to ask what was going on in the chemically complex brine-filled basin for nine-tenths of Zechstein time.

If, as it is reasonable to suppose, the abundant brine inclusions of the zoned halite crystals and the multiplicity of zones in any one crystal testify to repeated episodes of rapid crystal growth, these characteristics would argue against the halite crystals having grown beneath a deep body of brine. In such an environment one would expect physico-chemical conditions to have remained relatively uniform, and that those changes which did occur were gradual ones. In some of the chevron halites illustrated by Wardlaw and Schwerdtner[9] from the layers in the Prairie Evaporite there is an abundance of distinct zones within any one crystal. If each zone represents an episode of rapid growth, the assemblage of zones within each crystal would imply that the brine at the bottom of the basin was in a state of perpetual physico-chemical turmoil for the greater part of every year.

Differences between environments of formation of nodular anhydrite and layered halite

Nodular anhydrite and layered halite are among the most abundant rock types developed in ancient marine evaporites. All the present-day comparable sediments have been found forming in supratidal environments—namely nodular anhydrite in coastal sabkhas along the Trucial Coast and layered halite in salinas in Baja California. It may be pertinent to a better understanding of ancient evaporites to enquire into the reasons why nodular anhydrite with only accessory halite should form in the one locality, whereas layered halite with disseminated anhydrite occurs in the other.

The nodular anhydrite of the Trucial Coast is, with few exceptions, confined to the sediments above the rest level of the marine-derived groundwater, where the nodules grow by displacement from capillary brines. The halite of the layered halite rock of Baja California appears to have been precipitated initially from a standing body of brine and was subsequently modified texturally in the groundwater zone. Two differences are immediately apparent: first, nodular anhydrite requires the presence of a preexisting host sediment, whereas layered halite does not; and, secondly, nodular anhydrite develops in raised positions relative to sea level and the groundwater table. Layered halite, on the other hand, is formed in depressed environments where marine waters are impounded.

Accumulation of a thick succession of nodular anhydrite or layered halite necessitates a background of relative subsidence. However, the conditions which have to be superimposed on this background to generate nodular anhydrite, on the one hand, or layered halite, on the other, are mutually exclusive. Where there is a continuous supply of non-evaporitic sediment over and above that necessary to balance the subsidence, an emergent setting is created, and this would favour generation of nodular anhydrite. Such a continuous and massive influx of sediment is inimical to brine-pan development. It would be expected therefore that, in ancient evaporites, successions of nodular anhydrite should be relatively free from layered halite, and layered halite successions relatively free of nodular anhydrite. Further, one would predict that nodular anhydrite should be interbedded with non-evaporitic sediments, e.g. carbonates, clastics or organic matter, whereas non-evaporitic materials should be only an accessory at the most in layered halite successions. These predictions apply in the majority of instances, but there are, nevertheless, some exceptions.

Acknowledgement

Field work on the halite flats of Baja California, carried out in the spring of 1970, was financed by a grant from the Natural Environment Research Council, London. In 1968 the writer was able to visit Canada to study borehole cores of evaporites in the Provincial Sub-surface Geological Laboratories at Calgary and Regina by virtue of awards from the J. Vernon Hobson Bequest of the Institution of Mining and Metallurgy and the J. B. Tyrrell Fund of the Geological Society of London. The author expresses his thanks to all these bodies.

The author also expresses his gratitude to Dr. P. Roehl and Professor D. Gorsline for their companionship in the field, and to Dr. J. Bishof, Mr. P. R. Bush, Dr. G. P. Butler, Dr. J. G. C. M. Fuller, Dr. W. T. Holser and Miss M. Pugh for helpful discussion.

References

1. **Holser W. T.** Diagenetic polyhalite in Recent salt from Baja California. *Am. Miner.*, **51**, 1966, 99–109.
2. **Phleger F. B.** A modern evaporite deposit in Mexico. *Bull. Am. Ass. Petrol. Geol.*, **53**, 1969, 824–9.
3. **Thompson R. W.** Tidal flat sedimentation on the Colorado River delta, Northwestern Gulf of California. *Mem. geol. Soc. Am.* 107, 1968, 133 p.
4. **Smith J. T. Jr. ed.** *Manual of color aerial photography* (Falls Church, Virginia: American Society of Photogrammetry, 1968), 550 p. (Fig. 9.1.1, Gemini IV, 5, June, 1965)
5. **Dellwig L. F.** Origin of the Salina salt of Michigan. *J. sedim. Petrol.*, **25**, 1955, 83–110.
6. **Müller G. and Irion G.** 'Salt biscuits'—a special growth structure of NaCl in salt sediments of the Tuz Gölü ('Salt Lake', Turkey). *J. sedim. Petrol.*, **39**, 1969, 1604–7.
7. **Butler G. P.** Paper to be published in *Proc. 3rd symposium on salt* (Cleveland, Ohio: Northern Ohio Geological Society), in press.
8. **Kinsman D. J. J.** Modes of formation, sedimentary associations and diagnostic features of shallow-water and supratidal evaporites. *Bull. Am. Ass. Petrol. Geol.*, **53**, 1969, 830–40.

9. **Wardlaw N. C. and Schwerdtner W. M.** Halite–anhydrite seasonal layers in Middle Devonian Prairie Evaporite Formation, Saskatchewan, Canada. *Bull. geol. Soc. Am.*, **77**, 1966, 331–42.

10. **Jones C. L.** Petrography of evaporites from the Wellington Formation near Hutchinson, Kansas. *Bull. U.S. geol. Surv.* 1201–A, 1965, 70 p.

11. **Gottesmann W. von.** Eine häufig auftretende Struktur des Halits im Kaliflöz Stassfurt. *Geologie*, **12**, 1963, 576–81.

12. **Shearman D. J.** Origin of marine evaporites by diagenesis. *Trans. Instn Min. Metall.* (*Sect. B: Appl. earth sci.*), **75**, 1966, B208–15.

13. **Holliday D. W.** Contribution to discussion of reference 12. *Trans. Instn Min. Metall.* (*Sect. B: Appl. earth sci.*), **76**, 1967, B179–80.

14. **Richter-Bernburg G.** Zur Frage der absoluten Geschwindigkeit geologischer Vorgänge. *Naturwissenschaften*, **37**, 1950, 1–8.

15. **Borchert H. and Muir R. O.** *Salt deposits: the origin, metamorphism and deformation of evaporites* (Princeton, N. J., and London: D. Van Nostrand, 1964), 300 p.

Editors' Comments on Papers 22 and 23

22 **Anderson and Kirkland:** *Intrabasin Varve Correlation*

23 **Braitsch:** *Rhythmic Bedding and Influxes*

Temporal and Spatial Correlation

The almost ubiquitous presence of minor cycles in evaporite sequences, and the fact that those cycles are repeated over thicknesses of hundreds, even thousands of meters, pose a problem that ultimately bears on many aspects of the geology of saline deposits. Do these cycles represent seasonal accumulations similar to tree rings or varves in glacial lakes? To provide an answer to that question has been the goal of many geologists over the years, especially since German geologists first used the term "Jahresringe" to designate the minor repetitions.

To Richter-Bernburg, now director of the German Geological Survey, must go credit for demonstrating that evaporite complexes can be correlated over considerable distances and must therefore have regional significance. He measured the thickness of thousands of couplets from different cores from the second Zechstein cycle and, using these measurements as a basis for comparison, correlated one with the others (Richter-Bernburg, 1957, 1959, 1960), eventually covering a distance of 350 km (Richter-Bernburg, 1964).

In a paper included in this part, Anderson and Kirkland (1966) established definitive correlations of "Jahresringe" in the Jurassic Todilto Formation and the Permian Castile Formation, and demonstrated by statistical analysis that Richter-Bernburg's correlations over large distances in the Zechstein basin are, indeed, valid. It has since been shown that individual laminae in the Castile Formation can be correlated precisely for at least 113 km in an east–west direction and for at least 61 km in a north–south direction (Anderson et al., 1972).

R. Y. Anderson's interest in microstratigraphy has led to many diverse research efforts, from paleoecology of Pleistocene lacustrine sediments to investigation of rate of deposition in modern lakes in North America, as well as the study of evaporite "Jahresringe." He is now Professor of Geology at the University of New Mexico and his association with Kirkland began while the latter was a graduate student there.

With the knowledge that the couplets are features of regional dimension, the thought became compelling that some climatic control caused the mineralogical change during deposition. The similarity in form to tree rings is so striking that many investigators have been led to extend the similarity to the mode of origin, and to conclude that these couplets are, indeed, varves.

In the accompanying excerpt from his fundamental work on the geochemistry of evaporites, Braitsch echoes the caution of Lotze (1957) and warns against hasty and premature acceptance of all minor cycles as annual deposits. In particular, he indicates the difficulty of believing that the 11-year sunspot cycle can be identified by periodic (and subjectively identified) changes in thickness of the couplets. In fact, the presence of the sun-spot cycle in the "Jahresringe" of Richter-Bernburg has not been confirmed by objective techniques of harmonic analysis such as those used by Anderson and Koopmans (1963) on various varve-like series and by Bryson and Dutton (1961) on tree-ring data.

The other important aspect of Braitsch's work is in providing one of the few reasonable explanations of the origin of evaporite couplets. Recognizing the significance of the fact that the 37 ‰ salinity contour in the Mediterranean Sea fluctuates annually from 18°E to 2°E longitude, he reasons that a similar annual fluctuation of a contour of higher salinity in an evaporite basin could produce the precipitation of one salt during spring and summer, and another during autumn and winter.

Reprinted from *Geol. Soc. Amer. Bull.*, **77**, 241–255 (1966)

ROGER Y. ANDERSON *Dept. Geology, University of New Mexico, Albuquerque, New Mexico*
DOUGLAS W. KIRKLAND *Field Research Laboratory, Socony Mobil Oil Company, Inc., Dallas, Texas*

22 Intrabasin Varve Correlation

Abstract: Statistical correlation techniques have been applied to the stratigraphically correlated laminations in four sedimentary basins to demonstrate the degree of lateral variation of certain types of layers and to help determine their interrelationships. Laminae thicknesses in a varved sequence deposited in a small Pleistocene lake in Texas have a very high correlation coefficient (+0.90) over a distance of 275 m. Laminae in the Jurassic Todilto Formation of New Mexico were correlated stratigraphically over a distance of 5.3 km and would probably be traceable over a much greater distance except for the disturbing effects of currents on the bottom of the basin. Five contiguous Todilto sections over a distance of 3.7 km have correlation coefficients of +0.85 or greater. Couplets of bituminous calcite and anhydrite in the Permian Castile Formation of Texas correlate stratigraphically over a distance of 14.2 km with a coefficient as high as +0.99. In a section of 74 couplets 27.4 m above the base of the Castile, anhydrite laminae thicknesses have a correlation coefficient of +0.97; calcite laminae, +0.72; and organic-rich bands in the calcium carbonate laminae, +0.73. There is no relationship between the thickness of associated calcite and anhydrite laminae. Richter-Bernburg measured and subjectively correlated anhydrite laminae thicknesses in cores of the Permian Zechstein Formation in Germany. Correlation coefficients were obtained using his data which range from +0.85 to +0.32 over distances up to 290 km. Zechstein statistical correlation deteriorates with distance. If Castile correlation changes at the same rate, it should be possible to correlate laminae stratigraphically over the maximum length of the basin.

CONTENTS

Geological Society of America Bulletin, v. 77, p. 241–256, 8 figs., 5 pls., March 1966

INTRODUCTION

Part of the material deposited in most sedimentary basins is formed by processes that precipitate chemical or organic substances directly from water in the basin. In many cases this material is precipitated seasonally to form laminations that are preserved in a relatively undisturbed condition. The laminations provide a clue to the lateral continuity of this type of process as well as the associations of various kinds of substances including clastic deposits. This paper is mainly concerned with lateral variation and uses some statistical measures to help study the nature of this variation in several kinds of laminations.

The seasonal rain of particles onto the floor of a sedimentary basin can form laminations in which each repetition of a component represents 1 year (a varve). Lateral variations of glacial varves within a single basin have been investigated (Antevs, 1951), but nonglacial seasonal layers, considered in this report, are more common in the geologic record. They generally are composed of clastic or chemically and organically precipitated material which accumulates seasonally. *Seasonal* is used in this report to indicate that a layer or process occurs during a particular season of the year: it need not be annual. *Annual* is used to indicate that the process occurs almost every year.

Some groups of laminae formed from a widespread rain of precipitated particles are not strictly annual deposits: others are precisely annual: still others may be deposited in an unpredictable pattern. Proof of annual deposition will probably be obtained only through studies of the microstratigraphic position of fossils or by observing modern sedimentation. The microstratigraphic position of fossils has been used to demonstrate the varved nature of one of the examples in this study, and the writers are reasonably confident that the other examples are seasonal (and probably annual) accumulations: therefore, interpretations and explanations are phrased in these terms.

This work was carried out under Research Grant GP-742 from the Earth Sciences Division of the National Science Foundation. Kirkland's work was partly carried out under tenure as a National Science Foundation Postdoctoral Fellow at the University of New Mexico. The authors wish to thank John P. Bradbury, Walter E. Dean, Jr., and Henry I. Snider for helpful suggestions during the course of the study. They also wish to thank Mr. Jax Cowden, owner of the 9K Ranch, Culberson County, Texas, for permission to obtain a core of the Castile Formation. Thickness measurements used in computing coefficients of correlation are available from the ADI Auxiliary Publications Project.[1]

INTRABASIN STRATIGRAPHIC CORRELATIONS

Types of Examples

Four examples will be used to demonstrate the nature of lateral varve variation; they will be discussed in order of increasing distance correlated. The first (Rita Blanca lake beds, Blancan, Texas) demonstrates that the deposition of calcium carbonate was a seasonal and annual process. The second (the limestone of the Jurassic Todilto Formation of New Mexico) shows how potentially near-perfect statistical correlation can be destroyed by redistribution of precipitated material within a basin. The third (Castile Formation, Permian, Texas) has the highest coefficients of correlation and shows how statistical correlation techniques can be used as stratigraphic and interpretive tools. The fourth (analysis of the data presented by Richter-Bernburg on the anhydrite laminations in the Permian Zechstein basin in Germany) shows how varve correlation changes over long distances.

Pleistocene Rita Blanca Lake Beds

Rita Blanca Lake filled a small deflation basin on the High Plains of Texas (Fig. 1A) during a relatively moist interval near the beginning of the Pleistocene. The lake had a maximum diameter of 3 km. Recent erosion has exposed the laminated sediments which were formed in the deeper part of the lake. The water was apparently chemically stratified: oxygen depletion and high salinity or alkalinity in the bottom water eliminated a significant benthic fauna and was indirectly responsible for the preservation of the laminations. The varves consist of couplets of silty, light-colored laminae high in $CaCO_3$ (up to 60 per cent) and darker, relatively organic-rich clay laminae low in $CaCO_3$ (up to 20 per cent). Fossil insects, fishes, ostra-

[1] These have been deposited as Document number 8826 with the ADI Auxiliary Publications Project, Photoduplication Service, Library of Congress, Washington, D.C. 20540. A copy may be secured by citing the Document number and by remitting $1.25 for photoprints, or $1.25 for 35-mm microfilm. Advance payment is required. Make checks or money orders payable to: Chief, Photoduplication Service, Library of Congress.

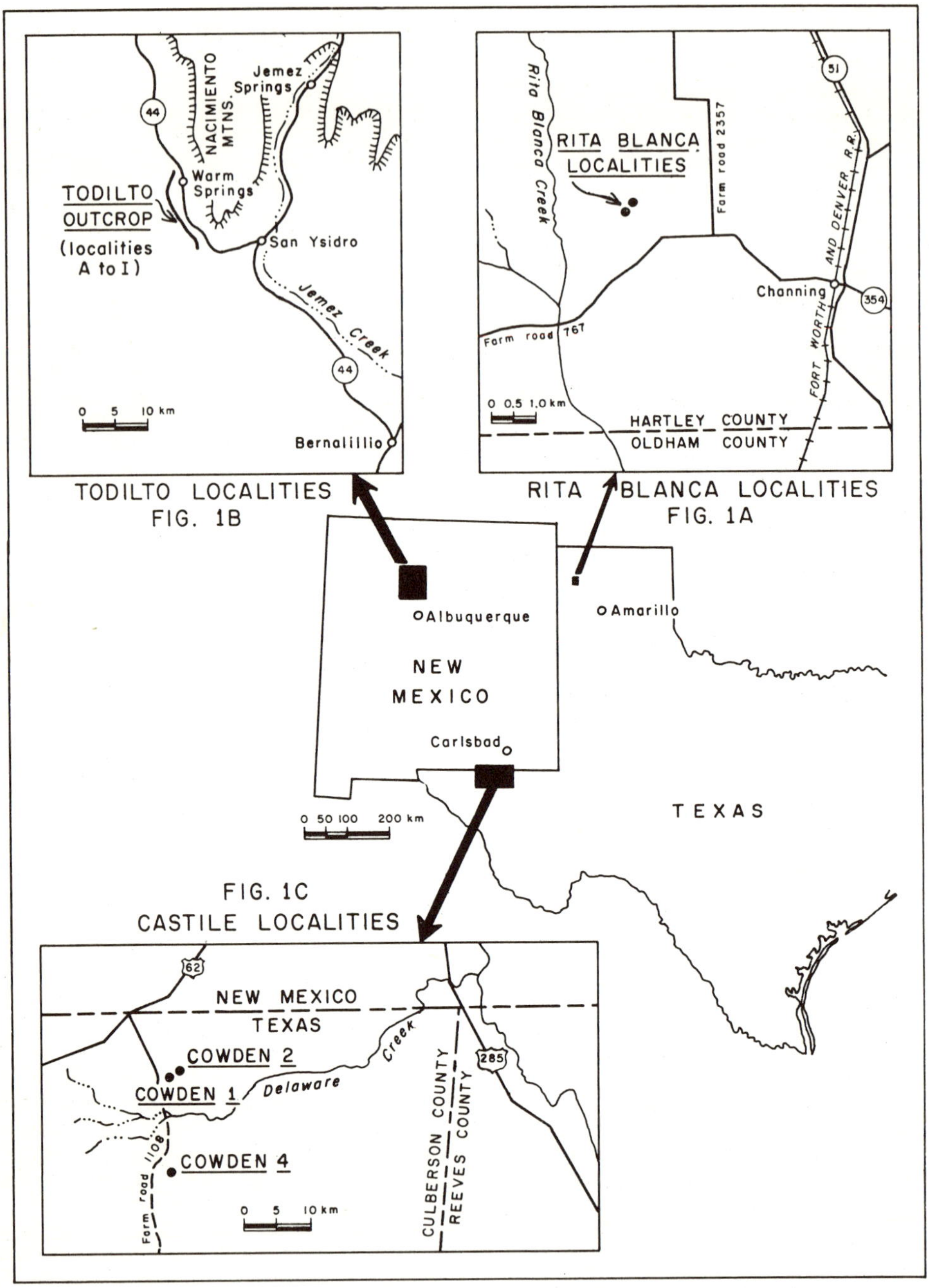

Figure 1. Location of Rita Blanca, Todilto, and Castile varve series

codes, algae, leaves, spores, and pollen are associated with the laminae, and the microstratigraphic position of seasonally introduced members of this population has established beyond a reasonable doubt that the couplets are varves. A comprehensive report on these deposits is now being prepared.

Two outcrops of laminated lake beds are exposed in arroyos. These are about 275 m apart and appear to be approximately at a right angle to the ancient shore line. The near-shore outcrop is 20 per cent thicker mainly because of greater amounts of $CaCO_3$. Seasonal precipitation of calcium carbonate is the principal reason why the varves are so easily distinguishable. Stratigraphic correlation of the varves was established subjectively by comparing the thickness and color of laminae in blocks collected from the two outcrops (Fig. 1A). Thickness measurements of the couplets were taken from enlarged photographs of the slabs (Pl. 1, fig. 1). The correlation coefficient[2] for couplet thickness was computed for a sequence of 21 couplets (+ 0.90). The excellent statistical correlation over a distance of 275 m suggests that stratigraphic correlation could be established over the entire "central plain" of the lake upon which laminated sediments were deposited.

Jurassic Todilto Formation

The Jurassic Todilto basin was a very large (minimum area during limestone deposition of at least 89,400 km²) and probably shallow depression mostly in northwestern New Mexico. The writers (1960) have discussed the basin and petrography elsewhere. The basin may have had a connection to the Curtis sea to the north, but a marine origin for the salts in the basin is not firmly established. A laminated limestone was deposited in the basin; assuming that the laminations are seasonal and the couplets are annual, this deposition phase lasted only about 20,000–30,000 years. The limestone grades upward into a gypsum member of less areal extent. In a section collected 51–58 inches above the base of the limestone, 4 miles south of Warm Springs (Fig. 2, near locality F), the average thickness of 1500 couplets was 0.13 mm. The couplets consist of laminae of calcium carbonate and sapropel. Silt grains were deposited within and on top of the sapropel layers. Rare micro-oölites with an average diameter of about 0.1 mm occur in pods in the low part of undulations in the sapropel layer.

The large size of the basin suggested that stratigraphic varve correlations of many kilometers might be obtainable, and the thickness of the limestone (approximately 2 m) made it possible to restrict the search to a zone of about 1 m (Pl. 1, fig. 2). Slabs were collected near the center of the basin from outcrops separated by 15 and 58 km. The slabs were polished, etched with dilute HCl, and stained so that sapropel laminae stood out in relief and were easily visible. Stratigraphic correlation on the basis of variations in calcite laminae thickness was then attempted by visual comparison of the slabs. Correlation was unsuccessful, and more collections were made over a distance of 10.3 km along a continuous outcrop extending south from Warm Springs along New Mexico Highway 44 (Fig. 2). The localities were from 170 m to 4.6 km apart, and stratigraphic correlation was ultimately established between these contiguous sections over a total distance of 9.7 km (Figs. 1B, 2). The validity of the stratigraphic correlations was corroborated by four observers except for some disagreement about the correla-

---

[2] Pearson's coefficient of correlation

$$r = \frac{N\Sigma XY - (\Sigma X) \quad (\Sigma Y)}{\sqrt{N\Sigma X^2 - (\Sigma X)^2}\sqrt{N\Sigma Y^2 - (\Sigma Y)^2}}$$

(where X and Y are variables and N is the number of units correlated) has been used as a estimator of the degree of perfection of correlation in this study. Perfect positive and negative correlations are expressed as +1.0 and −1.0, respectively.

The trends inherent in most natural time series present special problems in interpreting coefficients of correlation. Cyclic trends that by chance agree with a similar trend in another series might augment the correlation value and make it appear significant even when no stratigraphic correlation exists (*see* footnote 5). Eliminating the trend by the first-difference method (deviation of each thickness from the previous thickness in the series) is only a partial solution, because the long trend may be as important a factor in correlation as the shorter fluctuations. Rank correlation and an alternative technique recommended by Burnaby (1953) also fail to entirely eliminate errors caused by cyclic trends, and rank correlation does not lend itself to the sliding correlation technique used and discussed in a later section. Pearson's coefficient has been used in the study as an estimate of the correlation of time series without the use of significance tests and, when possible, with a large number of values in order to help minimize the influence of the shorter trends. In all cases, statistical correlations were computed for the same number of equivalent stratigraphic units. Variations in correlation coefficients are the result of relative variations in laminae thickness and not due to the occurrence of extra or missing layers.

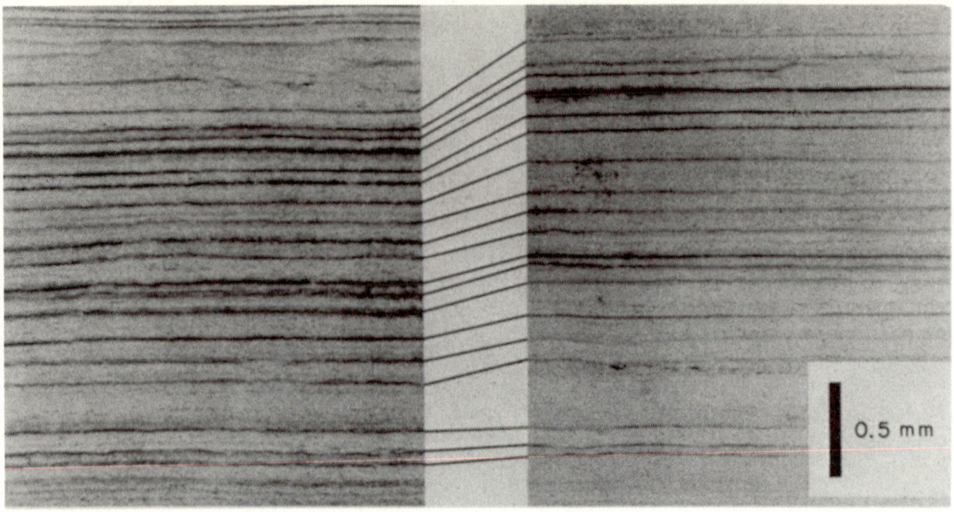

Figure 1. Correlative slabs of Rita Blanca laminations from localities separated by 275 m showing light carbonate-rich and dark organic-rich laminae

Figure 2. Typical exposure of Jurassic Todilto Formation along New Mexico Highway 44, near Warm Springs, New Mexico. The limestone member is about 2 m thick at this locality

CORRELATIVE VARVE SERIES OF RITA BLANCA LAKE BEDS: OUTCROP OF TODILTO FORMATION, NEW MEXICO

ANDERSON AND KIRKLAND, PLATE 1

Geological Society of America Bulletin, volume 77

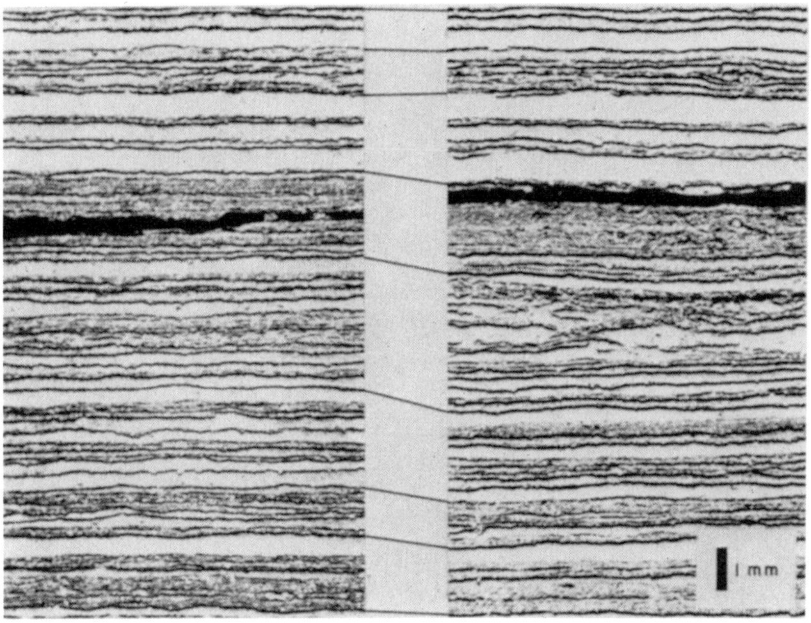

Figure 1. Etched and stained correlative slabs of Todilto limestone from localities D (left) and H, separated by 3.7 km. Note the disrupted sapropel layers in H. Light layers are calcite, dark layers sapropel

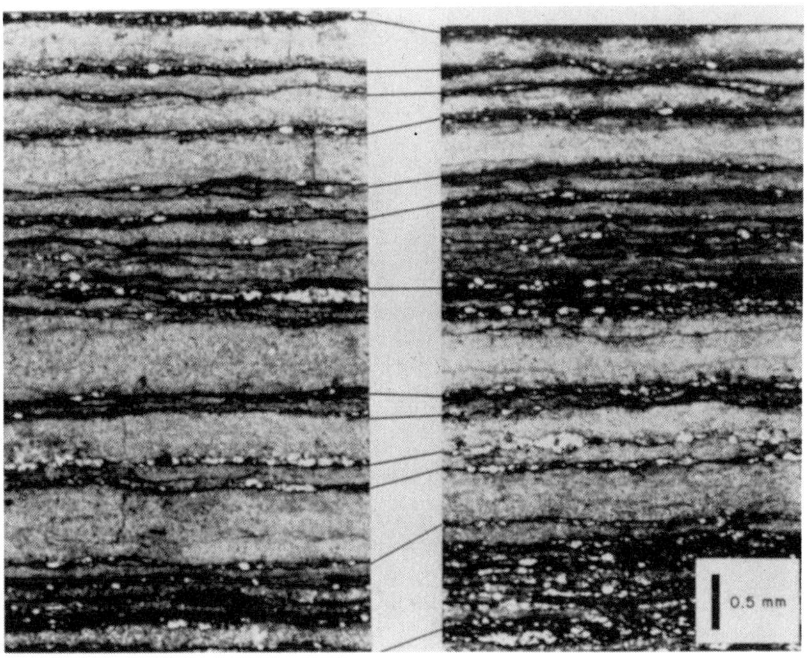

Figure 2. Correlative thin sections of calcite-sapropel laminae of Todilto limestone from localities F (left) and H, separated by 2.3 km. Light layers are calcite, dark layers sapropel

CORRELATIVE VARVE SERIES OF JURASSIC
TODILTO FORMATION, NEW MEXICO

ANDERSON AND KIRKLAND, PLATE 2
Geological Society of America Bulletin, volume 77

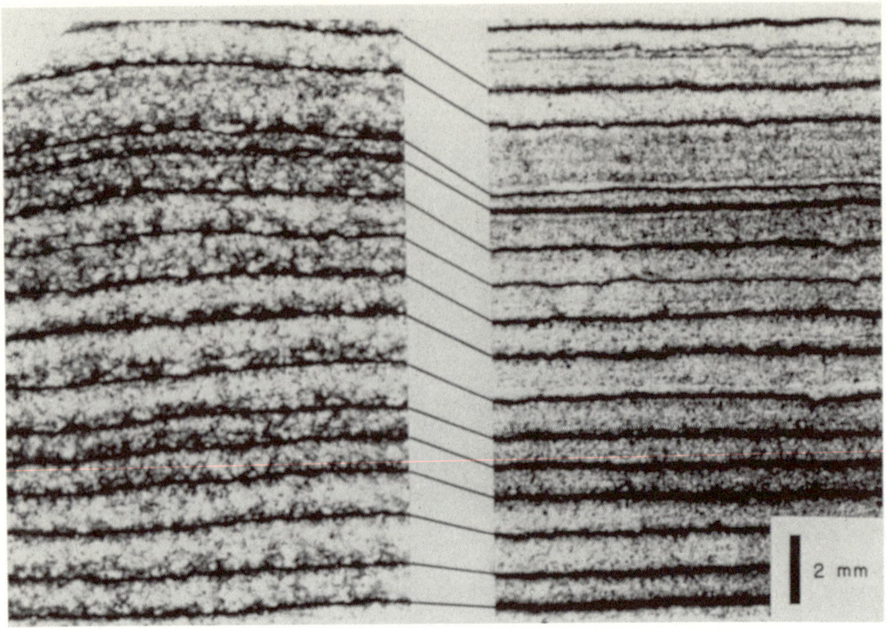

Figure 1. Correlative thin sections of typical Castile calcite-anhydrite varves from Cowden no. 2 (left) and no. 4, separated by 14.2 km. Light layers are anhydrite dark layers calcite

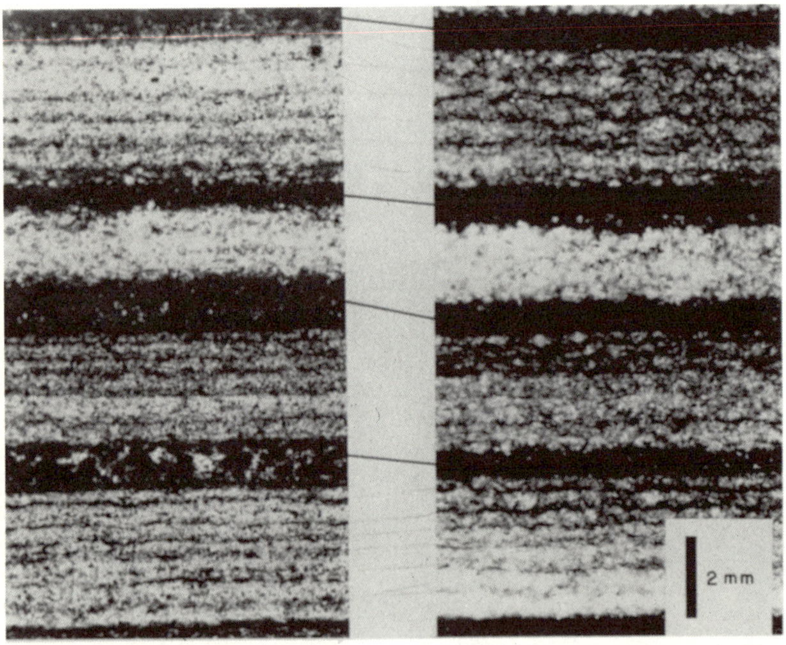

Figure 2. Correlative thin sections of atypical Castile calcite-anhydrite laminations from Cowden no. 2 (left) and no. 4 cores, separated by 14.2 km. Light layers are anhydrite, dark layers calcite

CORRELATIVE VARVE SERIES OF PERMIAN CASTILE FORMATION, SOUTHEASTERN NEW MEXICO AND WEST TEXAS

ANDERSON AND KIRKLAND, PLATE 3

Geological Society of America Bulletin, volume 77

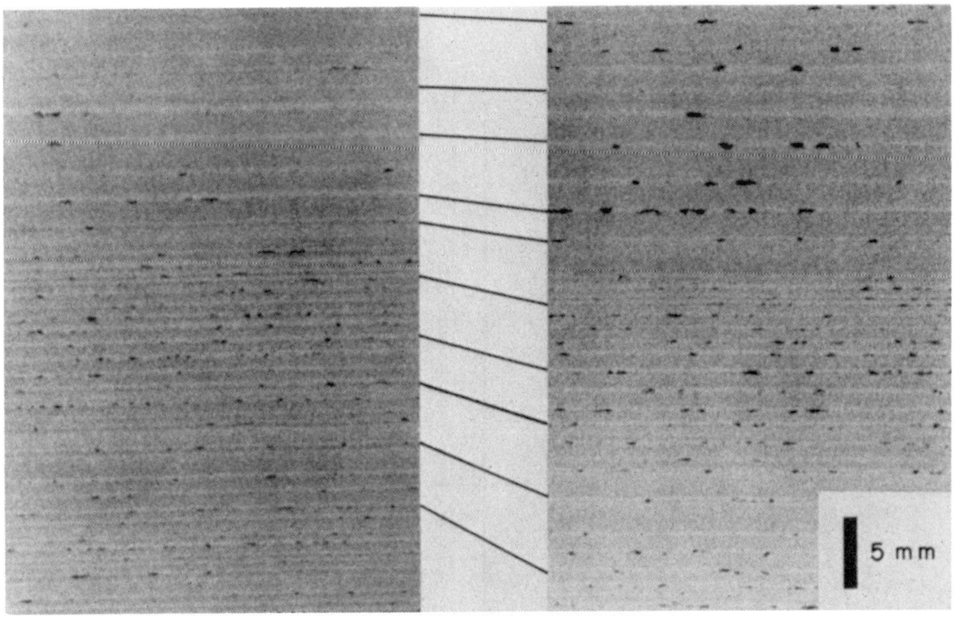

Figure 1. Correlative X-radiographs of 3 mm-thick slabs of Castile Formation from Cowden no. 2 (left) and no. 4 cores, separated by 14.2 km. The pyrite crystals (black) occur at the base of the calcite laminae; anhydrite laminae are thicker than calcite laminae and a darker gray

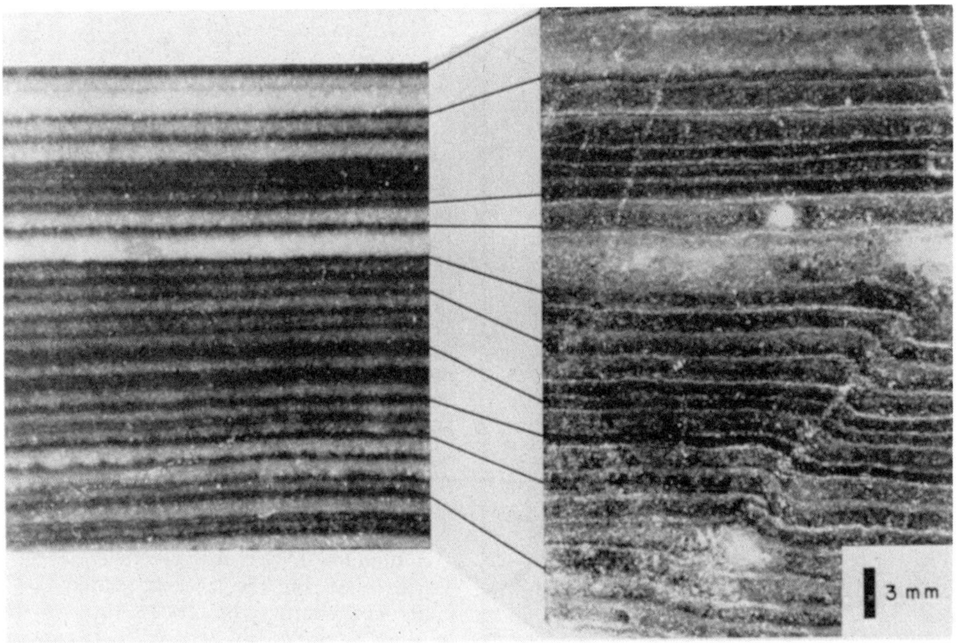

Figure 2. Correlative slabs of anhydrite (Cowden no. 2, left) and gypsum (Cowden no. 4) from the Castile Formation showing expansion of laminae due to hydration. The cores are separated by 14.2 km

CORRELATIVE VARVE SERIES OF PERMIAN CASTILE FORMATION, SOUTHEASTERN NEW MEXICO AND WEST TEXAS

ANDERSON AND KIRKLAND, PLATE 4

Geological Society of America Bulletin, volume 77

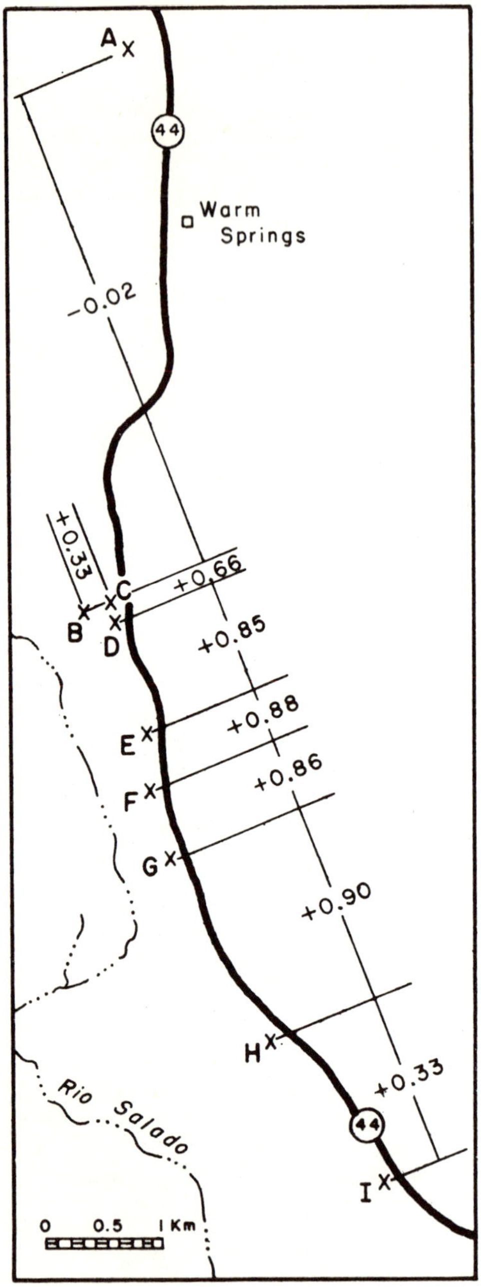

Figure 2. Todilto localities showing correlation coefficients of varve thickness between adjacent sequences. Note the low values at both the northern and southern ends of the outcrop.

tion of section A with the other sections. Correlation coefficients were computed using 35 thickness values which were obtained from a zone about 15 cm below the first gypsum laminae of the overlying gypsum member of the formation. In the area studied, the limestone member of the Todilto Formation consists of a lower limy siltstone and an upper blocky, silty limestone. The correlated zone is about three quarters of the way to the top of the blocky, silty limestone. The thicknesses of the couplets were scaled with the aid of a microscope from greatly enlarged photographs of the etched and stained blocks (Pl. 2, fig. 1). Photomicrographs of thin sections from this same zone (Pl. 2, fig. 2) illustrate the variation in relative thickness between sections F and H which are separated by 2.3 km. Contiguous stratigraphic correlations between sections D and H all have coefficients greater than +0.85 (Table 1). Both visual stratigraphic correlation and correlation coefficients deteriorate at each end of the outcrop, but on the basis of subjective visual correlations above and below the zone tested there is little question that sections C and I, at the north and south end of the outcrop respectively, were also correlated correctly. Correlation could not be established between sections B and D (correlation coefficient of best subjective stratigraphic correlation = +0.14), and the distance between sections had to be shortened to 180 m (distance between sections C and B) in order to bring about correlation (correlation coefficient of best subjective stratigraphic correlation = +0.33).

All statistical correlations between the sections studied appear in Table 1. There appears to be no relationship between the distance correlated and the value of the correlation coefficient. There is, however, a connection between the magnitude of the coefficient and the thickness of the zones correlated. The thickness of each of the sections was scaled from the photographs of etched slabs over a group of about 100 couplets which included the zone for which coefficients were computed. The amount of relative thinning in each of the sections was established by setting the thickest section (D) at unity and relating the other sections to this section. Sections C and B are 26 per cent and section I is 24 per cent thinner than section D. A "thickness index" between any two stratigraphically correlative sections was obtained by multiplying their relative thickness values. The thickness index was then plotted against the transformed[3] correlation coefficient in order to illustrate the relationship between thickness of the sections and statistical correlation. Figure 3 shows that thicker sections have higher correlation values than thinner ones. A plot of correlation coefficient against the index of the coefficient of variation

$$\left(\frac{\sigma a}{\overline{X}a} \times \frac{\sigma b}{\overline{X}b}\right),$$

where a and b are the two series correlated, was also constructed in order to determine the relationship between variability and statistical correlation. The more variable series have coefficients that are no higher or lower than the less variable series, and thickness remains the principal control on the value of the correlation coefficient.

The reason thinner sequences have lower coefficients may be related to the bottom configuration of the basin. High correlation values in the vicinity of sections D, E, F, G, and H probably reflect a low area in the basin. The specific mechanism could have been gentle "winnowing" of the fine calcite particles off high areas in the basin as they settled during their seasonal rain. The excellent preservation of fine laminations implies a stagnant body of water at depth and precludes strong current action. The gentlest currents, however, might have been capable of the slight movement required to bring about the changes in laminae thickness observed in the Todilto sections. Micro-oölites found on top of the sapropel laminae suggest a gentle current action during at least part of the year. The principal difficulty in this explanation is that the organic and carbonate particles would have to be differentially winnowed, but this could be explained by their different physical properties or by a difference in the strength of currents in different seasons.

The Todilto laminae in many places yield very high correlation coefficients over short distances (+0.90 over 1.7 km). If the same precipitation process operated in a basin less influenced by currents, laminae that correlate over a much greater distance would form. At least theoretically, the present laminae could be correlated over the entire basin if the sampling interval were short enough.

Table 1. Correlation Coefficients of Calcite-Sapropel Couplet Thickness from Nine Localities of Todilto Formation, $n = 35$

	Localities							
	B	C	D	E	F	G	H	I
A	−0.02	+0.03	+0.14	+0.08	+0.20	+0.09	+0.15	+0.01
B	..	+0.33	+0.13	+0.07	+0.16	+0.18	+0.13	+0.21
C	..	..	+0.66	+0.72	+0.66	+0.76	+0.69	+0.33
D	..	..	..	+0.85	+0.80	+0.81	+0.79	+0.47
E	..	..	..	..	+0.88	+0.89	+0.87	+0.42
F	..	..	..	..	..	+0.86	+0.87	+0.45
G	..	..	..	..	..	..	+0.90	+0.38
H	..	..	..	..	..	..	..	+0.33

Earlier studies by the writers (1960) and Anderson (1961) mentioned long-term, somewhat periodic changes in varve thicknesses in the Todilto Formation. Fortuitously, the section that was used for gathering these data came from the area where correlation coefficients between adjacent sections are high and thickness changes should reflect the amount of carbonate precipitated. This example clearly shows the desirability of determining the quality of a series, perhaps through studies of lateral variation, before interpretation is attempted.

Permian Castile Formation

The laminations of the Castile Formation are superbly preserved, and the Castile must be one of the most complete and continuously laminated evaporite series in the world. The laminated sequence of the Delaware basin, south-

[3] The correlation coefficient is not a linear measure of the "quality" of correlation. The coefficient can be converted to a proportional (linear) scale by the following transforming equation:

$$\text{Proportion } (p) = \frac{|r|}{\sqrt{1 - r^2} + |r|},$$

where r is the Pearson coefficient of correlation (Personal communication, Judah Rosenblatt, Univ. New Mexico).

eastern New Mexico and West Texas, encompases four geologic formations: the upper part of the Delaware Sandstone, the Lamar Limestone, the Castile Formation, and the Salado Formation. Along the western side of the Castile basin in the northwest part of Culberson County, Texas, laminations of several types are found in the underlying siltstone, Lamar Limestone, and Castile Formation.

A preliminary calibration of a core in the basal part of the anhydrite (Cowden no. 4) indicates that there are about 3300 couplets in the laminated portion of the basal siltstone (average 1320 per m), 44,000 in the Lamar Limestone (average 5880 per m), and about 46,000 in the cored part of the Castile Formation (average 715 per m). The laminations consist of anhydrite, several kinds of calcite, organic-rich calcite, silt, and layers in which iron sulfide or magnetite is enriched. The variety and excellent preservation of the laminations have made it possible to determine the nature of stratigraphic and statistical correlation between several of the constituents in the basal part of the anhydrite.

Three 2 1/8-inch (5.46-cm)-diameter cores were obtained from the Castile Formation near its western limit of outcrop (Fig. 1C).[4] The Cowden no. 1 core contained 31 m of anhydrite and gypsum. The Cowden no. 2, only 400 m from no. 1, contained 64 m of anhydrite and gypsum but did not penetrate the underlying limestone. The Cowden no. 3 was drilled in a disrupted zone and abandoned. The Cowden no. 4, 14.2 km south of the others (Fig. 1C), contained 63 m of anhydrite and gypsum, all the Lamar Limestone, and all the laminated part of the underlying siltstone (total depth = 71.8 m).

Stratigraphic correlation between the two cores separated by 14.2 km was easily established by visual comparison after locating a group of several hundred very thin laminae (average thickness about 0.3 mm) that served as a

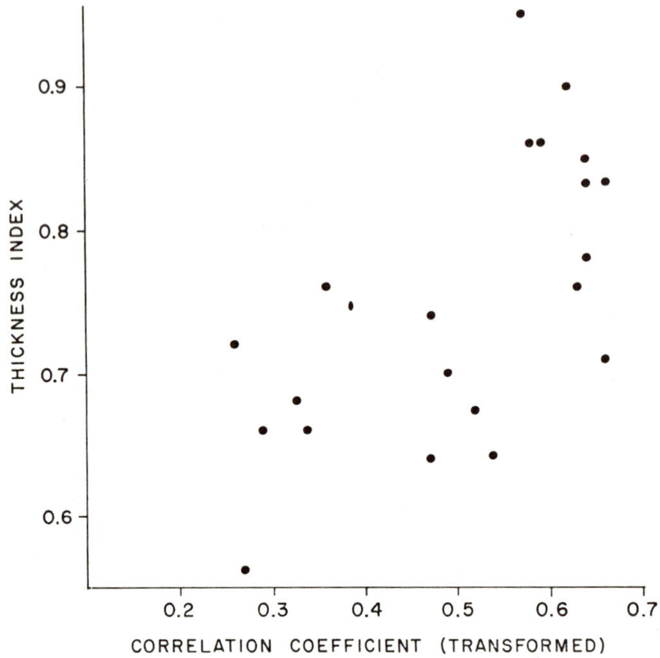

Figure 3. Relationship of thickness to correlation coefficient (transformed) in the Todilto Formation varves

[4] *Core-hole* *Locations*
Cowden no. 1: NW¼ NE¼ Sec. 34, Twp. 1, Blk. 62, T. & P.R.R. Co.
Cowden no. 2: NE¼ NE¼ Sec. 34, Twp. 1, Blk. 62, T. & P.R.R. Co.
Cowden no. 4: SW¼ SE¼ Sec. 33, Twp. 2, Blk. 62, T. & P.R.R. Co.

key unit. Without this distinctive zone, visual correlation would have been difficult. Stratigraphic correlation of cores 1 and 2 that are separated by 400 m has so far been unsuccessful.

PETROGRAPHY: The Castile Formation is basically a two-component (couplet) system of calcium carbonate and calcium sulfate. Udden's (1924) petrographic description of cores from the David Flood-Gresham and McAlpine well no. 1 (49 km southeast of the Cowden no. 4) typifies much of the Castile.

Two different patterns of calcite-anhydrite lamination are noticeable in the cores studied. Most of the sequence consists of calcite-anhydrite couplets in which thickness variation is relatively slight. In this type of lamination (Pl. 3, fig. 1) the calcite layers are 0.1–0.2 mm thick and are composed of rounded rhombohedrons of calcite which have an average diameter of about 0.02 mm (*see* Udden, 1924, Pl. 9, fig. 2). The rhombohedrons are concentrated into discrete calcite laminae and are also sparsely scattered within the anhydrite laminae. In this type of lamination all of each calcite layer is colored brown by organic matter. The anhydrite laminae are composed of a mosaic of rectangular crystals. The average thickness of the calcite-anhydrite couplets in the Cowden no. 4 core is 1.4 mm. Udden found that the couplets in the core he studied averaged less than 2 mm in thickness, and that the most common variation in thickness was between 0.5 and 2.0 mm.

Another, but atypical, type of calcite-anhydrite lamination occurs through short intervals of the core and consists of couplets of calcite and anhydrite that are extremely variable in thickness. In addition, the anhydrite layers in places contain very thin calcite laminae (Pl. 3, fig. 2). In this type of lamination the calcite laminae are about one third the thickness of each couplet. Some calcite layers appear to be essentially devoid of bituminous material, others are exceptionally dark, and still others have the organic matter confined to either the top or the bottom of the calcite layer. The relatively high organic content of the dark-brown layers in the calcite laminae was demonstrated by microkjeldahl analysis of material collected with the aid of a microtome. The dark-brown carbonate layers have an average nitrogen content of 0.01 per cent compared with an average of 0.003 per cent for the lighter colored calcite laminae.

X-radiographs at several stratigraphic levels in cores 2 and 4 reveal centers of pyrite-crystal growth along the basal contact of the calcite layer with the underlying anhydrite layer (Pl. 4, fig. 1). The pyrite crystals are small (average maximum length about 0.5 mm); only a few crystals were observed on polished sections, but many appear in X-radiographs of slabs 3 mm thick.

STRATIGRAPHIC CORRELATION: Subjective stratigraphic correlation between the two cores can usually be established by visual comparison unless there is a long search area. If the correlation coefficient is low or the varve series is uniform in thickness (low variance), the correct stratigraphic correlation is difficult to find. Fritts (1963) employed a method for correlating tree-ring series in which all possible statistical correlations between two series are computed. When plotted as a time series the highest coefficients stand above the others in the series, and tree-ring correlation can be established even when subjective correlation fails. This same technique has been modified for longer varve series by selecting a small test interval and computing all statistical correlations on a moving basis. A stratigraphic correlation based on couplet thickness from the Castile sequence was established in this manner where the master sequence was 94 units and the "slave" or sliding sequence was 20 units (75 correlations, Fig. 4; Table 2, series B). In this case the correct stratigraphic correlation coefficient falls well above the others. There are other places in the master sequence where the coefficient rises to a relatively high value. This is apparently due to the presence of thickness cycles or trends in the time series. If several high positive values are obtained they may be tested further by lengthening the "slave" interval.[5] When used with caution this method extends the range of the correlation coefficient technique in finding true stratigraphic correlations.

Data on the thickness of calcite and anhydrite layers from core no. 2 and core no. 4 were compiled from a series of overlapping thin sections representing a zone of erratic lamination (Pl. 3, fig. 2). Series A consists of 74 couplets approximately 27.4 m above the base of the formation. The thin sections were placed in an enlarger and projected onto photographic paper

[5] The higher values in the sliding correlation series as opposed to the +0.76 that appears in Table 2 are the result of using a shorter series when there is a trend (cycle) in the data. With $n = 20$, four positive correlations would be statistically significant at the 95 per cent level, but only the highest one is the correct stratigraphic correlation.

Figure 1. Folded laminae of anhydrite (light) and calcite in the Castile Formation (Cowden no. 2 core). The calcite layer is the incompetent unit

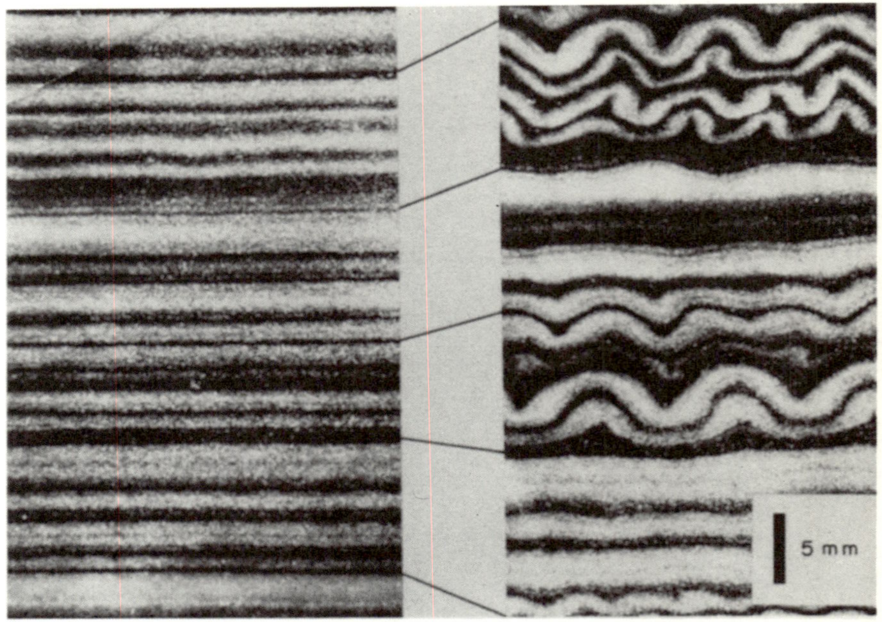

Figure 2. Folded and non-folded zones of calcite-anhydrite laminae in Cowden no. 4 (left) and no. 2 cores of the Castile Formation

CONTORTED LAMINAE FROM PERMIAN CASTILE FORMATION, SOUTHEASTERN NEW MEXICO AND WEST TEXAS

and measurements taken from the enlarged prints. This method allows measurement of both calcite and anhydrite layers and contains little error in determining laminae boundaries. Data from three other zones, series B, C, and D, approximately 22.0, 28.5, and 49.1 m above the base of the Castile, respectively, were taken from enlarged photographic prints of polished slabs which had been coated with a mixture of glycerine and water to improve visibility. The calcite-anhydrite boundary is not as distinct in this technique, and only couplet thickness was measured. Laminae in polished sections were also measured by means of a calibrated ocular using a stereoscopic microscope. The calcite-anhydrite boundaries could be distinguished but not as precisely as in thin section. This technique permitted the measurement of dark, organic-rich calcite bands within calcite layers. The same dark bands cannot be distinguished easily in thin section. Finally, data on iron sulfide distribution were taken from enlarged prints of X-radiographs (Pl. 4, fig. 1). The iron sulfide is organized into centers of pyrite crystallization that originated at the basal calcite layer contact. Estimates were made of the number of pyrite centers in unit length of each lamina.

Statistical correlations of couplet (combined calcite-anhydrite) thickness all have high positive correlation coefficients regardless of the measuring technique used (Table 2). The coefficients for anhydrite laminae thickness in series A are very high (greater than +0.97), and it is clear that in this part of the sequence they are the main contributor to the high coefficients of couplet correlation. The other components in the same section have lower values (calcite, +0.72; organic-rich layers in calcite laminae, +0.73) and verify couplet correlation. The number of pyrite centers in unit laminae length were statistically correlated between cores no. 2 and no. 4 in an 8-cm-thick zone (series C) and had a correlation coefficient of +0.61. Couplet-thickness coefficients for the atypical zone of erratic lamination are higher than for the typical zone with more regular laminae (+0.94 versus +0.76), a feature which may be inherent in the higher variance.

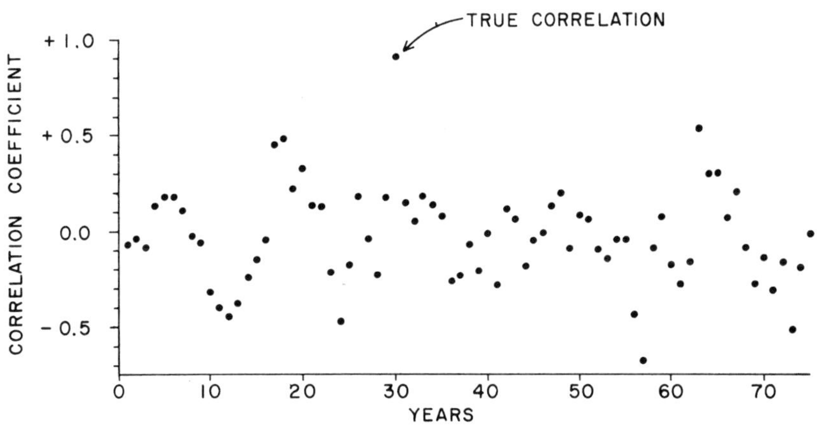

Figure 4. Correlation series of two Castile sequences separated by 14.2 km. The points represent the coefficients of correlation obtained by sliding a "slave" series of 20 units along a "master" series and computing every correlation.

Anhydrite in the Castile Formation is converted to gypsum by weathering in the upper 18.3 m of core no. 1, the upper 16.8 m of core no. 2, and the upper 17.4 m of core no. 4. A part of the Cowden no. 4 core was converted to gypsum, whereas the same part of the Cowden no. 2 core remained anhydrite (Pl. 4, fig. 2). There is a 44 per cent increase in the thickness of the zone after hydration. An underlying anhydrite zone was about 1.45 per cent thicker in core no. 4 than the same zone in core no. 2, so the actual thickness increase is 42.5 per cent. This figure was determined over a thickness of 9.2 cm containing 42 gypsum-calcite couplets. Allowing for the amount of calcite in the section, the thickness change upon conversion of anhydrite to gypsum is about 50 per cent. The theoretical volume change from the conversion of anhydrite to gypsum is 64 per cent. The correlation coefficient of the couplet thickness in the two cores is +0.99, indicating that little change in the relative thickness of the laminae took place as a result of hydration.

COMPONENT ASSOCIATIONS: The systems that controlled the precipitation of the various components in the Castile sequence might have been interrelated, and these components were tested for statistical independence by means of the correlation coefficient. In cores no. 2 and no. 4, coefficients were computed for associated calcite and anhydrite laminae, both overlying and underlying; for calcite and associated organic-rich layers in the calcite laminae; and for anhydrite and organic-rich layers in associated calcite laminae. The data used were gathered from the same zone of irregular couplet thickness used for lateral correlation (series A). In this zone there was no statistical correlation between the thickness of calcite laminae and either the overlying or underlying anhydrite laminae (correlation coefficient = +0.12 and −0.08, respectively). Richter-Bernburg (1960) reported a positive relationship between calcite and overlying anhydrite laminae thickness in the Permian Zechstein of Germany but did not test his observations. One might expect calcite and anhydrite laminae thickness to be related owing to a mutual dependence on temperature, evaporation, or dilution. The complete lack of correlation between these two components in the Castile sequence examined suggests that two independent precipitation systems may have been involved. Studies of other parts of the sequence might show a dependence in the long-term variations of calcium carbonate and calcium sulfate.

The correlation coefficient for calcite laminae thickness and the thickness of organic-rich bands in the same calcite layers in series A of core 4 is +0.33, n = 74, suggesting that a possible relationship exists. There is a ready mechanism in the relationship of plankton growth and carbonate precipitation to the carbon dioxide content of water. However, some of the thick carbonate layers in the zone of erratic thickness variation appear to be devoid or nearly devoid of organic matter.

Anhydrite laminae thickness and the thickness of the underlying organic-rich layers in calcite laminae in series A of core 4 have a negative correlation that may be meaningful (r = −0.36, n = 74). A dependence could be most easily explained if the same high or prolonged concentrations of sulfate or other associated ions in the water that produced a relatively thick anhydrite layer had a deleterious effect on the growth of the plankton. This might be related to the occurrence of periods of irregular and erratic plankton blooming.

TABLE 2. SUMMARY OF CORRELATION VALUES OF SEVERAL COMPONENTS IN THE CASTILE VARVES FROM THE COWDEN NO. 2 AND NO. 4 CORES

	Unit correlated	Distance correlated	Stratigraphic position	Number of units correlated	Correlation coefficient
Series A	Couplet thickness	14.2 km	Base of series 27.4 m above Lamar–Castile contact	74	+0.94
	Sulfate laminae thickness				+0.97
	Carbonate laminae thickness				+0.72
	Organic-rich bands in carbonate laminae				+0.73
Series B	Couplet thickness	14.2 km	Base of series 22.0 m above Lamar–Castile contact	94	+0.76
Series C	Couplet thickness	14.2 km	Base of series 28.5 m above Lamar–Castile contact	44	+0.93
	Pyrite centers				+0.61
Series D	Couplet thickness	14.2 km	Base of series 49.1 m above Lamar–Castile contact	49	+0.99

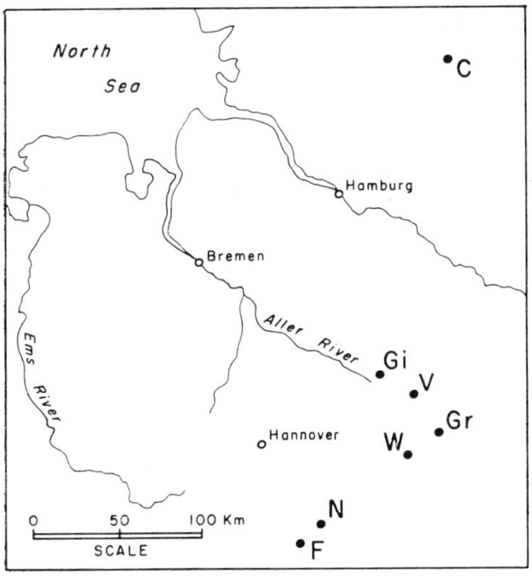

Figure 5. Core-hole localities in German part of Zechstein basin (*after* Richter-Bernburg, 1960)

FOLDED LAMINAE: The distinctive folds illustrated in Figure 1 of Plate 5 occur in the northernmost core (Cowden no. 2) about 42.4 m above the Lamar-Castile contact. Approximately 45 per cent of the laminae in a zone which extends from 42.0 to 43.8 m above the base of the Castile are folded; in one 47-cm part of this zone, 95 per cent of the laminae are folded. The same zone is involved in folding in both cores but in the equivalent zone of the southern core (Cowden no. 4) only 16 per cent (as opposed to 95 per cent) of the laminae are folded, and the amplitude of the folds is lower. Relatively few laminae are folded elsewhere in the cores. In 11 cases, the same laminae or groups of laminae are folded in both cores.

Folds of the same type were observed in outcrop a few miles north of the Cowden no. 2 locality in a large road cut in the Castile on U. S. Highway 62 near the Texas-New Mexico border. The folds observed in outcrop occur in "pods" usually not more than 1 m across and not involving more than several hundred laminae; the full linear dimension (parallel to the strike of the fold axes) is difficult to determine.

The axial planes of the folds dip at many angles, and some of the folds are overturned. As in outcrop, the strike of the axial planes in each core is unidirectional. The folds in core no. 2 have an average amplitude of 1.7 mm, and the layers have been folded in groups of 2–7 laminae. Parts of the folded zone in core no. 2 are thicker than nonfolded correlative parts of core no. 4 (Pl. 5, fig. 2). This difference in thickness may be accounted for in the sum of the amplitude of the groups of folds in the contorted zone. The bituminous carbonate laminae are the most deformed and were the incompetent layers: the calcium sulfate layers, which tended to maintain their thickness, are competent. A similar type of folding occurs in the Zechstein Anhydrite (Richter-Bernburg, 1958, Pl. 36). Kaufmann and Slawson (1950) observed analogous folds, although of much greater wave length, in the anhydrite-banded salt of the Salina Formation and believed they

TABLE 3. CORRELATION COEFFICIENTS OF COUPLET THICKNESS FROM SEVEN LOCALITIES IN PERMIAN ZECHSTEIN ANHYDRITE

(Series A, $T_o + 485$ to $T_o + 619$, $n = 134$; series B, $T_o + 760$ to $T_o + 926$, $n = 166$; series C, $T_o + 265$ to $T_o + 354$, $n = 89$: Richter-Bernburg, 1960, Fig. 3)

	Localities					
	W	C	N	Gi	V	F
Gr	$+0.62^{A}$ $+0.71^{B}$	$+0.55^{A}$ $+0.70^{B}$	$+0.56^{A}$ ..	$+0.60^{A}$ $+0.73^{B}$	$+0.59^{A}$ $+0.73^{B}$	$+0.57^{A}$ ..
W	.. ..	$+0.41^{A}$ $+0.60^{B}$	$+0.65^{A}$ $+0.74^{C}$	$+0.53^{A}$ $+0.67^{B}$	$+0.51^{A}$ $+0.68^{B}$	$+0.72^{A}$ $+0.78^{C}$
C	.. ..	.. ..	$+0.40^{A}$ ..	$+0.44^{A}$ $+0.65^{B}$	$+0.52^{A}$ $+0.69^{B}$	$+0.32^{A}$ ..
N	.. ..	.. ..	.. ..	$+0.45^{A}$ ..	$+0.44^{A}$ ..	$+0.83^{A}$ $+0.85^{C}$
Gi	.. ..	.. ..	.. ..	.. ..	$+0.42^{A}$ $+0.74^{B}$	$+0.48^{A}$ ..
V	..	..	..	..	..	$+0.46^{A}$

represented ripple marks. In some places the Castile folds are overturned, which precludes their formation as ripples.

Because the contortions occur in both gypsum and anhydrite, the structures were probably not produced as a result of the volume change due to hydration. The folds closely resemble the type of contortions produced experimentally under pressure in soft layers of differential competence (Bucher, 1956, Pl. 3). Their unidirectional strike and their occurrence in the same zone over a distance of 14.2 km suggest that the structures were all formed at about the same time, probably as a result of gravitational adjustment on the original slope of the basin.

Permian Zechstein Formation

Richter-Bernburg (1957; 1958; 1960) stratigraphically correlated the calcite-anhydrite couplets in many cores from the German part of the Permian Zechstein basin. The laminated bed at the base of the second Zechstein evaporite cycle is only about 2 m thick, and this simplified his search for correlation. The cores he used came mainly from the southern part of the basin: one locality was much farther to the north and nearer the center of the basin (Fig. 5). He found that he could correlate individual laminations over a distance of 290 km, that the laminations increase in thickness and variability toward the southern edge of the basin, and that anhydrite deposition began first in the south and moved northward into the basin.

Richter-Bernburg did not test the stratigraphic correlations he obtained other than to construct graphic plots of the data. Correlation coefficients were computed from his data, and all possible combinations between most of his cores were obtained (Table 3). All the coefficients appear to be significant with the possible exception of the one for the longest distance correlated (Cores C and F, 290 km, $T_o + 485$ to $T_o + 619$, $r = +0.32$, $n = 134$). However, high correlation values between these two most distant cores and cores of an intermediate distance indicates that the correct laminae have been correlated stratigraphically (core C with core V, $r = +0.52$, $n = 134$; core F with core V, $r = +0.46$, $n = 134$). A sliding correlation series using a "slave" interval of 20 units was computed for the cores separated by the greatest distance, cores C and F (Fig. 6). The correct stratigraphic correlation had the highest coefficient, +0.70, of the 115 coef-

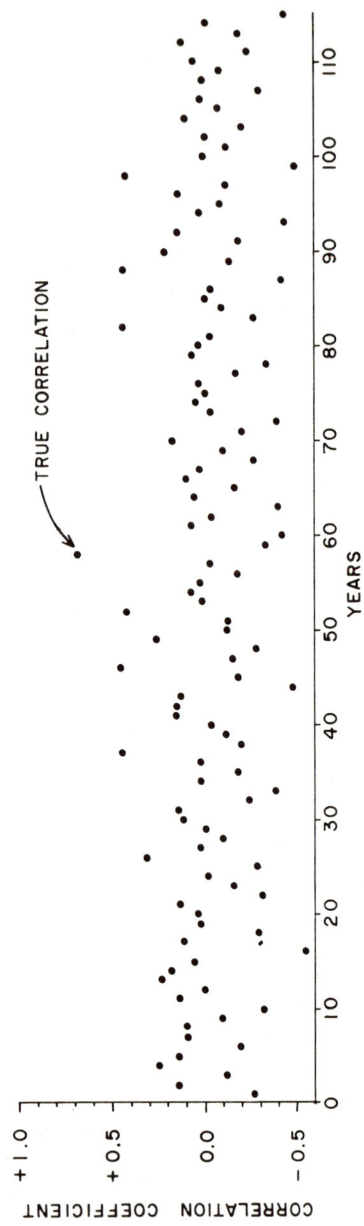

Figure 6. Correlation series of two Zechstein varve sequences (cores C and F, series A) separated by 290 km

ficients calculated. The next highest had a value of +0.45.

The number of cores studied by Richter-Bernburg make it possible to plot the transformed[6] correlation coefficient against distance between cores and determine the lowering of correlation coefficients with increasing distance (Fig. 7). The coefficients bear little relationship to distance in the region between 10 and 120 km, although there is a definite slope to the maximum values obtained for all distances. If the Castile coefficients decrease with distance at the same rate, it should be possible to correlate laminae stratigraphically throughout the entire Castile basin. The spread of values in the region between 10 and 120 km suggests that local and not regional changes in relative thickness of laminae may be involved. This was tested by constructing a plot of the transformed correlation value against thickness in the same manner as the plot for the Todilto varves. The thickness measurements were taken from the same series used in computing the correlation coefficients (T_o + 485 to T_o + 619, n = 134), and only stratigraphic correlations of less than 120 km were plotted (Fig. 8). The graph reveals that the coefficient is below +0.45 when the thickness index is below 0.7 (relatively thin). The thicker sections, however, tend to have higher coefficients. Although not as well defined as in the Todilto plot (Fig. 3), there is an alignment to the values, suggesting a positive relationship between thickness of the series and correlation coefficient. The variability of the series has no relationship to the value of correlation. The cores that were separated by less than 120 km are near the southern margin of the basin. The breakdown in high correlation in some of these cores may possibly be related to the local effects of cur-

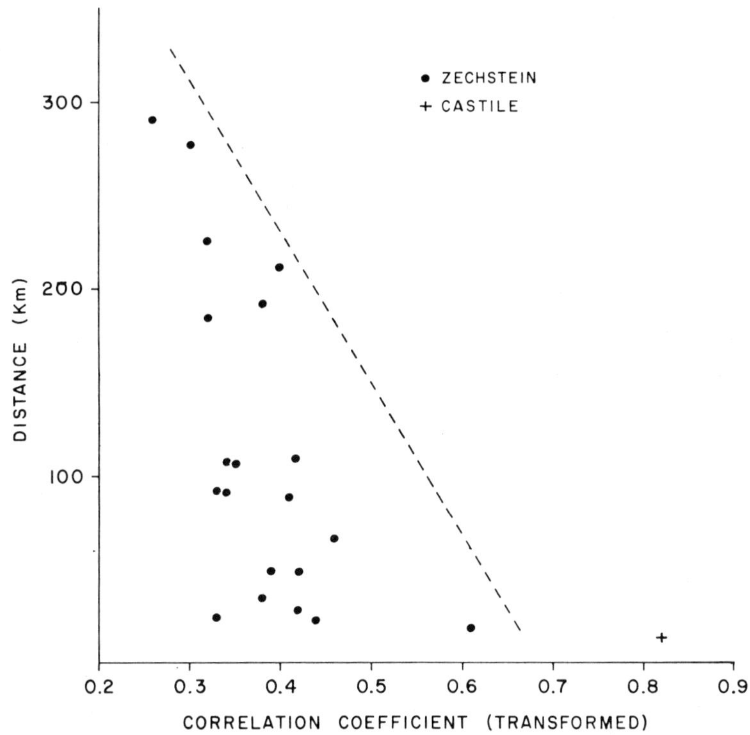

Figure 7. Relationship between distance and correlation coefficient (transformed) in Zechstein and Castile varve series. Zechstein correlation values should fall within the area bounded by the dashed line.

[6] *See* footnote 3.

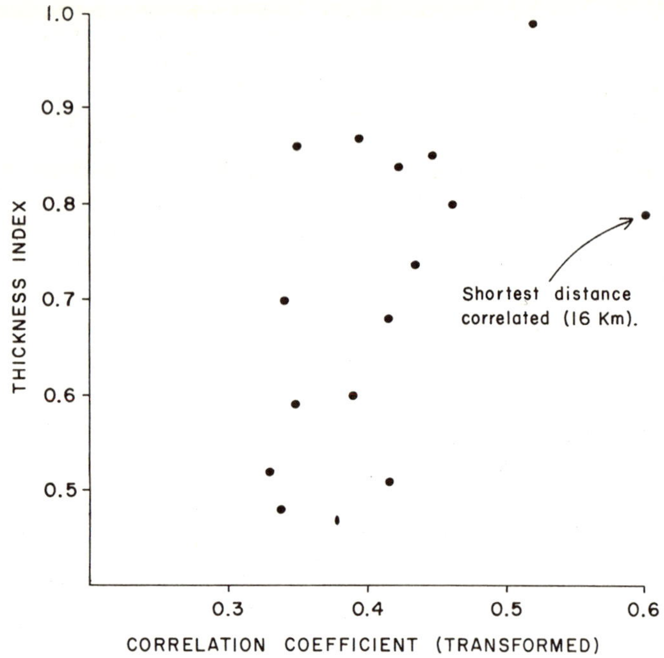

Figure 8. Relationship of thickness to correlation coefficient (transformed) in the Zechstein Anhydrite varves separated by less than 120 km, n = 134 (Data *from* Richter-Bernburg, 1960)

rents or the configuration of the basin as in the Todilto example.

CONCLUSIONS

(1) Organically or chemically precipitated laminae, known to be or suspected of being varves, are stratigraphically correlative over their area of formation. Stratigraphic correlations of laminae based on thickness variations frequently have correlation coefficients greater than +0.8.

(2) Correlation coefficients for correlative stratigraphic sections decrease gradually with increasing distance. The coefficients may also be lowered by local factors such as the bottom configuration of the basin and the redistribution of precipitated material by currents.

(3) Correlation coefficients derived from thickness measurements of stratigraphically correlative anhydrite laminae are higher than for associated calcite laminae, organic-rich layers, or pyrite crystals.

(4) Anhydrite and calcite laminae that are associated in the same couplet have changes in relative thickness that are statistically independent. There may be a negative association between anhydrite and organic-rich laminae.

REFERENCES CITED

Anderson, R. Y., 1961, Solar-terrestrial climatic patterns in varved sediments: New York Acad. Sciences Annals, v. 95, p. 424–439

Anderson, R. Y., and Kirkland, D. W., 1960, Origin, varves and cycles of Jurassic Todilto Formation, New Mexico: Am. Assoc. Petroleum Geologists Bull., v. 44, p. 37–52

Antevs, E., 1951, Glacial clays in Steep Rock Lake, Ontario, Canada: Geol. Soc. America Bull., v. 62, p. 1223–1262

Bucher, W. H., 1956, Role of gravity in orogenesis: Geol. Soc. America Bull., v. 67, p. 1295–1318

Burnaby, T. P., 1953, A suggested alternative to the correlation coefficient for testing the significance of agreement between pairs of time series, and its application to geological data: Nature, v. 172, p. 210–211

Fritts, H. C., 1963, Computer programs for tree-ring research: Tree-Ring Bull., v. 25, nos. 3–4, p. 2–7

Kaufmann, D. W., and Slawson, C. B., 1950, Ripple marks in rock salt of the Salina formation: Jour. Geology, v. 58, p. 24–29

Richter-Bernburg, Gerhard, 1957, Isochrone Warven im Anhydrite des Zechstein 2: Germany, Geol. Landesanst., Geol. Jb. Bd. 74, p. 601–610

—— 1958, Die Korrelierung isochroner Warven im Anhydrite des Zechstein 2 (zweiter Beitrag): Germany, Geol. Landesanst., Geol. Jb. Bd. 75, p. 629–639

—— 1960, Zeitmessung geologischer Vorgange nach Warven-Korrelation im Zechstein: Geol. Rund., Heft 1, Bd. 49, p. 132–148

Udden, J. A., 1924, Laminated anhydrite in Texas: Geol. Soc. America Bull., v. 35, p. 347–354

Manuscript Recieved by the Society September 28, 1964

Reprinted from *Salt Deposits: Their Origin and Composition,* 251–258 (1971)

23

2. Rhythmic Bedding and Influxes

The even rhythmic bedding of many anhydrite and halite rocks often for great thicknesses, is a characteristic feature. LOTZE (1957) has discussed in detail the bedding and the related general problem of its formation so that only one important special case will be touched upon here.

There is a general tendency to regard the regular rhythmic banding in the German Zechstein 2 as annual banding. Meanwhile it can be demonstrated that in the case of the Stassfurt halite series, at least,

1. the bedding is conditional upon the influx of fresh seawater and
2. the concentration of brine had strong, periodic variations.

Assumption 1. is common to all hypotheses which postulate a barrier of some kind. This is indispensable in explaining the quantity of salt precipitated and requires no further discussion. The desert hypothesis (J. WALTHER, 1910, p. 250/251) favoured earlier, for instance, by JÄNECKE (1923, p. 86) is not applicable to oceanic salt deposits for it explains neither the great salt thicknesses, nor the regular salt successions and in particular neglects the qualitative composition. It must be noted that the influx of fresh seawater onto a so-called "saturation shelf" in the absence of a barrier, as occasionally presented (RICHTER-BERNBURG, 1953, Fig. 2), does not lead to a stable increase in salt concentration because of the reflux of concentrated solutions.

Assumption 2. is common to all hypotheses which suppose periodic influxes. This idea first appeared in ZIMMERMANN (1915) and was based upon the fact than in fine banding a more and a less soluble salt alternate regularly, both in the Stinkschiefer (Ca 2 st) and in the Stassfurt halite (Na 2), and in the Basalanhydrit (A 2). In such cases there is usually a thin coating of clay, which sometimes contains carbonates, at the bottom of the less soluble layer.

According to LÜCK (1913), the rhythms in the anhydrite region in the Stassfurt anticline are on average 10 cm thick, the anhydrite "threads" being about $^1/_2$ cm. The halite grain size within the rhythm often increases upwards (cf. p. 155). In the polyhalite region the rhythms average 7 cm in thickness. The rhythms may also be irregular, particularly through the introduction of a thicker sulphate layer, or through the absence of bedding over a thickness of several meters. The halite layers also contain small quantities of loosely disposed sulphates. In hand-samples the separation into sulphate-rich (fine-grained, dull white, gray or dark) and halite-rich layers is clearly distinguishable. Between the anhydrite and polyhalite zones according to LÜCK (1913), there is a transition layer with about 1% glauberite. This zone was found by RÜHLE (1913) on the Bernburg anticline and later observed in several other localities in the central Zechstein basin.

In the Stassfurt section, the quantitative ratios in the anhydrite and halite layers correspond significantly enough to the theoretical quantities calculated for the static evaporation of seawater [theoretical: 3 mm anhydrite (4%) upon 10 cm NaCl out of $8^1/_2$ m seawater, or out of $2^1/_2$ m of a brine already saturated with respect to gypsum; observed: 4–5% anhydrite, Fig. 38]. This shows not only a regular change in the brine concentration, but, more importantly, that the formation of each couplet runs through the complete concentration range, at least from the beginning of gypsum saturation up to NaCl precipitation. Each couplet thus forms a small static evaporation cycle.

The alternative is to assume some form of continuous influx, as in Ochsenius' Bar theory; this, as so far presented, is physico-chemically and sediment-petrographically inadequate to explain the Zechstein 2. Neither the assumption of rhythmic precipitation (a detailed discussion in LOTZE, 1957) nor the assumption of temperature fluctuations can explain the double layers, especially as in an already saturated NaCl brine there is at most 0.7 g $CaSO_4$ per 100 g NaCl + $CaSO_4$ instead of the observed 4–5% (and in some places up to 9%). It is indeed possible to precipitate NaCl by cooling (p. 104) but this will not change the absolute amount of $CaSO_4$ in one and the same solution.

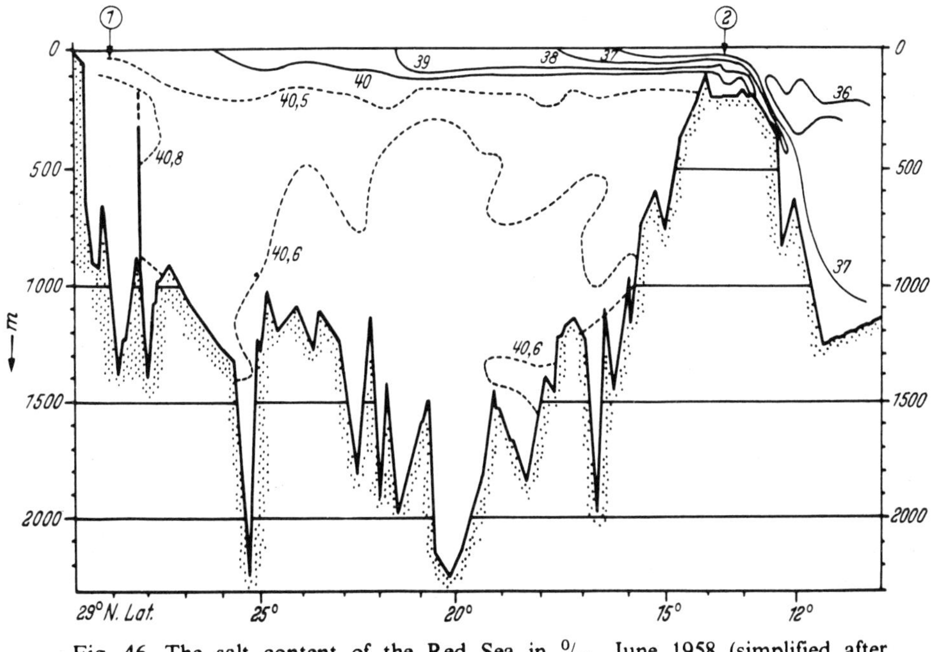

Fig. 46. The salt content of the Red Sea in ‰, June 1958 (simplified after NEUMANN and MACGILL, 1961). *1* Gulf of Aquaba; *2* Straits of Bab el Mandel in the Gulf of Aden

Assumption 2, that of periodic strong fluctuations in concentration in the upper brine layer, can be maintained even with continual brine influx. This possibility, which does not appear to have been discussed until now, will be developed below, using the Red Sea as an example.

The Red Sea, according to recent research (NEUMANN and MCGILL, 1961), shows a fairly continuous increase in salinity from about 36‰ at the bar in the Straits of Aden to over 40‰ in the Gulf of Suez (Fig. 46). The deep water throughout the Red Sea possesses the same salinity, about 40.6‰, up to the level of the bar (about 125 m below sea level). Immediately behind the bar there is thus a steep vertical concentration gradient (about 1‰/25 m) whilst the horizontal concentration gradient is very small (about 1‰/400 km). On account of the clear seasonal variation in evaporation (PRIVETT, 1959), upon which is superimposed a further variation due to latitude, a seasonal variation in the concentration gradient may be expected. In the Mediterranean such conditions are known to exist. In winter the 37‰ salinity line extends to about 2° E and in summer to 18° E (MCGILL, 1961, Fig. 3). In transposing these conditions to the Zechstein salt lake, a significantly greater horizontal gradient must first be assumed, which would be possible with a considerably shallower bar and longer or narrower straits than in the Red Sea, and in the case of the Kara Bugas Gulf is known to exist (SEDELNIKOV, 1958). As a consequence of seasonal fluctuations in the concentration gradient, the limit of gypsum or halite saturation in the surface layers at different times of the year must have lain at varying distances from the bar. The saturation limits must then have moved according to the season. There would thus be a transition zone in which there would be an alternation of gypsum and halite precipitation. All that is required is a pronounced variation in the horizontal concentration gradient in the surficial brine layers. From the data in Table 8 the magnitude of the necessary concentration variation from the beginning of gypsum precipitation to the precipitation of halite is evident, and this argues against the model. Should an annual rhythm, however, be demonstrable, it would appear to be most suitable as an explanation of anhydrite-halite banding and for that reason will not be totally rejected.

It is also possible to consider a continual seawater influx with a periodic influx from terrestrial sources containing Ca^{++} and HCO_3^- ions. This would lead to the precipitation of SO_4^{--} as calcium sulphate over a clay layer and at the same time bring about an impoverishment of the brine in $MgSO_4$ (p. 246). Although this can be excluded in the case of Zechstein 2, as then $MgSO_4$ depletion would occur in the central Zechstein basin, too, the model cannot be excluded in other cases.

The immediate cause of rhythmic bedding is certainly related to similar rhythmic fluctuations in the brine composition. The original bar theory model, which postulates that brine composition is more or less constant, or at least at any one spot in the salt lagoon excludes short-period variations in composition, cannot explain the results obtained from the Stassfurt Series, irrespective of whether or not a barrier is assumed to have separated subsidiary basins. If an annual rhythm is considered as the cause of the periodicity, then the available theories are either ZIMMERMANN'S (1915) hypothesis in which the seasonal influx occurred only during the Monsoon, or the rather more plausible hypothesis of the seasonal horizontal displacement of gypsum or NaCl saturation concentrations in the upper brine levels. In both cases, climate is the controlling influence. However, the second hypothesis is more consistent with oceanographic data than the first, and does not require a tongue of land as a barrier which has to be broken and reformed every year in the change from Monsoon to counter-Monsoon.

Yet the linking of the periodicity to an annual rhythm is an uncertain hypothesis. Apart from LOTZE'S (1955) views on the geologically short duration of salt formation, the following points must also be considered:

1. The normal thickness of the anhydrite layers (<1 mm) is quite compatible with the maximum annual evaporation, but this is not true of the not uncommon anomalies of 5 mm and more (up to >20 mm) in a single layer. To precipitate 5 mm anhydrite requires the evaporation of 14 m of normal seawater or 4 m of brine saturated in gypsum. At the present time the maximum evaporation from large water surfaces in very arid regions is 2 m (e.g. Red Sea; J. NEUMANN, 1952; PRIVETT, 1959). In more concentrated solutions this figure is never reached, but falls by about 1% for a 1% increase in salt concentration. In halite-saturated solutions the fall in evaporation is about 30% (HARDING, 1949, p. 65). Furthermore, in the absence of wind, halite precipitates at the water surface, reducing evaporation still further, so that the present-day evaporation from halite-saturated solutions may be taken as about $1^1/_2$ m per year. It is quite incompatible with well established physical principles to assume that a slight increase above an average temperature boundary will bring about much greater precipitation of $CaSO_4$ (RICHTER-BERNBURG, 1953). Only a marked negative temperature coefficient of gypsum solubility, which seems unlikely from the results of experiments so far carried out, under conditions of strong heating in shallow water could bring about the precipitation of gypsum from inflowing brines. This would require evaporation in the main basin to bring the brine to gypsum saturation, yet with the precipitation of gypsum itself concentrated over a much smaller area than the evaporation.

2. The precipitation of the observed anhydrite-NaCl couplets in Zechstein 2 from normal seawater likewise requires a higher annual

evaporation rate (about $8^1/_2$ m of normal seawater or 2.6 m of seawater concentrated to the point of gypsum evaporation). The Stassfurt halite cannot be directly compared with the average 10 cm halite precipitation per annum found in modern salt lakes such as Kara Bugas (FIVEG, 1954), as in these lakes the rhythmic banding with $CaSO_4$ is absent. The precipitation of a pure halite layer does not require strong periodic concentration variations nor does the precipitation of 10 cm NaCl demand greatly increased evaporation.

3. Regarding the anhydrite layer which sometimes replaces the Werra halite, JUNG (1959) established that it has 10 times as many "annual rings" as the halite itself. Of course, it could be argued that the halite was deposited only in the troughs without any equivalent on the ridges.

4. An attempt has been made to demonstrate statistically an 11-year cycle upon the basis of the anomalies in the fine bedding and to refer it to the sunspot cycles (RICHTER-BERNBURG, 1950, and later). Such a relationship is not yet well established. The anomalies of the sunspot cycles are a periodic fluctuation between a maximum and minimum, whilst the bedding thickness anomalies usually occur abruptly. One way of characterizing this is the determination of the periodicity from: $\frac{\text{number of varves}}{\text{number of anomalies}}$. Thus, even qualitatively, another rhythm may be suspected. Quantitatively the relationships are again not well established, quite apart from subjective factors in thickness measurements and counting and variations in the thickness of the actual beds. Even assuming that the measurements are reliable, the periodicity itself must be ascertained by harmonic analysis, as applied to all periodic geophysical processes (magnetic variations, sunspot numbers, etc.). Such an objective test is indispensable as, according to RICHTER-BERNBURG (1960), in counts of several hundred bands "no eleven-unit period is recognizable" or else the periods are of variable length. It remains to be seen whether under these circumstances there is a convincing dependence of the anomalies upon the rather constant solar periods. Apart from these methodological considerations, no effect of sunspot activity on the gross weather pattern has yet been detected (BAUR, 1961)[57], although the climate of the whole Earth has been recorded at a sufficient number of stations only over the last two or three sunspot cycles. Nevertheless,

[57] Page 65: "whilst sunspots and solar flares exhibit a clear, average 11-year variation, there is no 11-year periodicity in the general weather pattern. Rather in the rhythmic analysis of certain weather elements an 11-year cycle can be seen, but the amplitudes are always so small that they in no way explain weather anomalies. The 11-year period is thus well within the range of chance".

a connection between sunspot activity and the climate of the Earth is not impossible.

5. The enrichment of the clay layers in organic matter is regarded as a consequence of the plankton dying off in autumn (RICHTER-BERNBURG, 1960). This is open to more than one interpretation. Assuming a periodic influx of fresh seawater, a rich planktonic flora could be introduced from the adjacent open sea, but the increasing salinity due to evaporation would soon become too great for the majority of stenohaline plankton and their death would consequently be independent of season. The organic enrichment of the clay layers can thus be seen as a direct indication of the influx of fresh brine. This would involve evaporation on a scale so large it could not be explained by annual rhythm. Convenient and reasonable as the annual banding hypothesis may appear, it leads to difficulties at many points.

To be sure, the assumption of periodic influxes, resembling minor marine transgressions, runs into similar difficulties. In the first place it fails to explain the regularity of the influxes, and the same time applies to possible endogenic causes of rhythmic banding (LOTZE, 1957). Eustatic changes in sea level have also been suggested (HAM, 1961). Less probable is WILFAHRT'S great flood hypothesis, for as LOTZE (1957) noted it is highly improbable that the necessary amount of water could enter a basin as large as the Zechstein during a single high tide. The hypothesis requires that the evaporation surface of the salt sea should lie below sea level. This demands in turn an old depression, completely blocked off, with a depth at least equal to the salt thickness. This is quite feasible.

In this context the so-called "descendence" theory must be considered. It was introduced by EVERDING (1907, p. 31) for "salt beds which, immediately after precipitation of the sequence of mother salts and still during Zechstein time, were formed by redeposition or transformation of the mother material"[58]. This was clearly refuted by SCHÜNEMAN (1913) (see p. 178) for the Stassfurt area. For the younger Zechstein series (Z 3 and Z 4) it has been variously defended (D'ANS and KÜHN, 1940–1960), in particular as an explanation for the decreasing bromine curves. These arguments were disposed of on p. 142 upon the basis of a model with periodic seawater influx, although it was conceded that the D'ANS model (even with the solution of bromine-rich Stassfurt rock salt in rainwater) leads to a more rapid decrease in the bromine curve. KÜHN (in KÜHN and SCHWERDTNER, 1959; SCHAUBERGER and KÜHN, 1959) sought to use the large single crystals of NaCl (1–2 cm) occurring in the "Jüngeres Steinsalz" especially in zones Na 3 γ and Na 3 δ, as well as in Alpine

[58] Quoted from EVERDING as follows: "Salzablagerungen, die alsbald nach Abscheidung der Muttersalzfolge noch während der Zechsteinzeit aus der Umlagerung oder Umbildung des Muttermaterials hervorgegangen sind".

salt deposits, as an independent proof of the mechanical redeposition of coarse-grained older salts. In this context the so-called "Augensalze" were regarded as fragmentary salts or salt conglomerates. Excluding the sedimentary-petrographical weaknesses[59] of this explanation, geological objections have been raised, particularly as regards the amount of "descendence" (LOTZE, 1957; BORCHERT, 1959). It is certainly true that the character of the sedimentation of the younger halite horizons is disturbed with numerous progressive and recessive epochs. Only, however, in a fully isolated basin which is periodically flooded is basin shrinkage possible with, at least in places and at times, the redissolving of older salt precipitates in the dried-out marginal areas, either by a fresh influx of seawater or (less likely) by continental run-off or rainwater. On the other hand, the mechanical redeposition of sylvite and carnallite is quite certainly impossible, and that of NaCl is probably also excluded, so that this hypothesis is inapplicable to potash deposits. Yet, insofar as the level of the brine always lay at sea level, the redissolving of older salts is altogether unlikely, and the significance of such a process in the Zechstein salt sequences would generally have been slight.

The model of a completely isolated basin can only have developed late, if at all, either during or towards the end of chloride precipitation, possibly even following carbonate or gypsum precipitation in the bar region. If a return flow of concentrated brine was already excluded at the time of gypsum precipitation in the main basin, the amounts of halite and potash salts precipitated would have been much greater. There must thus have been an appreciable potential return flow of concentrated brines at the beginning of the saline phase, as is the case in the water layer above the barrier zone in the smaller enclosed seas at the present time (Fig. 46). To be sure, no quantitative calculation is ever likely to be possible for Zechstein 2, the total salt content of the basin being far from well known, because of the great variations in thickness and lateral changes in the facies.

In contrast, during the formation of the Hauptanhydrit, the greater part of the NaCl-solutions must have flowed back. In the present state of knowledge it appears that the Hauptanhydrit has a fairly uniform thickness over the whole basin while the thickness of the overlying "younger halite" (Jüngeres Steinsalz) is far too small in relation to the anhydrite thickness. Yet it is not known whether the highly concentrated

[59] From his examination of erosion in the salt stocks in the Persian Gulf, the author concludes that the transport of material is by pure solution weathering with the formation of sharp-angled solution channels. The only solids transported in salt water run-off streams are the insoluble residues. As well as the depositional mechanism required, the necessary flow rate is improbable within a salt lake; furthermore, such coarse-grained salt rocks occur too rarely in the Stassfurt halite and almost always as the result of secondary recrystallization.

residual solutions of the Stassfurt time were also washed out of the basin. This seems unlikely on account of the abnormally high bromine content of the "younger halite" (Jüngeres Steinsalz) immediately above the Hauptanhydrit (p. 204). It seems more likely that there was a very stable layering of the dilute influxes over a highly concentrated basal solution, so that the return flow did not carry away this old, residual brine. The characteristic radial structures of the Hauptanhydrit also suggest special precipitation conditions which can perhaps be explained by recrystallization of this concentrated basal solution. Admittedly, the Hauptanhydrit might, at least in part, be interpreted as a sedimentary precipitate (due to large influxes of $Ca(HCO_3)_2$ solutions). This would be consistent with the facies transition from Hauptanhydrit to Plattendolomit indicated by ZIMMERMANN (1915).

Weighing all the observed facts, it appears certain that during the formation of the fine, rhythmically banded carbonate and anhydrite rocks the surface level of the salt basin must have been the same as that of the open ocean, for otherwise the reflux of concentrated brine solutions would have been possible. This in turn would suggest periodic variations in the composition of the brines and be a point in favour of the annual banding hypothesis for the fine-banded layers. Yet whether the anomalously thick anhydrite layers could form in a single year remains very doubtful. Using the proposed mechanism of periodic horizontal displacements of the saturation limits for a given solid phase, the absence of an "annual ring" can be explained by the fact that, during the formation of these anomalously thick beds, the horizontal displacement was smaller and hence the surface brine layer remained saturated in the same solid phase over a longer period [60]. The same explanation could also be applied to the halite series, particularly where no strict regularity is found in the rhythmic bedding. It is certain that at the end of the halite precipitation, at least in the Stassfurt series, the salt basin became completely isolated. There were, nevertheless, several ingressions of fresh seawater, as witnessed by the intercalated rock salt banks (e.g. the Unstrutbänke, p. 212). This was early recognized by D'ANS (1915, p. 269).

It therefore seems right to adopt a compromise view of the bedding question, as LOTZE (1957) also does, since part of the sediments appear to be correlated with an annual rhythm. It is certainly an error to try to ascertain the duration of deposition simply by counting the bands, as an individual band may not uncommonly represent more than one year. Conversely, it is still uncertain whether individual bands always represent a complete year.

---

[60] BRAITSCH, O.: „Die Entstehung der Schichtung in rhythmisch geschichteten Evaporiten", Geol. Rundschau **52**, 405–417 (1962).

Editors' Comments on Papers 24 Through 27

Diagenesis and Alteration of Evaporites

It is a matter of simple experiment to show that evaporites are very soluble rocks. It is a matter of uncomplicated reasoning to consider, therefore, that they might be affected by profound diagenetic changes. The medium of their deposition, water, is frequently the agent of their transformation, and one longtime student of evaporites has termed them the "most vivacious and temperamental of rocks" (Lotze, 1938); temperamental, because the slightest change in the environment can induce complex changes in the rocks; vivacious, because as a result of their susceptibility to alteration, they display a varied, interesting, and lively face to those petrographers venturesome enough to become familiar with them.

The assertion that "most large evaporite bodies have suffered widespread metasomatic changes" (Stewart, 1963) can be supported by abundant petrographic and isotopic evidence. The descriptions of complex replacements in the British Permian (Stewart 1949, 1951a, 1951b) and the Salado Formation (Schaller and Henderson, 1932) are now classics in the field, and recent work has demonstrated that the alteration of evaporites may occur shortly after deposition (as in the case of dolomitization of aragonite in sabkhas) or many millions of years later (as in the case of the replacement of anhydrite by calcite associated with the formation of sulfur).

Many of the changes that affect evaporitic rocks after deposition are the result of what Borchert and Muir (1964), and many others before them, refer to as "geothermal metamorphism." It is, of course, the unstable mineral assemblages of potash deposits that are most seriously affected, and although details of the particular alterations are beyond the scope of this discussion, an important ingredient in those metamorphic changes is the water of crystallization of gypsum. This water is released when gypsum, the original precipitate, is itself affected by the higher temperatures resulting from burial, and converts to anhydrite, the common subsurface form in the $CaSO_4$–NaCl–H_2O system.

The most commonly encountered evaporites in which diagenesis has been studied in detail are calcium sulfate deposits. R. C. Murray, chairman of the Department of Geology, Rutgers University, and formerly with Shell Development Company, Houston, has described the diagenetic cycle of gypsum to anhydrite to gypsum, which has affected many of these deposits and which has been accepted by most students of evaporitic sediments. After reviewing the physical and chemical evidence, Murray (1964) concluded that calcium sulfate was almost always precipitated at the earth's surface as gypsum rather than as anhydrite. This conclusion did not stand unchallenged for long, however, for at about the same time that Murray's paper appeared, what was taken to be primary anhydrite was found in the sabkha sediments of the Persian Gulf (e.g., Kinsman, 1966, p. 312). Later work, however, confirmed Murray's contention that anhydrite nodules in carbonate sediments of the supratidal flat of the Trucial Coast are derived diagenetically from gypsum precursors (Butler, 1969, p. 81; Kinsman, 1969, p. 834).

The author of the second paper in this section, I. M. West, has traveled an unusual academic route. In the rigorous academic environment of the years immediately preceding entry into university in England, he failed to satisfy the minimum requirements for admission. With his great interest in geology, he took a post as laboratory technician in the Department of Geology at Southampton University, and it was there that he completed the detailed studies of diagenesis in the Purbeckian evaporites. One small indication of the unusual conditions surrounding his research is given by the acknowledgments in the paper we include here: ". . . I gratefully acknowledge the loan of a microscope from the Royal Society with which to carry out this work." Having demonstrated his aptitude for geologic investigations, he was allowed, sans B.Sc., to complete a Master of Science degree at Liverpool University, whence he returned to Southampton as a Lecturer in the Department of Geology.

West (1964) has deciphered the diagenetic history of the Lower Purbeck beds of Dorset, England, by careful petrographic observation. He unravels, in rather elegant fashion, the paragenetic and textural complexities within the general diagenetic sequence: gypsum replaced by anhydrite, subsequently replaced by gypsum. Some of the pseudomorphs are even visible in the field to the naked eye. Of particular interest is the rather extensive calcitization and partial silicification of gypsum and anhydrite in the units he examined.

F. H. Stewart, one of the most distinguished of living British geologists, has been Professor of Geology at Edinburgh University since 1957. He is a student of igneous and metamorphic petrology as well as evaporite petrology, and in the latter endeavor, he has been concerned chiefly with the Zechstein evaporites in the English part of that large basin. As one reads Stewart's paper, a summary of the mineralogy and mineralogical alterations in the British Zechstein, one cannot fail to be impressed by the number and complexity of diagenetic changes that have occurred. Many of the details of the changes have been documented in the remarkable series of papers on the famous Eskdale no. 2 boring (Stewart 1949, 1951a, 1951b); similar changes may be expected to have occurred in any large and complex evaporite sequence.

The problem of the origin of dolomite is one that falls naturally within the realm of evaporite diagenesis, for with the intuitive elucidation of "seepage refluxion" as a mechanism (Adams and Rhodes, 1960), and after the discovery of dolomite in Holocene evaporitic settings, it now appears that many dolomites may owe their origin to the reaction of calcium carbonate with magnesium-rich, hypersaline brines. Not only that, but many ancient dolomites may have formed as a result of a process akin to "seepage refluxion."

As gypsum precipitates from a marine brine, the value for Mg:Ca in the residual liquid increases, and the density of that liquid is raised to the point that it will displace "lighter" water within pore space in calcium carbonate sediments at the bottom of, and beneath, the basin. This brine, and the sediment through which it passes, react to form dolomite, and, at least according to the theory, will continue to do so until equilibrium between solid and liquid is attained.

In 1965 Deffeyes, Lucia, and Weyl reported on their investigation of a modern example of "evaporite reflux," on the island of Bonaire in the southern Caribbean. The brine from a hypersaline lake, the Pekelmeer, seeps through an extensive carbonate flat and has altered some calcite to dolomite. The hydrodynamics of the Pekelmeer may not be as simple as at first described, and with this in mind the reader should refer to the papers of Lucia (1968) and Murray (1969). The point of the paper by Deffeyes et al. is that it demonstrated the association of gypsum precipitation and dolomitization; if only as a prelude to later studies, this paper has considerable importance in that it provides sound field evidence for contemporary "evaporative reflux" or "seepage refluxion" as an agent of diagenesis.

Reprinted from *Jour. Sed. Petrol.*, **34**, 512–523 (1964)

ORIGIN AND DIAGENESIS OF GYPSUM AND ANHYDRITE[1]

R. C. MURRAY[2]

24

ABSTRACT

Despite data on the stability of anhydrite at high earth surface temperatures, the observation of Recent gypsum and the evidence for formation and preservation of metastable gypsum suggest that this form may be the common if not universal original calcium sulfate mineral. Gypsum beds can be produced by two mechanisms: (1) Precipitation and sedimentation in a standing body of water subjected to evaporation. Such deposits are original sedimentary facies, contemporaneous with other sedimentary facies. (2) Growth of abundant crystals by displacement of unconsolidated sediment or weathered rocks, which results in beds of gypsum which later is replaced by nodular anhydrite. These nodular anhydrite beds represent a diagenetic facies and postdate the host material. The distinction between sedimented and nodular is important in the interpretation of any given evaporite deposit.

These primary deposits act as a source of calcium sulfate for subsurface growth of replacement and void-filling anhydrite. The latter two secondary types of anhydrite can be distinguished in reflected light, since a dark color is imparted to replacement anhydrite by included material.

From observations in Recent sediments, older outcrops, and the subsurface, there appears to be a characteristic cycle in the diagenesis of gypsum-anhydrite minerals: surface or near-surface gypsum is replaced by anhydrite as a result of burial and is in turn replaced by gypsum if the anhydrite is thereafter brought close to the surface.

ORIGIN OF PRIMARY DEPOSITS

Sedimentary deposits of the calcium sulfate minerals (gypsum and anhydrite) are common throughout much of the geologic record. They are commonly associated with carbonate rocks, especially dolomite and ferruginous clastics. The object of this paper is to examine the evidence regarding which mineralogic form is the primary depositional product and to trace the diagenetic changes that accompany burial and re-exposure of these deposits. This has been the subject of much discussion in the literature with many authors favoring anhydrite as the original form.

Physical Evidence

In general, gypsum is found near the surface and in Recent sediments, whereas anhydrite is the common form of the subsurface. Indeed, it is not uncommon in large gypsum quarries to encounter interbedded anhydrite and gypsum followed by isolated gypsum patches and finally pure anhydrite with depth. Many authors, for example, Sund (1959), argue convincingly for the replacement of anhydrite by gypsum under near-surface conditions. The evidence for this replacement may be observed in all deep quarries and shallow cores where some anhydrite is still preserved. Although the transition commonly takes place within a few hundred feet of the present surface, the replacement of anhydrite by gypsum has been observed as deep as 3500 feet in the Permian San Andres Formation, Dune field, Crane County, Texas (fig. 1).

Bundy (1956) suggested that the gypsum may be the predominant or only form of calcium sulfate precipitated from sea water. Such a generalization is naturally at variance with interpretations of a primary origin of anhydrite. However, studies of Recent evaporite deposits—for examples, Fisk (1959), Morris and Dickey (1957), Phleger and Ewing (1962), Masson (1955), Talmage and Wootton (1937), and Wells (1962)—have emphasized the observations that gypsum is the common if not universal form present. It has generally been impossible to determine the specific composition and temperature of the solution at the time the gypsum was precipitated. However, gypsum is forming today in some of the hottest parts of the surface of the earth.[3] Recently, Curtis and others (1963) have reported Recent anhydrite above the free water level in the supratidal flats along the Trucial Coast of the Persian Gulf.

Stewart (1953) observed pseudomorphs of anhydrite after gypsum in the Permian evaporites of northeastern England. These pseudomorphs could only be seen when weathered sur-

[1] Manuscript received July 22, 1963.

[2] Shell Development Company (A Division of Shell Oil Company), Exploration and Production Research Division, Houston, Texas (Publication No. 354).

[3] Professor F. H. Stewart has correctly reminded the author that the argument of near universal Recent gypsum may not be applicable to deep water evaporites of the past which do not have Recent analogues. However, MacDonald (1953) suggests that increasing hydrostatic pressure favors gypsum.

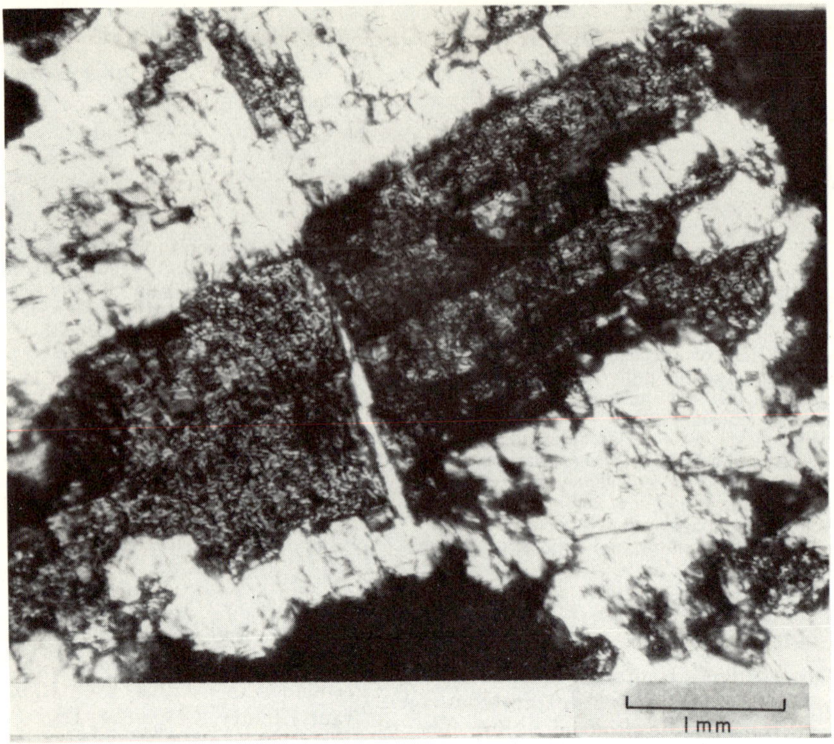

FIG. 1.—Gypsum replacing replacement anhydrite; San Anydres Formation, 3500 feet below the present surface. Gypsum (white) rims and indents a core of subhedral anhydrite. The anhydrite had originally replaced fine sucrose dolomite.

faces of the core were studied. Some of the original crystal outlines were flattened, and in extreme cases the structures were barely detectable. This direct type of observation has seldom been made in massive evaporite beds. The reason must, in part, be ascribed to imperfection of preservation. However, this example is prima-facie evidence of the metagypsum origin of these anhydrite beds.

These general observations suggest the generalization that a common diagenetic cycle exists in the calcium sulfate minerals—deposition of gypsum followed, during burial, by replacement of gypsum by anhydrite followed in turn, during uplift and erosion of overlying strata, by replacement of anhydrite by gypsum (fig. 2). Henderson (1954) argued that gypsum in the Stanford Range has never been buried to sufficient depth for replacement of gypsum by anhydrite. If his interpretation based on delicate preservation of textures is correct, these ancient deposits have never passed through the complete cycle.

Chemical Evidence

The physical-chemical environment that favors primary precipitation of gypsum or anhydrite or replacement of one mineral by the other has received much study since the early work of Van't Hoff in the early 1900's. Posnjak (1938) produced data on the stability of gypsum and anhydrite as a function of temperature. These data were obtained only by dissolution. The intersection of the solubility curves for the two minerals defines a temperature above which anhydrite is less soluble than gypsum and thus presumably more stable. The temperature of intersection for distilled water was 42°C; thus, anhydrite should be favored above this temperature. From his earlier work Posnjak recognized that solubility determinations in different concentrations of sea salt solutions could be helpful in interpreting the conditions of deposition of the two minerals from concentrated sea water. He made solubility determinations in solutions of sea salts at 30°C (Posnjak, 1940) and showed that an increase in salt concentration would lower the temperature of equal solubility. He concluded that sedimentary marine deposits of pure anhydrite either must have been partly derived from originally deposited gypsum or must have been formed near or above 42°C. However, from stability considerations some anhydrite could have been formed below 42°C.

During his experiments Posnjak observed that

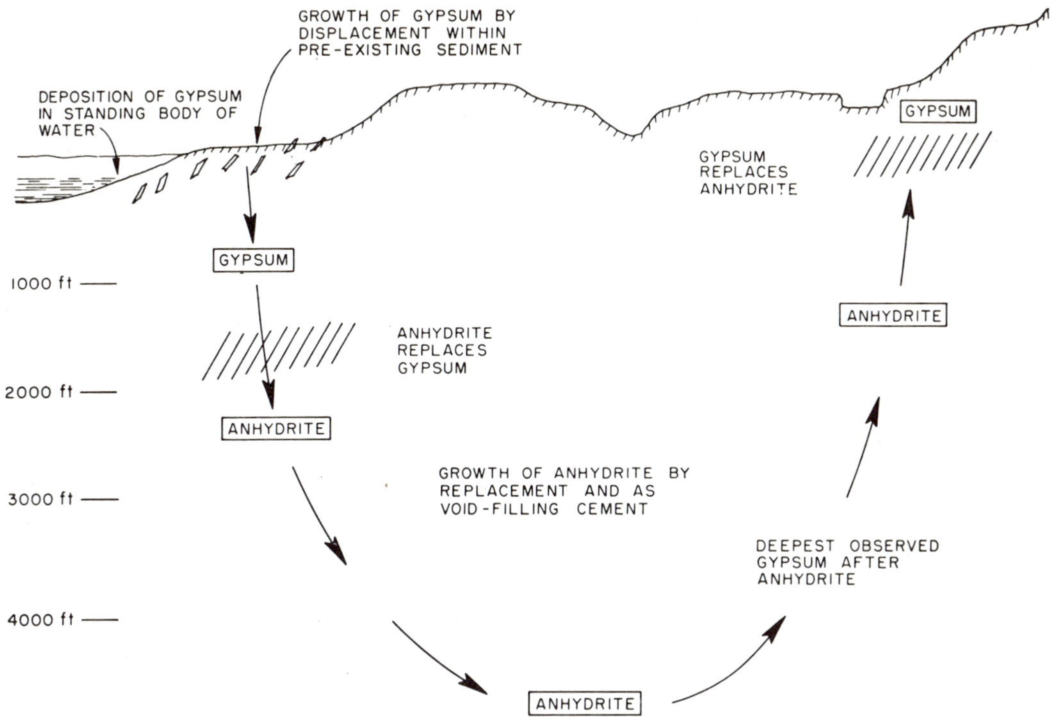

FIG. 2.—Schematic diagram illustrating gypsum-anhydrite-gypsum diagenetic cycle.

gypsum was very persistent and crystallized easily even in its metastable state. Indeed, evaporation of sea water in the temperature range 60°–80°C produces gypsum rather than anhydrite as a final product. This suggests that factors other than the solubilities of the two minerals may be involved in determining the conditions under which one or the other will be formed.

MacDonald (1953) approached the problem thermodynamically in order to study the effect of concentrated salt solutions and pressure on the gypsum-anhydrite transition temperature. From previously determined thermodynamic properties he calculated a transition temperature of 40°C for the reaction $CaSO_4 \cdot 2H_2O \rightleftarrows CaSO_4 + 2H_2O$ in pure water. This value corresponds closely with Posnjak's experimental data.

He also demonstrated the effect of pressure on the reaction in pure water. Under hydrostatic conditions the transition curve had a positive slope, indicating that an increase of total pressure would increase the transition temperature and would favor the formation or preservation of gypsum.

From further thermodynamic calculations, MacDonald (1953) produced a graph showing the transition temperature as a function of the concentration of NaCl, which was compared with sea water by corresponding chlorinities. This graph shows the transition temperature decreasing with an increased concentration of salts; thus, theoretically, one could determine the transition temperature of gypsum-anhydrite at any concentration of sea water. With an estimated concentration of 3.35 times that of normal sea water as the saturation point of $CaSO_4$ in the evaporation of sea water, a transition temperature of 34°C was determined. Therefore, if deposition is an equilibrium process, gypsum will be the stable phase and will precipitate out of sea water first at all temperatures less than 34°C, and anhydrite will precipitate out first at all temperatures greater than 34°C.

Henderson (1954) presented additional data on relative solubility of gypsum and anhydrite in salt solutions and concluded that gypsum should be stable to a maximum depth of 2000 feet.

Conley and Bundy (1958) studied the mechanisms involved in converting anhydrite to gypsum. By experimental work they further substantiated Posnjak's observations concerning the persistence and ease of crystallization of gypsum in its metastable state. They also determined that the reaction of gypsum going to anhydrite plus water is very slow or nonexistent below 42°C. Further, in contrast with Posnjak's and MacDonald's work showing that salt solutions lower the temperature at which anhydrite becomes more stable, they showed that some salts activate rather than inhibit the hydration process in the temperature range studied. In other words, these salts promote the conversion of anhydrite to gypsum, even under conditions theoretically favoring the formation of anhydrite. The best activators were found to be sodium and potassium sulphate.

From their data, Corley and Bundy postulated that it is highly improbable that anhydrite is ever precipitated primarily from sea water because (1) gypsum crystallizes relatively easily in its metastable state, and (2) as soon as crystallites of anhydrite would begin to form in sea water, the activating constituents in the sea water would bring about an almost immediate conversion to gypsum at temperatures below 42°C. They conclude that the thermodynamic applications which predict the formation of anhydrite from gypsum in concentrated salt solutions are incomplete in that they do not incorporate the effect of $CaSO_4$ activators which influence the reaction kinetics.

Despite the evidence from solubility and thermodynamic calculations that anhydrite is more stable in concentrated salt solutions below 42°C, the observation of common if not universal occurrence of Recent gypsum in brine and the influence of kinetic factors in allowing formation and preservation of metastable gypsum suggest that the primary deposit may almost always be gypsum.

Thus, there is a body of evidence that suggests that under most surface conditions gypsum should be the sedimentary product. Very high temperatures might locally cause formation of some anhydrite at the surface in the presence of gypsum. Indeed, this is happening today on the supratidal flats of the Trucial Coast. However, formation and preservation of gypsum may occur for kinetic reasons when surface temperatures are sufficiently high to make anhydrite the stable phase. With burial, gypsum should be replaced by anhydrite. The minimum depth for the transition should be a temperature of 42°C corrected for the pressure effect at the depth of the transition. Because the rate of the change of gypsum to anhydite with respect to the rate of subsidence is unknown, this value must remain a minimum depth. Theoretical values for the depth of this transition range around 2000 feet. However, the author is not familiar with any study of gypsum-to-anhydrite transformation in a formation undergoing initial burial. In th deeper subsurface, anhydrite is the almost uni versal form.

Environments of Formation

Studies of anhydrite in carbonate rocks and i association with carbonate and clastic rock suggest that a threefold distinction can be mad on the basis of the mode of formation. This dis tinction involves the recognition of (1) bedde anhydrite, (2) pore-filling anhydrite, and (3 replacement anhydrite. The anhydrite bed which represent the sedimentary unit of primar formation and thus generally metagypsum ar the obvious source of material for formation c the two secondary types.

From the previous discussion it is apparen that most of the sedimentary units of anhydrit observed in the subsurface are metagypsum. Thi may be true despite the fact that independen evidence of this replacement in individual ex amples is relatively rare. The original conditio of precipitation involved concentration throug evaporation of water, commonly sea water o ground water that has moved through pre existing evaporite deposits. Such deposits ma be of two general types, each with its own sig nificance with respect to paleoenvironmen Although an individual formation may exhib both types, a distinction between (1) sedimente and (2) nodular fabrics can commonly be recog nized.

The sedimented or laminated anhydrite o gypsum beds of which much of the Castile fo mation (King, 1947) and the Pleistocene Lisa formation of the Dead Sea are composed repre sents precipitation and sedimentation in a stan ing body of water. The concentration of th evaporating brine may be maintained within th field of $CaSO_4$ precipitation by limited flow o new sea water into the body of water and refl of the heavy brine out through the inlet (Kin 1947) or through the underlying sediments ar rocks (Adams and Rhodes, 1960; Deffeye Lucia, and Weyl, 1964). Such evaporite bed commonly show continuous laminations (fig. . and sedimentary structures and should be esse tially devoid of uniformly distributed mari fossils. They represent a depositional sedime tary facies, time-contemporaneous with oth facies. These other facies may be clastic sed ments or even carbonate. The latter is mo difficult because extensive carbonate depositic which depends on organic production is n possible in concentrated brines.

Nodular anhydrites or gypsum, although th occur in thick units relatively free of carbona and clastic material, present a different proble of genesis. All gradations exist, from isolated nodules in fossiliferous limestone or dolomite to thick beds of clustered nodules with only a thin matrix or sheath of limestone, dolomite, or siliceous clastic material. The crystal fabric of the anhydrite in nodules is similar to that of the sedimented anhydrite. Both exhibit a tightly interlocking fabric of needlelike to tabular crystals commonly preferentially oriented parallel or subparallel to bedding (fig. 4). This preferred fabric may change to an orientation subparallel to the nodule surface near the boundary.

Nodules of anhydrite commonly exist in fossiliferous limestones and fossiliferous limestone replaced by dolomite. However, because of general lack of foreign inclusions or ghosts and the draping of the host rock around the nodule, they do not appear to have formed by replacement of carbonate rock. The similarity of nodule shapes observed in rocks of different age and lithology and the absence of nonfilled vugs of similar shape and size argue against gypsum cementation of large vugs. The thin sheath or matrix of carbonate rock commonly observed between closely spaced nodules excludes the vug origin and suggests a displacement or compactional origin where one nodule has been squeezed against another. These anhydrite nodules appear to exist simply as ovoid masses within carbonate rocks whose origin is inconsistent with an evaporite environment.

Gypsum crystals similar to those found today on the shores of Lake Lucero near White Sands, New Mexico offer an explanation for the anhydrite nodule origin (fig. 5). These crystals have grown within the Recent alluvial sediments. During growth the crystals physically push aside the silt, sand, and gravel and form clear or nearly clear selenite crystals. When these sediments are buried, the gypsum will be replaced by anhydrite. If compaction of the sediment accompanies this change, draping of sediment both over and under the nodule will result. If the gypsum crystals grew in carbonate sediment, and they do today in most of the arid climate carbonate supratidal flats, the same relations should hold, and the product would be anhydrite nodules within, but not necessarily related by original depositional environment to the carbonate host rocks (figs. 6 and 7). Small gypsum crystals should produce the small pellet-like anhydrite nodules. The larger crystals should produce the larger anhydrite nodules.

The formation of selenite crystals and rosettes by displacement of soft sediments or weathered rocks appears to take place by at least two mechanisms: (1) Evaporation of ground water in the vadose or capillary zone (Talmage and Wootton, 1937). Crystals of this origin are indicative of subaerial exposure of the top of the host rock or sediment. (2) Evaporation of water on a playa or other standing body of water and down-

FIG. 3.—Laminated gypsum and calcite; Castile Formation, Permian, West Texas.

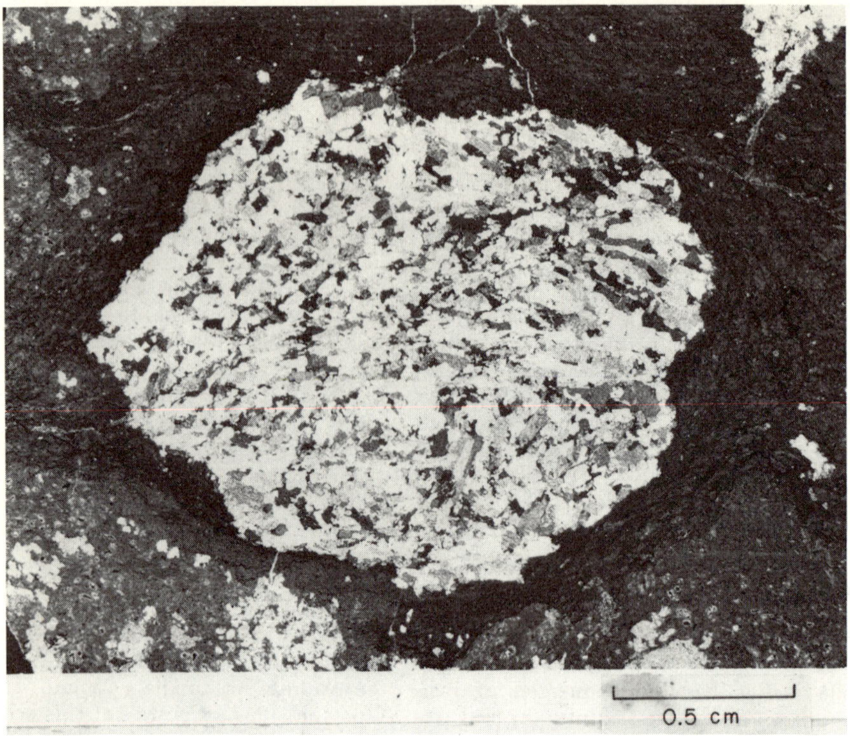

FIG. 4.—Small anhydrite nodule; transmitted light. Note the orientation of the individual anhydrite crystals subparallel to the bedding. The host San Andres dolomite is draped around the nodule.

FIG. 5.—Surface of valley fill sediments west of Lake Lucero, New Mexico. The clear gypsum crystals have grown within the sand and silt by displacement. The tops of some crystals have been exposed by erosion.

FIG. 6.—Bed of nodular anhydrite grading into fossiliferous sucrose dolomite; Clear Fork Formation, West Texas.

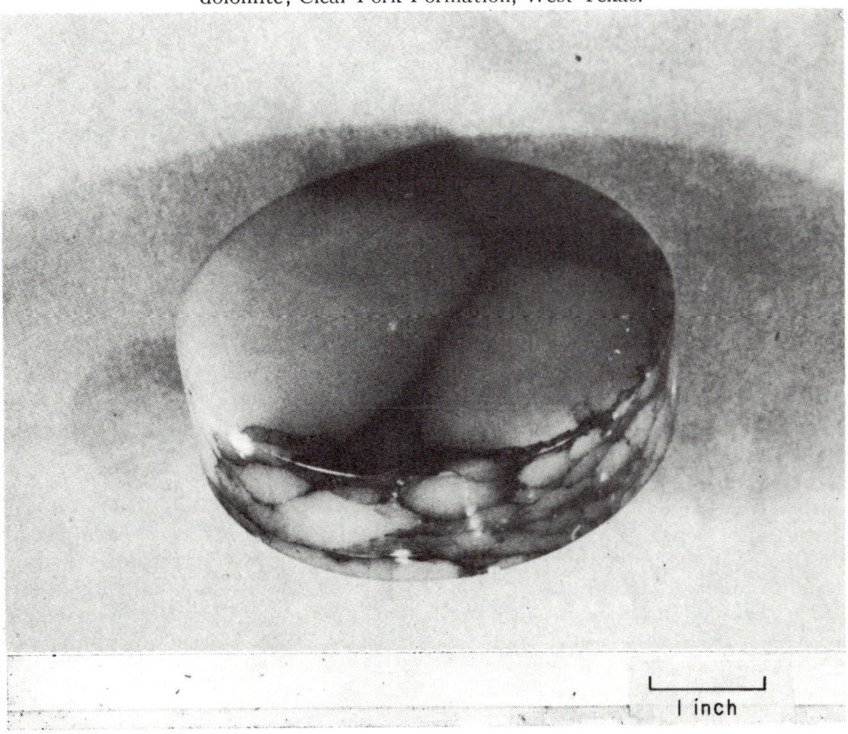

FIG. 7.—Bed of nodular anhydrite showing thin sheath of original sediment between pure anhydrite nodules; Madison Formation, Montana.

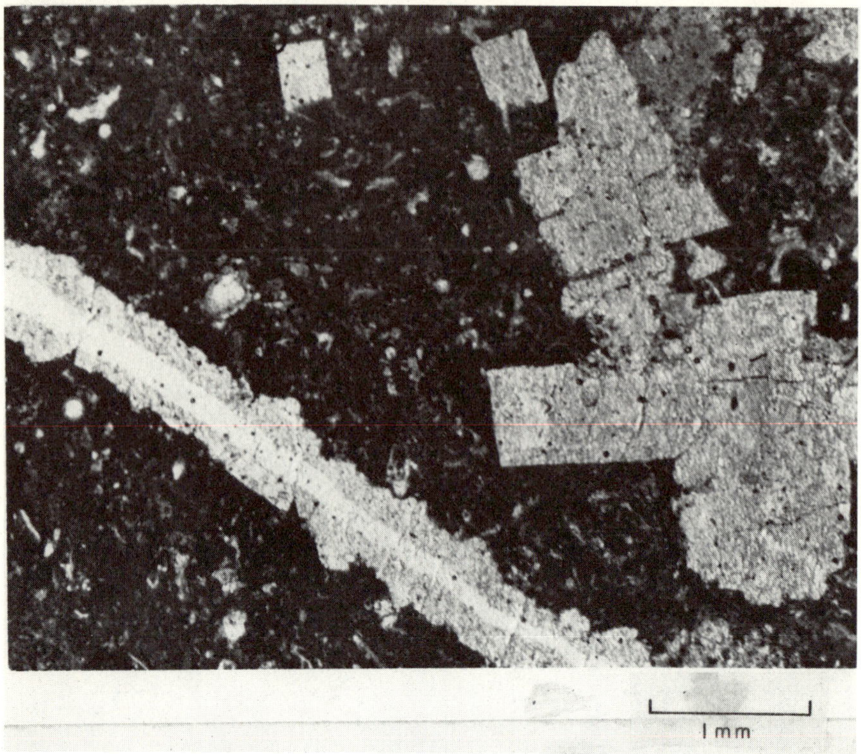

FIG. 8.—Replacement and void-filling anhydrite. The large blocky crystals have grown by replacement limestone. The fracture in the lower left has been filled with clear anhydrite crystals. These crystals ha developed overgrowths by replacement of the limestone wall rock. The same observation may commonly made in a carbonate sand where anhydrite fills the interparticle space and continues growth by replaceme of the particles.

ward movement of the dense hypersaline water into sediment below, with resulting formation of the gypsum crystals. Crystals of this origin are forming in the Laguna Madre tital flats (Masson, 1955). With either mechanism the presence of the anhydrite nodules is not necessarily indicative of hypersaline water during deposition of the host rock. Thus, the apparent inconsistency of coexisting marine organisms and rocks and hypersaline conditions is satisfied If the host rock existed at an unconformity that exposed sediments or weathered rocks of different age, the nodular bedded anhydrite would appear to be time-transgressive.

Recently, Kerr and Thomson (1963) have observed pseudomorphs of anhydrite nodules after gypsum crystals, indicating that the gypsum crystal form is not always destroyed in producing anhydrite nodules and presenting primafacie evidence of a metagypsum origin of the anhydrite. Using the Recent analogy of the Laguna Madre gypsum, they point out the association of tidal-flat sediments with nodular anhydrite and suggest that this environment of intermittent inundation with sea water offers an ideal environment for formation of nodula evaporites.

The nodular anhydrite beds thus represen growth of gypsum by displacement of soft sed ment after deposition of the host material an must be considered an early diagenetic facie rather than a depositional facies. Many of th bedded evaporites in the geologic record contai thick sections of the nodular form. Indeed, mos of the examples of Recent evaporites occur o supratidal flats and shallow lakes on supratida flats rather than in large, deeper-standing bodie of water.

Thus, the conclusion may be drawn that nodu lar anhydrite is indicative of subaerial exposure o soft sediments with evaporation of CaSO bearing water taking place within the sedimen Such conditions commonly exist today on th supratidal flats in arid climates or in desert pla yas. In the former the source of water is con monly the sea and transport into the supratida sediments may take place during a large storm o by seepage through the sediments. In the latte the source of the water may be ground wate that has passed through older evaporites an

SPECTROCHEMICAL ANALYSIS OF THE THREE TYPES OF ANHYDRITE*
(Weight Percent)

Metallic Elements	East Poplar Charles Formation			Wasson San Andres Formation			Sturgeon Lake Leduc Formation	
	Murphy #43 5791 Ft Replacement	Murphy #10 5634 Ft Pore-Filling	Murphy #10 5633 Ft Bedded	Roberts #1 4945 Ft Replacement	Roberts #1 4945 Ft Pore-Filling	Roberts #1 4944 Ft Nodules	SB-1-12 8346 Ft Pore-Filling	SB-1-12 8342 Ft Bedded
Aluminum	0.01	0.003	0.003	0.02	0.002	0.003	0.001	0.002
Antimony	0.0	0.0	0.0	0.0	0.0	0.0	0.0	0.0
Arsenic	0.	0.	0.	0.	0.	0.	0.	0.
Barium	0.002	0.001	0.001	0.002	0.0006	0.0003	0.00003	0.0008
Beryllium	0.000	0.000	0.000	0.000	0.000	0.000	0.000	0.000
Bismuth	0.00	0.00	0.00	0.00	0.00	0.00	0.00	0.00
Boron	0.0	0.0	0.0	0.0	0.0	0.0	0.0	0.0
Cadmium	0.0	0.0	0.0	0.0	0.0	0.0	0.0	0.0
Calcium	26.	30.	30.	33.	28.	29.	28.	29.
Chromium	0.000	0.000	0.000	0.000	0.005	0.000	0.000	0.000
Cobalt	0.00	0.00	0.00	0.00	0.00	0.00	0.00	0.00
Columbium	0.0	0.0	0.0	0.0	0.0	0.0	0.0	0.0
Copper	0.000	0.000	0.000	0.000	0.004	0.006	0.003	0.000
Gallium	0.00	0.00	0.00	0.00	0.00	0.00	0.00	0.00
Germanium	0.00	0.00	0.00	0.0	0.0	0.0	0.0	0.0
Gold	0.0	0.0	0.0	0.0	0.0	0.0	0.0	0.0
Indium	0.00	0.00	0.00	0.00	0.00	0.00	0.00	0.00
Iron	0.06	0.03	0.03	0.09	0.1	0.04	0.02	0.02
Lead	0.00	0.00	0.00	0.00	0.00	0.00	0.00	0.00
Magnesium	0.3	0.007	0.008	0.8	0.02	0.004	0.006	0.000
Manganese	0.000	0.000	0.000	0.000	0.000	0.000	0.000	0.000
Molybdenum	0.000	0.000	0.000	0.000	0.01	0.000	0.000	0.000
Nickel	0.09	0.09	0.09	0.1	0.08	0.09	0.07	0.1
Palladium	0.00	0.00	0.00	0.00	0.00	0.00	0.00	0.00
Potassium	1.	0.8	1.	0.0	0.0	0.0	0.0	0.0
Silicon	0.3	0.1	0.1	1.	0.2	0.1	0.06	0.08
Silver	0.000	0.000	0.000	0.000	0.000	0.000	0.000	0.000
Sodium	0.2	0.2	0.2	0.0	0.0	0.0	0.0	0.0
Strontium	0.3	0.5	0.2	0.3	0.1	0.1	0.1	0.2
Tantalum	0.	0.	0.	0.	0.	0.	0.	0.
Tin	0.00	0.00	0.00	0.00	0.00	0.00	0.00	0.00
Titanium	0.01	0.01	0.01	0.03	0.00	0.00	0.00	0.00
Tungsten	0.	0.	0.	0.	0.	0.	0.	0.
Vanadium	0.000	0.000	0.000	0.000	0.000	0.000	0.000	0.000
Zinc	0.0	0.0	0.0	0.0	0.0	0.0	0.0	0.0
Zirconium	0.0	0.0	0.0	0.0	0.0	0.0	0.0	0.0
Phosphorous	0.	0.	0.	0.	0.	0.	0.	0.
Mercury	0.	0.	0.	0.	0.	0.	0.	0.

*Ca may be in error as much as ± 8-10 percent absolute. Normal error of the other elements is ± 50 percent of the value reported.

FIG. 9.—Spectrochemical analysis of the three types of anhydrite.

thus represents a reconstructed water. Because nodules may form below standing bodies of water in earlier sediments and because shallow bodies of water may temporarily exist as lakes or supratidal flats or desert playas, both sedimented and nodular anhydrite may exist together. The relative abundance and distribution of the two types often permits the recognition of the environment of formation.

In addition, beds of both the sedimented and nodular calcium sulfate deposits probably act as the source material for many of the examples of pore-filling and replacement anhydrite observed in the subsurface. That is, they offer a source of $CaSO_4$ available for dissolution and reprecipitation in void space or growth by replacement in pre-existing rocks.

ANHYDRITE PETROLOGY

Void-filling Anhydrite

Free growth of anhydrite or gypsum takes place within previously existing voids and thus occurs in space formerly occupied by water. Several types of voids can be generated in carbonate rocks (Murray, 1960): interparticle space in carbonate or terrigenous clastic sand, intercrystalline space in dolomite, primary cavities in fossils, dissolution vugs or fractures. Anhydrite growing within such voids has little opportunity to include pre-existing rock and thus forms clear crystals. These crystals may occur as relatively large clear individuals that fill the pre-existing space or in larger spaces as clusters of clear tabular crystals. However, void-filling anhydrite in carbonate rocks commonly continues to grow by replacement into the rock margin of the pre-existing void (fig. 8).

Replacement Anhydrite

Replacement anhydrite crystals grow within the rock and occupy space previously occupied by other minerals and fine pore space. Bioclastic particles appear to be the most commonly replaced material. This process probably takes

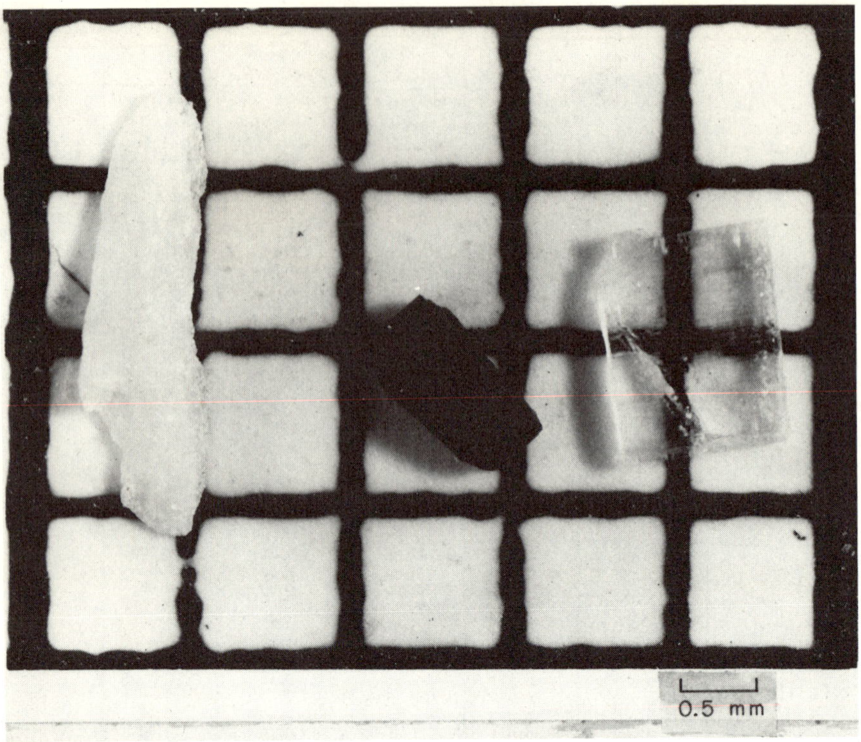

FIG. 10.—Anhydrite types in reflected light; Mississippian Charles Formation, Montana. Left—Nodular anhydrite from a thick anhydrite unit. Center—Replacement anhydrite from limestone 3 feet under the anhydrite bed. Right—Void-filling anhydrite from limestone 3 feet under anhydrite bed.

place by simultaneous dissolution of the parent rock and precipitation of the anhydrite, as demonstrated by the in-situ distribution of inclusion produced relict patterns. Dissolution is often incomplete, and the anhydrite continues to grow around the more resistant relict calcite, dolomite, and noncarbonate clastic grains, attempting to maintain its rectangular habit, but often the final shape is partially controlled by the shape of the material being replaced. These leftover particles remain within the replacement anhydrite crystal and, if they are large enough, can easily be seen when the anhydrite crystal is turned to extinction. Because of this imperfect replacement, resulting in the inclusion of relict particles within the anhydrite crystal, these crystals are usually cloudy to brown in reflected light. The color, of course, depends on the amount, distribution, and nature of the relict material.

Chemical analyses by emission spectrograph were made of anhydrite samples representing two sets of the three anhydrite types (fig. 9). These samples were prepared by removing all the surrounding carbonate rock with dilute hydrochloric acid. Within the limits of accuracy, the bedded nodular and the pore-filling anhydrites appear quite similar in trace element content. However, the replacement anhydrite from the San Andres dolomite and Charles limestone examples contains excess magnesium, silicon, and aluminum. This undoubtedly represents relicts of nonreplaced dolomite, quartz, and clay minerals within the replacement anhydrite crystals and confirms the nature of the relicts and the cause of the cloudy to brown color. Further confirmation was obtained by X-ray analysis of the three Charles anhydrites. The bedded and pore-filling anhydrite showed only anhydrite, whereas the replacement sample contained between 5 and 10 percent calcite and some quartz.

The three anhydrite types can be distinguished in reflected light because of the textural differences and the presence of inclusions in replacement anhydrite.

Bedded anhydrite is commonly granular to massive and translucent to opaque and may be white or light colored, depending on contained impurities (figs. 10 and 11). Pore- or vug-filling anhydrite commonly occurs as clear single crystals with well-defined cleavage (figs. 10 and 11). Rock-replacement anhydrite, because it often fails to digest completely the rock being replaced, appears as cloudy to brown single crystals (figs

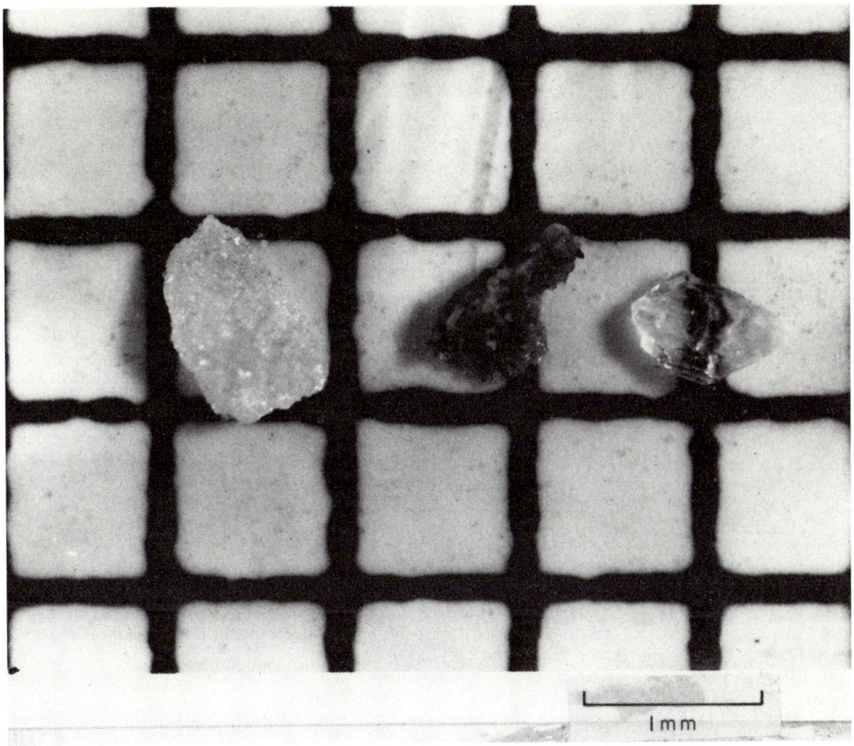

FIG. 11.—Anhydrite types in reflected light; Permian San Andres Dolomite West Texas. Left—Nodular anhydrite. Center—Replacement anhydrite from dolomite. Right—Void-filling anhydrite from dolomite.

10 and 11). Brown metasomatic anhydrite has been reported by Fuller (1956), and the presence of abundant inclusions has been noted by the author in all examples of replacement anhydrite studied.

Relationships of Anhydrite to Porosity

Bedded anhydrite seldom contains pore space in the subsurface. It is presumed that this anhydrite acts as a source of material for the two secondary anhydrite types. Commonly, void-filling and replacement anhydrite is concentrated near, and in association with, bedded anhydrite. In addition, the bedded anhydrite often appears to act as a barrier for further movement of hydrocarbons and thus may act as a seal for a hydrocarbon accumulation.

Replacement anhydrite in itself neither creates nor destroys porosity unless, as is quite common, earlier interparticle, intercrystalline, or vuggy void spaces are engulfed by a solid replacement crystal. However, replacement anhydrite is commonly dissolved later to produce anhydrite mold vugs (Murray, 1960).

CONCLUSIONS

The available evidence suggests that a common diagenetic cycle exists in the $CaSO_4$ minerals. The original material is commonly if not universally gypsum deposited either by precipitation in standing bodies of water or by growth of displacement gypsum crystals in pre-existing sediment or weathered rock. By either mechanism, beds of calcium sulfate minerals may be produced. With burial the gypsum is replaced by anhydrite, and with uplift and removal of overburden the anhydrite is replaced by gypsum. Pre-existing void space in rocks may be filled with anhydrite or gypsum and pre-existing carbonate rock is commonly replaced by anhydrite. The distinction between bedded, void-filling, and replacement anhydrite can be made in reflected light because of differences in texture between bedded and void-filling anhydrite and because of the abundance of included material in replacement anhydrite.

ACKNOWLEDGMENTS

The author is most appreciative for the assistance and comments generously given by his colleagues during this study and preparation of the manuscript, especially F. J. Lucia, J. W. Rasmussen, T. W. Donnelly, R. T. Buffler, and A. L. Diehl.

REFERENCES

Adams, J. E., and Rhodes, M. L. (1960), Dolomitization refluxion: Bull. Amer. Assoc. Petroleum Geol., v. 44, 1912–1920.

Bundy, W. M. (1956), Petrology of gypsum-anhydrite deposits in southwestern Indiana: J. Sed. Petrology, v. 26, 240–252.

Conley, R. F., and Bundy, W. M. (1958), Mechanism of gypsification: Geochimica et Cosmochimica Acta, v. 15, 57–72.

Curtis, R., Evans, G., Kinsman, D. J. J., and Shearman, D. J. (1963), Association of dolomite and anhydrite in Recent sediments of the Persian Gulf: Nature, v. 197, 679–680.

Deffeyes, K. S., Lucia, F. J., and Weyl, P. K. (1964), Dolomitization: Observations on the Island of Bonaire, Netherlands Antilles: Science, vol. 143, no. 3607, 678–679.

Fisk, H. N. (1959), Padre Island and the Laguna Madre flats, Coastal South Texas: Natl. Acad. Sci.-Nat Res. Coun., 2nd Coastal Geog. Conf., pp. 103–151.

Fuller, J. G. C. M. (1956), Mississippian rocks and oilfields in southeastern Saskatchewan: Saskatchewan Dept. of Min. Resources, Report 19.

Henderson, G. G. L. (1959), Geology of the Stanford Range of the Rocky Mountains: British Columbia Dep of Mines, Bull. 35.

Kerr, S. D., and Thomson, A. (1963), Origin of nodular and bedded anhydrite in Permian shelf sediment Texas and New Mexico: Bull. Amer. Assoc. Petroleum Geol., vol. 47, no. 9, 1726–1732.

King, R. H. (1947), Sedimentation in Permian Castile sea: Bull. Amer. Assoc. Petroleum Geol., vol. 31, 470–47

MacDonald, G. J. F. (1953), Anhydrite-gypsum equilibrium relations: Amer. J. Sci., v. 235-A, 247–272.

Masson, P. H. (1955), An occurrence of gypsum in Southwest Texas: J. Sed. Petrology, v. 25, 72–79.

Morris, R. C., and Dickey, P. A. (1957), Modern evaporite deposition in Peru: Bull. Amer. Assoc. Petroleum Geol., v. 41, 2467–2474.

Murray, R. C. (1960), Origin of porosity in carbonate rocks: J. Sed. Petrology, v. 30, 59–84.

Phleger, F. B., and Ewing, G. C. (1962), Sedimentology and oceanography of coastal lagoons in Baja California, Mexico: Geol. Soc. Amer. Bull., v. 73, 145–182.

Posnjak, E. (1938), The system, $CaSO_4$-H_2O: Amer. J. Sci., v. 235-A, 247–272.

——— (1940), Deposition of calcium sulfate from sea water, Amer. J. Sci., v. 238, 559–568.

Stewart, F. H. (1953), Early gypsum in the Permian evaporites of northeastern England: Proc. Geol. Assoc v. 64, 33–39.

Sund, J. O. (1959), Origin of the New Brunswick gypsum deposits: Canadian Mining and Met. Bull., v. 5 No. 571, 707–712.

Talmage, S. B., and Wootton, T. P. (1937), The non-metallic mineral resources of New Mexico and their economic features: New Mexico Bur. of Mines and Min. Resources, Bull. 12, 159 pp.

Wells, A. (1962), Recent dolomite in the Persian Gulf: Nature, v. 194, 274.

Reprinted from *Proc. Yorkshire Geol. Soc.*, **34**, 315–326, 330 (1964)

25

EVAPORITE DIAGENESIS IN THE LOWER PURBECK BEDS OF DORSET

BY I. M. WEST

(*Read at Durham, 14th March, 1964*)

CONTENTS

SUMMARY

Five stages have been determined in the diagenetic history of the calcium sulphate of beds of Lower Purbeck age in Dorset, particularly by reference to abundant pseudomorphs and other relict textures and structures preserved in late-formed gypsum and in secondary silica and calcite. The weight of overburden is thought to have controlled the changes, Occurrences are described of minerals, including celestite, calciostrontianite, lutecite and quartz, remaining after the removal in solution of gypsum. A description is also given of secondary limestones which occur particularly in the Caps and Broken Beds. They are shown to be almost entirely replacements of original sulphate deposits.

I. INTRODUCTION

The Lower Purbeck Beds of Dorset include a well-known breccia near the base—the Broken Beds—the origin of which has caused much discussion (see Arkell, 1938, for bibliography). In particular, Hollingworth (1938) suggested that gypsum or anhydrite was formerly present at this horizon and constituted a stratum of structural weakness in which the type of tectonic brecciation postulated by Arkell (1938) might have taken place. Following the recent report of celestite from these beds (West, 1960) which supports Hollingworth's hypothesis, a full petrographic investigation of the Caps and Broken Beds and of the gypsum of the Soft Cockle Beds was undertaken. Over four hundred thin sections were made and specimens from the Purbeck Beds of Sussex and elsewhere were used for comparison.

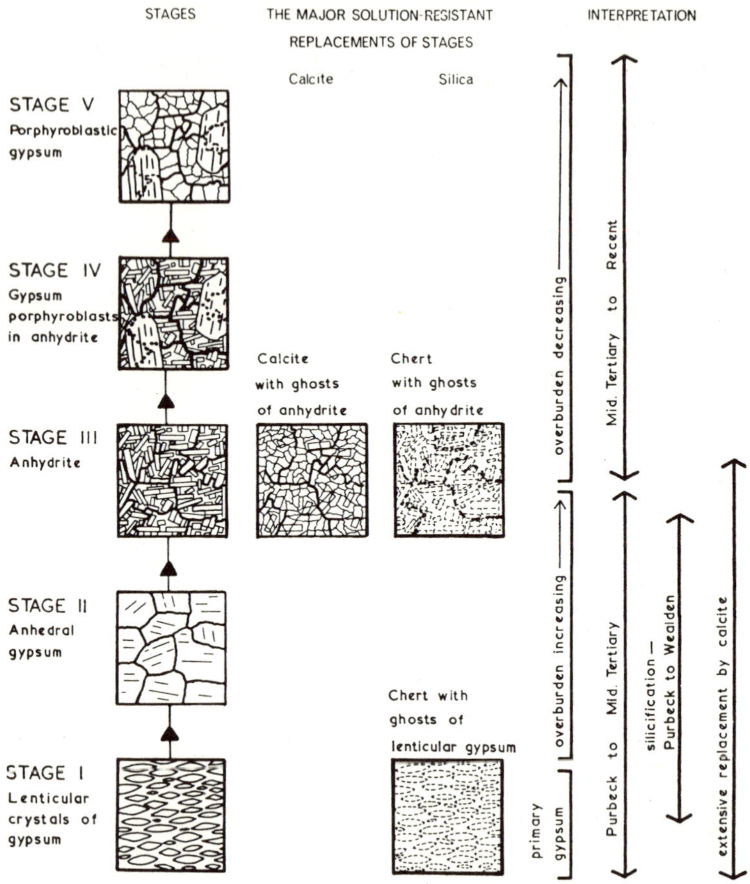

FIG. 1.—A diagrammatic representation of the five diagenetic stages of the Lower Purbeck calcium sulphate, together with three typical examples of replacement products. Postulated times of diagenetic change shown on the right are based on compaction effects and tectonic events.

There are many indications that gypsum and anhydrite were once present in quantity in the Caps and Broken Beds although they are almost completely absent today. Concurrently with the early diagenetic changes in the sulphate content of the rock there was replacement of it by silica and calcite. This replacement preserved the early stages of the sulphate diagenesis as palimpsests, and from these and from the textures in existing gypsum it has been possible to reconstruct the complete sulphate diagenetic history and to divide it into a series of five succeeding stages (Fig. 1).

II. Stratigraphy

The Lower Purbeck rocks in the type area rest conformably on the underlying marine Portland Beds and the succession at Stair Hole, Lulworth (Strahan, 1898, p. 104), which is typical of many exposures, is as follows:

6. Soft Cockle Beds
5. Hard Cockle Beds
4. '*Cypris*' Freestones
3. Broken Beds
2. Soft Cap
1. Hard Cap

(*Portland Stone*)

The Caps are the lowest division of the Purbeck Beds and in places attain thicknesses of up to twenty feet though an average of fifteen feet is more usual. They consist of laminated, lagoonal limestones deposited in shallow water under rapidly varying salinity conditions (West, 1961). Occasional subaerial exposure gave rise to thin "dirt beds"—ancient calcareous soils. Petrographically the limestones are mostly fine-grained pelsparites, pelmicrites, oosparites and biolithites; there is also crystalline limestone, considered below to be a replacement of evaporites.

The Broken Beds are brecciated limestones, sometimes nearly twenty feet thick. They follow roughly the level of apparent maximum evaporite precipitation in the lower part of the Lower Purbeck Beds, and both the upper and lower limits of the brecciation cut across beds of different ages (Woodward, 1895, p. 248). The lower part of the Broken Beds consists mostly of crystalline limestones only distinguishable from those of the Caps by their brecciated condition; a porous matrix of crystalline calcite is often present in this part of the breccia. The upper part usually consists of brecciated pelsparites, highly laminated and with abundant ostracods. There are only minor indications of evaporites and the beds are petrographically similar to the '*Cypris*' Freestones above. A recent study by Brown (1963) of the basal Purbeck limestones has shown the presence of algal remains compatible with high salinity. Some have undergone partial replacement by anhydrite which has been since removed in solution.

Chert nodules occur in the Caps and Broken Beds. Some of the chert is a replacement of limestone (see Strahan, 1898, p. 83) but most, when weathered or etched, shows abundant ghosts of gypsum or anhydrite crystals. These are visible both in the hand specimen and in thin section. Pseudomorphs, in calcite of diagenetic origin, are abundant in the limestones, and traces of residual gypsum are still present in the Hard Cap at Bacon Hole (SY/*839797*).

Gypsum occurs in the Soft Cockle Beds, higher in the Lower Purbeck sequence, notably at Durlston Bay and Worbarrow Tout, while in the same beds north of Portland Bill (SY/*676692*) celestite and relics of evaporites have been found. Bedding is rarely visible in the gypsum, and nodules and enterolithic veins of gypsum are extensively developed; these in turn are cut by later satin-spar veins. Chert also occurs in the gypsum and exhibits ghost textures.

III. The Diagenetic Stages

(a) General

From pseudomorphs, ghosts and present gypsum textures the sequence of diagenesis which the calcium sulphate of the Caps, Broken Beds and the Soft Cockle Beds has undergone can be interpreted. The evidence is similar at widely different localities and horizons within the Purbeck Beds of Dorset suggesting that the calcium sulphate at each locality underwent the same general type of diagenesis from gypsum to anhydrite to gypsum. This diagenesis can be divided into five stages (Fig. 1), the first three of which were extensively replaced by calcite and silica. Definite evidence of the later stages is present in the textures of the existing secondary gypsum of the Soft Cockle Beds. It is inferred that the calcium sulphate of the Caps and Broken Beds passed through the same later stages before being removed in solution leaving only siliceous and calcareous replacements.

(b) Stage I

Chert in the Caps and Broken Beds often contains minute (up to two millimetres) lens-shaped bodies of chalcedonic silica (Plate 36, fig. 1) which closely resemble the lenticular crystals of

lenses. ×46. Lower Purbeck Caps, Worbarrow Tout (L.P. 148). *c*, A secondary limestone of anhedral crystalline calcite with net texture. The walls of this consist of microcrystalline calcite. Numerous lutecite spherulites are present. ×17. Lower Purbeck Beds, trench on Bayard Hill, Upwey (L.P. 123). *d*, A lutecite spherulite and subhedral calcite crystals of the net-texture in stage V anhedral gypsum. Crossed polarizers. ×154. Soft Cockle Beds, Worbarrow Tout (L.P. 163). *e*, Stage V gypsum pseudomorphs after stage III prismatic anhydrite, in a matrix of microcrystalline calcite. ×60. Soft Cockle Beds, Worbarrow Tout (L.P. 161). *f*, Stage V gypsum with the outlines of stage II gypsum shown by minute inclusions, mainly of calcite and quartz forming net-texture. Larger inclusions consist of relics of stage III anhydrite and chalcedonic silica, calcite and quartz crystals. ×52. Soft Cockle Beds, Durlston Bay (L.P. 125). Fig. *d* is shown as between crossed polarizers, the other five being in plane polarized light.

CA, calcite other than microcrystalline calcite; CL, celestite; G, gypsum; L, lutecite; M, microcrystalline calcite; Q, quartz. The prefix L.P. denotes specimens in the author's collection.

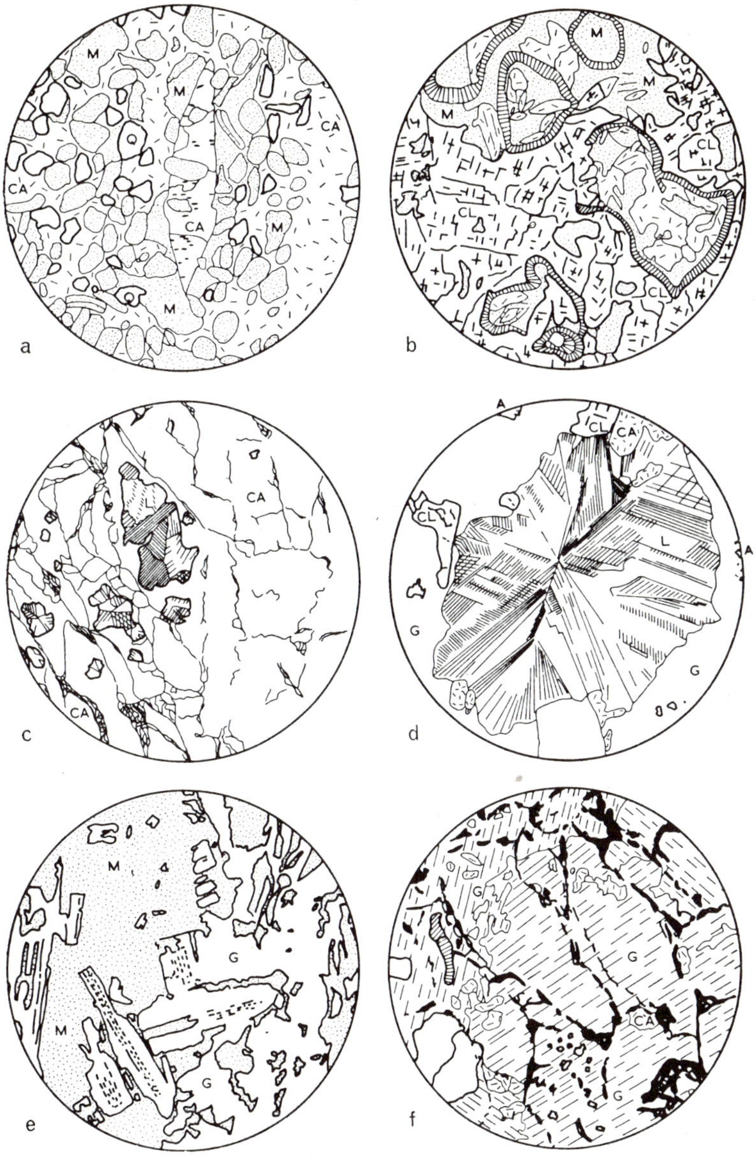

FIG. 2.—Sections of Lower Purbeck rocks. *a*. Sandy pelsparite, partly composed of pellets, with an originally poikiloblastic gypsum lens since replaced by calcite. ×48. '*Cypris*' Freestones, Stair Hole, Lulworth (L.P. 115). *b*, Anhedral celestite replacing limestone. This contains oolitically coated nodules with calcite pseudomorphs after gypsum

gypsum[1] figured by Lacroix (1897, pl. 8, fig. 10), Masson (1955, fig. 3) and by others. These lenses also occur in the chert in the gypsum of the Soft Cockle Beds at Worbarrow Tout.

In the Caps in the old quarries at Upwey (SY/*671851*), near Weymouth, Strahan (1898, p. 108) listed a four-foot bed of "white limestone with nests of brown chert weathering into a white impalpable powder". The chert consists of lutecite pseudomorphs, almost certainly after gypsum crystals, having a circular to oval periphery and attaining two millimetres in diameter. Chevron-like striations on the surfaces may by traces of crystal edges. In thin section numerous lanceolate sections are seen of the pseudomorphs—brown in transmitted light, white in reflected light. Scattered circular basal sections show outlines of the original (010) cleavage. Sometimes smaller crystals have been incorporated in the larger ones. Often the replacement of a gypsum crystal commenced at its periphery but was incomplete; a quartz mosaic later replaced the central part. Hollow siliceous pseudomorphs after gypsum, reaching fifteen millimetres in diameter, in the Lower Purbeck Beds near Dancing Ledge (SZ/*011768*) are thought to be of similar origin.

Another type of chert with pseudomorphs, common in the basal Purbeck Beds, is similar to that at Upwey except that the gypsum is replaced by a different variety of chalcedonic silica; the chert of the Soft Cap in the Lulworth district is an example. In this, etching of the thin section with hydrofluoric acid is necessary to make the pseudomorphs visible.

In evaporitic limestones, hollow moulds, up to five millimetres long, left by the solution of gypsum lenses are common. They are often orientated with the original *c*-axis approximately parallel to the bedding and are abundant, for example, in the Soft Cap, east of Lulworth Cove. Above the gypsum of the Soft Cockle Beds of Durlston Bay moulds are abundant in marly limestone which overlies a bed containing composite pseudomorphs of quartz, celestite and lutecite after gypsum. Similar pseudomorphs occur in the Soft Cockle Beds north of Portland Bill, together with pseudomorphs solely of celestite after gypsum.

Of the minerals replacing gypsum lenses, calcite is the most common and these pseudomorphs were found in all major sections of the basal Purbeck Beds of the Dorset mainland. Perfect specimens form the nuclei of ooliths, as for instance, in the Hard Cap near Lulworth Cove (Plate 36, fig. 2) which suggests a syngenetic origin for some of the gypsum lenses. Others truncate, or poikiloblastically enclose, calcareous pellets (Fig. 2a; Plate 36, fig. 6), for example in the '*Cypris*' Freestones at Stair Hole, Lulworth (Fig. 2a), in such a way as to suggest occasional development of the gypsum below the surface of the sediment. The well-defined

[1]Referred to hereafter simply as lenses for brevity.

stratigraphical limits and considerable lateral continuity of the evaporite beds indicate, perhaps, that such penecontemporaneous gypsum only developed immediately beneath the original sediment surface. Beds with pseudomorphs rarely contain faunal remains incompatible with high salinity. At Portesham, however, chert with both pseudomorphs after gypsum lenses and freshwater fossils is a notable exception (West, 1961).

(c) Stage II

On weathering many Purbeck evaporitic rocks exhibit a cellular texture. The gypsum of the Soft Cockle Beds and the quartz-calcite rock of the Caps at Worbarrow Tout are good examples. The cellular texture is ubiquitous in thin sections of the evaporites and their replacements (Figs. 2c, f). It is seen as a sparse network of impurities, mostly microcrystalline calcite and argillaceous material; epigenetic crystals of quartz, calcite and celestite have often developed on it. Its texture resembles that figured by Brown (1931, fig. 22) in anhydrite of salt-dome cap rock and described by him as "net-texture".

The net-texture in the Purbeck Beds is formed typically of sub-lenticular units about two millimetres in length. The present-day secondary gypsum crystals cross the "walls" indiscriminately while ghosts of earlier anhydrite crystals have a similar relationship or alternatively project into and displace the walls. Thus the net-texture is older than either the secondary gypsum or the original anhydrite crystals and it seems likely that, in the Purbeck Beds, net-texture arose from the compaction, and perhaps crystal growth, of stage I gypsum lenses, followed by the conversion of the mass into anhedral gypsum with considerably reduced porosity. Illustrations of this process occur, as in the Caps at Portesham, where there is a lateral transition from rock with calcite pseudomorphs after gypsum lenses to rock with crystalline calcite with net-texture—the transitional stages showing the compression of the lenses, with the loss of the perfect lenticular shape and the reduction of the interstitial matter to thin walls.

(d) Stage III

Pseudomorphs and ghosts of lath-shaped crystal sections, usually exceeding two millimetres in length and sometimes with a fibroradiate texture, are present in thin sections of chert and limestones from the Caps and Broken Beds (Plate 36, fig. 3). They are common in the gypsum of the Soft Cockle Beds (Fig. 2e) where they occur associated with small relics of anhydrite and appear to be pseudomorphs, in this instance, of gypsum after anhydrite. In addition to cryptocrystalline silica, calcite and gypsum, quartz also has replaced anhydrite; pseudomorphs of quartz after tabular anhydrite crystals are abundant in the soft debris of the Broken Beds a hundred yards north of Durlston Head. At Mountfield in Sussex similar pseudomorphs are present in Purbeck anhydrite rock, and in gypsum formed by its hydration. Most of the anhydrite at this locality is, however, fine-grained with only occasional relics of the larger tabular type of crystals.

In parts of the gypsum of the Soft Cockle Beds small anhydrite relics of variable optical orientation suggest that comparable finer-grained anhydrite was also originally present in this rock. In the rocks examined, anhydrite, or its replacements, usually displays net-texture, and indications of the less compact stage I texture commonly occur in rocks from the same exposures. Thus the Purbeck anhydrite appears to be secondary and to have formed after stage II. This agrees with the theoretical conclusions of Conley and Bundy (1959) that primary anhydrite is rarely deposited in marine evaporites.

(e) Stage IV

With a limited supply of water the hydration of anhydrite usually commences with the growth of gypsum porphyroblasts. This has been recognized in many deposits (for example, by Dunham, 1948; Ham, 1962). The present porphyroblastic texture of the Dorset gypsum strongly suggests that it was generated in such a manner. This is confirmed by the presence of anhydrite with porphyroblasts of gypsum at the Mountfield mine, Sussex (Allen, 1960). Similarly the calcium sulphate once present in the Caps and Broken Beds probably passed through the hydration sequence of this and the succeeding stages but apart from minute traces has since been removed in solution.

(f) Stage V

The gypsum of the Soft Cockle Beds differs only from the gypsum-anhydrite of stage IV in that the matrix between the porphyroblasts has also been hydrated. This is now a mosaic of small, nearly equant gypsum crystals with consertal (Niggli, 1954, p. 244) intergranular boundaries. The porphyroblasts are predominantly anhedral and small relics of anhydrite are often present. Net-texture is usually visible, as in all the stages later than the first.

IV. Interpretation of the Diagenetic Evidence

All the main evaporite occurrences in the Purbeck Beds of Dorset suggest that the diagenetic changes took place in the same order and through the same series of textural stages (Fig. 1) except that occasionally there was passage from stage III to stage V either directly or without any trace of stage IV remaining, the other stages taking place unchanged. In the sequence considered, there is a loss of pore water in the change from stage I to stage II; of water of crystallization from stage II to stage III, while stages IV and V result from the rehydration of the calcium sulphate. These changes in water content could have been dependent on variations in the weight of overburden. The weight initially increased with the burial of the gypsiferous sediments and provided the power to express the pore water, and, as pressure became greater, to weld the gypsum lenses, aided by the ambient diagenetic conditions. Theoretical considerations (MacDonald, 1953) and evidence of other sulphate deposits (Ogniben, 1957) suggest that the change to anhydrite took place under an overburden of not more than two thousand feet, gypsum rarely being found below this depth. In east Dorset a sufficient thickness of overburden would have been achieved by the end of Wealden deposition, but since no major erosion took place until mid-Tertiary times, considerably more than this amount accumulated. From mid-Tertiary times the thickness has been progressively reduced by erosion until the present day. If therefore the dehydration stages are directly related to overburden it is possible to fix the major change—of gypsum to anhydrite—in Cretaceous times when the thickness was enough, and similarly the reverse change, in late Tertiary or Quaternary when it was reduced.

Stages I, II and III only are affected by silicification and the process is presumed to have begun early in the diagenetic history and ended before the change to stage IV. That some of the chert was formed soon after the deposition of the limestones is implied by the compaction and deformation of the pelsparites round chert nodules, which thus are deduced to have been formed before final lithification; in addition the nodules do not follow the joints but are broken by them, and thus presumably were earlier. On the other hand the pseudomorphs after anhydrite that are also preserved in the chert are a strong indication that some was formed relatively late, during or after the first appearance of anhydrite in stage III. The occurrence of blocks of chert containing these pseudomorphs within the Broken Beds shows, however, that chert formation preceded the brecciation (Bristow and Fisher, 1857).

V. Residual Minerals and Secondary Limestones

(a) Residual minerals from the solution of the evaporites

There has been much relatively recent removal in solution of calcium sulphate from the Purbeck Beds, particularly from the Caps and Broken Beds of the Dorset mainland and from the Soft Cockle Beds of the Isle of Portland. The process has in places been preceded by the replacement of much of the sulphate by calcite to form secondary limestones of considerable magnitude. Solution alone left the less soluble minerals which include celestite, calciostrontianite, the lutecite variety of chalcedonic silica and euhedral quartz.

Celestite was first found at Durlston Head (West, 1960) in beds now known to contain abundant pseudomorphs after anhydrite. Amongst other localities, it occurs in the Caps at Worbarrow Tout where it partly replaces limestone containing calcite pseudomorphs after gypsum lenses (Fig. 2b). A little has been found at Stair Hole, Lulworth, in the Soft Cap, a bed containing chert with pseudomorphs after gypsum. In the Mountfield mine, where gypsum still exists in the basal Purbeck Beds, celestite occurs immediately above it (James in discussion of Howitt, 1964). As in the Dorset Purbeck Beds, celestite in Miocene beds of Poland has been found in limestone replacements of calcium sulphate (Laszkiewicz, 1957).

A mineral from the beds containing celestite at Durlston Head was described by West (1960, p. 396) as spherulitic calcite. Mr. D. Salter (personal communication) has since found by X-ray diffraction that it is calciostrontianite. It is abundant in the upper and lower celestite-bearing beds at Durlston Head (West, 1960, p. 393) both as a vein material and as a matrix to the celestite crystals.

The lutecite variety of chalcedonic silica occurs abundantly as spherulites in the evaporite beds (Figs. 2c, d) to which it appears to be confined. It is noteworthy that similar spherulites are present in an evaporite collapse breccia in Spitsbergen (Hutchins, 1962, p. 66).

Authigenic quartz often occurs as euhedral, doubly terminated crystals in the gypsum of the Purbeck Beds and other deposits (see, for example, Tarr, 1929). Such crystals are abundant in the calcite-quartz rock of Worbarrow Tout (Plate 36, fig. 5), considered below to be a replacement of calcium sulphate. Small authigenic quartz crystals occur in the Broken Beds wherever evaporites were once present.

(b) Secondary limestones

In the Caps and Broken Beds there are present unusual, unfossiliferous, crystalline limestones. In many places they are brecciated or contain acute folds and contortions. At Durlston Head secondary limestones form most of the beds from the top of the Portland Stone to the basal bed of the '*Cypris*' Freestones (West, 1960, fig. 1), and at this locality the original presence of evaporites, easily affected by tectonic forces, explains the extent of local contortion. Secondary limestones are present at the base of the Broken Beds at almost all localities; they usually form the matrix of the breccia and some of the blocks within it.

There is the following evidence that these limestones are replacements of anhydrite. First, they usually contain chert with pseudomorphs after gypsum or anhydrite (Plate 36, fig. 3). Second, net-texture is almost invariably present (Fig. 2c) and, third, ghosts of anhydrite crystals are often preserved. In addition they contain quite commonly celestite, lutecite spherulites and euhedral quartz. The calcite-quartz rock of Worbarrow Tout described by Watts (*in* Strahan, 1898, pp. 98-9) as a silicified tufa is a typical example of such a secondary limestone (Plate 36, fig. 5). Similar limestone replacement of calcium sulphate has been described by Laszkiewicz (1957) from Miocene rocks of Poland but regarded as a replacement of gypsum rather than of anhydrite.

The limestones range from a soft, white, saccharoidal, porous rock which can be crumbled with the fingers, to hard laminated limestone. In the Broken Beds north of Durlston Head they consist of barely cemented aggregates of calcite pseudomorphs after anhydrite. Often the texture is an anhedral calcite mosaic with ghosts of anhydrite crystals. Some porous varieties, as in the Soft Cockle Beds of Portland, consist almost entirely of calcite rhombs. Yet others are porphyroblastic; only the coarsely-crystalline nature and net-texture are constant characteristics. Replacement of anhydrite by limestone took place partly before and partly after the brecciation that formed the Broken Beds. Such limestone occurs both as blocks and as matrix.

VI. Acknowledgements

I am indebted to Professor F. Hodson for advice and encouragement. I am most grateful to Professor P. Allen and Dr. D. Moore for reading an early draft of the typescript. I would like to express my thanks to the Directors and staff of Gyproc Limited for permission to collect from their mines, and I gratefully acknowledge the loan of a microscope from the Royal Society with which to carry out this work.

VII. References

Allen, P., 1960. Geology of the Central Weald: the Hastings Beds. *Geol. Ass. Guide*, No. 24, 28 pp.

Arkell, W. J., 1938. Three tectonic problems of the Lulworth district: studies on the middle limb of the Purbeck Fold. *Quart. J. Geol. Soc. Lond.*, **94**, 1-54.

Bristow, H. W. and O. Fisher, 1857. Comparative sections of the Purbeck strata of Dorset. *Vertical Sections, Geol. Surv.*, sheet 22.

Brown, L. S., 1931. Cap-rock petrography. *Bull. Amer. Ass. Petrol. Geol.*, **15**, 509-22.

Brown, P. R., 1963. Algal limestones and associated sediments in the basal Purbeck of Dorset. *Geol. Mag.*, **100**, 565-73.

Conley, R. F. and W. M. Bundy, 1959. Mechanism of gypsification. *Geochim. et Cosmoch. Acta*, **15**, 57-72.

Dunham, K. C., 1948. A contribution to the petrology of the Permian evaporite deposits of north-eastern England. *Proc. Yorks. Geol. Soc.*, **27**, 217-27.

Ham, W. E., 1962. Economic geology and petrology of gypsum and anhydrite in Blaine County. *Bull. Okla. Geol. Surv.*, **89**, 100-51.

Hollingworth, S. E., 1938. The Purbeck Broken Beds. *Geol. Mag.*, **75**, 330-2.

HOWITT, F., 1964. Stratigraphy and structure of the Purbeck inliers of Sussex (England). *Quart. J. Geol. Soc. Lond.*, **120**, 77-113.
HUTCHINS, P. F., 1962. Authigenic minerals in Carboniferous sediments from Central Vestspitsbergen. *Geol. Mag.*, **99**, 63-8.
LACROIX, A., 1897. Le gypse de Paris et les minéraux qui l'accompagnent. *Nouv. Arch. Mus. Hist. Nat., Paris* (sér. 3), **9**, 201-96.
LASZKIEWICZ, A., 1957. Siarka i celestyn Tamobrzega i Szydlowa. *Arch. Min. Warsz.*, **20**, 95-120.
MACDONALD, G. J. F., 1953. Anhydrite-gypsum equilibrium relations. *Amer. J. Sci.*, **251**, 884-98.
MASSON, P. H., 1955. An occurrence of gypsum in south-west Texas. *J. Sediment. Petrol.*, **25**, 72-7.
NIGGLI, P., 1954. *Rocks and mineral deposits.* 559 pp. W. H. Freeman, San Francisco.
OGNIBEN, L., 1957. Petrografia della Serie Solifera Siciliana e considerazioni geologiche relative. *Mem. Carta Geol. Ital.*, **33**, 268 pp.
STRAHAN, A., 1898. The geology of the Isle of Purbeck and Weymouth. *Mem. Geol. Surv.*, xi + 278 pp., London.
TARR, W. A., 1929. Doubly terminated quartz crystals occurring in gypsum. *Amer. Min.*, **14**, 19-25.
WEST, I. M., 1960. On the occurrence of celestine in the Caps and Broken Beds of Durlston Head, Dorset. *Proc. Geol. Ass., Lond.*, **71**, 391-401.
—— 1961. Lower Purbeck Beds of Swindon facies in Dorset. *Nature*, **190**, 526.
WOODWARD, H. B., 1895. The Jurassic rocks of Britain. Vol. 5. The Middle and Upper Oolitic rocks of England (Yorkshire excepted). *Mem. Geol. Surv.*, xiv + 499 pp., London.

Revised manuscript received:
21st November, 1963.

I. M. WEST,
Department of Geology,
The University,
Southampton.

* * * * * * *

EXPLANATION OF PLATE 36

FIG. 1.—Polished surface of a chert nodule showing sections through lutecite pseudomorphs after stage I gypsum lenses. ×16. Lower Purbeck Caps, Upwey (L.P. 114).

FIG. 2.—A calcite pseudomorph, after a stage I syngenetic gypsum lens, with an oolitic coating of fibrous calcite in a partly oolitic pelsparite. ×106. Hard Cap, Dungy Head, near Lulworth Cove, Dorset (L.P. 51). Plane polarized light.

FIG. 3.—Ghosts of stage III prismatic anhydrite crystals in etched chert; the larger ghosts consist of fibrous lutecite. ×87. Broken Beds, Durlston Head (L.P. 290). Plane polarized light.

FIG. 4.—Net-texture, a relic of stage II, consisting of an aggregate of subhedral to euhedral calcite crystals (dark patches) in stage V anhedral gypsum. ×32. Soft Cockle Gypsum, Worbarrow Tout (L.P. 273). Plane polarized light.

FIG. 5.—Euhedral quartz crystals in secondary limestone, probably a replacement of anhydrite. Pellets and impurities of microcrystalline calcite form a net-texture. ×48. Quartz-calcite rock, Caps, Worbarrow Tout (L.P. 153). Plane polarized light.

FIG. 6.—Calcite pseudomorphs after penecontemporaneous gypsum lenses in a pelsparite. ×43. Lower Purbeck Beds, Portesham (L.P.351). Crossed polarizers.

The prefix L.P. denotes specimens in the author's collection.

[Owing to the loss of the original photographs, this plate has been prepared again from the original samples, under the same optical conditions, and at the same magnifications. There are slight differences from the original in the areas shown in Figures 1 and 6.]

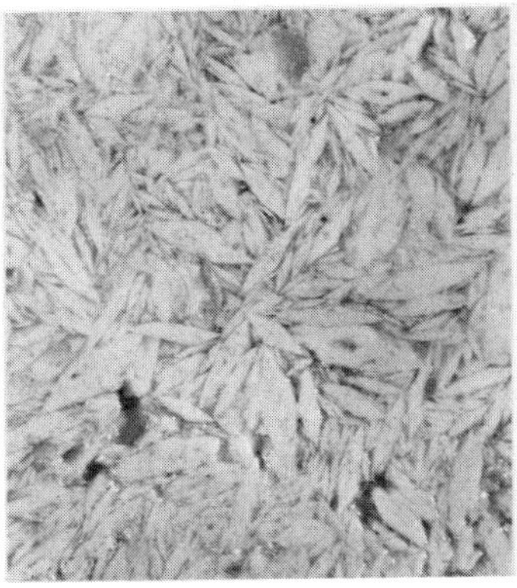

Fig. 1

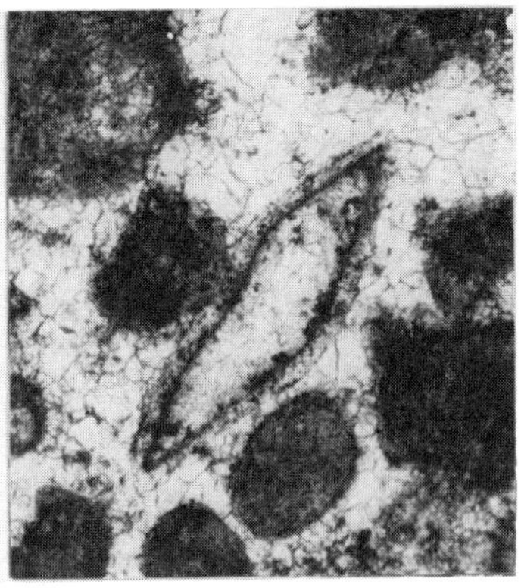

Fig. 2

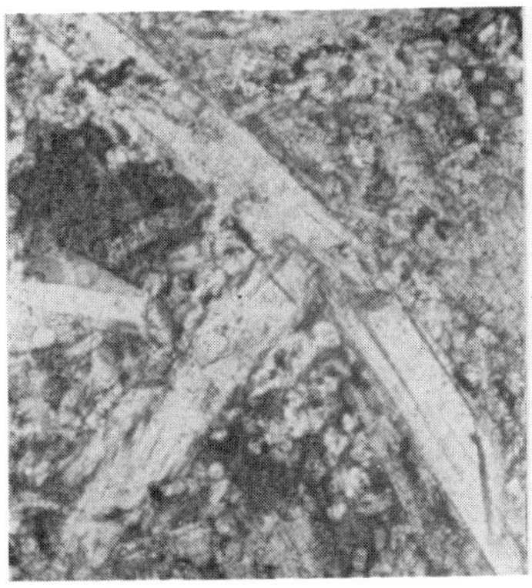

Fig. 3

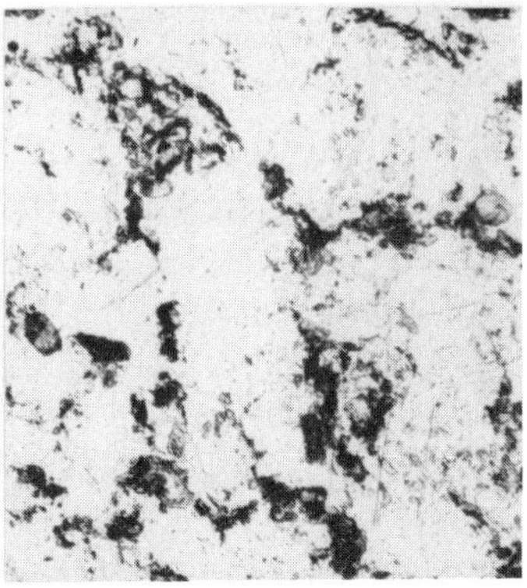

Fig. 4

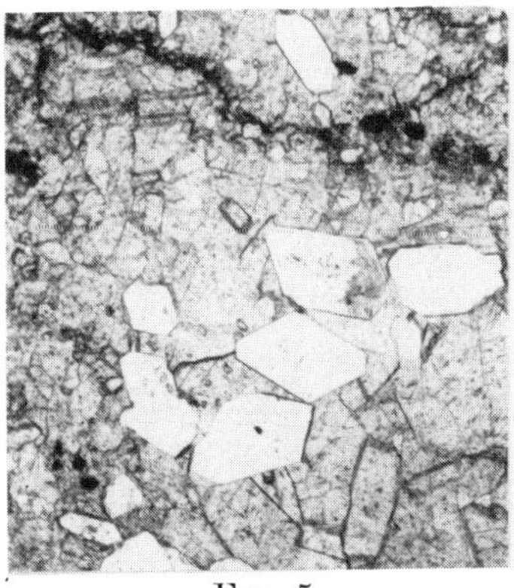

Fig. 5

Fig. 6

Pseudomorphs and Replacements of Gypsum and Anhydrite

Reprinted from *Mineralog. Mag.*, **34**, 460–470 (1965)

26

The mineralogy of the British Permian evaporites

By F. H. Stewart

Department of Geology, University of Edinburgh

Summary. The occurrence of the minerals is reviewed. Consideration of their genesis leads to a tentative list of primary minerals, and of early diagenetic, later diagenetic and geothermal metamorphic, and late near-surface changes.

In 1943 Professor Tilley recorded the occurrence of polyhalite in the rocks of the D'Arcy Exploration Company's E.2 borehole at Aislaby, near Whitby; this was the first published record of a potassium-bearing salt in British evaporites. The occurrence of rocks containing sylvine (Lees and Taitt, 1946) prompted further exploration by Imperial Chemical Industries and Fisons Ltd. after the war, and this showed the presence of large deposits of potassium ores in the Whitby region. There is now a considerable literature related to these, and it seems timely to give a short review of the mineralogy of the British Permian evaporites. The writer is most grateful to Fisons Ltd. who have kindly allowed him to use unpublished information from his study of their boreholes F.1 at Robin Hood's Bay, F.2 at Staintondale, F.3 at Little Beck, and F.4 at Hawsker, in the Whitby–Scarborough district. The position of these and the other boreholes mentioned in this paper can be found in the map of Dunham (1960).

Distribution of the minerals

The evaporites occur in two principal areas—east and west of the Pennine Hills. In the western area (the Vale of Eden, west Cumberland, and the Furness district of Lancashire), where four evaporite beds have been recognized (Hollingworth, 1942), the mineralogy is relatively simple; carbonates, anhydrite, and gypsum are the main constituents. In the eastern area the mineralogy is much more complex. Here there are three main evaporite beds, each in general becoming thicker from the north, west, and south in County Durham, Yorkshire, and Nottinghamshire towards a central area in the Whitby–Scarborough region. In the margins of the area carbonates are the only major constituents, and these are joined successively in each evaporite bed by anhydrite (with or without secondary gypsum), halite, and potassium-bearing salts towards the Whitby Scarborough potash field. In this potash field, the Top Anhydrite is probably the start of a fourth evaporite cycle.

The potash field. Table I shows the stratigraphy of the potash field, and the distribution of the minerals so far recorded, with an indication of their relative abundance, excluding detrital and clay minerals which need further study. The table is based on publications by Armstrong, Dunham, *et al.* (1951), Beevers and Stewart (1960), Dunham (1960), Fleck (1950), Guppy (1944), Raymond (1953, 1962), Sabine (1953), Stewart (1949, 1951*a* and *b*, 1953, 1954, 1956, 1963), and Stewart, Chalmers, and Phillips (1954), together with the writer's study of the four Fisons boreholes.

North-eastern England outside the potash field (Dunham, 1948, 1960; Edwards, 1951; Forbes, 1958; Fowler, 1943, 1944; Hollingworth, 1942, 1948; Napier, 1948; Raymond, 1959, 1962; Sherlock, 1921; Sherlock and Hollingworth, 1938; F. H. Stewart, 1954; G. E. Stewart, 1947; other references are given in Dunham, 1960). Dolomite, anhydrite, gypsum, and halite are major constituents of certain zones. Calcite is a fairly common minor constituent, and locally a major constituent of the Magnesian Limestones. Pseudomorphs of anhydrite after early gypsum are abundant in the Upper Evaporites at Hayton. In the sulphate and chloride zones of the Upper and Middle Evaporites hematite is associated with clays and silts but is less abundant in the salts than it is in the potash field, while quartz and talc appear to be much less common than in the potash field and there is only one record of magnesite (Forbes, 1958): traces of pyrite are not uncommon and celestine has been found rarely in sulphate rocks. In the Magnesian Limestone, dolomite is predominant, calcite, gypsum, and anhydrite are common minor and occasional major constituents, accessory pyrite is widespread, and sulphur is a local minor constituent; Raymond (1962) also records traces of halite, hematite, and talc and gives the only record of magnesite.

A suite of metalliferous minerals occurring mainly as veins and cavity fillings in dolomite of the Lower Magnesian Limestone (Dunham, 1948; Dunham, Claringbull, and Bannister, 1948; Fowler, 1943, 1944; Raymond, 1962) includes galena, sphalerite, chalcopyrite, pyrite, and malachite, associated with anhydrite, gypsum, barite, celestine, fluorite, ankerite, goethite, collophane, dickite, calcite, and quartz. One vein carried epsomite and sylvine (Raymond, 1962), and the only other record of sylvine is from the Billingham Anhydrite Mine in County

TABLE I. Mineralogy of the Yorkshire potash field

		Aphthitalite	Boracite	Calcite	Carnallite	Pseudomorphs after carnallite	Celestine	Dolomite	Fluorite	Glauberite	Pseudomorphs after gypsum	Hematite	Halite	Kieserite	?Langbeinite	Magnesite	Magnetite	?Northupite	Picromerite	Polyhalite	Pyrite	Quartz (authigenic)	Rinneite	Sulphur	Sulphoborite	Sylvine	Talc	p-Veatchite
TOP ANHYDRITE	A	—	—	—	—	—	—	B	—	—	—	B	D	—	—	—	—	—	—	—	D	D	—	—	—	—	—	—
SALT CLAY	B	—	—	—	—	—	—	B	—	—	—	B	A	—	—	—	—	—	—	—	—	B	—	—	—	—	—	—
UPPER EVAPORITES																												
Upper halite	B	—	—	—	—	—	—	B	—	—	—	B	A	—	—	—	—	—	—	—	—	B	—	—	—	D	B	—
Potash zone	B	—	—	—	B	—	—	D	—	—	—	B	A	—	—	B	—	—	—	—	—	B	D	—	—	A	B	—
Lower halite	B	—	D	—	B	—	—	D	—	—	D	B	A	—	—	B	—	—	—	—	—	B	B	—	—	B	B	—
Anhydrite	A	—	—	—	B	—	—	D	—	—	A	B	B	—	—	B	—	—	—	—	D	B	B	—	—	B	—	—
Carbonate	A	—	—	—	B	—	—	C	—	—	A	B	B	—	—	A	—	—	—	—	D	B	B	—	—	B	—	—
CARNALLITIC MARL	B	—	—	—	A	—	—	—	—	—	D	B	A	—	—	B	—	—	—	B	—	B	B	—	—	B	D	—
MIDDLE EVAPORITES																												
Upper halite	B	—	—	—	B	C	—	—	—	—	—	B	A	—	—	B	—	—	—	—	D	B	B	—	—	B	—	—
Potash zone	BC	D	D	—	B	A	—	—	—	—	—	B	A	D	—	B	—	—	D	—	D	B	B	—	—	A	—	—
Lower halite	B	—	?D	—	B	D	—	—	—	—	D	B	A	D	—	B	—	—	—	—	D	B	D	—	—	B	D	—
Anhydrite-halite & anhydrite	A	—	?D	D	D	—	D	B	—	—	C	D	A	—	—	B	D	—	—	—	D	B	—	—	—	—	D	—
UPPER MAGNESIAN LIMESTONE	B	—	—	BC	—	—	—	A	D	—	—	—	D	—	—	B	—	—	—	—	B	—	—	—	—	—	—	—
LOWER EVAPORITES																												
Upper halite-anhydrite	A	—	—	—	—	—	D	B	—	—	D	—	A	—	—	D	—	—	—	—	D	D	—	D	D	—	B	D
Upper anhydrite	A	—	—	—	—	—	—	B	—	—	C	—	B	—	—	D	—	—	—	—	D	—	—	D	D	—	—	D
Halite-kieserite-anhydrite	B	D	—	—	—	—	—	—	—	—	—	—	A	A	—	—	—	—	—	B	—	—	—	—	—	—	—	—
Halite-polyhalite-anhydrite	B	—	—	—	—	—	—	—	—	—	—	—	A	—	—	B	—	—	—	A	—	—	—	—	—	—	D	—
Halite-polyhalite-kieserite	—	—	—	—	—	—	—	—	—	—	—	—	A	B	—	B	—	—	—	A	—	—	—	—	—	—	D	—
Halite-polyhalite	—	—	—	—	—	—	—	—	—	—	—	—	A	—	—	B	—	—	—	A	—	—	—	—	—	—	D	—
Polyhalite-anhydrite-halite	A	—	—	—	—	—	—	—	—	—	C	—	A	D	—	B	—	—	—	A	—	—	—	D	—	—	B	—
Anhydrite-polyhalite	A	—	—	—	—	—	—	B	—	D	C	—	B	B	D	B	—	—	—	A	D	—	—	D	—	—	B	—
Lower halite-anhydrite	A	—	—	—	—	—	—	B	—	—	C	—	A	—	—	D	—	—	—	B	D	—	—	—	—	—	B	—
Lower anhydrite	A	—	—	D	—	—	D	BC	D	—	C	—	B	—	—	—	—	—	—	—	D	—	—	—	—	—	—	—
LOWER MAGNESIAN LIMESTONE	BC	—	—	A	—	—	—	A	D	—	—	—	—	—	—	—	—	D	—	—	B	D	—	D	—	—	—	—

A. A major constituent of the zone.
B. Widespread minor constituent (may sometimes be major constituent of certain layers in the zone).
C. A major constituent of the zone in certain localities.
D. Scarce minor constituent.

Durham (Raymond, 1959) where it occurs in efflorescences resulting from upward seepage of water from the Magnesian Limestone, and is associated with halite, thenardite, gypsum, epsomite, pyrite, sulphur, lepidocrocite, ? ilsemannite, ? tamarugite, talc, and other minerals. Efflorescences of trona are also found.

North-western England (Dunham and Rose, 1949; Hollingworth, 1942, 1948; Sherlock and Hollingworth, 1938; Stewart, 1954). Dolomite, anhydrite, and gypsum are major constituents; calcite is not abundant. Halite is practically absent, but pseudomorphs of gypsum after halite have been found in the St. Bees Shales (Hollingworth, 1942). Traces of hematite and pyrite occur, and the gypsum of the 'Daisy Bed' of the Stamp Hill Mine contains siderite. Mirabilite was found in gypsum at Kirkby Thore (Trechmann, 1901).

Genesis of the minerals

Carbonates. Experimental work suggests that dolomite would not be expected as a primary precipitate, and work on some modern evaporites such as those of the Persian Gulf (Curtis *et al.*, 1963; Shearman, 1963) shows that dolomite can replace aragonite and calcite at a very early stage in diagenesis. However, in the water of some Australian lakes and lagoons calcite, dolomite, and magnesite have been collected as suspended sediments (Alderman, 1959; Alderman and Skinner, 1957; Alderman and von der Borch, 1961).

In the British Magnesian Limestones there is direct evidence of some replacement of calcite by dolomite, including dolomitization of fossils and ooliths, and the distribution of the carbonates provides indirect evidence of large-scale replacement (e.g. Raymond, 1962), but it is impossible to prove how much, if any, dolomite here is of primary origin. It seems probable that most of it has replaced primary calcite and aragonite and that much of this replacement was effected early in diagenesis by the action of brines percolating from above. Later, some calcite has replaced dolomite, and this is a large-scale replacement in the concretionary limestones of County Durham.

In the calcium sulphate zones and parts of the halite zones dolomite may have been a primary constituent, but here again there is a lack of reliable criteria. Small-scale late replacement of anhydrite by dolomite has been recorded (Stewart, 1963).

Magnesite is a widespread minor constituent of the potash field, but is very scarce elsewhere. It is the normal carbonate of the potash zones of the Upper and Middle Evaporites, and the only carbonate of the zones carrying ? primary polyhalite and kieserite in the Lower Evaporites (Stewart, 1963). It is therefore possible that it was a primary constituent in the later stages of evaporation. Its distribution in some of the calcium sulphate and carbonate zones of the potash field suggests replacement of earlier carbonates by descending solutions during the formation of higher zones. This would explain its virtual absence outside the potash field.

Small quantities of secondary magnesite accompany other salts in veins in the Carnallitic Marl (Armstrong, Dunham, *et al.*, 1951) and are also found in a few veins towards the top of the Middle Evaporites in the F boreholes. Magnesite in the Middle Potash Zone has recrystallized adjacent to clay (Stewart, 1956).

The siderite in the 'Daisy Bed' of Westmorland forms rhombohedra in secondary gypsum. The traces of ankerite found outside the potash field are of secondary origin and are associated with metalliferous minerals.

Sulphates. It is almost certain that all the existing gypsum in the British Permian is secondary, and most of it has formed by near-surface hydration of anhydrite. It forms porphyroblasts, veins, and patches in anhydrite, dolomite, and clastic rocks, and massive beds which pass into anhydrite with increasing depth of cover (e.g. Dunham, 1948; Forbes, 1958; Hollingworth, 1942; Sherlock and Hollingworth, 1938).

Although no gypsum is found in the potash field, pseudomorphs of anhydrite, halite, polyhalite, and other minerals after gypsum indicate that it was an abundant early mineral (probably largely primary and partly early diagenetic) in most of the sulphate zones and some halite–anhydrite zones (e.g. Stewart, 1949, 1953), and that it was a minor constituent of the Carnallitic Marl (Armstrong, Dunham, *et al.*, 1951).

Although much of the anhydrite in the sulphate zones and anhydrite–halite zones has replaced early gypsum, some of it may have been primary. Reliable criteria are rare, but textural relationships suggest that anhydrite was the primary deposit in the varved rocks of the anhydrite–polyhalite zone and higher zones of the middle cycle of the Lower Evaporites at Forden near Scarborough (Stewart, 1963). It is conceivable that the sulphate was precipitated as gypsum from surface brines and changed to anhydrite while settling through the water. In the greater part of the halite zones and in the potash zones of the Upper and Middle Evaporites there is no petrographic evidence to suggest the presence of early gypsum.

There is evidence of large-scale replacement of dolomite by anhydrite, especially in the Lower Magnesian Limestone and towards the base of the Middle and Lower Evaporites (Dunham, 1948; Fowler, 1944; Stewart, 1949, 1951, 1954), and although there was presumably some primary sulphate in the Magnesian Limestone (Raymond, 1962), it seems likely that much of it grew during diagenesis, perhaps like the anhydrite and gypsum of recent sediments in the Persian Gulf (Curtis *et al.*, 1963; Shearman, 1963).

The writer believes that the major replacements of gypsum by anhydrite, and dolomite by anhydrite, took place in early diagenesis, by the action of interstitial waters and brines percolating through the crystal mush from above. Evidence from Forden suggests that polyhalite replaced anhydrite in this way (Stewart, 1963), and in other localities (e.g. E.2, Stewart, 1949) the anhydrite → polyhalite change followed the gypsum → anhydrite change. On a minor scale, secondary anhydrite formed at later stages, as a constituent of veins and patches in many parts of the succession.

Polyhalite, in exceptionally large quantity in the Lower Evaporites, is almost absent from the rest of the succession, although it is a minor constituent of the Carnallitic Marl (Armstrong, Dunham, *et al.*, 1951). It has been suggested that at Forden about half the polyhalite is primary; and that the large amount of calcium required has been derived from the pene-contemporaneous replacement of anhydrite below by a great deal of secondary polyhalite, through downward percolation of brines rich in potassium and magnesium which displaced brines rich in calcium (Stewart, 1963). The traces of polyhalite in the halite–kieserite–anhydrite subzone are of late secondary origin (post-consolidation). At other localities (E.2, Stewart, 1949; E.3 at Sleights and E.5 at Eskdalegate, Armstrong, Dunham, *et al.*, 1951, and Raymond, 1953), the polyhalite of the Lower Evaporites is largely or entirely secondary, having replaced anhydrite and halite after the gypsum → anhydrite change.

Some kieserite has replaced secondary polyhalite in the Lower Evaporites of E.3 (Armstrong, Dunham, *et al.*, 1951). Greater quantities of kieserite at Forden may have replaced pre-existing primary magnesium-bearing sulphates such as epsomite, after consolidation. The mineral is scarce in the Middle Evaporites (F.2 and F.3).

Among the scarce sulphates, secondary aphthitalite and picromerite accompany kieserite in the Middle Evaporites of the F.3 borehole. At Forden aphthitalite replaces kieserite and halite, and glauberite replaces anhydrite and polyhalite, and these are late-post-consolidation effects.

At least some celestine is secondary, replacing anhydrite at Forden (Stewart, 1963) and occurring in veins in carbonate and sulphate rocks elsewhere (Fowler, 1944; Forbes, 1958). Epsomite is secondary (Raymond, 1962). The occurrence of langbeinite (E.3, Armstrong, Dunham, *et al.*, 1951) needs confirmation. Mirabilite (Trechmann, 1901) is presumably of late secondary origin.

Chlorides. Halite has suffered widespread recrystallization with some movement, so that early textures have often been obliterated. Nevertheless there is evidence of widespread early replacement of gypsum by halite in the pseudomorphous layers of the anhydrite zone of the Upper Evaporites, the halite–anhydrite zone of the Middle Evaporites and the polyhalite-bearing rocks of the Lower Evaporites. Halite has also replaced sulphates and carbonates in oolitic rocks at Forden and elsewhere. There are many records of halite replacing anhydrite in minor quantity in all the chloride zones, after movement of halite under pressure. A complex series of replacements in the potash zone of the Middle Evaporites involves three generations of halite and two of carnallite, rinneite, and anhydrite (Stewart, 1956), and here halite has replaced fairly large amounts of carnallite, perhaps during leaching of the primary potash zone soon after its formation. Veins of various ages, containing halite, are found in all chloride and many sulphate zones and in the Carnallitic Marl, and occasionally in the Magnesian Limestones.

Pseudomorphs indicate that carnallite was the main primary potassium salt of the potash zone of the Middle Evaporites, and that most of the sylvine here is secondary (Stewart, 1956). However, the writer believes that sylvine may have been primary in a layer of halite–sylvine rock, with some anhydrite and very little other impurity, which occurs below the part of the potash zone which carries pseudomorphs after carnallite in the F boreholes. Small amounts of sylvine and carnallite in the lower halite zone may have formed by crystallization partly from original interstitial brine, and partly from brines percolating from above.

The upper halite zone of the Middle Evaporites, and to a much greater extent the Carnallitic Marl, are veined and impregnated with halite, carnallite, sylvine, and rinneite (Armstrong, Dunham, *et al.*, 1951). Some similar veins cut the overlying carbonate and anhydrite zones of the Upper Evaporites, along fractures which are sometimes slickensided, and lead upwards to a subsidiary development of these potassium-bearing minerals towards the base of the lower halite zone of the Upper Evaporites: this potassium enrichment is presumably entirely secondary and at least partly of fairly late origin, resulting from upward migration of brines from below.

Sylvine appears again in force, with scarce carnallite and rinneite, in the potash zone of the Upper Evaporites, where there is no evidence of primary carnallite.

In summary, halite, carnallite, and possibly some sylvine were primary constituents, but these minerals have also taken part in large-scale pene-contemporaneous and post-consolidation changes. Rinneite is almost certainly entirely secondary, and it is likely that the iron required for its formation has been derived from reaction of metasomatic brines with clastic material and hematite.

Borates. Boracite forms isolated crystals and crystal groups in well cuttings from the Middle Evaporites of the E.2 borehole (Guppy, 1944; Stewart, 1951), and radiating nodules in halite in the Upper Evaporites of E.5 (Sabine, 1953). Isolated crystals in halite, seen by the writer in slices of rocks from the Middle Evaporites (potash zone) of the F.1 borehole, show no sign of secondary origin.

p-Veatchite has been found only as isolated crystals in well cuttings of the Lower Evaporites of E.2 (Stewart, Chalmers, and Phillips, 1954; Beevers and Stewart, 1960). Sulphoborite replaces anhydrite in the Lower Evaporites of Forden (Stewart, 1963). No borates have been recorded outside the potash field.

Other minerals. Hematite is widespread in clastic rocks and salts (especially chlorides) above the Upper Magnesian Limestone. The common occurrence of flakes of this mineral in secondary sylvine and other salts, especially adjacent to clastic material, and also in some veins, suggests that much of the hematite in the salts has been formed by leaching of the clastic material by brines, during diagenesis and later geothermal metamorphism. The same applies to quartz, which is common in the Upper and Middle Evaporites of the potash field, especially in chloride zones, often fringing patches of clay or silt. There is definite evidence of late replacement of halite and anhydrite by quartz (Stewart, 1951).

Talc (Raymond, 1962; Stewart, 1949, 1951, 1963) is a widespread minor constituent of the potash field and occurs sporadically outside it. Its common occurrence along cleavages of recrystallized halite, and in veins with halite cutting halite and silt (in the Upper Evaporites of the F boreholes), suggests formation during geothermal metamorphism.

A biogenic origin for the small quantity of sulphur found in patches and veins has been suggested by Raymond (1962). Sulphur has replaced some anhydrite (Stewart, 1949; Raymond, 1962).

In the potash field traces of fluorite have been recorded in anhydrite rock (Lower Evaporites, E.2; Stewart, 1954), in dolomite in the Upper Magnesian Limestone filling rods of the fossil *Filograna permiana* (E.2, Trechmann in discussion on Raymond, 1953), and in dolomite veins in the Lower Magnesian Limestone, with traces of a mineral tentatively identified as northupite (E.5, Raymond, 1962). The origin of the suite of metalliferous minerals, with fluorite, barite, etc., found in the Lower Magnesian Limestone outside the potash field is discussed by Fowler (1943) and Dunham (1948, 1960). Dunham points out that their emplacement appears to be epigenetic, but that they might originate from Hercynian mineralizing fluids from depth or from fluids squeezed out of the Permian Marl Slate during diagenesis.

Conclusions

Our present knowledge of these evaporites suggests the following tentative history.

Primary minerals: Probably calcite, dolomite, magnesite, gypsum, anhydrite, polyhalite, epsomite, halite, carnallite, sylvine, boracite; possibly aragonite, celestine, fluorite, *p*-veatchite.

Early diagenetic (*pene-contemporaneous*) *changes*, resulting largely from reaction with brines percolating downwards during evaporite formation: Large-scale replacement of calcite (and aragonite?) by dolomite, dolomite by anhydrite (and gypsum?), gypsum by anhydrite and halite, anhydrite and halite by polyhalite, and carnallite by halite and secondary sylvine; smaller-scale replacement of dolomite by magnesite; leaching of magnesium sulphates by fresher brines in the Lower Evaporites; development of some hematite and quartz by alteration of clastic material.

Later diagenetic and geothermal metamorphic changes, involving rise of temperature and pressure during burial: Some dolomitization of calcite, conversion of remaining gypsum (if any) to anhydrite, widespread but moderate scale replacement of anhydrite by halite, and development of kieserite from earlier magnesium-bearing sulphates; veining and impregnation of upper part of Middle Evaporites and Carnallitic Marl with halite, carnallite, sylvine, rinneite, magnesite, and quartz, and introduction of these minerals from below into the lower part of the Upper Evaporites; development of quartz, hematite, rinneite, talc, partly through reactions with clay and silt; widespread migration of halite, anhydrite, and other minerals forming veins and patches in both evaporites and overlying sediments; many minor changes including replacement of anhydrite by dolomite, of polyhalite and anhydrite by glauberite, of glauberite, anhydrite, polyhalite, and magnesite by halite, of halite and kieserite by aphthitalite, and development of scarce late polyhalite; large-scale recrystallization of halite, with some movement. The suite of metalliferous minerals and their associates, galena, chalcopyrite, sphalerite, barite, fluorite, etc., were probably formed by upward migration of fluids from below the evaporites.

Near-surface changes, involving the action of meteoric water, and fall of temperature and pressure on erosion of the overburden: Large-scale hydration of anhydrite to gypsum (with release of some celestine?); solution and removal of halite and gypsum; the replacement of dolomite by calcite in the concretionary limestones may belong here; development of mirabilite.

References

ALDERMAN (A. R.), 1959. J. Geol. Soc. Australia, vol. 6, p. 1.

—— and SKINNER (H. C. W.), 1957. Amer. J. Sci., vol. 255, p. 561.

—— and VON DER BORCH (C. C.), 1961. Nature, vol. 192, p. 861.

ARMSTRONG (G.), DUNHAM (K. C.), HARVEY (C. O.), SABINE (P. A.), and WATERS (W. F.), 1951. Min. Mag., vol. 29, p. 667.

BEEVERS (C. A.) and STEWART (F. H.), 1960. *Ibid.*, vol. 32, p. 500.

CURTIS (R.), EVANS (G.), KINSMAN (D. J. J.), and SHEARMAN (D. J.), 1963. Nature, vol. 197, p. 679.

DUNHAM, (K. C.), 1948. Proc. Yorks. geol. Soc., vol. 27, p. 217.

—— 1960. *Ibid.*, vol. 32, p. 229.

—— CLARINGBULL (G. F.), and BANNISTER (F. A.), 1948. Min. Mag., vol. 28, p. 338.

—— and ROSE (W. C. C.), 1949. Proc. Geol. Assoc., vol. 60, p. 11.

EDWARDS (W.), 1951. The concealed coalfield of Yorkshire and Nottinghamshire. Mem. geol. Surv.

FLECK (A.), 1950. Deposits of potassium salts in north-east Yorkshire. Chemistry and Industry (separate, 17 Oct.), 15 pp. (This work was issued and is paginated independently of the volume for 1950.)

FORBES (B. G.), 1958. Proc. Yorks. geol. Soc., vol. 31, p. 351.

FOWLER (A.), 1943. Geol. Mag., vol. 80, p. 41.

—— 1944. *Ibid.*, vol. 81, p. 193 and p. 254.

GUPPY (E. M.), 1944. Min. Mag., vol. 27, p. 51.

HOLLINGWORTH (S. E.), 1942. Proc. Geol. Assoc., vol. 53, p. 141.

—— 1948. Proc. Yorks. geol. Soc., vol. 27, p. 192.

LEES (G. M.) and TAITT (A. H.), 1946. Quart. Journ. geol. Soc., vol. 101, p. 255.

NAPIER (E.), 1948. Proc. Yorks, geol. Soc., vol. 27, p. 210.

RAYMOND (L. R.), 1953. Quart. Journ. Geol. Soc., vol. 108, p. 283.

—— 1959. Min. Mag., vol. 32, p. 172.

—— 1962. Quart. Journ. geol. Soc., vol. 118, p. 39.

Sabine (P. A.), 1953. Petrog. Dept. Report. Summ. Prog. Geol. Surv. for 1951, p. 60.
Shearman (D. J.), 1963. Abs. Proc. Geol. Soc., no. 1607, p. 63.
Sherlock (R. L.), 1921. Rock salt and brine. Spec. Rep. Min. Resources, G.B., Geol. Surv., no. 18.
—— and Hollingworth (S. E.), 1938. Gypsum and anhydrite and celestine and strontianite. Spec. Rep. Min. Resources, G.B., Geol. Surv., no. 3.
Stewart (F. H.), 1949. Min. Mag., vol. 28, p. 621.
—— 1951*a*. *Ibid.*, vol. 29, p. 445.
—— 1951*b*. *Ibid.*, vol. 29, p. 557.
—— 1953. Proc. Geol. Assoc., vol. 64, p. 33.
—— 1954. Proc. Yorks. geol. Soc., vol. 29, p. 185.
—— 1956. Min. Mag., vol. 31, p. 127.
—— 1963. Proc. Yorks. geol. Soc., vol. 34, p. 1.
—— Chalmers (R. A.), and Phillips (R.), 1954. Min. Mag., vol. 30, p. 389.
Stewart (G. E.), 1947. Trans. Instn. Min. Metall., vol. 56, p. 113.
Tilley (C. E.), 1943. Min. Mag., vol. 26, p. lvii.
Trechmann (C. O.), 1901. *Ibid.*, vol. 13, p. 73.

Reprinted from *Dolomitization and Limestone Diagenesis*, Soc. Econ. Paleon. Mineralog. Spec. Publ. 13, 71–88 (1965)

DOLOMITIZATION OF RECENT AND PLIO-PLEISTOCENE SEDIMENTS BY MARINE EVAPORITE WATERS ON BONAIRE, NETHERLANDS ANTILLES[1,2]

K. S. DEFFEYES,[3] F. JERRY LUCIA, AND P. K. WEYL[3]
Shell Development Company, Houston, Texas

27

ABSTRACT

The carbonate rocks and sediments on the island of Bonaire in the Netherlands Antilles contain interesting examples of early dolomitization. A large flat area of Recent supratidal sedimentation exists on the south end of Bonaire, and in this area evaporation of sea water is depositing calcium carbonate and gypsum, which produces dense brines having large Mg/Ca ratios. Dolomite is found in most of the Recent supratidal sediments, and carbon-14 dates on the dolomite establish that the time since dolomitization has been less than 2200 years. Textural evidence indicates that some of the dolomite was formed by replacing lime sediments.

The dense brines produced by evaporation tend to flow downward into the permeable sediments, and an analysis of the chemistry and hydrology of a hypersaline lake chosen for detailed study shows that downward drainage of brine having an Mg/Ca ratio of about 30 must be taking place today. Examination of marine Plio-Pleistocene rocks on the north end of Bonaire shows large areas of dolomitization whose boundaries cut across the bedding. The field evidence is consistent with the hypothesis that this dolomite has been produced by the flow of dense brines from a supratidal area. The time required to produce the estimated volume of dolomite found in the Plio-Pleistocene rocks would be of the order of 10^5 years.

INTRODUCTION

In spite of the existence of an extensive literature on dolomitization, the process itself is not yet clearly understood. For a recent review on dolomitization and an extensive bibliography, the reader is referred to Fairbridge (1957). Part of the difficulty has been that the mechanism of dolomitization had to be inferred from the evidence in the ancient rocks. Although dolomites which give Recent radiocarbon dates have been found in several places (Skinner et al., 1963; Graf et al., 1961), these examples have shed little light on the mechanisms that have produced most of the dolomite rock in the geologic record.

In order for a limestone to become dolomitized, the following requirements must be fulfilled:

1) A water having a Mg/Ca ratio larger than the ratio that would be in equilibrium with both calcite and dolomite must be produced (Halla and Ritter, 1935).

2) This water must flow through the limestone, because magnesium transport by diffusion is inadequate to explain most dolomite occurrences.

3) The rate of production and flow of the high-magnesium water must be adequate to dolomitize the rock in the time available.

A mechanism which meets these requirements has been proposed by several writers to explain ancient dolomite deposits. In particular, Newell et al. (1953, p. 179), in their study of the Permian Reef Complex, state:

> "Theoretical considerations and the ubiquitous dolomitization of lagoonal sediments, however, point to shelf brines as the most effective agent of dolomitization. Not only were these waters much enriched in magnesium by evaporation; their magnesium-calcium ratio must have exceeded that of normal marine waters because of withdrawal of calcium in the lagoon, in the form of gypsum and anhydrite.
>
> "Lagoonal waters may have entered the reef rock in several ways. They may have flooded the reef in times of storm. They may have seeped from lagoonal sediments through the reef and into the basin under the pressure of hydrostatic head caused by their relatively high specific gravity."

More recently, Adams and Rhodes (1960) have also discussed the process of dolomitization by the reflux of hypersaline brines.

It was important to find a location where the process of evaporation and reflux is taking place at the present time. What we have needed as a clue to the past is to find a geochemical engine operating today that can supply adequate volumes of high-magnesium waters to carbonate sediments or limestones. We believe that we have found such an engine in operation on the island of Bonaire in the Netherlands Antilles. The processes operating today on the southern end of Bonaire are largely the consequences of evaporation of sea water on an exposed carbonate flat. Because the combination of sediment surfaces very near sea level and a

[1] Manuscript received July 8, 1964.

[2] EPR Publication 348, Shell Development Company (A Division of Shell Oil Company), Exploration and Production Research Division, Houston, Texas.

[3] Present address: Department of Oceanography, Oregon State University, Corvallis, Oregon.

permanently or seasonally dry climate has probably occurred repeatedly in the geologic record, we are interested in the diagenetic effects that could be caused by evaporite brines. In this paper we are reporting observations on the chemistry and hydrology of south Bonaire as a model for processes that probably have taken place many times in the past.

On the north end of Bonaire are outcrops of Plio-Pleistocene limestones that contain large dolomitized areas which cut across the bedding. We shall show how these dolomites can be explained by the model process established for south Bonaire.

The material for the data set forth below was collected on a three-week field study by the writers early in 1961. Information furnished by Mr. P. H. de Buisonje from his study of the Pleistocene rocks of Aruba, Curacao, and Bonaire was valuable in carrying out the work.

GEOLOGIC SETTING OF BONAIRE

Aruba, Curacao, and Bonaire are located about 50 to 100 miles north of the Venezuelan coast (fig. 1). They are small islands, each about 30 miles long and 10 miles wide, and are spaced about 60 miles apart. All three are composed of a folded core of Cretaceous volcanic and intrusive rocks covered by Eocene and Pleistocene carbonate rocks. On Bonaire the oldest rocks are flows, tuffs, and igneous intrusions of Upper Cretaceous age which constitute the Washikemba Formation (Pijpers, 1933). These rocks were folded before the deposition of the conglomerates and limestones of the Rincon and Soebi Blanco formations, also believed to be of Cretaceous age. These formations are presently exposed in the hills and valleys of the northwestern and central parts of the island (fig. 2).

The carbonate rocks which overlie the Cretaceous rocks are primarily Plio-Pleistocene in age, with two small areas of Eocene rocks. The late Pliocene or Pleistocene age is determined by the identity of the Recent biota with the fossils in the rocks. The Plio-Pleistocene carbonate rocks are similar in many ways to the sediments being deposited today adjacent to the island. On the windward side of Bonaire several Plio-Pleistocene terrace levels exist, and they were probably formed as combination wave-cut and wave-built surfaces similar to that being formed today on the windward coast. On the leeward coast the Recent sediments consist of skeletal lime sands with widely scattered coral heads which are being deposited on slopes of 10 to 30 degrees. The Plio-Pleistocene rocks on the leeward side of the island show similar seaward dips, and the au-

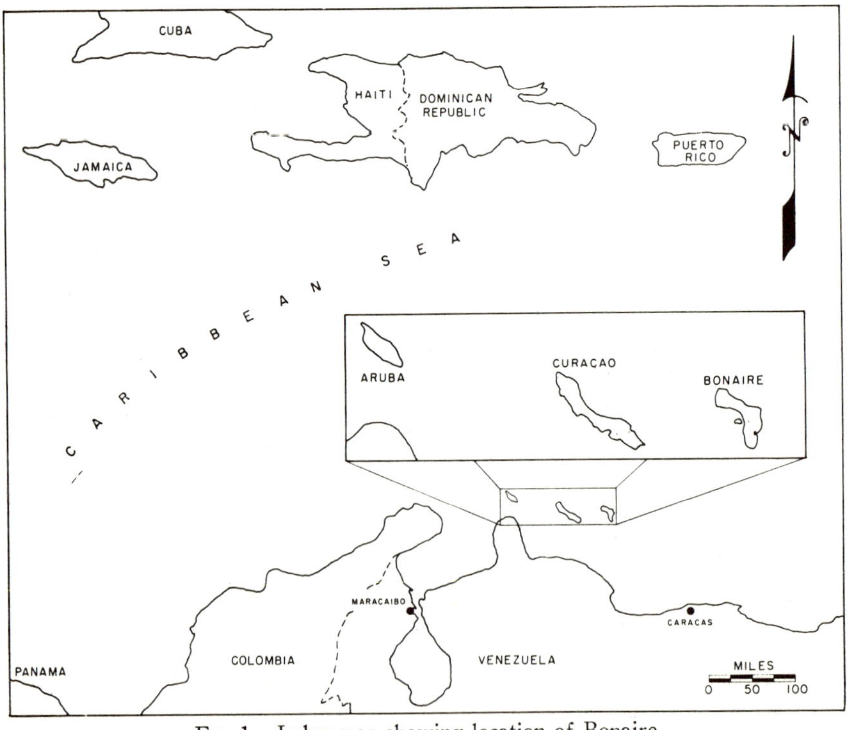

FIG. 1.—Index map showing location of Bonaire.

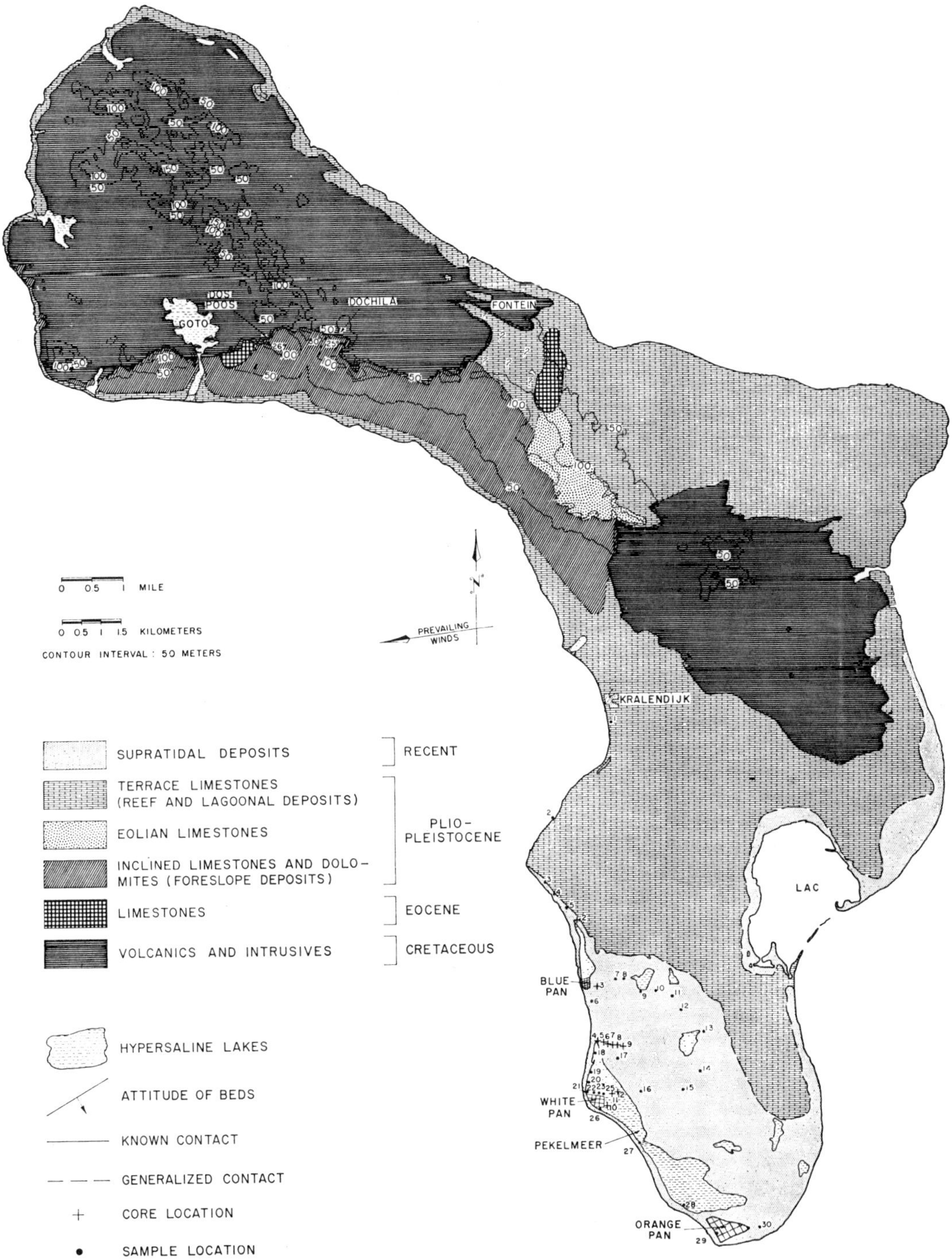

FIG. 2—Geologic map of Bonaire (after Westermann and Zonneveld, 1956).

thors are in agreement with the conclusion reached by Martin (1888), Schaub (1948), and de Buisonje (personal communication) that the dip of these "inclined" beds is depositional.

In addition to the open ocean environments on the leeward and windward sides of Bonaire, two other distinct depositional environments exist today. On the windward coast is a sizable reef-protected lagoon, the Lac, in which *Thalassia, Halimeda, Goniolithon,* and *Porites* are the common organisms living on the bottom. Although this lagoon is ecologically different from the open ocean, the water chemistry and diagenetic environment are very similar to those in the open sea.

In contrast, a large area in south Bonaire is isolated from the sea by a wave-built coral rubble ridge at the coast line, and the internal area receives sea water only by seepage and rare storm tides. Inside of the ridge is an extremely flat area of about 13 square miles where Recent carbonate sediments are being deposited on subaerially exposed flats, in low dunes, and in a number of shallow hypersaline lakes. The largest lake, the Pekelmeer, has an area of about 1 square mile and is located along the leeward coast just inland from the coral rubble ridge.

The interior of south Bonaire is similar to the supratidal environment of Logan (1961, p. 527). Specifically, the supratidal environment can be defined as an area which is not in daily tidal communication with the ocean but receives the majority of its sediment from the ocean as particles washed up by the storm tides and (or) as minerals precipitated from ocean water which has seeped in through the sediment or washed in by higher-than-normal tides.

Within the supratidal area the most common sediments are pelleted lime muds and silts, algal mats in intermittently wet areas, and gypsum on the bottoms of the hypersaline lakes. Diagenetic modification of the Recent sediment has two common forms. Cementation which forms hard, lacy limestone crusts is quite common. A more interesting diagenetic feature in the supratidal area is the occurrence of Recent dolomite.

RECENT DOLOMITE

Recent dolomite is found in crusts near the surface and in the unconsolidated sediment of the supratidal area. The dolomite crusts are in the top few inches of a pelleted lime sand and mud. The crystal size is about 2 microns, and the Ca/Mg ratio in the dolomite crystals, as measured by X-ray spacing (Goldsmith et al., 1961), ranges from $Mg_{46}Ca_{54}$ to $Mg_{44}Ca_{56}$. The ordering peaks 10·1, 10·5, and 20·1 of dolomit are visible on the X-ray powder pattern. A well developed example of this crust was found o the northeast shore of the Pekelmeer. It contains between 80 and 95 percent dolomite, an the remainder is aragonite. The crust is eithe faintly laminated or massive, and where th top of the crust is exposed, it contains widel spaced polygonal cracks. In some areas th crust is being eroded, and clasts of the crust ar being scattered in the pelleted lime sand an mud. If no further dolomitization occurred this mixture would form a limestone containin dolomite lithoclasts. Most of the unconsolidate sediments in the supratidal area contain som dolomite.

The maximum time that could have elapse since the dolomitization of the crusts has bee established by radiocarbon analyses. Two samples of the crust from different localities, bot of which contained more than 95 percent dolomite, gave carbon-14 ages of 1480 ±140 an 2190 ±150 years.

Petrographic examination shows that th dolomite was formed in the sediment after was deposited. Gastropod shells composed c dolomite in the dolomite crust (fig. 3), and th selective occurrence of dolomite in the mud matrix rather than in the sand size pellets an shells in the partially dolomitized sediment show that replacement of pre-existing calciu carbonate has occurred. The occurrence dolomite filling pre-existing pore space is suggested by the fact that both the porosity of th dolomite crust and the porosity of the undolomitized equivalent pelleted lime mud are 50 pe cent, whereas the pore type distribution different. The porosity of the dolomite cru was determined in the laboratory, and the porosity of the undolomitized equivalent was dete mined by water loss from a known volume water-saturated sediment. In the pelleted lim mud the porosity is between the mud size part cles, but in the dolomitized equivalent the sa size pellets and shells are commonly leached form vuggy porosity. To balance the amount vuggy porosity, matrix porosity must be r duced. The fact that leached shells and pelle are found only in the dolomite crust sugge that the dissolution is part of the dolomitizati process. The carbonate ions produced by t dissolution of shells and pellets were probab used for the growth of the dolomite crysta This apparent balance between the amount carbonate in the sediment and the amount dolomite formed supports the hypothesis Weyl (1960) and Murray (1960) that low-c

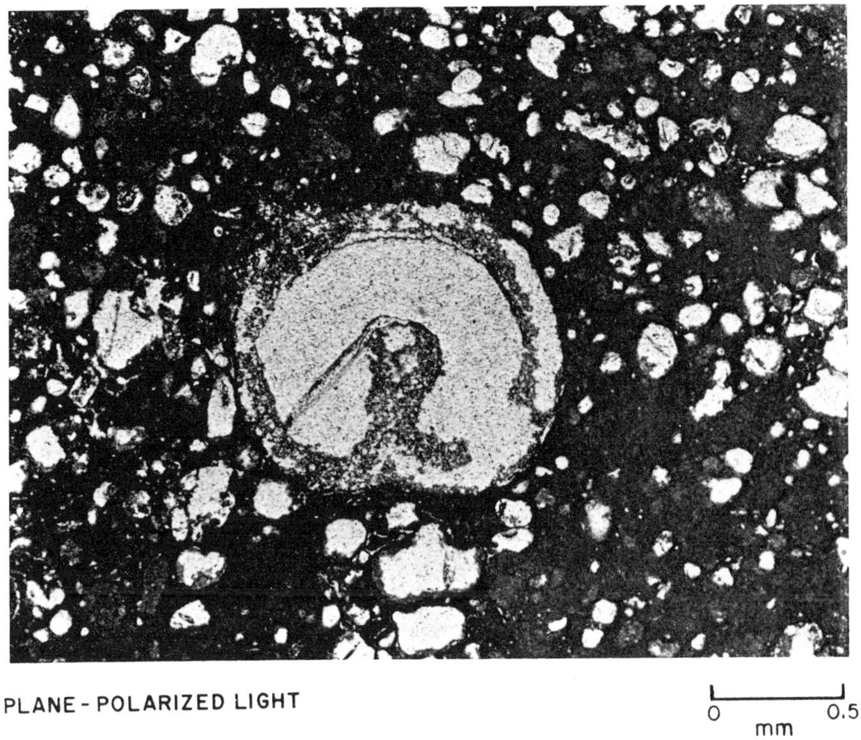

Fig. 3.—Recent dolomite crust containing gastropod shell replaced by dolomite.

bonate waters can form dolomite most effectively by using the local carbonate.

CHEMISTRY OF THE WATERS ON SOUTH BONAIRE

A large number of water samples were collected from surface waters on Bonaire and from seepage water into core and test holes. Chemical analyses of many of these waters are given in table 1. For the waters that are under-saturated with sodium chloride, the concentrations of the important constituents are plotted in figure 4 against the chloride (total halides) concentration. If these waters resulted exclusively from the evaporation of sea water, the ratio of chloride to other concentrations should remain constant. Thus, on the log-log plot of figure 4 the data points should fall on a

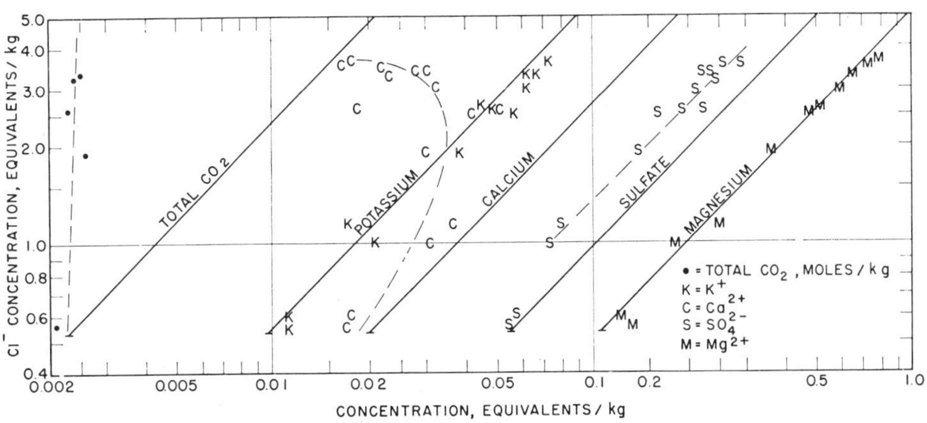

Fig. 4.—Analysis of evaporite waters.

TABLE 1.—*Water Analyses*

Sample No.	Sample Description	Density, g/cc	Concentration, equivalents/kg					Ratio		(equivalents/kg)²	Concentration, millimoles, kg		
			Cl^-	Ca^{2+}	Mg^{2+}	K^+	SO_4^{2-}	Cl^-/Cl^-_{sea}	Mg^{2+}/Ca^{2+}	$Ca^{2+} \times SO_4^{2-}$	HCO_3^-	CO_3^{2-}	Total CO_2
4	Atlantic sea water	1.023	0.558	0.017	0.113	0.011	0.054	1.04	6.7	0.0009	1.92	0.15	2.07
71	Sea water spring Pm	1.026	0.561	0.017	0.123	0.012	0.057	1.05	7.2	0.0009			
82	From core hole No. 9	1.045	1.02	0.031	0.181	0.021	0.074	1.91	5.8	0.0023	3.00	0.01	3.01
25	Below crust, loc 19	1.040	1.14	0.036	0.225	0.017	0.079	2.13	6.2	0.0028			
33	Salt pond, orange pan	1.057	1.25	0.026	0.284	0.028	0.12	2.34	10.9	0.0031			
77	From core hole No. 8	1.087	1.92	0.030	0.361	0.038	0.14	3.59	12.0	0.0042	2.45	0.15	2.60
86	Below crust, loc 27	1.115	2.48	0.042	0.480	0.056	0.16	4.63	11.4	0.0067			
80	Goto Lake	1.121	2.60	0.050	0.510	0.048	0.22	4.86	10.2	0.011			
55	Pekelmeer	1.122	2.62	0.018	0.506	0.044	0.19	4.90	28	0.0034	1.70	0.60	2.30
68	From core hole No. 4	1.142	3.00	0.033	0.598	0.061	0.22	5.60	18.4	0.0072	1.71	0.67	2.38
43	Pekelmeer	1.152	3.25	0.018	0.633	0.057	0.24	6.06	35	0.0043	1.95	0.55	2.50
31	Pekelmeer, bottom layer	1.159	3.36	0.023	0.644	0.061	0.23	6.27	28	0.0053			
66	From core hole No. 7	1.160	3.37	0.030	0.635	0.065	0.22	6.29	21	0.0066			
63	From pit near salt pan	1.162	3.36	0.028	0.630	0.089	0.22	6.28	23	0.0062			
64	From core hole No. 5	1.167	3.48	0.022	0.731	0.062	0.26	6.49	33	0.0057			
59	From core hole No. 6	1.176	3.62	0.017	0.778	0.072	0.29	6.76	47	0.0048			
29	Salt pan	1.180	3.70	0.018	1.01	0.092	0.36	6.90	56	0.0065	2.50	1.50	4.00
63	Interior salt lake	1.216	4.37	0.013	0.997	0.089	0.33	8.16	75	0.00098			
57	Salt pan	1.238	4.31	0.004	2.14	0.18	0.68	8.05	535	0.0027	5.4	2.1	7.5

straight line with a slope of 1. Based on the standard analysis of sea water, such ideal evaporation-only lines are indicated in the figure.

Magnesium was determined titrimetrically. The magnesium concentration falls on the theoretical line, indicating that not more than a few percent of the magnesium present is precipitated.

Potassium was determined flame-photometrically. Although the potassium concentrations scatter rather widely about the theoretical line, no significant deviation is apparent.

Sulfate was determined turbidimetrically. The sulfate concentrations of waters that have more than doubled their chloride content are significantly lower than the theoretical line. Thus, sulfate must have been precipitated.

Calcium was determined titrimetrically. The calcium ion concentration is significantly lower than the theoretical curve, indicating precipitation of calcium.

Total CO_2. For eight samples the total CO_2 content of the waters was determined by a potentiometric titration. The total CO_2 content is plotted as moles per kg. The total CO_2 concentration in the waters remains essentially constant as the salinity increases, indicating loss of CO_2 from the water. Total carbonate can be lost either by the precipitation of a carbonate mineral or by the loss of CO_2 gas. These losses can take place either inorganically or by the organic precipitation of crystalline carbonate and by photosynthetic absorption of CO_2.

Taking the composition of Atlantic Ocean water as being representative of the original water, we can calculate, from the data of table 1, the loss of carbonate and CO_2 per kg of original sea water. This is indicated in figure 5. It can be seen that most of the loss takes place as the chloride content increases to 2 equivalents per kg. At the salinity of the Pekelmeer about 0.9 millimole of carbonate has been precipitated per kg of original sea water.

The water analyses discussed above indicate a significant loss of calcium and sulfate from the evaporating sea water. X-ray diffraction analysis of the sediments on south Bonaire indicates that calcium sulfate was precipitated as gypsum. In figure 6 the concentration product of calcium times sulfate in (equivalents/kg)² is plotted against the chlorine content. A line of slope 2 on this log-log plot indicates the concentration product that would be obtained if no precipitation took place. Also shown is the concentration product obtained experimentally by Posnjak (1940) for a water closely approximating sea water of various salinities that is saturated with gypsum. Above a chlorine concen-

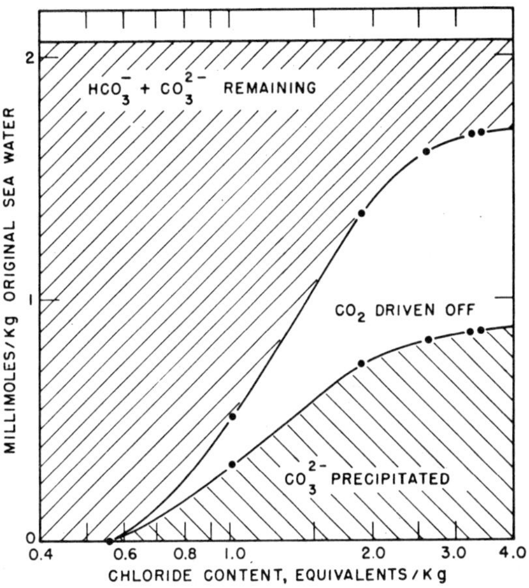

FIG. 5.—Change in total carbonate content of sea water upon evaporation.

tration of 1.5 equivalents/kg, the concentration product of $Ca^{2+} \times SO_4^{2-}$ scatters about the data of Posnjak.

Using the data of Posnjak and extrapolating them to higher salinities, we can then calculate the number of equivalents of gypsum precipitated per kg of original sea water. This is shown in figure 7. Precipitation begins when the chloride content has been increased to 1.3 equivalents per liter. At this stage about 1 meq of calcium carbonate has been precipitated. For comparison, the results of a laboratory evaporation by Usiglio (*in* Clarke, 1924) are indicated. Apparently, Usiglio's solution was considerably supersaturated before gypsum precipitated. At the salinity of the Pekelmeer about 8 millimoles of gypsum per kg of original sea water was precipitated.

DOLOMITIZATION BY HYPERSALINE WATER

Some insight into the question of which of the many types of natural water could cause the replacement of calcite by dolomite can be gained from an examination of the reaction for the replacement process:

$$2CaCO_3 + Mg^{2+} \leftrightarrows CaMg(CO_3)_2 + Ca^{2+}.$$

The activity ratio of Mg^{2+} to Ca^{2+} ions is fixed at some constant value if the water is in equilibrium with ideal dolomite and with pure calcite at any given temperature and pressure (Halla and Ritter, 1935). Waters which have larger Mg^{2+}/Ca^{2+} ratios will dissolve calcite and precipitate dolomite until they have lost enough magnesium and picked up enough calcium to approach the equilibrium ratio. Thus, as a first approximation, we are interested in those waters which have large Mg^{2+}/Ca^{2+} ratios.

The precipitation of a large fraction of the calcium, both as gypsum and as aragonite, and the retention in solution of essentially all of the magnesium cause a very large rise in the magnesium-to-calcium ratio in the brines resulting from sea water evaporation. This increase is shown in figure 8, which compares the ratios measured in the waters from south Bonaire with the predicted ratio computed from the gypsum and aragonite precipitation curve. At the time of the field work, much of the hypersaline water had a magnesium-to-calcium ratio in excess of 20 to 1, which is considerably larger than the 5-to-1 ratio in sea water. Thus, one might ask whether this brine could be capable of producing dolomite. If the hypersaline water could replace calcite with dolomite, one would expect to find dolomite in the carbonate sediments containing the hypersaline water. As described above, partially dolomitized sediments are present in most of the localities sampled on south Bonaire, and completely dolomitized layers occur in some places. The presence of dolomite in many of the Recent samples from south Bonaire causes us to become increasingly interested in the chemistry of the hypersaline water, although we must realize that the association of

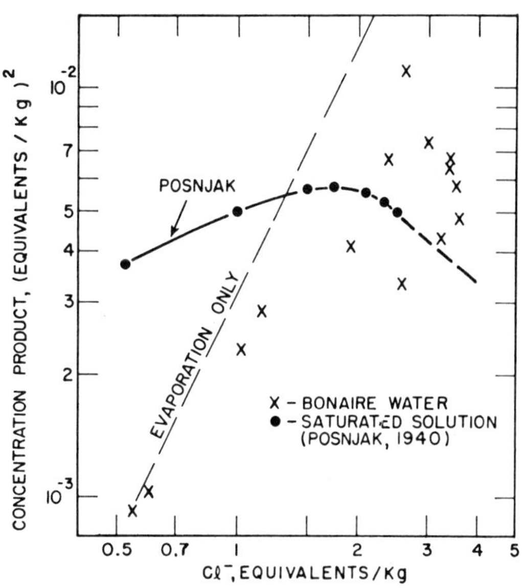

FIG. 6.—$Ca^{2+} \times SO_4^{2-}$ concentration product for Bonaire waters.

this brine with Recent dolomite is only circumstantial evidence that the hypersaline water is capable of producing dolomite.

Typical hypersaline water contains about 0.25, 0.01, and 0.002 mole of Mg^{2+}, Ca^{2+}, and total CO_2 per kg of solution, respectively. It thus contains one hundred times as much magnesium ion as total CO_2 carbon. Weyl (1960) has shown that such waters should dolomitize by utilizing the carbonate in the limestone and replacing half its calcium by magnesium. The petrographic study of the Recent dolomite on south Bonaire (see Page 74) indicates "local source" dolomitization, and Murray (1960) concludes that such "local source" dolomitization is often found in dolomite rocks.

HYDROLOGY OF THE PEKELMEER

We have shown that sea water that has been concentrated by evaporation is likely to dolomitize lime sediments. It therefore becomes important to study quantitatively the rate of production of these brines and their subsequent flow history. We therefore made some observations on the largest of the hypersaline lakes on south Bonaire, the Pekelmeer.

A hypersaline lake such as the Pekelmeer is a dynamic system whose hydrology depends upon the geometry of the lake, the volume of sea water inflow, the volume of rain water inflow, the rate of evaporation, and the volume of outflow through the bottom of the lake (reflux). The volumes of sea water and rain

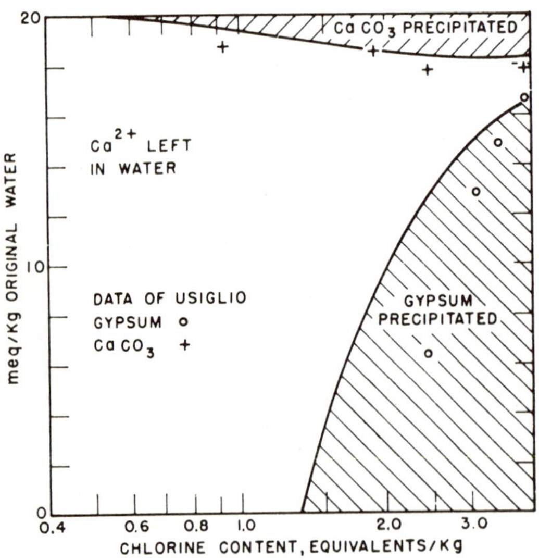

FIG. 7.—Gypsum precipitation during evaporation (data of Usiglio *in* Clark, 1924).

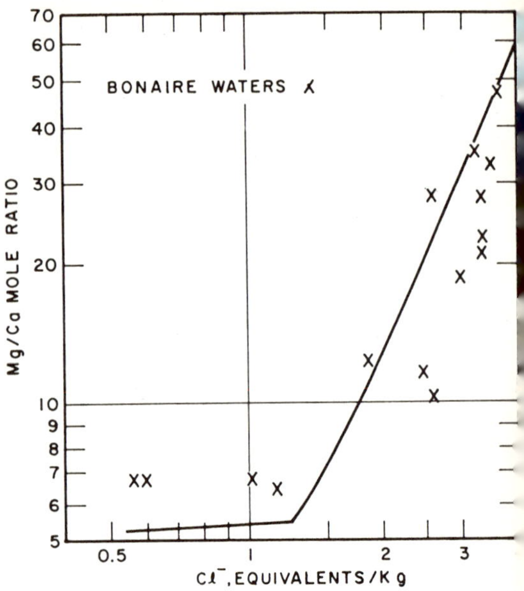

FIG. 8.—Magnesium-to-calcium ratio during evap- oration. (Curve is computed from gypsum an aragonite precipitation curves.)

water inflow and the rate of evaporation fo the Pekelmeer have been observed and mea sured. The conclusion that reflux of hypersalin water exists is inferred from the fact that th salinity of the Pekelmeer is not high enough t precipitate sodium chloride. Permanent salt de position takes place only in the artificially con structed salt pans. These salt pans have been i operation intermittently for the last 200 year The fact that their construction was necessar indicates that large amounts of sodium chlorid were not precipitated naturally in the Pekelmee or elsewhere on Bonaire. [No halite was foun in the sediment below the Pekelmeer (corin depth, 1 meter maximum).] The only natur halite was an ephemeral crust about 1 centim ter thick near the shore of the Pekelmeer whic was redissolved after a rainstorm during th field work. If the Pekelmeer acted as a simpl evaporating pan for sea water, the sedimen should have contained about twenty times a much halite as gypsum. The absence of halite i the presence of 1 meter of gypsum indicate that the hypersaline water must leave the lak before it becomes saturated with sodium chlo ride.

In order to study the dynamic hydrologic be havior of a hypersaline lake of the Pekelme type, we shall investigate a model based on ob servations from the Pekelmeer.

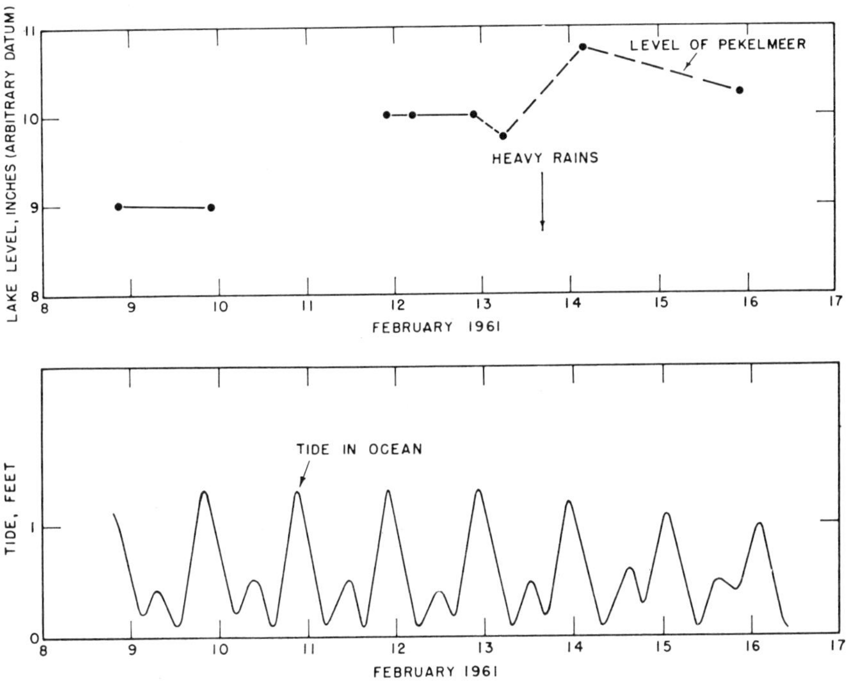

FIG. 9.—Tide data, Pekelmeer and open ocean (ocean data from tide tables).

Volume of Sea Water Inflow

The lake surface was about 20 centimeters below a beach platform believed to be at low tide level. The fact that seepage into the Pekelmeer through the coral ridge was active both at high tide and at low tide confirms the fact that the level of the Pekelmeer is below low tide level. Although the tidal range was about 1 foot, the fluctuation in the lake was less than 1/4 inch, with major fluctuations correlating with rainfall (fig. 9).

A number of areas of sea water inflow into the Pekelmeer were observed. All sea water inflow was through the sediments. The largest area was found south of the White Pan salt ponds. The geometry of this inflow is illustrated in figure 10. Upon entering the lake, the sea water formed a colder, less dense layer on top of the warm hypersaline water of the Pekelmeer. The following temperatures and salinities were observed:

	Top Water	*Bottom Water*
Temperature, °C	25.6	29.0
Salinity, °salometer[5]	35	95

[5] The salometer scale is a density scale which takes distilled water as 0 and saturated NaCl solution as 100. Normal sea water is about 13° salometer.

In the inflow area itself a salinity of 13° salometer was observed.

To obtain a rough estimate of the volume of sea water inflow, the following procedure was employed: The sill over which the water entered the Pekelmeer was subdivided into 1-meter intervals. The depth of the water at the center of each interval was measured. A dye marker was introduced at middepth in the water, and its flow was timed over a 50-centimeter interval. The flow was then computed for each interval and was summed over the width of the sill. Flow measurements at high and low tide gave total flow rates of 5 and 7 cubic meters per minute, respectively. The average inflow for this area is thus of the order of 10^{10} cubic centimeters per day.

Volume of Rain Water Inflow

The rainfall in Bonaire shows a considerable annual variation; 66 percent of the rain falls from October through January. The average rainfall is 51 centimeters. Considerable variation also exists from year to year. During the period 1940-1955, there were two years in which the rainfall was less than 30 centimeters and two years in which the rainfall was over 90 centimeters. Figure 11 is a plot of the mean monthly rainfall during the period 1903-1933 and of the excess of Pekelmeer evaporation over precipitation. We shall assume that the evaporation rate is constant during a year.

In addition to the direct rainfall on the Pekelmeer itself, a certain amount of rain water drains into the Pekelmeer. In the absence of data, we shall assume that about half of south Bonaire, or 15 square kilometers, drains into the Pekelmeer. For hills in a desert area in Israel, Evenari et al. (1961) have estimated that the coefficient of runoff is 15 to 20 percent. Since the relief on Bonaire is considerably less, we shall assume a runoff of 10 percent. The annual runoff into the Pekelmeer is therefore 15 $\times 10^{10}$ cm^2 $\times$ 51 cm $\times$ 0.1 = 77 $\times 10^{10}$ cm^3. This runoff is distributed over the year in proportion to the rainfall.

Evaporation rate

Bonaire has a mean annual temperature of 27°C, with an average minimum and maximum of 24.5°C and 30.0°C, respectively. An almost constant trade wind blows from N 85°-95° E with an average velocity of 3.9-4.9 meters per second.[6] The average relative humidity is 75 percent. The fresh water evaporation rate, as measured on Curacao, is 230 centimeters per year. Sling psychometer measurements by the present authors indicate that the air was 64 percent saturated with fresh water and 77 percent saturated with respect to the vapor pressure of water from the Pekelmeer. We shall

[6] The meteorological data were taken frcm Westermann and Zonneveld (1956).

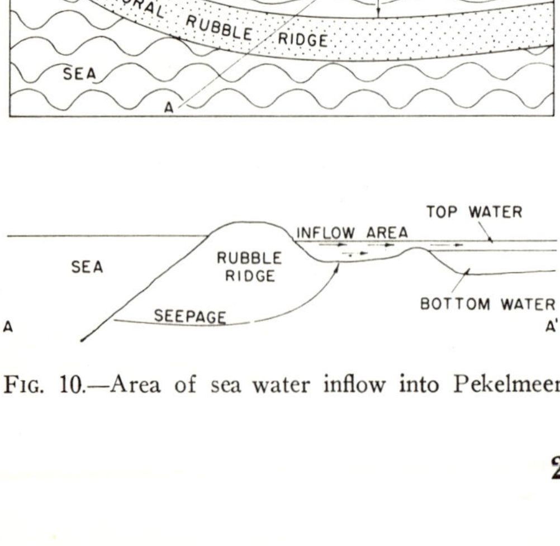

FIG. 10.—Area of sea water inflow into Pekelmeer.

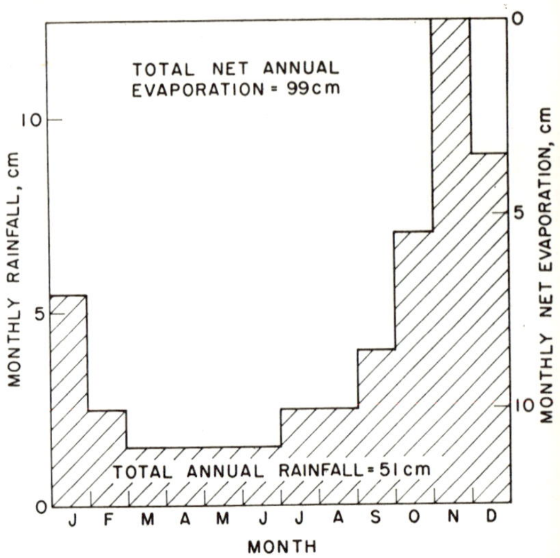

FIG. 11.—Monthly average rainfall on Bonaire, Netherlands Antilles (from Westermann and Zonneveld, 1956) and net evaporation rate.

therefore estimate that the rate of evaporatior from the Pekelmeer is 150 centimeters per year

Reflux through Bottom of Pekelmeer

Assuming an average area of 3 $\times 10^{10}$ cm^2 fo the Pekelmeer, we obtain the following exces of evaporation over precipitation:

Total evaporation	450 $\times 10^{10}$ cm^3	
Rain water, direct		153 $\times 10^{10}$ cm
Rain water, runoff		77 $\times 10^{10}$ cr
	450 $\times 10^{10}$ cm^3	230 $\times 10^{10}$ cm
Excess of evaporation	220 $\times 10^{10}$ cm^3	

In order for the level of the lake to remai constant over the years, this deficit in wate must be equal to the difference between the se water influx and the brine reflux. In order fo the salinity of the lake to remain constant, th reflux must carry out of the lake a quantity o salt equal to the inflow. Since the concentratio of the lake water is about seven times that o sea water, the volume of reflux must be one seventh of the sea water influx. We therefor obtain the following annual volumes:

Sea water influx	257 $\times 10^{10}$ cm^3
Brine reflux	37 $\times 10^{10}$ cm^3
Net water addition	220 $\times 10^{10}$ cm^3

For comparison, the measured sea water influ is 365 $\times 10^{10}$ cm^3 per year.

The dynamics of the Pekelmeer requires th about 36 $\times 10^{10}$ cm^3 of brine reflux annuall Since the Pekelmeer is six times as long as it

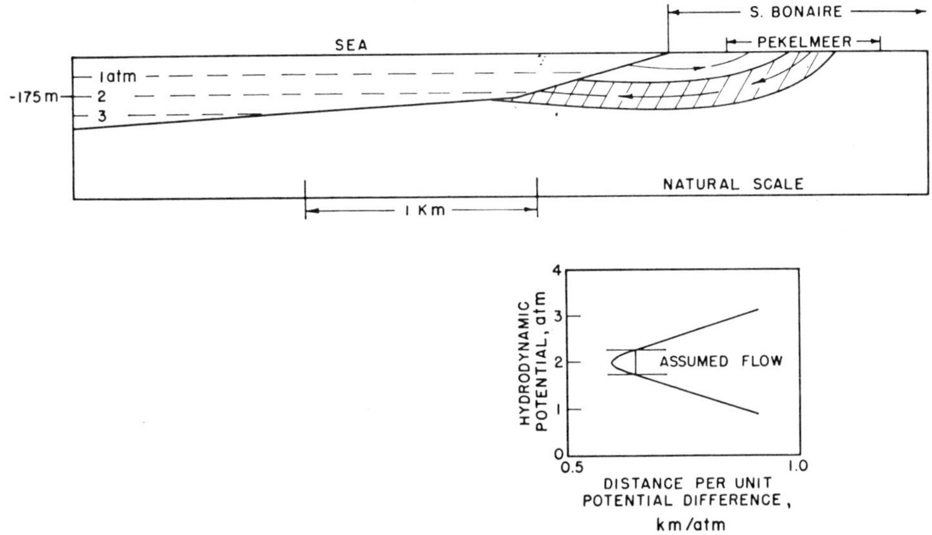

FIG. 12.—Hydrology of brine reflux.

wide, we can approximate the outflow as a two-dimensional pattern. Figure 12 is a cross section through south Bonaire perpendicular to the long axis of the Pekelmeer. Per cm thickness, 6×10^5 cm^3 per year must flow through this cross section. This is about 2×10^{-2} $cm^3/sec \cdot cm$ thickness.

The driving force behind this outflow is the difference in density between the Pekelmeer water and the sea water. The densities are 1.14 and 1.02 g/cc, respectively, giving a difference of 0.12 g/cc. The refluxing brine will first fill up the available pore space beneath the Pekelmeer down to the first impermeable layer. After the pore space is largely filled, an approximation to the steady-state flow is given by the following analysis.

Since the average level of the Pekelmeer is about 30 cm below mean sea level, 2.85 m of sea water is required to balance 2.55 m of Pekelmeer water. Below this depth the hydrodynamic potential of the Pekelmeer water will increase at a rate of 1 atm per 86 m of depth in the sea. This is indicated in figure 12. To determine where the maximum flow will occur, the distance from the center of the Pekelmeer to the point at which the equipotentials intersect the sediment has been divided by the potential. It can be seen that the minimum distance per atm potential difference occurs at a potential of 2 atm. We shall assume that the flow is uniformly distributed in a 100-m zone about a line joining the Pekelmeer to the 2-atm potential at 175 m below sea level.

The rate of the reflux is 2×10^{-6} $cm^3/sec \cdot cm^2$. The brine will thus require about 2000 years to flow from the Pekelmeer back to the sea. The viscosity of the brine is about 2 cp. Thus, the average permeability of the sediments through which the reflux flows must be about

$$\frac{2 \times 10^{-6}\ \text{cm/sec} \times 1.2 \times 10^5\ \text{cm} \times 2\ \text{cp}}{2\ \text{atm}} = 0.24\ \text{darcy}.$$

For comparison, the permeability of the loose offshore sediment, after repacking in the laboratory, ranged from 70 to 45 darcies. (The permeability of the Plio-Pleistocene inclined limestones ranged from 5 darcies down.)

The actual flow pattern and permeability distribution in the area of reflux are likely to be complex. The purpose of the calculation presented above is merely to obtain an estimate of the order of magnitude of permeability required to maintain the reflux at the required rate.

Geometry of the Pekelmeer

The dynamic behavior of the lake will depend to a large extent on the variation of surface area with depth. Because of the low relief, the area increases rapidly as the lake level rises. We shall approximate the area A as a simple function of the depth z:

$$\frac{1}{A} = [0.875 - 0.011z\ (\text{cm})] \times 10^{-10}\ \text{cm}^{-2}.$$

For the later computations, it is convenient to use a linear function in 1/A. The function was obtained by measuring the area of the lake on the map and estimating the increase in area with a 20-cm rise from the topography. The

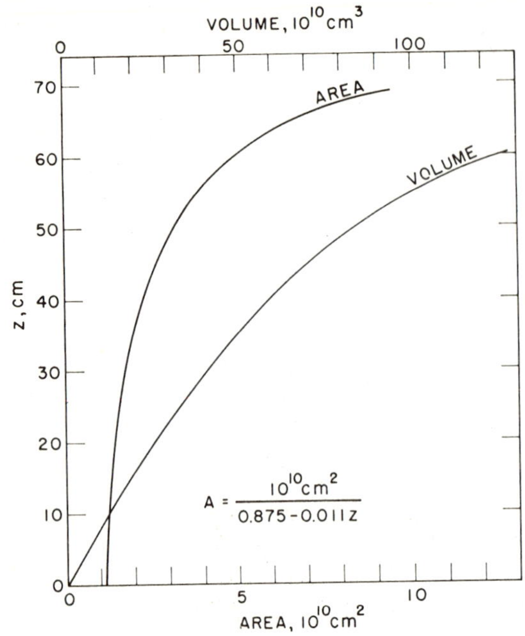

FIG. 13.—Geometry of Pekelmeer model, interrelating lake depth (z), area, and volume.

area and volume as a function of z are plotted in figure 13. We shall assume that the lake level during our visit to Bonaire was 50 cm and that mean sea level was 80 cm above the bottom of the lake.

Dynamics of the Lake Level

The time rate of change of the level of the Pekelmeer dz/dt (in cm per month) is the sum of the following terms:

1) The negative of the excess of evaporation over precipitation $-E(t)$. This is a function of the month of the year only and is independent of the level of the lake.

2) The volume of the monthly runoff $R(t)$ divided by the area of the lake $R(t)/A(z)$. This is a function of both the month and the level of the lake.

3) The negative of the volume of brine reflux divided by the area of the lake. The volume of reflux should be independent of time and the level of the lake. We shall assume it to be 3×10^{10} cm^3 per month, giving for the term due to reflux

$$\frac{-3 \times 10^{10} \text{ cm}^3}{A(z) \text{ cm}^2}.$$

4) The influx of sea water. The rate of sea water influx should be proportional to the difference between mean sea level and the level of the lake. Since we assumed that mean sea level is 80 cm above the bottom of the lake, the monthly influx of sea water is $0.7 \times (80-z) \times 10^{10}$ cm^3. The change in lake level due to this influx is then

$$\frac{0.7\,(80-z) \times 10^{10} \text{ cm}^3}{A(z) \text{ cm}^2}.$$

Adding all these terms together and remembering that

$$\frac{1}{A} = (0.875-0.011z) \times 10^{-10} \text{ cm}^{-2},$$

we obtain

$$\frac{dz}{dt} \text{ (cm per month)} = -E(t) + [R(t)-3](0.875-0.011z) + 0.7\,(80-z)\,(0.875-0.11z),$$

where $E(t)$ is in cm per month, and $R(t)$ is in units of 10^{10} cm^3. An initial value of z must be so chosen that, after integrating this expression over 1 year, one returns to the original value of z. Once z has been obtained as a function of

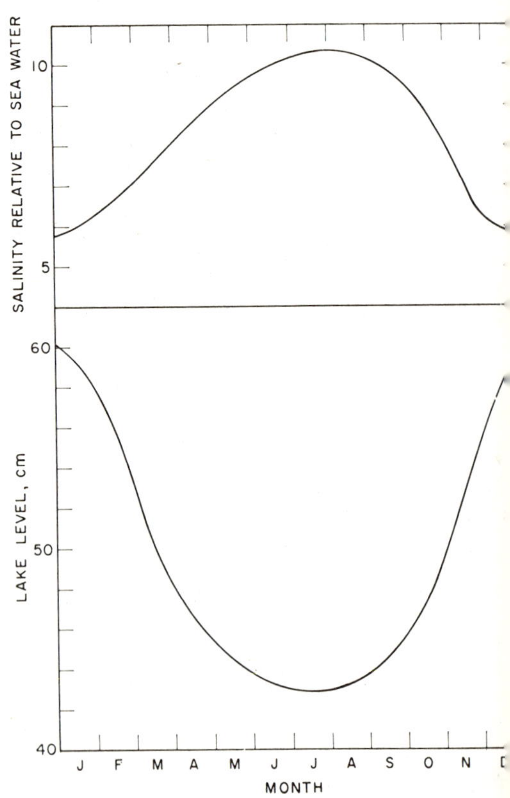

FIG. 14.—Dynamics of Pekelmeer model.

time, a similar procedure can be employed to determine the concentration as a function of time. The above calculations have been carried out approximately, and the results are shown in figure 14. In view of the crude nature of the model, a more exact numerical analysis is not warranted. The data indicate that, for the assumptions made, the lake level fluctuates between 43 and 60 cm, and the salinity varies from six to ten times that of normal sea water. The model shows that the lake varies considerably in response to annual rainfall variation. The chemical processes of such a lake will depend on these variations.

Although we have singled out the Pekelmeer for study, it is clear that evaporation and the production of hypersaline water are not restricted to the lakes of south Bonaire. Evaporation also takes place from the capillary water held in the exposed sediment. As a result, gypsum is precipitated in the supratidal sediments, and the hypersaline brines that are produced reflux downward to be replaced by less dense water.

To provide a qualitative insight into reflux hydrology, flow model experiments were conducted in a 1- by 2-foot plastic tank ½ inch wide. The model presented here (fig. 15) consists of the open "sea" (dark gray), a permeable sediment composed of cryolite sand (initial-

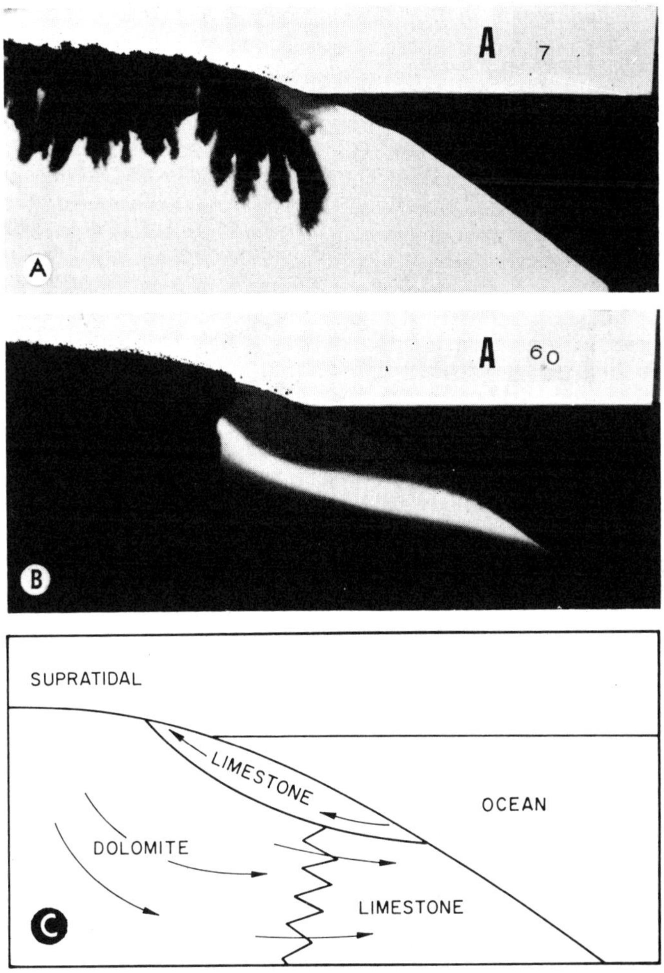

Fig. 15.—Model reflux experiments.

A. Initial downward displacement of dense water after 7 minutes.
B. Steady-state flow after 60 minutes.
C. Hypothetical relationship between supratidal, refluxing hypersaline water, influxing ocean water, and dolomitization, based on model experiments and observations on Bonaire.

ly white, and an area of evaporation or "supratidal area" slightly above sea level. The effect of evaporation is duplicated by sprinkling the surface of the supratidal area with chromium nitrate novahydrate [$Cr(NO_3)_3 \cdot 9H_20$], whose saturated solution has a density of 1.45. As the water is drawn to the surface of the supratidal area by capillary forces, it dissolves the chromium nitrate novahydrate, increasing the density of the water, similar to the increase in density due to evaporation of sea water. The salt also dyes the water black; thus, the flow of water which has reached the surface can be seen. In figure 15A the experiment is 7 minutes old and shows the black tongues of dense water beginning to flow down and the inflow of light gray ocean water to replace the downward flow. In figure 15B the experiment is 60 minutes old and has reached a steady state. The black water is flowing down and out into the "ocean" while the light gray ocean water is flowing up and into the area of evaporation. Figure 15C shows the model flow pattern and the hypothetical dolomitization pattern which we would expect if this flow pattern were to persist for a sufficient length of time.

RATE OF PRECIPITATION IN THE PEKELMEER

The mineral gypsum is the main component of the sediment in the hypersaline lakes and is common in other sediments. The gypsum in the Pekelmeer is at least 1 meter thick, the depth limit of the coring equipment. It is present in typical prismatic crystals which range from about 0.1 mm to about 12 mm in length (fig. 16). In some cases the crystals are oriented parallel to the bedding, whereas in others they are randomly oriented or are oriented perpendicular to the bedding. In some cases they are clumped together into pelletoids. The gypsum sediment is laminated in some samples and is churned in others. Several thousand flamingos live on the shore of the Pekelmeer and feed by pumping its sediment through their mouths (Allen, 1956). What effect the flamingos have on the gypsum other than reworking the sediment is unknown.

From the chemical considerations, we find that 0.9 millimoles of aragonite and 8 millimoles of gypsum are precipitated per kg of sea water evaporated. In our hydrological model the yearly rate of evaporation of sea water was about 90 cm, resulting in a yearly precipitation of 0.08 and 0.7 millimole of aragonite and gypsum, respectively, per cm^2. If a porosity of 50 percent is assumed, this precipitate would form a yearly layer 1.1 mm thick.

A core taken near the center of the Pekelmeer permits us to compare the above estimate with the core material. Core 11B, which was 53 cm long, was cut into seven pieces. Sections 2, 4, and 6 were analyzed for porosity and for carbonate and gypsum content. In addition, an age determination on the combined sections 1-2 and 6-7 was made with the carbon-14 method on the carbonate fraction in the core. The results are given in figure 17. The gypsum-to-carbonate mole ratio was found to be somewhat smaller (5.7, 7.1, and 6.5 rather than 9) than that predicted from the chemical considerations. The excess of aragonite may be due to two causes:

1) Calcium carbonate other than that precipitated locally may be added. As this carbonate may be old, the addition would result in C^{14} age that is older than the carbonate precipitate.

2) After deposition, the brines may precipitate further carbonate in the sediments. This addition of more recent cement would give an age that is younger than the actual sediment.

If an appreciable amount of old carbonate is added to the sediment, the age of the surface should be older than expected. Actually, the date of sections 1 and 2 was 110 $\pm$130 years, indicating that at least in that section of the core the addition of old carbonate was not significant. The age of the bottom two sections was 710 $\pm$130 years. If we accept that age as correct, the average rate of sedimentation is 0.6 mm per year, compared with our estimate of 1.1 mm. If we assume that the excess carbonate in the bottom section is later cement having an average age equal to one-half the age of the sediment, we obtain a rate of sedimentation of 0.5 mm per year.

It is of interest to compare the rate of gypsum precipitation in the Pekelmeer with inferred rates in the geologic past. Udden (1924) has studied the laminated anhydrite in the Castile Formation of the Permian in west Texas. He concludes that the laminations, which consist of broad anhydrite bands alternating with dark thin bituminous layers of calcite and dolomite represent yearly cycles. The thickness of the layers ranges from 0.2 to 7 mm and generally is between 0.5 and 2 mm. The average thickness is 1.63 mm, of which about 1.3 mm is anhydrite.

If the bands represent a yearly cycle, 2.8 millimoles of calcium sulfate was precipitated per year per cm^2, which is about four times the rate of precipitation in our model and eight times the observed rate in the Pekelmeer. If this precipitate is the result of evaporating normal sea water, a net average rate of evaporation of 360 cm of sea water per year would be implied.

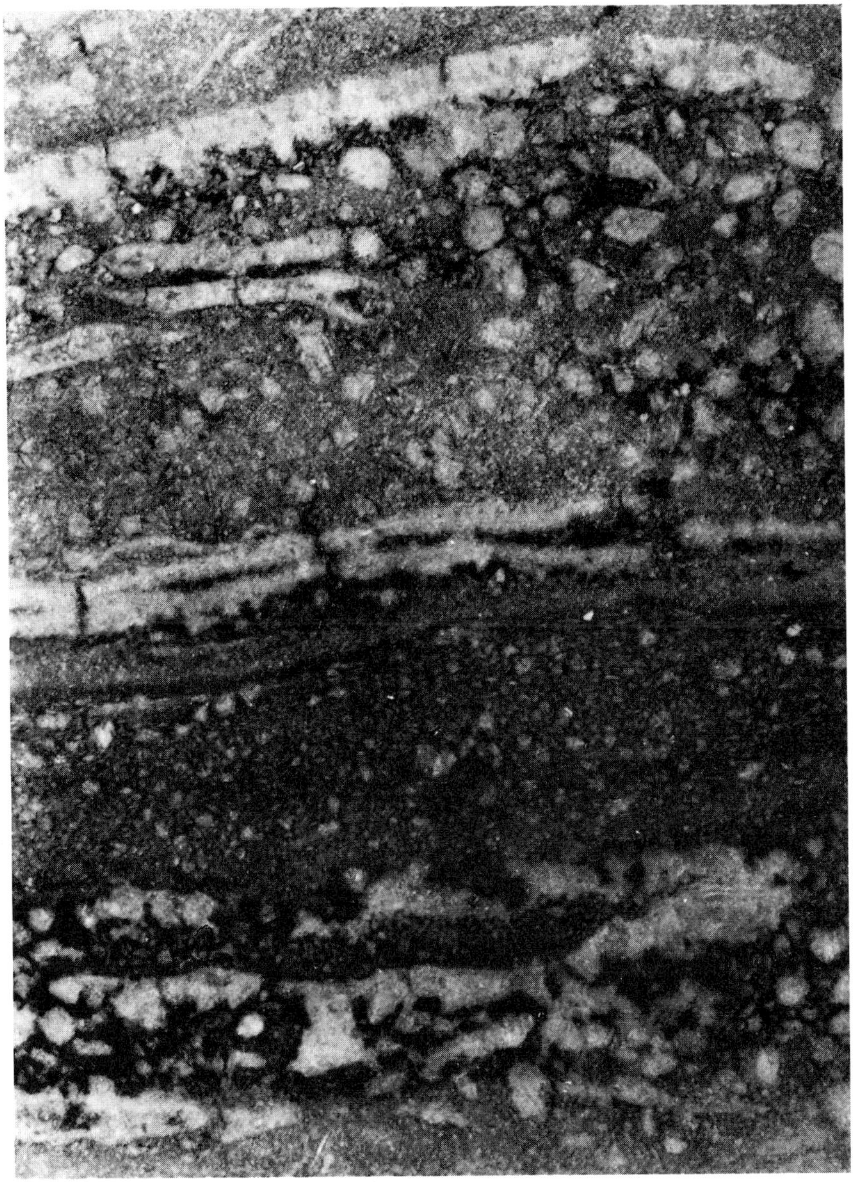

FIG. 16.—Impregnated core from Pekelmeer (location 11, Fig. 2) showing crusts and loose crystals of gypsum.

THE PLIO-PLEISTOCENE DOLOMITE

The study of south Bonaire showed that dense brines having a high magnesium-to-calcium ratio are produced by the evaporation of marine water in the supratidal environment. These brines must reflux down through the underlying sediments. The fact that these brines can dolomitize limestones is suggested by chemical arguments and by the occurrence of Recent dolomite on south Bonaire. It is then of interest to inquire if the same mechanism could explain the dolomitization in the inclined beds of north Bonaire.

Outcrop and thin-section examination shows that the dolomite in the inclined beds is of secondary origin, as shown by replacement of cor-

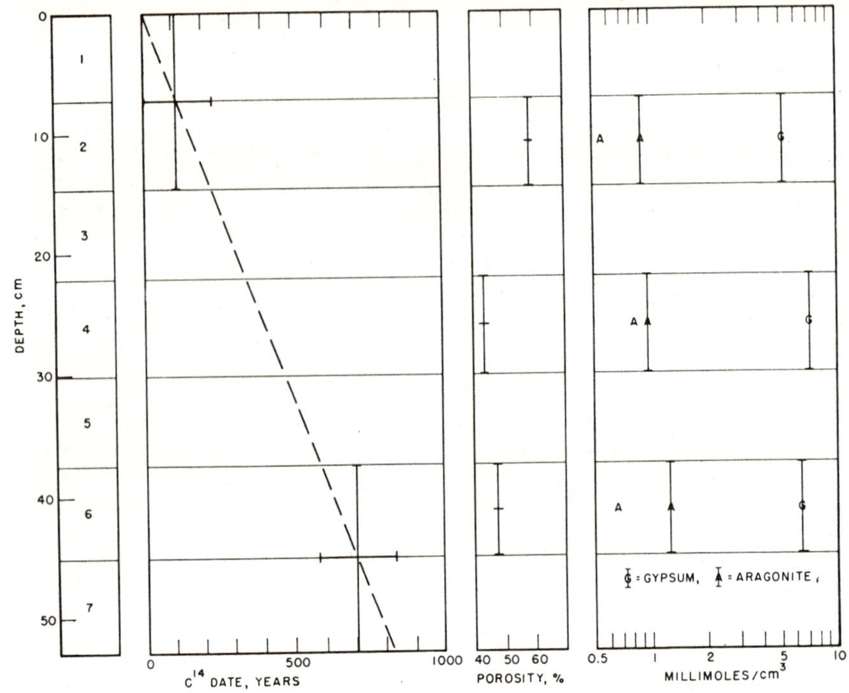

FIG. 17.—Analysis of core 11.

als. The average size of the dolomite rhombs is about 75 microns. The distribution of dolomite in the inclined beds was mapped in two areas, Dos Poos and Dochila (see figure 2). Although the outcrops were an oblique section, mapping indicated that dolomite changes to limestone in a downdip direction. This suggests that the dolomitizing waters moved downdip through the inclined marine lime sands some time after deposition.

Unfortunately, erosion has destroyed the sediments above and inland from these inclined deposits, but it can be reasonably inferred that an island existed at the time of deposition of these sediments which was somewhat smaller than the present island but of the same general type. This island could have had a supratidal environment similar to that of south Bonaire today, as reconstructed in figure 18. The source of the magnesium for the dolomite was probably the magnesium in the ocean, because that was the only significant source present. However, the sea water must have been altered; otherwise all the inclined sediments would have beer dolomitized.

The Pleistocene topography of north Bonaire could have had an evaporating surface of 10 km². A yearly evaporation rate of 100 cm would evaporate a total of 10^{10} kg of sea water per year. This would supply 5×10^8 moles of magnesium ion to the reflux. If this water is used to convert calcite to dolomite, changing the Mg/Ca ratio from about 30:1 in the hypersaline water back to a ratio of 5:1 would produce about 5×10

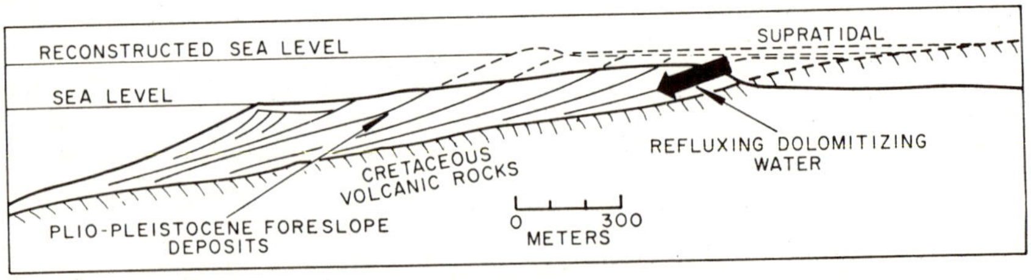

FIG. 18.—Cross section of north Bonaire showing reconstructed Plio-Pleistocene supratidal area.

cm^3 of dolomite per year. A rough estimate indicates that the inclined beds contain about 0.5 cubic kilometer (5×10^{14} cm^3) of dolomite. Therefore, the time required to form the estimated volume of dolomite is of the order of 10^5 years.

Although these estimates do not prove that the inclined beds were dolomitized by the proposed mechanism, they suggest that the mechanism is adequate for the time available. We are not aware of any other mechanism of dolomitization that can explain the observed distribution and quantity of dolomite.

CONCLUSION

The Recent dolomites of south Bonaire are similar to many ancient dolomites described by most authors as "penecontemporaneous" or "primary." These ancient dolomites are characterized by some or all of the following criteria: very restricted faunas, fine grain size, an association with gypsum, the presence of thin extensive dolomite beds, and dolomite clasts broken from these beds. The Plio-Pleistocene dolomites of Bonaire are similar to rocks often described as "secondary" or "replacement" dolomites. These are usually characterized by a normal marine fauna, grain size in the range from 40 to 80 microns, and a pattern of dolomitization which cuts across the depositional bedding. We have put forward the hypothesis that the Plio-Pleistocene dolomites on Bonaire resulted from dolomitization by downward-flowing hypersaline waters. The descriptions given in the literature of many ancient replacement dolomites are not inconsistent with the hypothesis that these dolomites were also formed by the reflux of hypersaline waters produced by the partial evaporation of sea water. In fact, this was the explanation put forward by Newell et al. (1953) for the dolomite in the Permian Reef Complex. In order to increase significantly the magnesium-to-calcium ratio of the hypersaline brines, gypsum must first be precipitated. The gypsum deposits so formed are susceptible to dissolution either by sea water or by fresh water and are therefore not always preserved in the geologic record. The dolomites in the cores from the Pacific atolls can be explained by the hypersaline reflux mechanism if the lagoons were temporarily converted to areas of evaporation. The high-magnesium waters so formed would dolomitize the underlying sediments, and the gypsum precipitated would later be removed by dissolution or erosion.

We certainly do not wish to suggest that all ancient dolomites were formed by the interaction of sea water concentrated by evaporation with lime sediments. The hypersaline reflux mechanism combines a source of high-magnesium water with natural driving force to move this water through the sediments. The only environmental requirements are a seasonally or permanently dry climate, so that evaporation may exceed precipitation, and a nearly flat sediment surface near sea level to provide a supratidal environment of sufficient areal extent. It is highly probable that these requirements have been met many times in the geologic past.

ACKNOWLEDGMENTS

The authors wish to acknowledge the help of the staff of Shell Curacao, N.A., in arranging the field work. In particular, the assistance of Mr. J. W. de Vries in making arrangements and the field guidance of Mr. P. A. van der Werf on Curacao are sincerely appreciated. The work of Mr. L. D. Gerharts in arranging for the field work on Bonaire was most helpful. The field assistance of Mr. Eric Weyl was invaluable. The carbon and dates were determined by Dr. E. L. Martin, Jr., and the chemical analyses were supervised by Dr. H. F. Young. Discussions with Dr. R. C. Murray and his review of the manuscript were of great help in preparing this paper.

REFERENCES

ADAMS, J. E., AND RHODES, M. L., 1960: Dolomitization by seepage refluxion, Am. Assoc. Petroleum Geologists Bull., v. 44, no. 12, p. 1912-1920.

ALLEN, R. P., 1956, The Flamingos: Their Life History and Survival, National Audubon Society: Research Report No. 5, New York, 285 pp.

CLARKE, F. W., 1924, Data of Geochemistry: U.S. Geol. Surv., Bull. 770.

EVENARI, M., SHANAN, L., TADMOR, N., AND AHARONI, Y., 1961, Ancient agriculture in the Negev, Science: v. 133, p. 979-996.

FAIRBRIDGE, R. W., 1957, The dolomite question: *in* Regional Aspects of Carbonate Deposition—A Symposium (LEBLANC, R. J. and BREEDING, J. G., editors), Soc. Econ. Paleontologists and Mineralogists Spec. Pub. No. 5, p. 125-178.

GOLDSMITH, J. R., GRAF, D. L., AND HEARD, H. C., 1961, Lattice constants of the calcium-magnesium carbonates: Am. Mineralogist, v. 46, p. 453-457.

GRAF, D. L., EARDLEY, A. J., AND SHIMP, N. F., 1961, A preliminary report on magnesium carbonate formation in Glacial Lake Bonneville: Jour. Geology, v. 69, p. 219-223.

HALLA, F., AND RITTER, F., 1935, Eine Methode zur Bestimmung der freien Energie bei Reaktionen des

Typus A(s) + B(s) = AB und ihre Anwendung auf das Dolomitproblem: Z. Physik. Chem. (Leipzig), vol. 175, 465-467.
Logan, B. W., 1961, *Crypotzoon* and associate stromatolites from the Recent, Shark Bay, Western Australia: Jour. Geology, v. 69, p. 517-533.
Martin, K., 1888, Bericht über eine Reise nach Niederländisch Westindien und darauf gegründete Studien: Leiden, Geologischer Teil.
Murray, R. C., 1960, Origin of porosity in carbonate rocks: Jour. Sed. Petrology, v. 30, p. 59-84.
Newell, N. D., et al., 1953, The Permian Reef Complex of the Guadalupe Mountains Region, Texas and New Mexico—A Study in Paleoecology: W. H. Freeman, San Francisco, 236 p.
Pijpers, P. J., 1933, Geology and Paleontology of Bonaire (N.W.I.): Dissertation, University of Utrecht.
Posnjak, E., 1940, Deposition of calcium sulfate from sea water: Am. Jour. Sci., v. 238, No. 8, p. 559-568.
Schaub, H. P., 1948, Geological observations on Curacao, N.W.I.: Am. Assoc. Petroleum Geologists Bull., v. 32, No. 7, p. 1275-1291.
Skinner, H. C. W., Skinner, B. J., and Rubin, M., 1963, Age and accumulation rate of dolomite-bearing carbonate sediments in South Australia: Science, v. 139, p. 335-336.
Udden, J. A., 1924, Laminated anhydrite in Texas: Geol. Soc. America. Bull. v. 35, No. 2 p. 347-354.
Westermann, J. H., and Zonneveld, J. I. S., 1956, Photo-Geological Observations and Land Capabilities and Land Use Survey on the Island of Bonaire (Netherlands Antilles): Royal Tropical Institute, Amsterdam.
Weyl, P. K., 1960, Porosity through dolomitization: Conservation-of-mass requirements: Jour. Sed. Petrology, v. 30, no. 1, p. 85-90.

Geochemistry of Evaporites

III

Gypsum–Anhydrite Relationships

One of the more interesting and prolonged arguments in geology has been that concerning the formation of gypsum and anhydrite. Ever since Usiglio identified both hydrous and anhydrous forms of calcium sulfate in the precipitates from his famous experiment, geologists have indulged in lengthy, repeated, and occasionally acrimonious discussions on the nature of the original form of those compounds. But, in the manner of all conflicts in science, new research arose from the fragmentary remains of early theories, and today we have a much more reasonable grasp of the chemical and geological relationships between gypsum and anhydrite, and are, therefore, better able to explain their origin.

This part is unlike all the others in this volume, for no individual papers were selected for inclusion in it. This is an unfortunate result of the nature of much of the research into this problem: the otherwise excellent papers on the gypsum–anhydrite equilibrium by Hardie (1967) and E-An Zen (1965) provide us with the chemical information that we need in order to understand the rocks, but give insufficient attention to the sedimentological features of those rocks, and the many, many papers dealing with the field occurrences of gypsum and anhydrite pay scant regard to the rigors of chemistry in the calcium sulfate system. It seemed more useful, therefore, to cull the relevant information from the mass of detail and provide the reader with introductory reading on the nature of the initial deposit in the calcium sulfate evaporites.

The conclusions reached by examination of the rocks in the field and in the laboratory seemed to depend as much on the mineralogy itself as on sound geological reasoning: where gypsum was found alone, no one questioned for a moment that it was the original precipitate (McKee, 1938; Brill, 1952); where the gypsum could be shown to be a weathering product of deeper anhydrite, anhydrite was the original deposit (Rogers, 1910; Muir, 1934; Sellards and Baker, 1934; Jones, 1935; Dorheim and Campbell, 1958); and the occurrences of anhydrite at depth were the result, too, of precipitation of anhydrous calcium sulfate (Udden, 1924; King, 1947; Dellwig, 1955). Few geologists, indeed, considered the possibility that under different circumstances of environment, either one or the other might form (see, however, Kulstad et al., 1956, Douglas and Goodman, 1957); fewer still were aware that to demonstrate that gypsum was derived from anhydrite was not tantamount to proving that anhydrite was the original precipitate. Both Bundy (1956) and Murray (1964) have demonstrated that a cycle from gypsum to anhydrite to gypsum is the most likely history of surface gypsum deposits, and this conclusion is widely accepted.

If geology was proved inadequate for the task, perhaps chemistry would be able to resolve the dispute. The exact relationships among the calcium sulfate minerals have, therefore, been the subject of intensive experimentation and calculation. The simple binary system $CaSO_4$–H_2O was studied by van't Hoff et al. (1903), Partridge and White (1929), Toriumi and Hara (1934), Hill (1934), Posnjak (1938), and D'Ans et al. (1955); the effect of the addition of NaCl to that system was considered in the work of van't Hoff et al. (1903), D'Ans et al. (1955), Madgin and Swales (1956), Bock (1961), and Zen (1965); the gypsum–anhydrite equilibrium in seawater was studied experimentally by Toriumi et al. (1938) and Posnjak (1940); and the transition temperature has been calculated by Kelley et al. (1941) and MacDonald (1953). Although, as expected, there is no universal agreement, certain regularities in the gypsum–anhydrite relationship are accepted:

1. The transition temperature in pure water is between 40 and 55°C, anhydrite being the stable form above that temperature; gypsum the stable form below (Hardie, 1967).

2. The transition temperature for gypsum–anhydrite in a brine decreases continuously with increasing concentration so that at 5 or 6 times the seawater concentration, the transition temperature may be less than 35°C (Toriumi and Hara, 1934; D'Ans et al., 1955; Madgin and Swales, 1956).

3. Gypsum is extremely persistent in the metastable state so that if conditions of temperature and concentration fall within the anhydrite field, this means only that anhydrite will not convert to gypsum, and not that anhydrite will precipitate, even when seeding of solutions has taken place (Zen, 1965; Cruft and Chao, 1970). Field data, however, do indicate that anhydrite can be formed, probably diagenetically, from brines whose composition falls just barely within the anhydrite stability field (Kinsman, 1966; Butler, 1970a, 1970b).

It is, therefore, by no means certain that anhydrite can precipitate from concentrated marine brines. However, in general the *conversion* of gypsum to anhydrite is favored by increasing temperature and increasing salinity. These conclusions are summarized in the phase diagram of Fig. 1, and in Fig. 2, which is based on the work of Hardie (1967) and the field data of Kinsman (1966).

In the study of the calcium sulfate evaporites, much attention should be paid to the petrographic features that they display. Chemical data indicate that it would have to be an unusual environment to produce anhydrite, yet there are certain indications that anhydrite may sometimes have been a primary mineral (Ogniben, 1957;

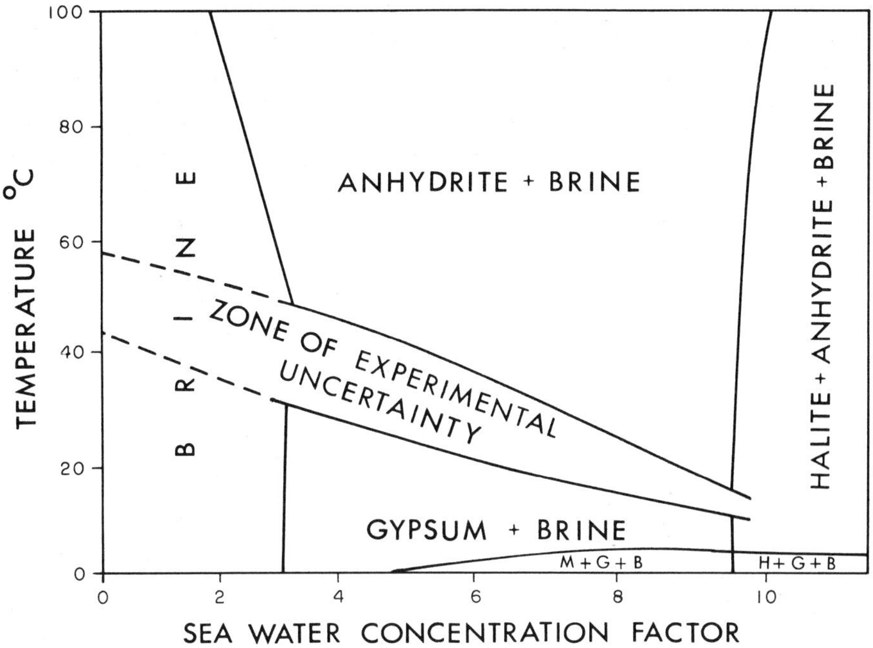

Figure 1. Gypsum–anhydrite equilibrium relations. The upper limit of the "zone of experimental uncertainty" is the boundary defined by the temperature and activity $_{H_2O}$ data of Hardie (1967); the lower limit is that drawn on the basis of solubility measurements in pore water and sea water. M, mirabilite; G, gypsum; B, brine; H, halite.

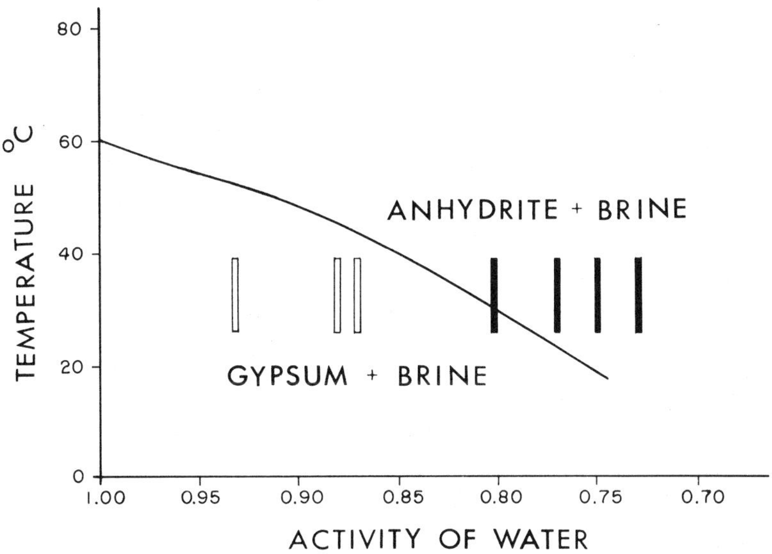

Figure 2. Temperature and activity$_{H_2O}$ relations of brines coexisting with gypsum and anhydrite in sabkhas. The equilibrium curve is that of Hardie (1967), and the field data are from Kinsman (1966). In seawater, activity$_{H_2O}$ is a function of salinity, so that in the diagram above, high salinities are represented by the lower values of activity$_{H_2O}$. Open bar, gypsum; solid bar, anhydrite; length of bar represents annual temperature range of brine within sabkha sediments.

Stewart, 1949). The wedding of chemistry and geology may supply the progeny that will finally resolve the problem, as it has in so many other controversial subjects within our science.

Editors' Comments on Papers 28 and 29

Bromine

From the moment that they are exhaled in volcanic gases until the time that they accumulate in seawater, chlorine and bromine accompany each other in the geochemical cycle. In the formation of evaporites, however, these elements go separate ways because chlorine is more easily fixed in crystallizing solids than is bromine, which tends to accumulate in the residual brine. The timely fractionation of these

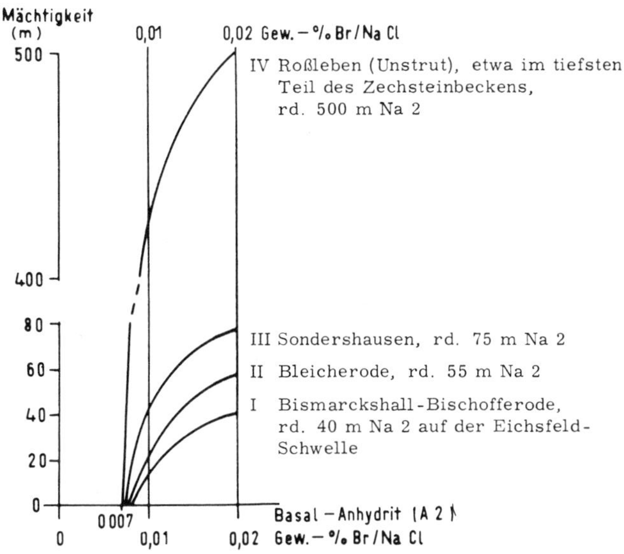

Figure 3. Normal bromine profile in the Stassfurt halite (Na 2) beneath the Stassfurt potash layer (K 2) south and southeasterly of the Harz; after Baar (1954).

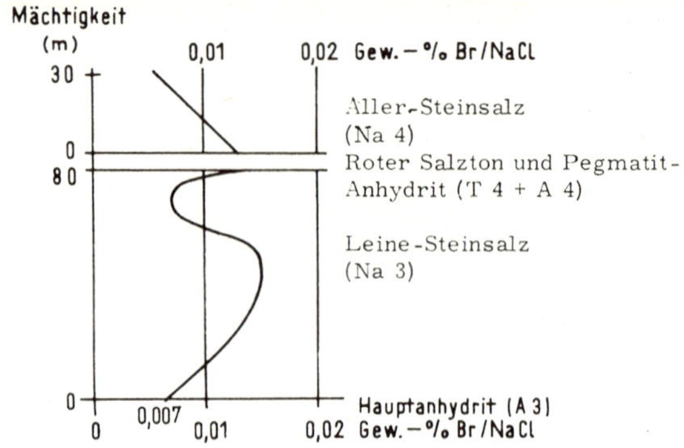

Figure 4. Normal bromine profile in the Leine halite (Na 3) and of the Aller halite (Na 4) in the Zechstein in the region of Schöneback (Elbe); according to information from Schulze (1960).

halogens is of value to geologists because the ratio of Cl^- to Br^- in chlorides can be used to provide information about deposition and alteration of evaporites.

Since 1908, when Boeke reported on the halides of sodium and on the presence of bromine in potash minerals, the leaders in the study of bromine in evaporitive chlorides have been German and Soviet geologists. Much of the work was stimulated

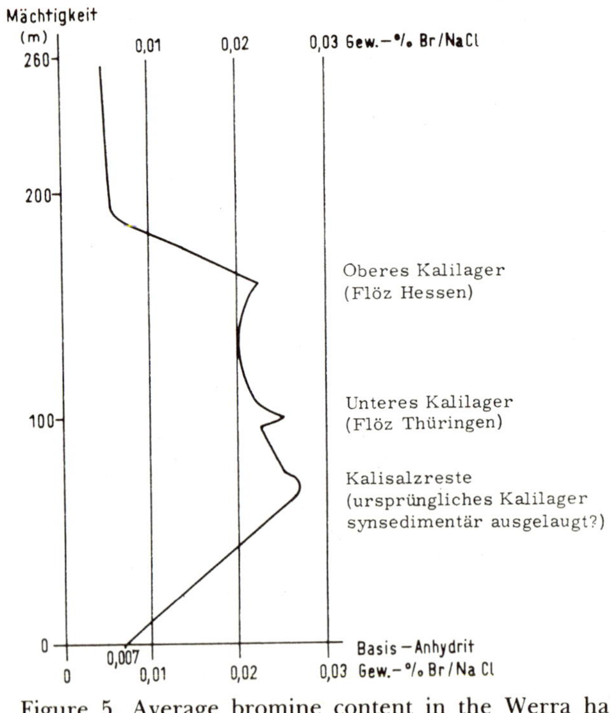

Figure 5. Average bromine content in the Werra halite (Zechstein 1) of the Werra basin. After D'Ans and Kühn (1940) (from the base to the upper potash layer), and Dittrich (1962).

by economic necessity, in Germany to overcome problems of stratigraphy in deformed salt domes, in the U.S.S.R. to locate new deposits of potash. Only recently have North American geologists begun to use bromine values in research on deposits in the Western Hemisphere (Schwerdtner and Wardlaw, 1963; Raup, 1966; Holser, 1966; Adams, 1969).

The basic theory behind the inclusion of a trace element in a chloride is well summarized in both papers included in this part. To gain an appreciation of the stratigraphic value of the systematic variation of value of Br in NaCl, it is instructive to compare the theoretical profiles in Holser's paper with the "real" profiles obtained from Stassfurt halite, in the second Zechstein cycle, reprinted here from Baar (1963) (Fig. 3); then to consider the fluctuations in profiles taken from halite associated with or adjacent to potash units (Figs. 4, 5, and 6).

The inclusion of only one paper by M. G. Valyashko in this volume is an inadequate tribute to a man who has studied the geochemistry of saline lakes and deposits for more than 35 years. Since he graduated in 1930, he has had a distinguished career, serving for 20 years as the chief of the Geochemical Laboratory of the All-Union Scientific Research Institute in Leningrad, and acting as Vice-Principal of the Geochemistry Department of the Geological Faculty at the M. V. Lomonosov Moscow State University since 1957. Among his 180 publications on various aspects of geochemistry are four monographs on saline lakes and evaporites, including the 1962 volume *Geochemical Laws Governing the Formation of Salt Deposits.*

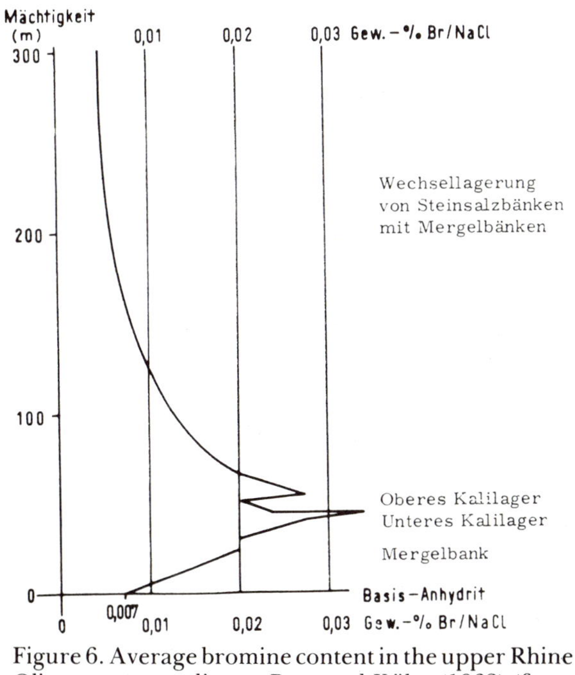

Figure 6. Average bromine content in the upper Rhine Oligocene; according to Baar and Kühn (1962) (from the base to the upper potash layer), and Manger (1961).

Included in Valyashko's paper is a very emphatic outline of how the content of bromine in halite may be used in prospecting for potash. His confidence, of course, is the confidence of any successful prospector or explorer. Holser's paper, more academic in thrust, concerns itself with the problems yet to be solved in this branch of geochemistry, and the tone of his final statement is one of bewilderment and humility in the face of 50 years' work that has not been able to supply all the answers.

Reprinted from *Geochemistry*, **6**, 570–589 (1956)

GEOCHEMISTRY OF BROMINE IN THE PROCESSES OF SALT DEPOSITION AND THE USE OF THE BROMINE CONTENT AS A GENETIC AND PROSPECTING CRITERION*

28

M. G. Valyashko

All-Union Institute of Halurgy, Leningrad

Bromine occupies a special place among the elements composing the salts of the ocean. Unlike the other elements, some of which are present in even smaller amounts (for example, boron), bromine does not form independent minerals when water is evaporated, but is precipitated as a diadochic admixture in chlorides.

It is well known that halite is the first chloride to be precipitated; as concentration proceeds, it is joined by sylvite, which later passes into carnallite or kainite. Kainite also eventually passes into carnallite. At the point of most intensive crystallization (eutonic point) these minerals are joined by bischofite. All these chlorides capture bromine and bring it into the solid phase.

The data of Table 2 show that, as the concentration of sea water proceeds, the percentage of bromine in it increases steadily in spite of the precipitation of the chlorides which transfer bromine into the solid phase.

During the evaporation of sea water, from the very beginning of crystallization of the first chloride, halite, bromine is distributed between the solution and one or several solid phases. This distribution is such that the amount of bromine passing into the solid phase is less than the amount remaining in the solution.

A number of investigators have studied the distribution of bromine between the crystals and the solution. These researches, begun by Boecke in 1908 [1], were continued and brought to completion by Yu. V. Morachevskii (1928-1950), V. I. Nikolaeev (1932-1936), N. A. Shlezinger, F. P. Zorkin, A. V. Novozhenova and others (1935-1939), S. K. Chirkov (1935-1946), V. P. Ilinskii, V. M. Filippeo and D. I. Sapirshtein (1928-1948), M. G. Valyashko, E. M. Petrova and T. V. Mandrykina (1943-1949).

In 1946 Chirkov summarized the existing knowledge on the subject and developed a theory of distribution of the bromine ion between the crystals and the solution. Omitting, therefore, a review of the history of the problem and of the work done in this field, we shall begin with the basic conclusions concerning the principles governing this process. The most important of these are:

*Paper read at the XXth International Geological Congress in Mexico.

Table 1. Composition of the Ocean, Seas and Some Lakes Reduced to Approximately the Same Concentration (12 B) [1] (% by wt.)

Basins	Sp.G.	°B	$Ca(HCO_3)_2$	$CaSO_4$	$MgSO_4$	$MgBr_2$	$MgCl_2$	KCl	NaCl	Sum of salts	Cl^-	Br^-	$\frac{Br^- \cdot 10^3}{Cl^-}$
Ocean	1.090	12.0	0.040	0.47	0.77	0.025	1.13	0.20	9.40	12.00	6.65	0.022	3.30
Mediterranean Sea	1.086	11.5	0.016	0.39	0.71	0.023	0.93	0.14	8.49	10.73	6.51	0.0204	3.14
Black Sea	1.076	10.2	0.030	0.33	0.67	0.019	0.93	-	8.18	10.16	5.69	0.016	2.82
Gulf	1.092	12.2	0.030	0.35	0.89	0.022	1.044	0.26	9.39	11.95	6.61	0.019	2.90
Lake No. 8	1.090	12.0	0.028	0.31	0.94	0.029	1.93	0.36	7.76	11.36	6.31	0.025	3.96
Lake No. 5	1.090	12.0	0.031	0.54	0.86	0.025	1.38	0.23	9.09	12.08	6.67	0.022	3.30

1. Bromine* enters diadochically into solid chlorides.

2. The content of bromine in a solid chloride is determined by the bromine content in the crystallizing solution.

3. Each chloride is characterized by its own coefficient of distribution

$$J = \frac{C_{solid}}{C_{s.r.}}$$

where C_{solid} is the bromine content in the crystals (wt.%); $C_{s.r.}$, the concentration of bromine in the dry residue (salt) of the remaining liquid phase (wt.%).

It was shown by Chirkov that for the concentrations of the bromide in natural brines and sea water, which are very low as compared with the concentrations of chloride, the distribution coefficient is constant.

In the distribution curve $C_{solid} = f(C_{s.r.})$, $J = \frac{C_{solid}}{C_{s.r.}}$ is the slope of the tangent to the distribution curve.

4. Experiments have shown that for all chlorides crystallizing from sea water, $J < 1$. The consequence of this is that, although the crystallization of a bromine-carrying chloride removes some bromine into the solid phase, the concentration of bromine in the solution increases.

5. If two chlorides into which bromine enters diadochically crystallize simultaneously, bromine is distributed as if only one chloride were present.

In other words, the bromine content in each chloride is determined by the distribution coefficient of each chloride and by the content of bromine in the solution.

These principles lead to the following conclusions of practical importance:

1. As sea water becomes concentrated, the bromine content in it increases

*Here and in what follows the bromine ion (Br^-) is meant.

gradually; and to each stage of concentration a definite bromine content corresponds.

2. During the crystallization of a chloride which always occurs over a definite interval of concentration, the bromine content in the first chloride precipitate is low and increases gradually with further crystallization, since the concentration of bromine is increasing in the sea water while the distribution coefficient remains constant.

3. With simultaneous crystallization of two chlorides from sea water the bromine content in each of them will be determined by the distribution coefficient of each chloride and the concentration of bromine in the solution.

4. The subsequent recrystallization of the chloride in the absence of the mother liquor will lead to the impoverishment of bromine in the chloride, since $J < 1$.

Thus, the bromine content in salt deposits and minerals is determined by their history and is an indicator of the conditions of their formation.

It is well known that the majority of at least the larger salt deposits formed as a result of evaporation of sea water. Therefore, all constituents of salt deposits may be regarded as products of crystallization from sea water. If the content of bromine characteristic of each stage of concentration of sea water is determined, then, knowing the distribution coefficients for each chloride, it is possible to determine

Table 2. Bromine Content and the Value of Bromine-Chlorine Ratio in the Products of Evaporation of Sea Water in Lake No. 5

Sp.G.	Degrees Baumé	Cl^-, %	Br^-, %	$\frac{Br \cdot 10^3}{Cl^-}$	Solid phase
1.01	1.0	1.01	0.0035	3.48	
1.075	10.0	5.49	0.018	3.28	
1.091	12.0	6.62	0.022	3.33	
1.108	14.0	7.77	0.026	3.22	
1.162	20.0	11.83	0.039	3.29	
1.199	24.0	14.33	0.047	3.28	
1.220	26.1	15.72	0.051	3.24	NaCl
1.231	27.0	15.57	0.083	5.34	NaCl
1.252	29.0	15.26	0.137	9.00	NaCl
1.273	31.0	15.29	0.179	11.70	NaCl
1.291	32.6	15.63	0.214	13.70	NaCl
1.300	33.4	15.77	0.226	14.30	NaCl
1.308	34.0	15.93	0.236	14.80	$NaCl + MgSO_4 \cdot 7H_2O$ (KCl)
1.315	34.6	17.93	0.296	16.50	
1.323	35.3	19.10	0.334	17.50	
1.325	35.5	19.23	0.342	17.80	$NaCl + MgSO_4 \cdot 6H_2O + KCl \cdot MgCl_2 \cdot 6H_2O$
1.331	35.9	20.97	0.421	20.01	
1.336	36.5	21.91	0.469	21.40	
1.350	37.5	23.10	0.516	22.40	
1.359	38.3	23.68	0.520	22.00	

the normal concentration of bromine which must exist in the chlorides separating from evaporating sea water and to establish the so-called *scale of normal bromine concentrations in the chlorides.*

This will make it possible, on the basis of the bromine content and its ratio to chlorine, to determine the stage at which a given salt was precipitated and also to discover whether recrystallization has taken place. The bromine content in halite may serve also as a guide to potash deposits.

The $\frac{Br^- \cdot 10^3}{Cl^-}$ ratio in the ocean and the seas is about 3.30 (Table 2).

The content of bromine in seas and oceans fluctuates, and occasionally lower values are recorded for individual stations and samples; but these fluctuations are probably caused by local conditions. The above value of the $\frac{Br^- \cdot 10^3}{Cl^-}$ ratio may be regarded as constant and characteristic of sea water [7], not only at present but over a long period of geologic time.*

BROMINE IN THE PRODUCTS OF CONCENTRATION OF SEA WATER

To determine the behavior of bromine during the process of concentration of sea water and its content at the different stages of the process, we used the materials of the Crimean salt station and our own experimental data on the natural evaporation of the waters of the Black and Azov Seas and of the brines of the coastal lakes fed by marine waters.

The Br^-/Cl^- ratio and the chemical composition of these waters is very near those of sea water, as can be seen from Table 1. The data selected from all materials investigated are presented in Table 2 and the graph (Fig. 1).† An examination of these data leads to the following conclusions:

1. During the evaporation of sea water, bromine accumulates in the liquid phase. At first its content is a linear function of the concentration of sea water.

2. The first break in the bromine content-brine concentration curve occurs at the moment when NaCl begins to crystallize. The second break marks the beginning of crystallization of epsomite, $MgSO_4 \cdot 7H_2O$; the third break corresponds to the appearance of carnallite, $KCl \cdot MgCl_2 \cdot 6H_2O$; and the last one, to the eutonic point (E), the beginning of crystallization of bischofite.

It is interesting to note that the intensity of accumulation of bromine during the process of evaporation of sea water increases continuously

*The last assumption is very important for further deductions. An analysis of the data on the bromine content in salt deposits (see below) shows that during a long period of geological history this ratio had a value close to the present.

†For details, see [2].

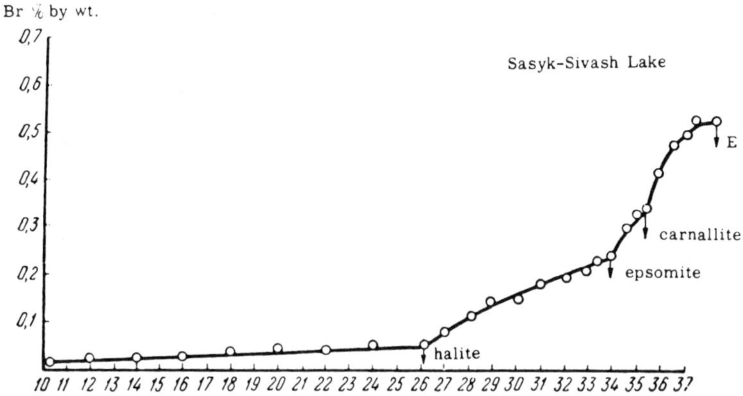

Fig. 1. Curve of bromine accumulation in the products of concentration of sea water.

and the curve becomes steeper with the appearance of each solid phase. This is due to the fact that the appearance of each new solid phase does not terminate the separation of the preceding phase. The new phase joins the preceding one hastening the decrease in volume of the liquid phase and increasing its salt content.

For our purpose it is very important to determine the bromine content at the inflection points on the curve. A comparison of data for these points for different lakes shows that there is a certain difference in their bromine content. Different also are the lengths of the segments of the curve corresponding to the separation of the individual solid phases. This is due to the slight differences in the composition of the lake brines and in the amounts of salts separating during the different stages of concentration.

The bromine content corresponding to the inflection points on the crystallization curves of these lakes is given in Table 3.

Table 3. Bromine Content at Inflection Points (wt. %)

Lake	Beginning of precipitation of halite		Beginning of precipitation epsomite (sylvite)		Beginning of precipitation carnallite			
	Sp.G.	Br^-	Sp.G.	Br^-	Sp.G.	Br^-	Sp.G.	Br^-
Lake No. 8	1.224	0.062	1.308	1.222	1.319	0.259	1.361	0.431
Lake No. 5	1.220	0.051	1.308	0.236	1.325	0.342	1.359	0.520
Sea of Azov	1.221	0.044	1.308	0.223	1.321	0.317	1.357	0.608
Gulf	1.220	0.033	-	-	-	-	-	-

We see that the maximum differences in bromine content occur at the final stages of crystallization and the minimum difference at the point of the first separation of carnallite. At the beginning of crystallization of epsomite, the bromine content is very nearly the same in all waters. Somewhat greater differences are observed between the points marking the beginning of crystallization of halite.

There is no need to average these data, since all of them characterize the concentration of sea water under slightly different conditions of interaction with waters of the land. This interaction always occurs in nature, and therefore these data will give us a natural scatter of points which we must take into consideration in establishing the scale of normal bromine content in salt deposits of marine origin.

Before passing on to the solution of our main problem we shall draw another conclusion from these data which may be useful in proving the marine origin of a given salt basin.

The value of the bromine-chlorine ratio, according to Vinogradov's suggestion [7], is a convenient genetic criterion for natural waters. However, the constancy of this ratio for waters of marine origin is preserved only until the moment of crystallization of the first chloride, halite. From that moment on, the $\frac{Br^{-} \cdot 10^{3}}{Cl^{-}}$ ratio increases sharply. Therefore, this value may be used only for a definite interval of salt formation.

More widely applicable is a graph showing the relation between the $\frac{Br^{-} \cdot 10^{3}}{Cl^{-}}$ ratio and the content of the chlorine ion, which is an index of the stage of salt deposition. The chlorine ion content is plotted on the axis of the abscissas. The graph is shown in Figure 2. The points below and to the right of the average curve for waters of marine origin represent waters poorer in bromine, and those above and to the left of the curve, waters richer in bromine.

During the sinking of a borehole in Siberia, a mineral water was encountered at the depth of 1410 m, with a bromine-chlorine coefficient of 1905 and a chlorine content of 19.3%. Comparison of this coefficient with that for sea water shows clearly that this water is enriched in bromine. When this coefficient is plotted on the graph, it becomes evident that the water is a brine of marine origin. Point (x) falls exactly on the curve.

Lakes of marine origin, marked with ● and ○ on the graph, fit the curve well. Continental lakes usually fall below and to the left of the curve. Above the curve there are several points representing salt lakes of the Kerch peninsula, which probably receive waters of high bromine content from the mud volcanoes. Judging by these data, the ocean is the place of maximum bromine concentration. Circulation of land waters disperses bromine. Another locus of its concentration is living matter, and this explains the high bromine content in the mud volcanoes and some oil field waters.

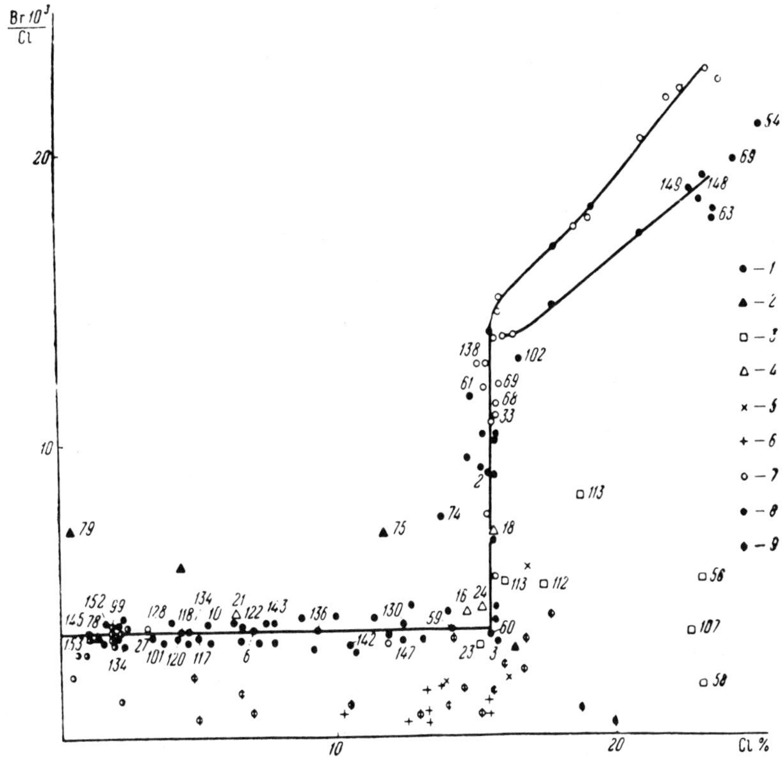

Fig. 2. The value of the bromine-chlorine ratio as a function of chlorine content in the brines of marine origin: 1—Black Sea and lakes fed by sea water; 2—lakes of Kerch Peninsula; 3—Perekop lakes; 4—Genichensk lakes; 5—Lake Elton; 6—lakes of Middle Asia; 7—artificial salt pans and lakes fed by sea water; 8,9—lakes of the Aral-Caspian depression.

DISTRIBUTION COEFFICIENTS OF BROMINE

The determination of accurate values of the coefficient of distribution of bromine between the solid chlorides and the solution $J = \frac{C_{solid}}{C_{s.r.}}$ is very important for the purposes indicated above.

For this reason, we undertook a special experimental investigation to determine the values of these coefficients and the effect of the mode and environment of salt formation on them. This investigation was made in association with E. M. Petrova for the two principal minerals, halite and sylvite.

In choosing the final values of the distribution coefficient, all available data from the literature were taken into consideration, and for halite the results of calculations based on our determinations of bromine in fresh halite deposits and in brines from a number of salt lakes, as well. The selection of the coefficient for carnallite and bischofite was based entirely on the data from the literature. The details of these calculations are given in the paper by Valyashko and Mandrykina [2]. Here, only the results are presented (Table 4).

The above values of the coefficients are the basis for all further computations and discussions.

BROMINE IN CHLORIDES OF MARINE ORIGIN

Knowing the bromine content of brines concentrated from sea water and the coefficients of distribution, $J = \frac{C_{solid}}{C_{s.r.}}$, for the individual chlorides, it is possible to compute the bromine content (C_{solid}) which must be present in a chloride crystallizing from sea water during its concentration. This bromine content we shall call normal.

We compute it from the equation:

$$C_{solid} = J.C_{r.s.}$$

If, for example, the bromine content in the brine, at the moment when NaCl begins to crystallize, is 0.051% and the total salt content is 27.49%, then

$$C_{s.r.} = \frac{0.051 \cdot 100}{27.49} = 0.185\%, \text{ and } C_{solid} = (0.037)(0.185) = 0.0068\%.$$

By the same method, but taking the corresponding coefficients, the normal bromine content in sylvite, carnallite and bischofite may be computed. Let us consider the data for the individual minerals.

Halite begins to crystallize from sea water when 26-27% of the total salt content (d = 1.220) has separated and continues to crystallize until the very end of concentration. This makes halite a very convenient mineral for determination of the stratigraphic position of individual groups of unfossiliferous salt beds.

It follows from Table 5, which gives bromine content and the values

Table 4. Coefficients of Distribution of Bromine Between the Solid and Liquid Phases in the Principal Chlorides

Chloride	$J = \frac{C_{solid}}{C_{s.r.}}$
Halite (NaCl)	0.037
Sylvite (KCl)	0.200
Carnallite ($KMgCl_3 \cdot 6H_2O$) . .	0.320
Bischofite ($MgCl_2 \cdot 6H_2O$) . . .	0.420

Table 5. Normal Bromine Content and the Value of the $\frac{Br^- \cdot 10^3}{Cl^-}$ Ratio in Halite, Sylvite and Carnallite Precipitated from Sea Water (selected data [2])

Liquid phase	Halite		Sylvite		Carnallite	
$\frac{Br^- \cdot 10^3}{Cl^-}$	Br^-, wt.	$\frac{Br^- \cdot 10^3}{Cl^-}$	Br^-, wt.	$\frac{Br^- \cdot 10^3}{Cl^-}$	Br^-, wt.	$\frac{Br^- \cdot 10^3}{Cl^-}$
3.24	0.0068	0.11				
5.34	0.011	0.18				
9.00	0.017	0.288				
11.70	0.022	0.36				
13.70	0.025	0.41				
14.30	0.026	0.43				
14.80	0.027	0.44	0.137	2.9		
16.50	0.033	0.54	0.176	3.70		
17.50	0.037	0.61	0.190	4.00		
17.80	0.037	0.60	0.199	4.20	0.318	8.3
20.01	0.045	0.74	(0.242)*	(5.10)	0.387	10.1
21.40	0.049	0.81	(0.266)	(5.60)	0.425	11.3
22.40	0.052	0.86	-	-	0.523	13.7
26.7	0.063	1.03	-	-	0.544	14.1

*The values for sylvite assumed to separate in the carnallite zone are placed in parentheses.

of the bromine-chlorine coefficient in halite, that the normal bromine content in the first crystals of halite is 0.005–0.006%, whereas in the halite separating at the end of concentration, it is 0.05-0.06%, i.e., 10 times as high. This is shown graphically by the curve of Figure 3. The breaks in the curve mark appearances of new solid phases in the precipitate.

It is even more convenient to consider not the bromine content but its ratio to the chlorine content $\left(\frac{Br^- \cdot 10^3}{Cl^-}\right)$, which is shown by the curve of Figure 4.

Knowing the distribution of the fields of crystallization of potassium and magnesium salts on the concentration scale, and therefore the position of the individual zones, the bromine content, or better, the value of the bromine-chlorine coefficient of the halite beds may be used as an index of the stratigraphic position of a given group of beds and as a prospecting guide to the potassium-magnesium salts. If the bromine content in halite (computed for 100% halite) is near 0.02%, or the value of the bromine-chlorine coefficient is $\sim$0.4, it follows that the halite crystallized near the beginning of the potassium salt zone (sylvite zone) and its presence indicates the probability of finding potash salts in the locality.

Sylvite. The crystallization of sylvite from oceanic waters, as shown by investigation [3], begins when the brine has the following composition: Cl^- - 16.28%, SO_4^{-2} - 6.03%, Mg^{2+} - 5.8% and K^+ - 1.94%, which is very near the beginning of separation of epsomite at summer temperatures. Therefore, we may consider this point with sufficient accuracy as the beginning of crystallization of sylvite. The normal bromine content in sylvite was calculated by using the data on bromine content in the brines between this concentration and the beginning of crystallization of carnallite, and by taking $J = 0.20$ (Table 5).

It follows from these computations that the normal bromine content in sylvite separated from brines of marine origin will vary from 0.13% to 0.20%. The bromine-chlorine coefficient for sylvite ranges from 2.9 to 4.2. Lower bromine content in sylvite will indicate that it had crystallized from brines poor in bromine or that recrystallization has occurred.

According to the determination by D'Ans and Kühn [8], the bromine content in sylvite from a northern potash deposit is 0.21-0.28%, and in sylvite from the Werra region, 0.13%. These values indicate primary crystallization of the salt.

According to Mandrykina's determination, the sylvite from Solikamsk contains 0.06 to 0.008% bromine; this is much less than the normal content and indicates that recrystallization has occurred (after disappearance of mother liquor). Sylvite, as a rule, does not form monomineralic beds but is most commonly found in association with halite and much less commonly with carnallite and kainite. This will be discussed below.

Carnallite. At a definite moment when the brine has accumulated a sufficient amount of magnesium chloride, carnallite begins to separate

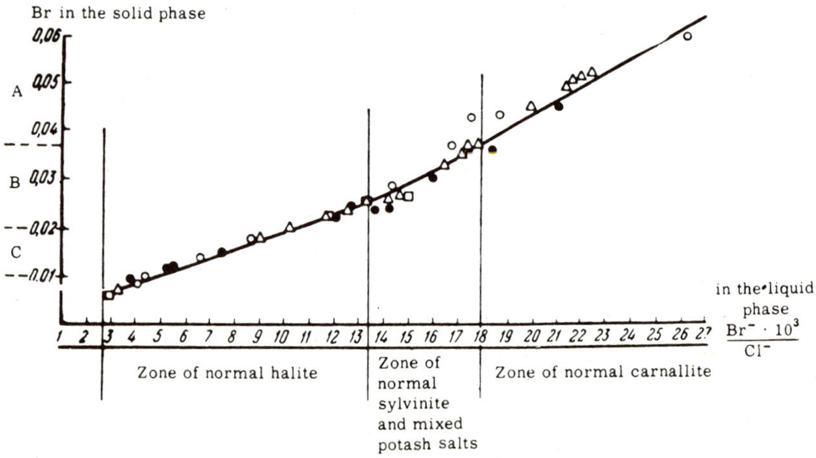

Fig. 3. Normal bromine content in solid halite as a function of the bromine-chlorine ratio in the liquid phase during concentration of sea water. A—occurrence of carnallite possible; B—potash deposits are probable; C—finding of potash and magnesium salts improbable. Halite zone.

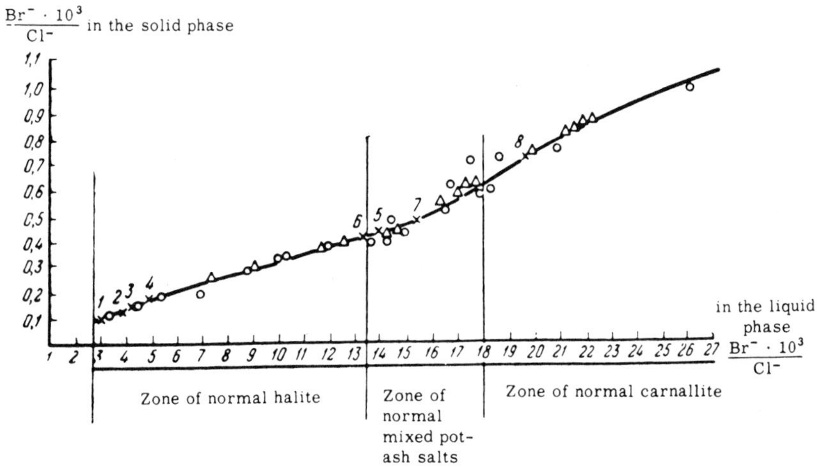

Fig. 4. Normal bromine-chlorine ratios of solid halite condensed from sea water, as a function of the bromine-chlorine ratio in the liquid phase.

out. In brines of marine origin, this occurs when the density of the brine is from 1.32 to 1.325 (salt content 34-35%). Using the distribution coefficient, 0.32, and the bromine content in the brines of marine origin, the normal bromine content in carnallite must range from 0.24 at the beginning of crystallization to about 0.6% at the first appearance of bischofite (Table 5). The corresponding bromine-chlorine coefficient for carnallite ranges from 6 to 15. Carnallite may have a higher bromine content when it crystallizes at the very end of the concentration process.

In fossil salt deposits, carnallites are found with a rather variable bromine content, which, however, does not exceed the normal content. The highest bromine content is found in the Inder carnallites (borehole 244). German carnallites have normal bromine contents. The carnallites of the Solikamsk, Ozinki and a number of other deposits have bromine contents somewhat below normal. The reasons for this will be touched upon later.

Bischofite is the last chloride crystallizing from the evaporating sea water. Being very hygroscopic, it is seldom found in salt deposits.

On the basis of bromine content in brine of marine origin, at the end of concentration and $J = 0.42$, the normal bromine content in bischofite must be from 0.5 to 0.7%, and with further evaporation of the brine it may reach 1%. The bromine-chlorine coefficient for bischofite is 14 to 20 and higher.

Among bischofites from USSR deposits, the mineral from Inder has a normal bromine content and that from the Ozinki deposit a very low content.

Kainite. Kainite is widely distributed in a number of potash deposits and especially in the Galician deposits. Unfortunately we have no determinations of the distribution coefficient J for it and are unable, therefore, to compute its normal bromine content. Bromine determinations have been made by D'Ans and Kühn [8] for the German and Galician localities; by Valyashko for the kainite from Inder; and by Morachevskii, Valyashko and Mandrykina for the Kalusz kainite. According to these data, the bromine content in kainite is lower than in the associated sylvite. But it is 1.5 times higher than it would be if the KCl in kainite had retained its distribution coefficient J = 0.2. Thus, the distribution coefficient based on these data may be considered to be 0.3 for the KCl contained in the kainite, or 0.1 for the whole molecule of kainite. These, however, are assumptions requiring experimental confirmation.

Salt deposits. Of the numerous types of salt deposits we shall review only the main ones, those of halite, sylvite and carnallite.

The essential mineral of rock salt deposits is halite, usually accompanied by anhydrite, less frequently by polyhalite and a small amount of calcium and magnesium carbonates. Neither sulfates nor carbonates contain bromine, and therefore their mixture with halite only lowers the percentage of bromine in the deposit, but is not in any way reflected in the value of the bromine-chlorine coefficient. Therefore, in using bromine content in halite deposits to establish the stratigraphic position of a given bed for prospecting and genetical studies, the graph of the normal values of the bromine-chlorine coefficient may be used (Fig. 4).

The essential minerals in the sylvite deposits are halite and sylvite. Both minerals are carriers of bromine. The value of the bromine-chlorine ratio in the rock consisting of a sylvite-halite mixture is determined by the composition of the mixture and the bromine content in the two minerals.

In order to utilize the bromine content in sylvinite for genetic studies without separating its component minerals or determining bromine by an indirect method (D'Ans and Kühn, 1940; Valyashko, 1944), it is necessary to determine the normal bromine content in sylvinite. This is most conveniently done graphically.* A straight line segment taken as 100% represents the composition of halite-sylvite mixtures from 0 to 100%. The values of the bromine-chlorine ratio are plotted on the vertical axis (Fig. 5).

Crystallization of sylvite from sea water begins when about 32% of

*There is another method proposed and applied in our laboratory by N. S. Spiro, which consists in computing from the data of a chemical analysis halite equivalents for sylvite, carnallite, etc. This method is convenient for deposits of complex composition.

the total salt content has separated and the bromine content in the solid residue from the remaining brine $C_{s.r.} = 0.69$. The bromine-chlorine coefficient of sylvite is 2.9 and of the associated halite, about 0.4. Using these data, we computed the bromine and chlorine content in the mixtures of halite and sylvite and the values of the bromine-chlorine ratio. The latter are plotted on the graph. Connecting the plotted points, we obtained the curve $H_1 S_0$ representing normal values of the

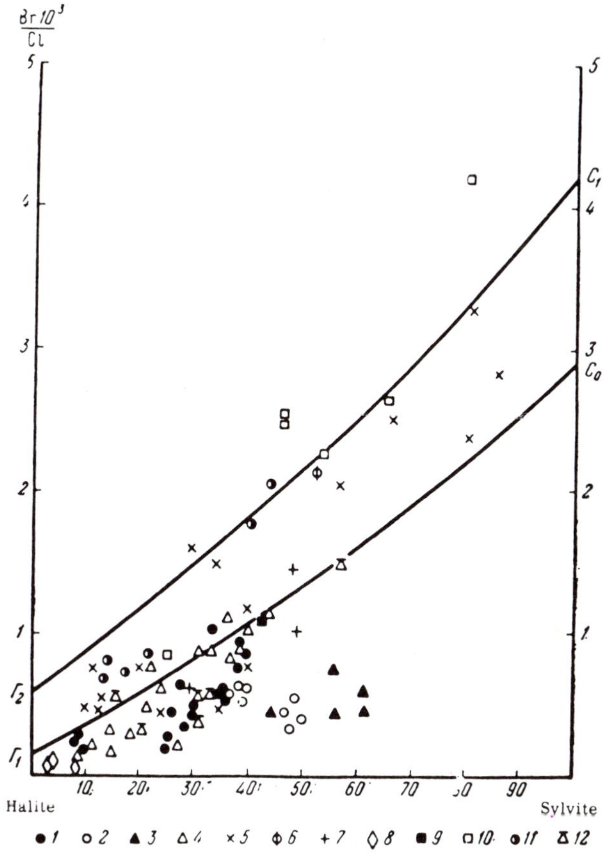

Fig. 5. Normal values of the bromine-chlorine ratios for sylvinites formed from marine brines. 1—Solikamsk, red sylvinites; 2—Solikamsk, variegated sylvinites; 3—Inder, variegated sylvinites; 4—Inder, sylvinites; 5—Kalusz; 6—Stebnik; 7—Gaurdak; 8—Ozinki; 9—Stassfurt; 10—Werra; 11—Utah; 12—New Mexico.

bromine-chlorine coefficient in the mixtures of halite and sylvite at the beginning of crystallization of the latter.

The separation of sylvite ends when $C_{s.r.} = 1.0$. The values of the bromine-chlorine coefficient for sylvite and halite crystallizing at this moment are 4.2 and 0.57, respectively. After computing the bromine-chlorine coefficients for the mixtures of these two minerals and plotting them on the same graph we obtain curve H_2S_1, which gives the normal values of the bromine-chlorine coefficient for the halite-sylvite mixtures at the end of the crystallization of sylvite. The field included between the two curves, H_1S_0 and H_1S_2, contains normal values of the bromine-chlorine coefficient for all possible mixtures of halite and sylvite separated from sea water during its concentration.

In nature, rocks often contain other minerals besides these two, minerals which do not carry bromine (kieserite, polyhalite, etc.) and which lower the bromine concentration in the rock but not the value of the bromine-chlorine coefficient. Before plotting such rocks on the graph, it is necessary to compute the percentage of KCl with respect to NaCl and plot this value on the axis of compositions on the graph.

A graph of normal values of the bromine-chlorine coefficient for carnallite rocks (halite-carnallite) is shown in Figure 6. The field between the curves H_2S_0 and H_3C_1 contains normal values of the bromine-chlorine coefficient for all possible compositions of

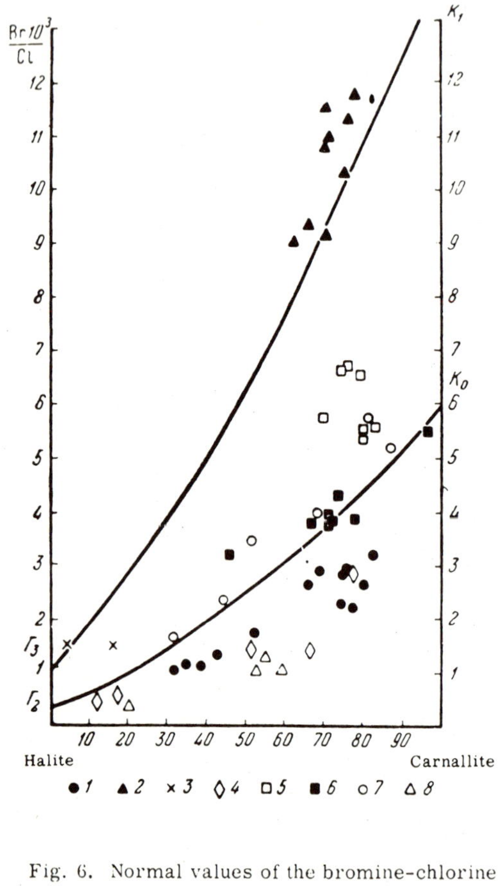

Fig. 6. Normal values of the bromine-chlorine ratio for carnallite rock separated from brines of marine origin. 1—Solikamsk; 2—Inder; 3—Kalusz; 4—Ozinki; 5—Stassfurt; 6—Werra; 7—Utah; 8—New Mexico.

carnallite rocks separated from sea water during the period of crystallization of carnallite.

Carnallite, like halite, is a so-called congruent mineral and continues to crystallize during the bischofite stage. For such carnallite rocks the value of the bromine-chlorine coefficient is higher than that shown on the graph.

These graphs permit the use of the bromine content, or, more accurately, the value of the bromine-chlorine coefficient, in a variety of genetic investigations.

It has already been mentioned that salts other than chlorides do not contain bromine. Our investigations have shown that adsorbed bromide may be present in finely dispersed clay material, which commonly contaminates salt beds; and in the Galician deposits, for example, the clay fraction contains as much bromine as sylvite. Therefore, in dealing with salts containing clay, it is necessary to consider clays as bromine carriers.

It follows from the above discussion that an analysis of the data on the content and distribution of bromine in salts made from a certain point of view enabled us to establish a scale of normal bromine content and normal values of the bromine-chlorine coefficient for the minerals of salt deposits and the main salt rocks and to develop methods of utilizing them.

In conclusion, let us consider a few examples of the application of the bromine-chlorine coefficient.

DETERMINATION OF THE STRATIGRAPHIC POSITION OF A HALITE BED

It has already been stated that the bromine content in halite will depend on the stage of concentration of sea water during which it crystallized. This makes it possible to determine the stratigraphic position of a halite bed by the value of its bromine-chlorine coefficient. For this purpose the $\frac{Br^- \cdot 10^3}{Cl^-}$ coefficient is computed for the given sample of halite and plotted on the $H_0 H_1 H_2 H_3$ curve (Fig. 4). The position of the coefficient on this curve determines the stratigraphic position of the halite bed. If the point lies between H_0 and H_1, it is in the halite zone; if in the H_1-H_2 segment, in the sylvite zone; and if in the H_2 - H_3 segment, in the carnallite zone. If it is nearer to H_0, it is near the beginning of the halite zone. If it is nearer to H_1, it is in the upper part of this zone. For example, in borehole No. 6 in the Verkhnekamsk deposit, Mandrykina determined the following values for the bromine-chlorine coefficient:

depth in meters	528	491	458	430
$\frac{Br^- \cdot 10^3}{Cl^-}$	0.08	0.12	0.14	0.17
No. of point	1	2	3	4

Plotting these values on the graph (Fig. 4), we find that the borehole penetrated the lower part of the halite zone.

At the Inder deposit the following values of the bromine-chlorine coefficient were determined by E. M. Petrova for the halite beds:

Borehole No.	1134	1137	1141	1989
$\frac{Br^- \cdot 10^3}{Cl^-}$	0.422	0.40	0.46	0.69
No. of point	5	6	7	8

Points 5, 6 and 7 lie near H_1 and correspond to the uppermost part of the halite zone (point 6) and the beginning of the sylvite zone. Point 8 is in the carnallite zone.

The positions of points 5, 6, 7 and 8 may serve as guides to potash salts. Indeed, not far from boreholes 1134 and 1141 sylvinite has been found and in borehole 1989, inclusions of carnallite.

There are cases of such low bromine content in halite that the value of the bromine-chlorine coefficient falls to the left of H_0. This indicates erosion and subsequent redeposition of halite.

It must be remembered that if mistakes are to be avoided, no conclusions must be made on the basis of the analysis of a single sample.*

PRIMARY AND SECONDARY NATURE OF SYLVITE AND CARNALLITE DEPOSITS

Using the bromine-chlorine coefficient, let us attempt to decipher the history of sylvite and carnallite rocks of a number of localities. To do this we shall plot the values of the coefficient for sylvinites on the graph of Figure 5 and of carnallite rocks, on the graph of Figure 6.

An examination of the graph for sylvinites shows that among the "normal" sylvites are those from Stassfurt, the Werra district, Kalusz and Stebnik. Some of the Inder sylvinites (boreholes 273 and 221) lie at the base of the normal zone. Most of the Inder sylvinites, and especially those from borehole 238, called "variegated" by analogy with the Solikamsk sylvinites, undoubtedly have undergone solution and recrystallization, as indicated by the low value of the bromine-chlorine coefficient and high KCl content.

The Solikamsk sylvinites lie mostly below the lower boundary of the zone, suggesting that the deposit is secondary, redeposited. The variegated Solikamsk sylvinites have also been recrystallized and lost bromine. An especially low position is occupied, according to the value of the bromine-chlorine coefficient, by the sylvinites from the

*It was recently suggested by Kuhn [35] that the variation in bromine content of halite be used to determine the depth of the basin in which salts were deposited. In our opinion, the intensity of accumulation of bromine in halite deposits is related not so much to the depth of the basin as to its chemical inertness—the supply of brine.

Ozinki deposit. The sylvinites of Gaurdak also lie below the lower boundary of the field of normal sylvinites.

Above the upper boundary of the normal sylvinites lie some of the German sylvinites. It is probable that they were formed as the result of decomposition of carnallite.

An examination of the graph for carnallite rocks from different localities shows that the Stassfurt and Werra carnallites are normal. The carnallites of Inder also are primary. They lie mostly above the upper boundary. This indicates that they separated out from nearly fully concentrated brines. This is confirmed by the presence of borates among the Inder carnallites. It is interesting to note that borates have been found also in the German deposits, which according to all data are primary, of marine origin and nearly unchanged by later processes.

The Solikamsk carnallites, according to the values of the bromine-chlorine coefficient, lie noticeably below the lower boundary of the normal bromine content. The carnallite and bischofite rocks of Ozinki also occupy this low position.

Some data from the analyses of the carnallite-bearing sequence of Kalusz lie in the field of normal bromine content.

Thus the construction of the graphs gives us an objective method of distinguishing between primary and secondary salts and deposits.

The correctness of these data and their correspondence to reality rest upon careful documentation by the distribution coefficients determined for each rock on hand, and by the abundant material used to derive the normal bromine content in the products of concentration of sea water.

Plotting of the bromine-chlorine coefficients for all the potash deposits of the world for which we had data showed that they fall either in the field of normal values of the coefficient or below it. This distribution is perfectly regular and follows from the nature of the probable processes affecting salt deposits. Very few points fall above the boundaries of the normal bromine content.

Values in excess of normal are found in the sylvinites of Stebnik and Kalusz, which are poor in sylvite and chlorides in general but are enriched in clay.

The high values of the coefficient have been found in some German deposits. It is probable that these are sylvinites formed from carnallites with high bromine content decomposed by the action of solutions also with a relatively high bromine content. Experiments performed by the author in collaboration with Petrova showed that in this case the bromine content in sylvite exceeds normal. This deviation from the normal, therefore, finds an explanation confirmed by experiment.

Finally, the carnallite graph shows that the Inder carnallites also lie above the curve. Their enrichment in bromine is explained by separation from the highly concentrated solutions of the last stages of crystallization.

The method developed in the present work of utilizing the values of the bromine-chlorine coefficient in salt beds provides an objective way of determining the changes which have taken place in the deposits, of discovering potash salts in halite deposits and of establishing the stage at which individual halite beds were precipitated. The application of this method enabled the author and his associates to predict the existence of potash salts in certain localities, subsequently confirmed by drilling.

Still another conclusion can be drawn from the above discussion. The distribution of the values of the bromine-chlorine coefficient for sylvinites and carnallite rocks in the field of normal and low values points out the important fact that the value of the bromine-chlorine coefficient for ocean water during the part of geologic time for which salt deposits are known (beginning with the Cambrian) has been close to the modern value.

The data reviewed here lead to the conclusion that the processes tending to lower the value of the bromine-chlorine coefficient are widespread in nature, whereas the processes of concentration of bromine are much more rare. Principal among the latter processes are salt deposition and concentration of bromine by living matter and, in some cases, adsorption.

REFERENCES

1. Boecke, H. E. Uber der Kristallisationschema der Chloride, Bromide und Jodide von Na, K, Mg u.s.w., Z. f. Kristall, 41, p. 4, 1908.
2. Valyashko, M. G. and T. V. Mandrykina. Brom v Solyanykh Otlozheniyakh kak Geneticheskii i Poiskovyi Priznak (Bromine in salt deposits as a genetic and prospecting guide). Trudy Vsesoyuznogo nauchno-issled, in-ta galurgii, Vyp. 23, L. Goskhimizdat, 1952.
3. Valyashko, M. G. and E. F. Solov'eva. Issledovanie Metastabil'nykh Ravnovesii v Sisteme $2Na^-$; $2K^-$; Mg^{2-}; So_4^{2-}, $2Cl^-/H_2O$. Fiziko-Khimicheskie Issledovaniya Solyanykh Sistem (An investigation of metastable equilibrium in the system: $2Na^- - 2K^- - Mg^{2-} - SO_4^{2-} - H_2O$. Physicochemical studies of salt systems). Trudy Vsesoyuzn. nauchno-issled. in-ta galurgii, Vyp. 21, L. Goskhimizdat, 1949.
4. Vant Hoff, J. Okeanicheskie Solyanye Otalozheniya (Zur Bildung der ozeanischen Salzablagerungen, 1909). Khimteoretizdat, M, 1936.
5. Vilnyanskii, Ya. and V. Zelyanskii. Ob Izvleckenii broma iz Silvinita (Extraction of bromine from sylvinite). Kalii, Vyp. 2, p. 12, 1935.
6. Vilnyanskii, Ya. and V. Zelyanskii. O Predele Nakopleniya Broma v Silvinitovykh Schelokakh (On the limit of bromine accumulation in sylvinite brine). Kalii, Vyp. 3, 1936.
7. Vinogradov, V. P. O Khlor-Bromnom Koeffitsiente Podzemnykh Vod (The chlorine-bromine ratio in ground waters). Doklady, Vyp. 44, Izd, Akad. Nauk, p. 2, 1944.
8. D'Ans, I and R. Kühn. Uber die Bromgehalt von Salzgesteinen Kalisalzlagerstatten. Kali, 4, 5, 6, 1940.
9. Efremov, N. N. and A. A. Veselovskii. O Spderzhanii Broma v Solikamskikh Karnallitakh (The content of bromine in the Solikamsk carnallites). Izv. In-ta fiz-khim. analiza, tom. 4, Vyp. 2, Izd. Akad. Nauk, p. 99, 1930.
10. Illinskii, V. P. Poluchenie Khloristogo Natriya v Protsesse Ispareniya Morskoi Vody (Formation of sodium chloride in the process of evapor-

ation of sea water). Gos. in-t prikl. khim., Sbornik pabot in-ta, Vyp. 20, 1948.

11. Ilinskii, V. P. Poluchenie Khloristykh i Sernokislykh Solei Magniya i Kaliya i Rassolov dlya Proizvodstva Broma. (Extraction of chlorides and sulfates of magnesium and potassium and preparation of brine for the extraction of bromine). Gos. in-t prikl. khim., Sbornik rabot In-ta, Vyp. 40, 1948.
12. Kurnakov, N. S. and V. I. Nikolaev. Izvestiya In-ta fiz. khim. analiza, Vyp. 10, Izd. Akad. Nauk, p. 333, 1938.
13. Kurnakov, N. S., and I. N. Lepeshkov. Inderskoe Ozero. Bor i Kalii v Kazakhstane (Lale Inder. Boron and potassium in Kazakhstan). Izd. Akad. Nauk, 1933.
14. Kurnakov, N. S., I. N. Lepeshkov and N. I. Buyalov, Kaliinye Mestorozhdeniya Yuzhnogo Basseina Permoskogo Morya i Solyanye Ozera Zapadnogo Kazakhstana (potash deposits of the southern basin of the Permian sea and the salt lakes of the western Kazakhstan). Izvestiya, seiya khim., Vyp. 1, Izd. Akad. Nauk, p. 13, 1938.
15. Lepeshkov, I. N. Kaliinye Soli Volga-Emby i Prikarpat'ya (Potash salts of the Volga-Emba region and Galicia). Izd. Akad. Nauk, M.-L, 1946.
16. Morachevskii, Yu. V. and A. N. Fedorova. Brom v Solikamskom Karnallite (Bromine in Solikamski carnallite). Vestnik Geologicheskogo Komiteta, No. 4, 1928.
17. Morachevskii, Yu. V. O Khimicheskom Sostave Solikamskikh Solyanykh Otlozhenii (Chemical composition of the Solikamsk salt deposits). Izv. in-ta fiz.-khim. ahaliza, tom 4, Vyp. 2, Izd. Akad. Nauk, 1930.
18. Morachevskii, Yu. V. and A. N. Fedorova. Rezul'taty Oprobovaniya na Brom Srednikh Prob Soley Solikamskogo Mestorpzhdeniya (Assays for bromine on the average samples from the Solikamsk salt deposits). Izv. Vsesoyuzn. geologo-razvedochn. ob''edineniya, Vyp. 52.
19. Morachevskii, Yu. V. Altsessornye Element i Nerastvorimye Ostatki Solikamskogo Karnallita (Accessory elements and insoluble residuesi in Solikams carnallites). Sbornik Solikamskie karnallity. (Solikamsk carnallites. Collection), 1935.
20. Morachevski, Yu. V. Ocherki Geochimii Verkhnekamskikh Solyanykh Otlozhenii (Essays on the geochemistry of the Upper Kama salt deposits). Trudy Vsesoyuzn. nauchno-issled. in-ta galurgii, Vyp. 17, 1939.
21. Nikolaev, V. I. Khloristyi Natrii Permskikh Solyanykh Otlozhenii kak Istocknik Broma (Sodium chloride of the Permian salt deposits as a source of bromine). Priroda, Vyp. 6, p. 91, 1936.
22. Nikolaev, V. I. and E. M. Petrova. K Voprosu o Poluchenii Broma is Silvinitovykh Shchelokov (On the extraction of bromine from sylvinite brines). Kalii, Vyp. 8, p. 18, 1935.
23. Nikolaev, V. I. and I. N. Lepeshkov. Bromnye Ravnovesiya Broma v Silvinitovykh Shchelokakh (Bromine equilibria in Lake Inder and the question of the limit of accumulation of bromine in sylvinite brines). Kalii, Vyp. 9, p. 29.
24. Sapirshtein, D. I. Zh. prikl. khim, Vyp. 2, 3-4, p. 210, 1925.
25. Usiglio, J. Ann. de. chim. et phys., 27, 92, 172; 1849. Kali, 5-6, 1924.
26. Chirkov, S. K. Raspredelenie Broma v Vodno-Sil 'vinitovoi Sisteme (Distribution of bromine in the system: sylvinite—water). Kalii, Vyp. 10, 1935.
27. Chirkov, S. K. and M. S. Shnee. K Voprosu o Vliyanii "Primesei" na Raspredelenie Veshchestv pri ikh Kristallizatskii (The effect of "admixtures" on the distribution of substance during crystallization). Kalii, Vyp. 9, 1936.

28. Chirkov, S. K. K Voprosu ob Izvlechenii Broma iz Solikamskikh Sil'vinitovykh Shchelokov (On the extraction of bromine from the Solikamsk sylvinite brines). Kalii, Vyp. 3, p. 35.
29. Chirkov, S. K. Soderzhanie Ioda i Broma v Solikamskikh Karnallite, Sil' vinite i Galite (Iodine and bromine content in the Solikamsk carnallite, sylvinite and halite). Kalii, Vyp. 9, 1937.
30. Chirkov, S. K. Raspredelenie Broma Mezhdu Kristallami i Rastvorom KCl i KBr (The distribution of bromine between the crystals and solution of KCl and KBr) .J. prikl. khim., t. 12, Vyp. 2, 1939.
31. Chirkov, S. K. Kristallizatsiya Isomprfnykh Veshchestv is Vodnykh Rastvorov (Crystallization of solid solutions from aqueous solution). Trudy Ural'skogo filiala, t.V, Sverdlovsk, Izd. Akad. Nauk.
32. Shlezinger, N. A., F. P. Zorkin and L. V. Novozhenova. O Raspredelenii Broma Mezhdu Kristallami i Rastvorom Khloristogo i Bromistogo Kaliya (The distribution of bromine between the crystals and solutions of KCl and KBr). Zh. prikl. khim., t. 11, Vyp. 9, p. 1259, 1938.
33. Shlezinger, N. A. and F. P. Zorkin. O Vydelenii Smeshannykh Kristallov Khloristogo i Bromistogo Natriya iz Vodnykh Rastvorov i ob Aktivnosti v Nikh Bromistogo Natriya (On the separation of mixed crystals of NaCl and NaBr from aqueous solutions and on the role of NaBr in them). Uchenye Zapiski Saratovskogo gos, un-ta.
34. Selivanov, L. S. Geokhimiya i Biogeokhimiya Rasseyannogo Broma (Geochemistry and biogeochemistry of dispersed bromine). Trudy Biogeokhimicheskoi laboratorii, Vyp. 8, Izd. Akad. Nauk, 1940.
35. Kuhn, R. Tiefenberechnung des Zechsteinmeeres nach dem Bromgehalt der Salze. Z. Deutsch, geolog. Geselsche., B. 105, 4, 1953.

Received for publication
September 15, 1956

Reprinted from *Second Symposium on Salt*, **2**, 248–275 (1966)

Bromide Geochemistry of Salt Rocks

29

William T. Holser
Chevron Research Company
La Habra, California

ABSTRACT

Starting with the experimental distribution coefficient and various geological models, theoretical profiles of bromide content against salt thickness are derived. For a constant regime of inflow, evaporation and reflux all profiles are monotonic logarithmic functions, without extrema or inflections.

A Recent salt pan in Baja California has bromide between the 75 ppm predicted by the distribution coefficient of Braitsch and Herrmann and the 38 ppm in the evaporation experiments of Bloch and Scherb. The value to be expected in basal salt is still indeterminate within this range and may be a function of crystallization conditions. Most basal salts of earlier geological ages fall in this range and show no trend with geological time that would imply a change in the Cl/Br of the oceans.

Published profiles of the Stassfurt salt are regular, with two inflection points interpreted as simple changes of evaporation-inflow regime. Most bromide profiles are much more irregular. The most common type (Wellington, Castile, Salina, Prairie) have irregular profiles with most values in the range 30-80 ppm and occasional spikes up to 150 ppm. In the Salado salt the halite is mostly below 70 ppm even in the potash zone, and sylvite is also low -- almost in paragenetic ratio. Apparently most evaporite basins do not represent a simple inflow-evaporation-reflux cell, but have suffered multiple incursions of sea water with recycling (surface or alteration) of earlier salts and nearly complete loss of bitterns. Meteroric water may also be important in the recycling, as indicated by the extremely low bromide content throughout the Lower Elk Point and at the top of the Wellington.

The theory of profiles permits explicit calculation of brine depth from the basal halite/initial slope ratio, independent of the evaporation-flow regime. To the extent that the lowermost profiles may be primary, this relation suggests initial brine depths in the range 20-200 meters for various evaporite basins, but these depths are somewhat less than the final salt thicknesses.

INTRODUCTION

This paper reviews a several years' reconnaissance, to find out what sort of geological information could be derived from the distribution of a solid-solution element in chemical precipitates. When this work was begun in 1959, there was already available more information about the distribution of bromide in salt rocks than about any other evaporite mineral. Since Boecke (1908), German geologists had intensively studied the distribution of bromide, both experimentally and in the Zechstein rocks. During the past 20 years, Soviet geologists, particularly M.G. Valiashko (1958), applied bromide geochemistry in their own area, particularly in prospecting for potash. I have tried to develop, from first principles, a general model for the distribution of trace elements in precipitates, and to use this model to interpret analyses of bromide in salt rocks throughout North America. The emphasis in this study has been on the distribution of this single element in one mineral -- halite, as a model for application to other mineral-element pairs. This paper reviews the results on mainly marine evaporites; other work on nonmarine evaporites of the Western Interior United States will be published elsewhere.

This study was greatly aided by numerous companies and individuals who furnished samples or provided access to them: J. E. Adams, Standard Oil Co. of Texas, Midland; L. I. Briggs, University of Michigan; L. F. Dellwig, University of Kansas; T. L. Gambill, United States Borax and Chemical Corp., Carlsbad, New Mexico; W. B. Heroy, Geotechnical Corp., Garland, Texas; R. W. Hougland, International Mining and Chemical Co., Carlsbad, New Mexico; S. B. Horrell, Carey Salt Co., Hutchinson, Kansas; A. A. Ivanov, VSEGEI, Leningrad; J. R. Kramer, University of Western Ontario, London, Ont.; F. W. Lines, California Standard Co., Calgary; W. R. Muehlberger, University of Texas; F. L. Parker, Oak Ridge National Laboratory, Oak Ridge, Tenn.; the Alberta Oil and Gas Conservation Board; and Sifto Mines, Goderich, Ontario. I am particularly indebted to various individuals for stimulating discussion of the problems uncovered in these samples: Otto Braitsch, C. L. Jones, Robert Kühn, R. W. Rex, W. F. Schwerdtner, and N. C. Wardlaw.

I begin with a review of the general geochemistry of bromide, give the results of a general model developed for trace element precipitation, present bromide profiles measured in a wide variety of evaporite basins, and, finally, review some of the general geological and geochemical problems in which bromide geochemistry has been informative. In describing the evaporite profiles, I will make only summary reference to the geological and petrographical relations that have been considered in their evaluation.

BROMIDE GEOCHEMISTRY

Bromide is a unique element because it is not very reactive, and ordinarily follows chloride through the geochemical cycle. Practically the only situation in which there is fractionation of the chloride/bromide ratio is in the crystallization of evaporite minerals, where generally chloride is removed in the solid phase in preference to bromide. Figure 1 is an over-all view of the distribution of bromide in all kinds of rocks. There you see four general groups in terms of the ratio Cl/Br: most igneous rocks, sediments, and meteorites, with a Cl/Br of about 100, brines, atmospheric halides, and connate waters, Cl/Br = 200-400, thermal waters, Cl/Br = about 900, and salt rocks, Cl/Br > 5,000. From this simple diagram, one can draw several interesting conclusions: (1) neither mafic igneous rocks nor meteorites can represent primordial halides in the earth (the meteorite analyses should be re-examined); (2) the intermediate position of thermal waters between the two main reservoirs of halides in the crust, sea water and salt rock, suggest that volcanism represents average halides, either because they represent juvenile halides degassed from the crust, or because they represent an average of halides regenerated from both connate waters and salt rocks during deep crustal metamorphism; (3) if thermal waters are representative of average halide in the crust, then the mass balance requires that there be about twice as much chloride present in salt rocks as there is in the present sea (this is not unreasonable, inasmuch as the volume of evaporites in the Gulf Coast Jurassic alone represents five per cent of the chloride in sea water); (4) second cycle (descendent) waters formed by the solution of salt rock are unique in their very high Cl/Br ration, and second cycle salt rocks deposited from such waters have an even higher ratio; (5) most connate waters are neither second-cycle solution waters nor are they bitterns residual to the crystallization of salt rock.

TRACE ELEMENT PRECIPITATION

Fractionation in simple evaporation. Now let's consider how a trace element will distribute in a series of solid solutions crystallized from a brine. A distribution coefficient

$$K = \frac{r_c}{r_l} = \frac{B_c/A_c}{B_l/A_l} \qquad (1)$$

is strictly defined as a thermodynamic constant (at a given temperature) in terms of activities of the components, but for the present we will consider these ratios equivalent to those involved in

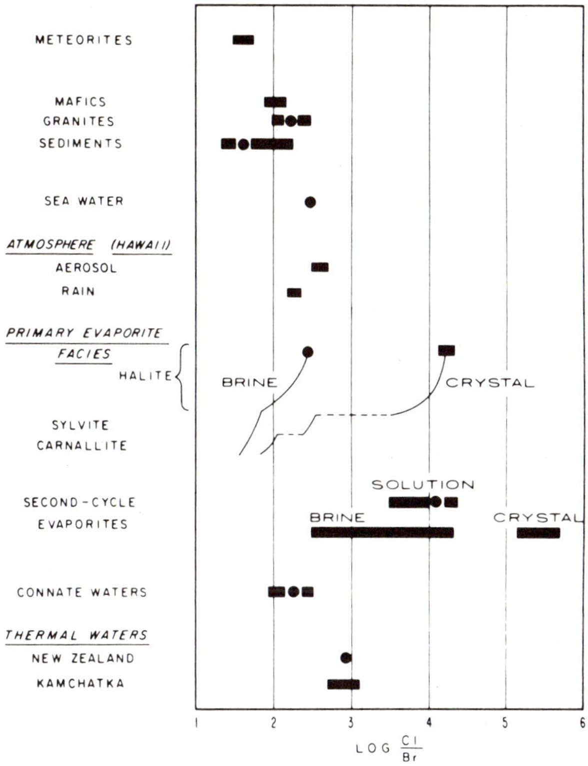

Figure 1. General geochemistry of bromine.

concentrations, as enumerated in Fig. 2. It is well known that in a closed system this relationship leads to a logarithmic decrement or increase of a trace element, depending upon the value of $\underline{K}$:

$$\ln \left[\frac{r_c}{r_{co}} \right] = (K - 1) \ln \left[\frac{M_o - M_c}{M_o} \right] \qquad (2)$$

where $\underline{m}_o$ and $\underline{m}_c$ are the total mass of crystallizing component and the mass crystallized at any given time respectively.

While this equation is perhaps applicable to crystallization in a beaker, or even to some magmatic minerals, a model that will be of general application to evaporite situations must account for the inflow of sea water into a basin, and for the reflux of concentrated brine back to the sea. For this purpose, I assume in Fig. 3 a weighted average of the inflow and reflux, and characterize this as a net flow $\underline{dW}_f$, with a concentration of main component $\underline{A}_f$ and trace element ratio $\underline{r}_f$. The relative importance of all factors in the regime, including net flow, $\underline{dW}_f$, water evaporation $\underline{dW}_e$, and saturation concentration $\underline{A}_l$, is characterized in a way as to best simplify the equations that follow, as the flow-crystallization factor,

$$g = \frac{\text{change of amount of main component A in reservoir}}{\text{change of amount of main component A in crystals}}$$

$$= \frac{A_l\,(dW_e + dW_f)}{A_l\,(dW_e + dW_f) - A_f dW_f} \qquad (3)$$

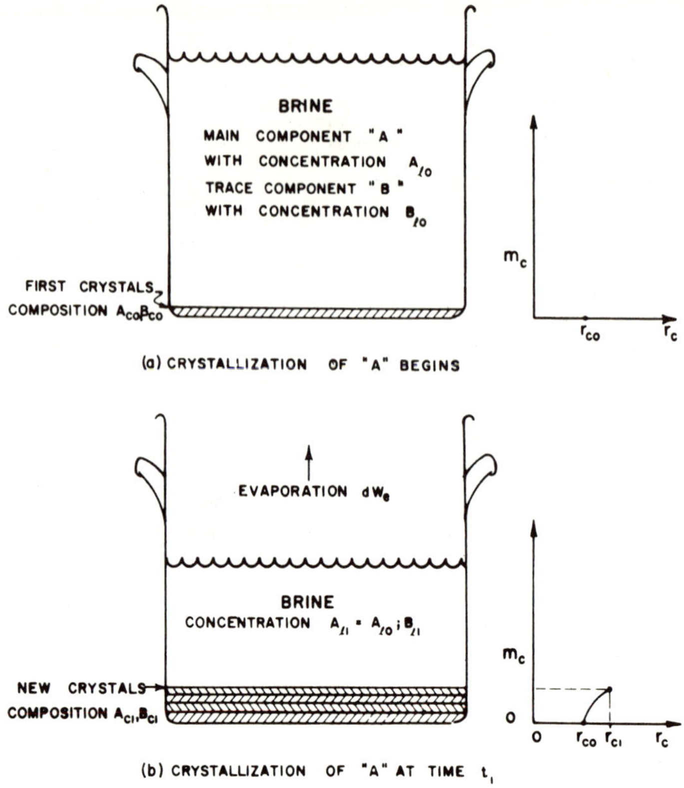

Figure 2. A simple evaporation model for the crystallization of a solid solution of a trace component B in A, showing the development of a trace element profile.

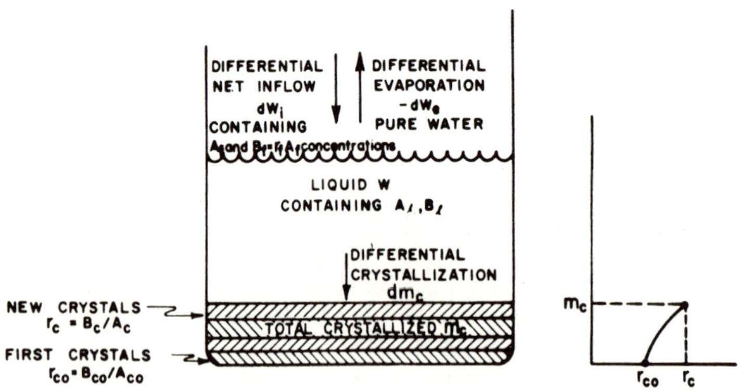

Figure 3. Model for a general open system analogous to Figure 2. Note that signs for flows dWe and dW_f are both positive for increase in basin volume, that is, dWe is always negative.

Specifically, $\underline{g} = 0$ for simple evaporation, $1 > g > 0$ for basins in which evaporation exceeds net inflow, $\underline{g} = 1$ for the important geological case of constant brine volume in the basin, and $g > 1$ if net inflow exceeds evaporation. The consequent general relation between trace element content and the mass of salt crystallized is

$$\ln \left[\frac{r_c\ (K - g) + (g - 1)\ Kr_f}{r_{co}\ (K - g) + (g - 1)\ Kr_f}\right] = \frac{K - g}{g} \ln \left[\frac{m_o - gm_c}{m_o}\right] \quad (4)$$

The full development of this mathematical model will be published elsewhere; I here give only the results. This rather formidable general equation reduces, for the case of straight evaporation ($\underline{g} = 0$), to equation (2). The constant volume case ($\underline{g} = 1$) is a singular point at which the particular solution is

$$\ln \frac{r_c - r_f}{r_{co} - r_f} = K \frac{m_c}{m_o} \quad (5)$$

Figure 4 sketches the shapes of trace element profiles for various $\underline{k}$'s at a particular $\underline{g}$, while Fig. 5 shows curves for various values of $\underline{g}$ at constant $\underline{k}$.

An important generalization can be derived from this general model of trace element precipitation. If the regime ($\underline{g}$) of net inflow and evaporation remains constant during the deposition of any particular sequence of rocks, then the trace element profile will be a logarithmic curve with no maxima, no minima, and no inflections, that is, no changes in the signs of slope or

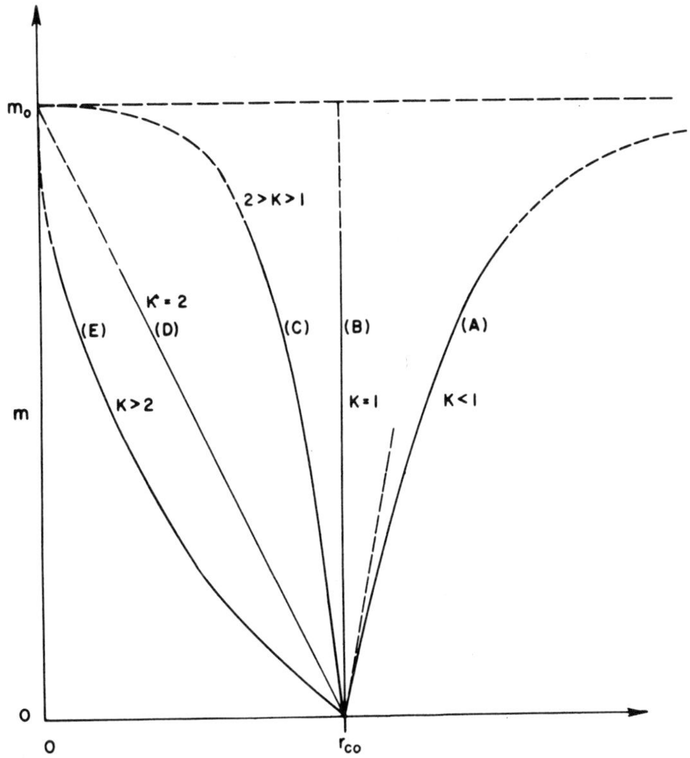

Figure 4. Effect of the distribution coefficient, $\underline{K}$, on profiles for simple evaporation (Eq. 2). Dashed upper portions of the curves will generally not be attained due to the intervention of a new facies or change of conditions.

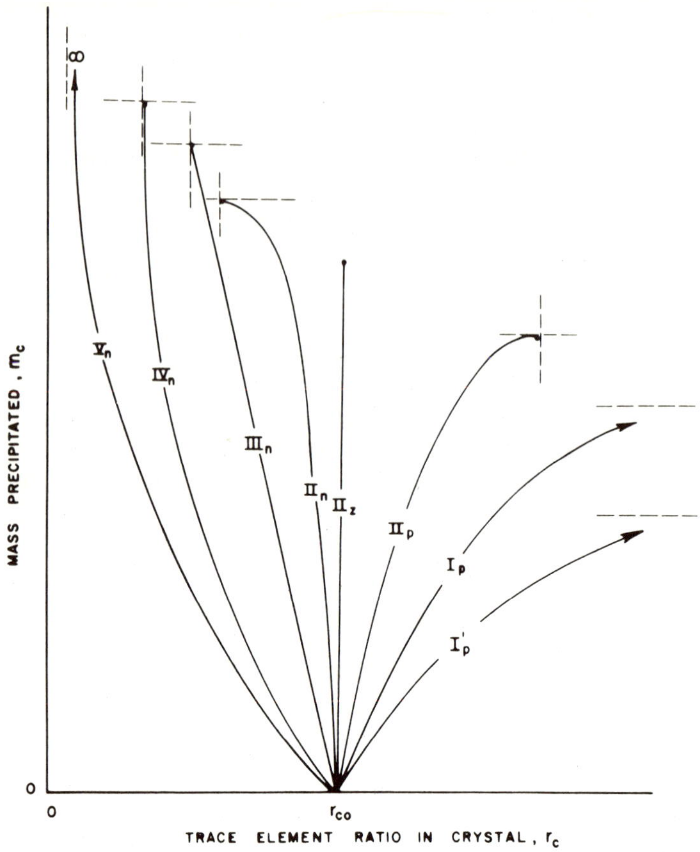

Figure 5. Profiles in an open system with low distribution coefficient, K, and low trace element concentration of inflow r_f. The curves illustrate the effect of increasing importance of inflow vs. evaporation (the g factor) from right to left.

curvature. Conversely, extrema or inflections positively signify a change in the net inflow-evaporation regime of an evaporite basin, regardless of the geological conditions.

Bromide as a trace element. The distribution coefficient K for bromide in halite has been extensively studied in various laboratories recently, and definitively by Braitsch and Herrmann (1963). The essential results are shown in Fig. 6, where the distribution coefficient, stated now as k in terms of bromide weight concentrations, varies from 0.16 in a pure NaCl solution, decreasing to 0.14 when the magnesium concentration reaches that at the beginning of the halite facies (evaporation of normal sea water), and continuing to decrease to a value of 0.07 by the beginning of the potash-magnesia facies. At that time, other chlorides may begin to precipitate, and the situation gets more complicated.

The general model discussed above did not allow for variations in k, but it is still applicable for short ranges of evaporation of sea water. If account is taken of the variation of k, theoretical bromide profiles will be like those shown in Fig. 7. The first salt to be expected in an evaporite basin, as found by applying k = 0.14 to the level of bromide in modern sea water at the beginning of the halite facies, gives 75 ppm Br in the halite.

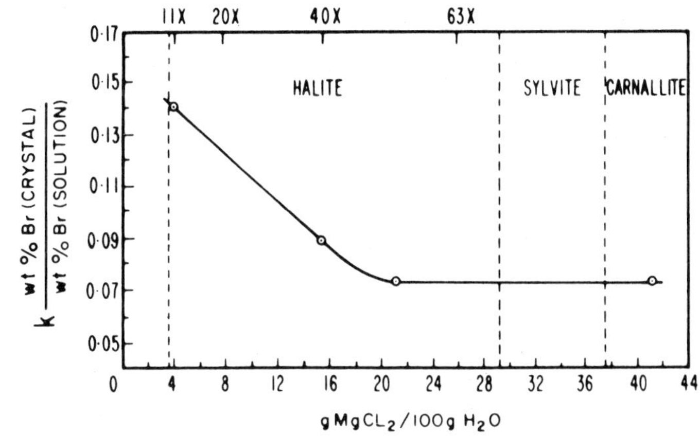

Figure 6. Distribution coefficient k, for bromide in halite, as a function of $MgCl_2$ concentration (lower scale) (Braitsch and Herrmann, 1963, p. 371), or corresponding sea water concentration (upper scale).

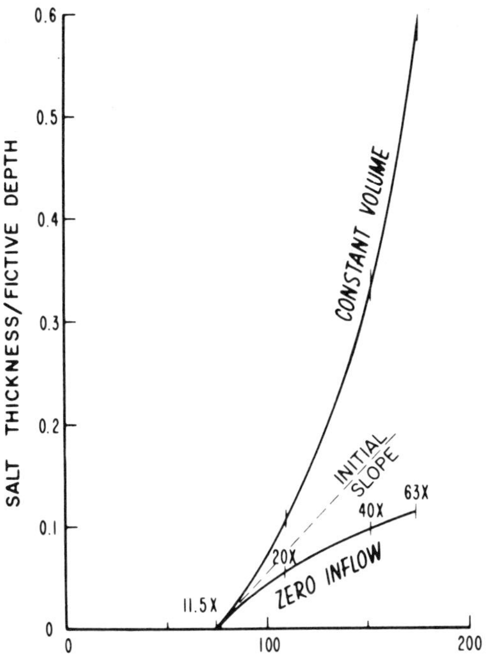

Figure 7. Theoretical bromide profiles in halite, corrected for the variation in distribution coefficient with brine concentration. The ordinate represents the fraction to which an original fictive depth of brine has been filled by solid salt at any point during the sequence of salt crystallization.

It is important to note that the bromide content in halite is constant for any net amount of evaporation accumulated by the brine remaining in the basin, regardless of the inflow-evaporation regime. One hundred and ten ppm bromide means a 20X concentration of sea water [= (total mass of water, present and evaporated)/(Mass of water present)], no matter whether there has been inflow, constant volume, or any other regime. At constant volume, as shown in Fig. 7, the thickness of salt in which this bromide profile is distributed is simply stretched out. The ordinate in Fig. 7 is the ratio of the thickness of salt deposited (at any stage) to the fictive depth of brine, here defined as the depth the brine would have had in the beginning if the basin were cylindrical.

Analysis for bromide. In order to put all analyses on a constant basis of ppm bromide in halite, in the easiest way, samples were redissolved in water, filtered, and recrystallized as NaCl, then carefully remixed before weighing for analysis.

Early in the program we settled on a colorimetric procedure for the determination of bromide that was sufficiently free of interferences and sensitive enough to determine bromide in second-cycle salts and in relatively fresh waters. The method is adapted from that published by Balatre (1936), in which bromide is oxidized with chloramine-T in the presence of phenol-red and a pH 5 buffer. Color development is carefully timed and the complex measured at 595 nm in a Bausch & Lomb Spectronic colorimeter with a 1-inch test tube.

The analytical method was developed with Mr. Gale Baker of Chevron Research Company, and he made most of the analyses reported here. A standard sample ("analytical reagent" KCl, with a bromide content of 150 ppm) was included in each batch of analyses, and over a 5 year period, including several recalibrations, the standard deviation on this sample was $\pm$ 4%. Replication is a little better, about $\pm$ 3%.

Where contamination by potassium salts was important, two salt samples were picked visually and both analyzed for bromide and for their mineralogy by quantitative X-ray diffraction. Bromide concentrations in halite and, say, sylvite, were then calculated by the method of moments (Kühn, 1962). If contamination by potash salts was merely suspected, perhaps because of a high bromide content, the potassium content of the sample could be quickly checked on a Fisher clinical flame photometer, using the same solution that had been analyzed for bromide. Where contamination by fluid inclusions was suspected, the total water content of the sample was determined in an apparatus designed for other studies of water in salt: melting in a vacuum, freezing in a cold finger, reducing to hydrogen over hot uranium, and measurement in a calibrated-volume manometer. This result, combined with extensive experience in the chemical analysis of brine inclusions (Holser, 1963), served as criteria for confidence in the bromide analysis of halite.

Colored soluble material interferes with the analysis, but it is usually either organic material or iron in solution. Either one is removed by roasting the recrystallized sample at 500°C in air, redissolving, and filtering. Standard addition tests indicate that this treatment does not lose any bromide.

BROMIDE DISTRIBUTION IN SALT ROCKS

Recent salt. Halite and gypsum are now being deposited on extensive tidal flats at the upper end of Laguna Ojo de Liebre, Baja California (Phleger and Ewing, 1962; Holser, 1962). They are underlain by up to 2 meters of porous bedded evaporites, interbedded with sands, and permeated by bitterns whose stage of evaporation increases with distance from the lagoon. Frequent tidal- and wind-driven inflows continually evaporate and begin to crystallize both salt and gypsum about 2 km from the beach, both on the surface of the brine and also on the bottom beneath the few centimeters of brine. Halite from such a locality has about 50 ppm Br.* Comparable data from other Recent salt deposits do not seem to be published. This level is considerably below the 75 ppm calculated from a distribution coefficient carefully measured in the laboratory on synthetic solutions, as described above. However, experiments with the direct evaporation of Mediterranean water (with the same Cl/Br as modern ocean water) gave a first halite of only 38 ppm (Bloch and Scherb, 1953, p. 156). Comparable data from other Recent salt deposits do not seem to have been published. Apparently the value for the first halite from modern sea water is still indeterminate by a factor of two, either because some important component has not been included in the laboratory determination of distribution coefficients, or, even more likely, because the value actually varies within the rate or other conditions of crystallization. Hard salt flats 4-15 km from Laguna Ojo de Liebre apparently experience inflows only during occasional storm tides. In that area, one ordinarily finds several centimeters below the surface a brine that is very highly concentrated, in some cases nearly into the potash-magnesia facies (concentration 60× sea water), and gypsum is pseudomorphically replaced by polyhalite. Halite from such localities have the highest bromide in this deposit, but this is still only about 90 ppm, about half the amount expected from equilibrium with such a brine (Fig. 7). Capillary evaporation may have been an important part of the history of this deposit, resulting in a higher stage of evaporation of the brine, with the deposition of only a small quantity of late-stage, high-bromide halite around the pores. Whether or not this is the case, the remarkable thing is that this concentrated brine has not been able to raise the bromide content of the porous salt to its equilibrium level, even after hundreds of years. The possible importance of these processes in ancient evaporites should not be overlooked. Work is continuing on a large collection of samples from surface and cores in this area.

Normally regular profiles. Bromide distribution has been studied intensively in the Permian Zechstein of Germany (Kühn, 1962; Braitsch, 1962, pp. 151-159; geology reviewed by Borchert and Muir, 1964, pp. 44-57; Richter-Bernburg, 1957). This first work on bromide was, in some

**Footnote added 15 June 1972:* Later analyses of 25 samples of halite from cores in this area indicate a range of 100–195 ppm Br, which represents, in general, a smaller discrepancy from theoretical values than the earlier analyses (the single earlier analysis at 50 ppm may have been an isolated case of recrystallized halite).

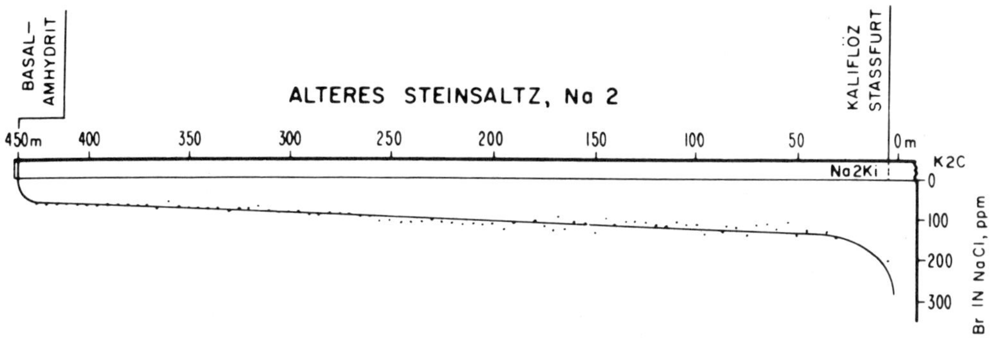

Figure 8. Typical "normally regular" bromide profile from the second salt of the Zechstein series, after Schulze (1960, p. 99). Thirty meters below the top of the profile, the slow, regular climb of bromide content curves sharply to the right, probably because of a change to simple evaporation, finally leading to the Stassfurt potash bed at the top.

respects, a lucky choice, in that at least some of the bromide profiles resemble the ideal profiles of Fig. 7. I choose just one illustration, from a large number published by Schulze (1960, p. 99). Using the principles outlined above, this profile can evidently be divided into three sections, differing from one another in their inflow-evaporation regimes. The short section at the base has the characteristics of a constant volume reservoir of rather small depth (as discussed in a later section). The sharp rightward curvature of the upper section suggests that most or all of the inflow of sea water was cut off while evaporation proceeded rapidly to finally deposit the potash bed near the top of the section. Interpretation of the long steady rise in the main center section is not so clear: possibly it represents a shift to a more efficient reflux of concentrated brine back to the sea, that is, a smaller $\underline{r}_f$. If my interpretation of this bromide profile is correct, the implication is that final formation of the potash bed at the top of the sequence was almost an accident that depended upon cutting off access to the sea, after a moderate concentration level had been built up in the brine over hundred of years (see also d'Ans and Kühn, 1960). Indications of the high brine concentrations necessary for potash crystallization are evident only above this final break in the curve; no hint of them is given in the bromide distribution of the main part of the salt section.

Similar curves are found in many localities for this second of the four main evaporite cycles of the Zechstein series. Data, from near and in the potash facies only, in the Paradox Formation of Utah (Lindberg, 1946; Raup, 1965) and the Oligocene of the Upper Rhine valley (Barr and Kühn, 1962, p. 298) suggests that these profiles may also be fairly regular. Some other parts of the Zechstein salts also have a regular development of this sort, but much of the salt rock in the upper two evaporite cycles has very irregular profiles. These are well displayed in the literature, and my results from many basins in the United States and Canada suggest that irregular profiles are actually more usual than the ones found for the Stassfurt series.

A regular and low bromide content also prevails through the small sampling available from the Triassic-Jurassic salt of the Gulf Coast. General geology of this evaporite province has been ably surveyed by Murray (1961), pp. 202-287). The undisturbed Louann salt is seldom seen, in a few wells along the north edge of the basin, and no cores of this were available to me. However, many sections of the salt are well displayed in two extensive belts of salt domes. Kupfer (1963, 1965) and Muehlberger (1959, 1965) have graphically described the intensely folded vertical structures of these piercement domes. In several mines samples were taken using this kind of structural data so that, insofar as possible, they represented a section across the bedding. Although the sequences covered distances of hundreds of feet, it is impossible to guess the original stratigraphic thickness to which they correspond. The sequence at Grand Saline, Texas, is typical: 44, 50, 47, 46 ppm bromide probably on one fold from bottom to top; on other folds: 39, 39, 42, 38, 39 ppm. At Weeks Island, Louisiana, a sequence is 49, 51, 47, 45 ppm; single values from other domes are 34, 49, and 89 ppm bromide. The original Louann salt in this basin apparently covered 150,000-200,000 square miles, with a thickness up to 5,000 feet (Murray, 1965), so what we have here doesn't represent much of a sample. It is sufficient, however, to suggest a low and regular bromide content for part of this formation.

Commonly irregular profiles. A profile through the Hutchinson salt member of the Wellington Formation, in the Permian of Kansas, illustrates the highly irregular form that is more common than any other in bromide profiles. The distribution of the Hutchinson salt is described by Kulstad (1959), and its textural features are interpreted in terms of environment by Dellwig (1963). The samples analyzed in this study were collected in the Carey Mine, and from a continuous core cut 8 miles south at the Naval Air Station. The petrography of this core has been described in detail recently by Jones (1965), and at the previous Symposium I discussed our chemical analyses of fluid inclusions from this locality (Holser, 1963).

Much of the Hutchinson salt is nicely interbedded with thinner layers of clay, anhydrite, or carbonate, and exposures in the Carey Mine show that many of these are continuous for at least hundreds of feet. Occasional masses of white to clear, very coarsely crystalline salt were probably, to judge by the chemistry of their included brines (Holser, 1963), formed by recrystallization in place (see below). Much of the Hutchinson salt shows the chevron textures indicative (Dellwig, 1955, 1963; Wardlaw, 1964) of primary crystallization. Samples were chosen to represent a variety of these textures and stratigraphic situations in the sequence of salts and interbeds.

The profile, as shown in Fig. 9, starts out in a normal fashion, gradually rising to almost 60 ppm at about the Carey mine level. Small-scale sampling in the mine indicates no substantial variations in a horizontal or vertical direction, within a short sequence of salt layers. Most of the rest of the core has bromide values that are all low, but very irregular in their sequence, until they decrease generally near the top of the section. The last salt at the top of the section has only 10 ppm bromide, indicating that it must have suffered recrystallization and probably redeposition -- a "second cycle" salt.

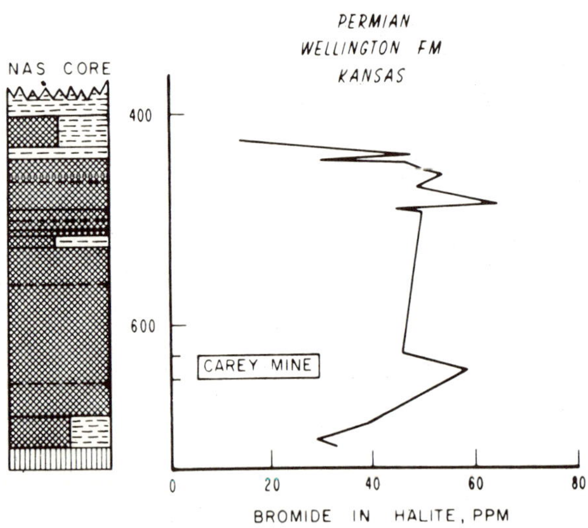

Figure 9. Bromide profile through the Hutchinson salt member of the Wellington Formation, Permian of Kansas. Crosshatching is mainly salt, vertical lines are anhydrite, and dashed lines are shale, all highly generalized. A detailed section has been published by Jones (1965, pp. 55-65).

The Salina Formation of Michigan and Ontario is another example of this sort of profile. The general distribution of salt is shown in Fig. 10, and the stratigraphic sequence of salt beds is

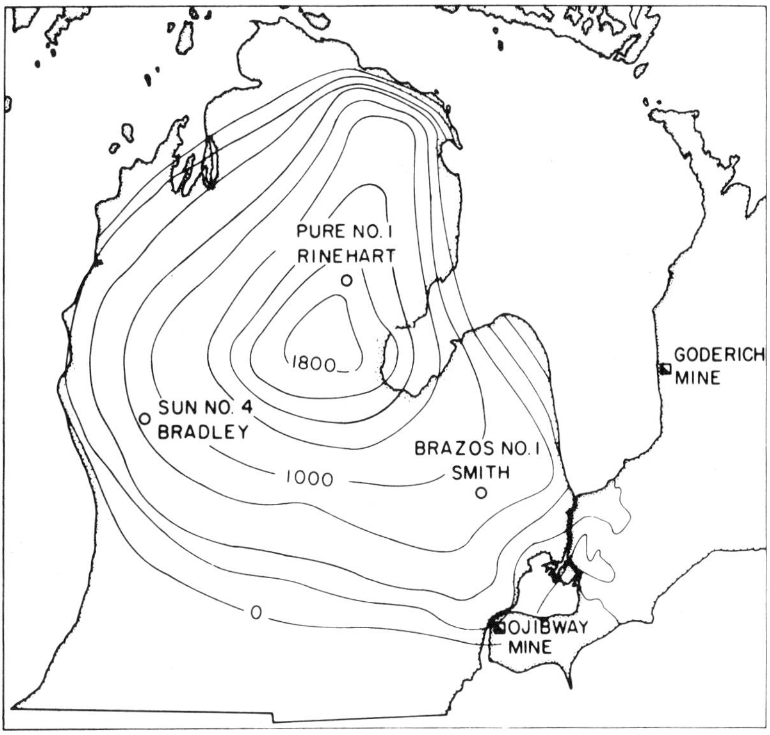

Figure 10. Aggregate thickness of all salt in the Salina Formation of Michigan and Ontario (Landes, 1945). Sample localities are indicated.

indicated in Fig. 11. The paleography and general evaporite conditions have been discussed fully by Alling and Briggs (1961), and Dellwig (1955) gives considerable detailed petrography and interpretation from a core out of the Sun #4 Bradley well in the western part of the basin (Fig. 10). My samples came from two mines very close to the eastern edge of the basin (Hewitt, 1962), and two wells on the slope and in the center of the basin. Where exposed in the mines, salt in both the bottom and top beds shows excellent and continuous bedding, mainly accentuated by anhydrite concentrations. At the basin center the first, or A_1, salt starts out with the usual 40 ppm bromide, but rises near the top of this bed to values high enough to suggest the presence of potash in this part of the basin. The rest of this profile is irregular, and, like that at Hutchinson, rather low. For the most part, salt toward the edge of the basin is even lower in any given bed. In connection with these profiles, a piece of core from Pure No. 1 Rinehardt in the Devonian Lucas salt (2500 feet above the top of the profile in Fig. 12) analyzed 69 ppm Br. Evidently this salt is marine and cannot have any substantial second cycle contribution by solution of the Salina salt, as hypothesized by Landes (1951, p. 19).

Many large salt sections apparently have a bromide content that lies generally below the minimum value of 75 ppm expected from the crystallization of sea water, and many of these profiles are very irregular in the long run, while being homogeneous on a small scale. This requires

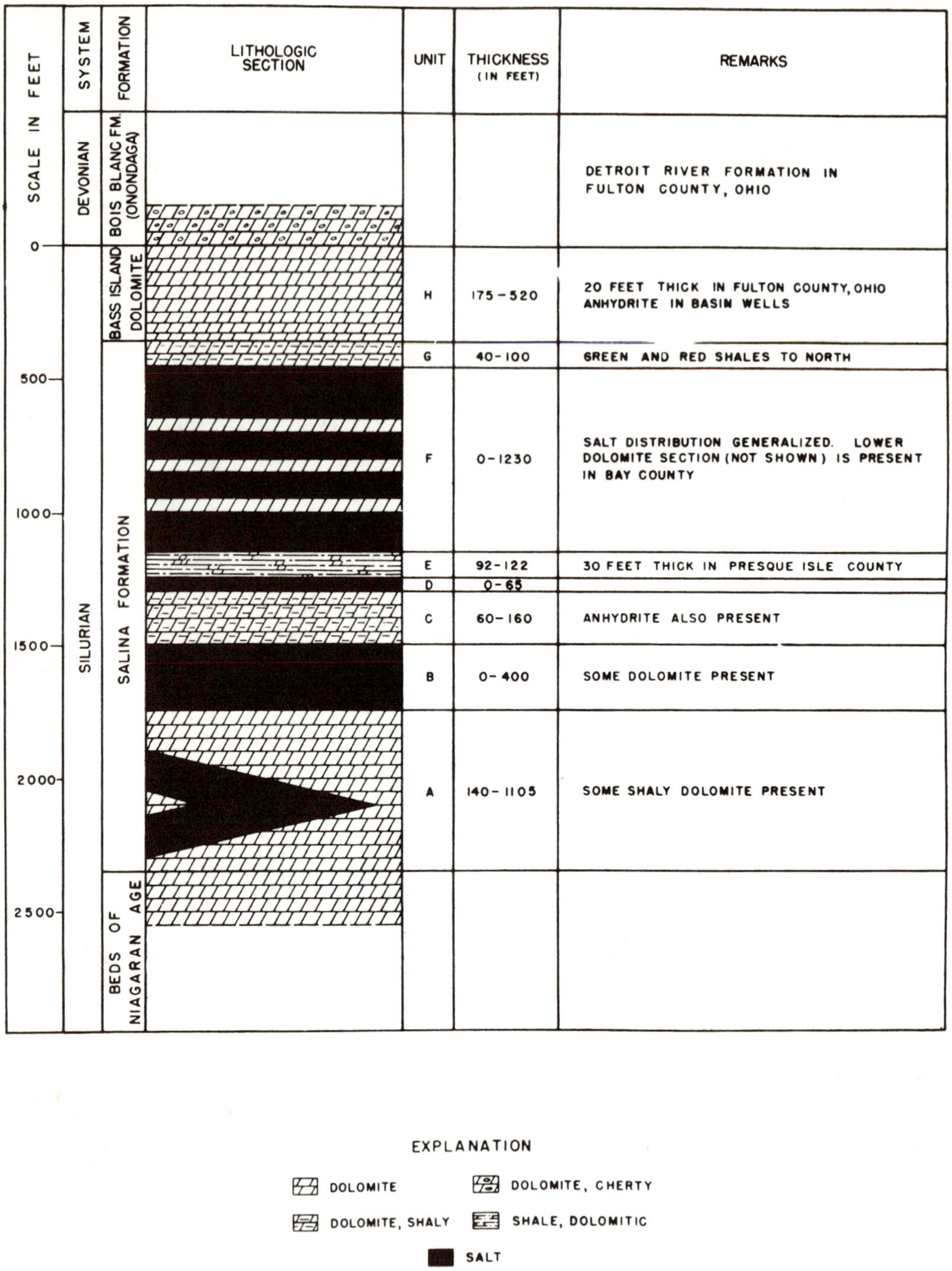

Figure 11. Generalized stratigraphic section of the Salina Formation (Landes, 1945).

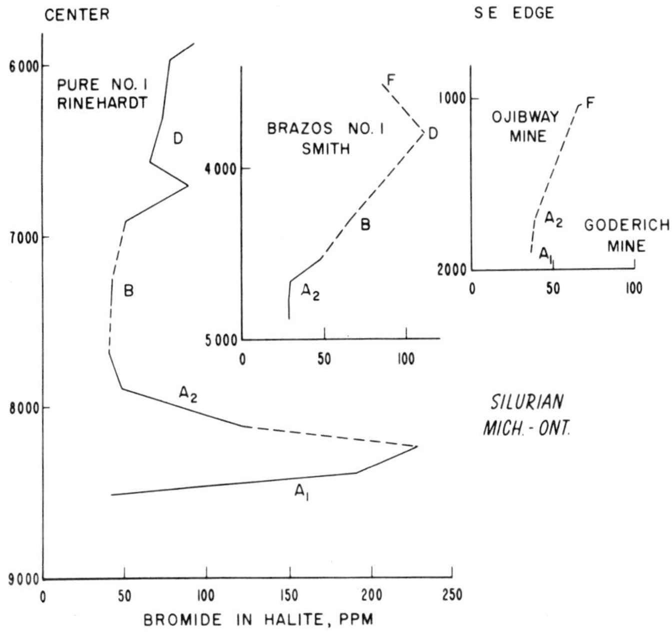

Figure 12. Bromide profiles in the Salina salt in various parts of the Michigan basin.

a modification of the conventional picture of an evaporite basin with uniform inflow and reflux. Perhaps these basins suffered multiple separate incursions of sea water, each one of which may have dissolved and redeposited some of the salt previously deposited, and each of which lost most of its bitterns and consequently its bromide by subsequent outflow or flushing. It is difficult to see how the irregularities observed could be perpetuated in the presence of the flywheel-effect of a body of brine continually overlying the depositing body of salt.

Uncommonly problematical profiles. As irregular as the above profiles may be, they do not present such startling problems as do profiles from the two largest evaporite basins in North America: the Permian Delaware basin in Texas-New Mexico, and the Devonian Alberta basin in Canada.

The geology of the Delaware basin was reviewed by Adams (1963) in the first Salt Symposium and the Carlsbad potash district was described in this symposium by Linn and Adams (also Jones, 1954, 1962). Figs. 13 and 14 show the great horizontal and vertical extent of this series. The Delaware basin itself is a deep basin in front of the high Capitan reef. This pre-existing basin was filled by evaporites of the Castile Formation, including thick sections of primary laminated anhydrite, as well as salt of undetermined texture. The overlying Salado evaporites covered the Castile and extended north and east over the back-reef and shelf areas for considerable distance, beyond the edges of Figs. 13 and 14. Anhydrite here was subordinate to salt, but the anhydrite beds are very persistent throughout the basin. In an area around Carlsbad, New Mexico, both in front of and behind the reef (Figs. 13, 14), potash mineralogy is developed through a considerable part of the Salado evaporite section. Most of the salt in the Salado Formation, both within the potash zone and outside of it, is coarsely crystalline and never shows the chevron textures that would prove primary deposition.

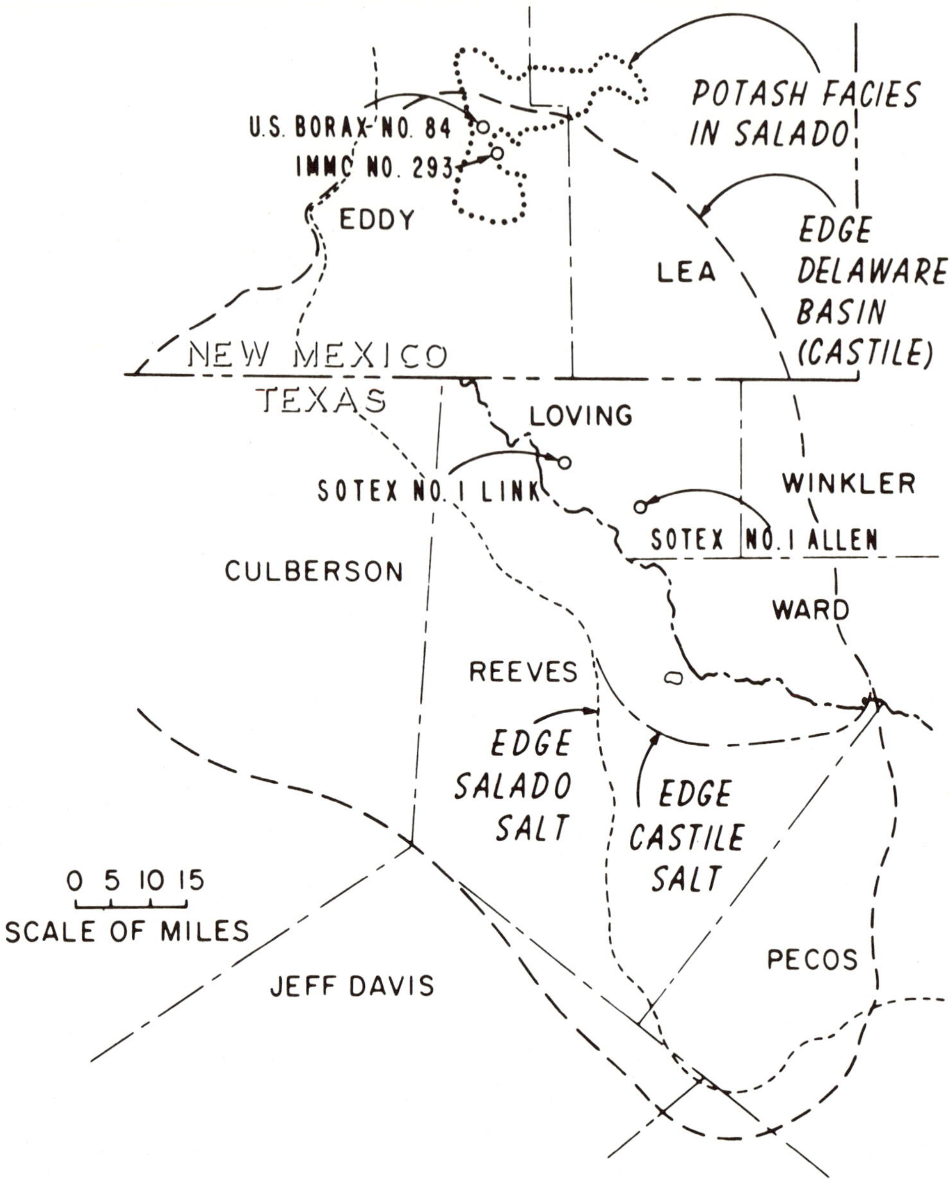

Figure 13. Evaporites of the Ochoa series in the Delaware basin, Texas, and New Mexico, showing sample localities.

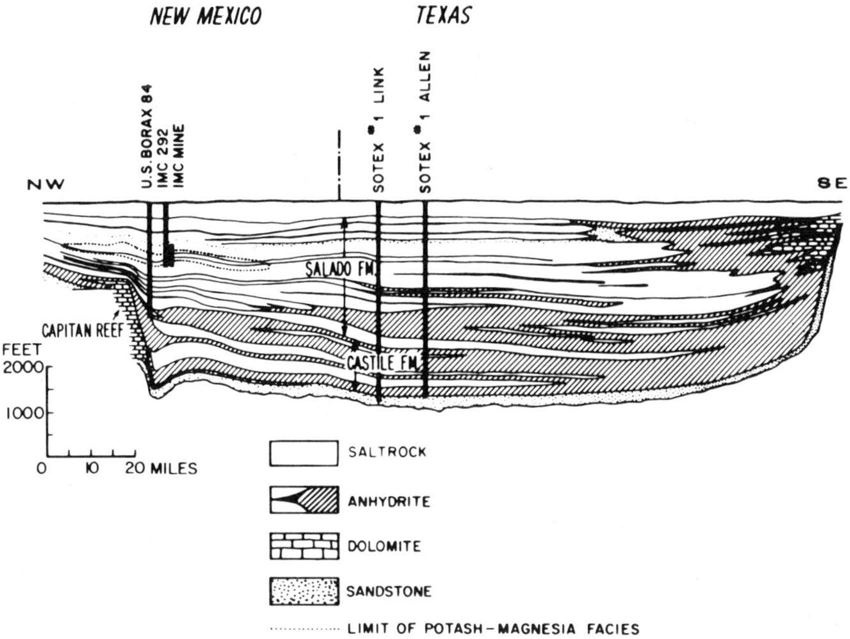

Figure 14. Generalized cross section through the Delaware basin, Texas, and New Mexico.

Some profiles from the Delaware basin are shown in Fig. 15. In the Castile salt no cores were available, so cuttings from two old churn-drill holes were analyzed, at locations indicated on Fig. 13. The lowermost Castile salt might well be classified with the above "commonly irregular profiles," including the characteristics of a normal beginning and a continuing irregular curve in the range 30-50 ppm. A second Castile salt, as well as the base of the Salado Formation (far below the potash zone) is even lower in bromide, in the range of 20 ppm. Although my main interest was in halite, these results posed the interesting question as to just how far up the section one had to go before finding the rise in bromide content expected to accompany potash deposition (Figs. 7, 8). The answer turned out to be, as you see in the upper section of the Salado profile (Fig. 15), that such high values never are attained. Halite continues to show the irregular values of a common profile even through the potash zone, with most values below 70 ppm. This raised the further question as to the bromide content of the potash minerals. Four sylvite/halite pairs were found to have Br_{syl} = 200, 450, 740, and 780 ppm, with ratios Br_{syl}/Br_{hal} of 10, 12, 22, and 30, respectively. The bromide in sylvite is far below the more than 1500 ppm expected by either metastable crystallization from the sea water brines or by thermal decomposition of carnallite (Braitsch, 1962, pp. 105-113; Kühn, 1962), while the ratios approach the theoretical value of 10 ± 1. Generally low bromide in the salts from carnallite, sylvite, and halite in Carlsbad was already recognized in the earliest analyses (Lindberg, 1946; Lang, 1954). These data suggest a rather closely related origin for the halite and sylvite minerals that we see in Carlsbad today. Fluid inclusions from Carlsbad salts are also very low in bromide, with Cl/Br near that of sea water, but with higher magnesium content (Holser, 1963, pp. 90-91).

These data all indicate that the evaporites in the Carlsbad area were (most recently) crystallized from a second cycle of new water, low in bromide. The bromide content is not so low but that this new cycle of crystallization could have been caused by sea water, but there is no question that it must have been a fresh cycle of sea water, with the earlier concentrations of bromide removed. There is a further question as to whether the second cycle crystallization occurred

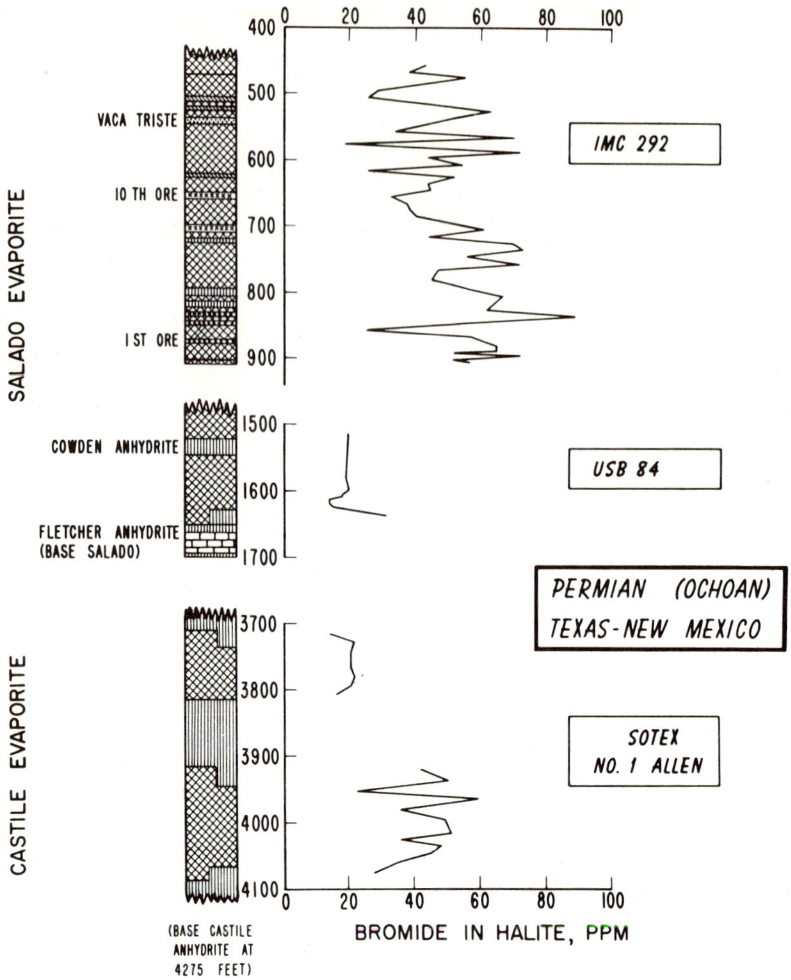

Figure 15. Bromide profiles through the Ochoa series in the Delaware basin, at localities indicated in Figures 13 and 14. Symbols for the generalized cross section are as in Figure 9; crosses indicate the thickest potash ore beds.

in place, or whether the salt was dissolved elsewhere by the water and newly crystallized here. Detailed mapping and petrography (Jones, 1962) seem to support the first idea, although in that case one might have expected a much smoother bromide profile, as well as an even greater incidence of solution-collapse than is found in the area (Maley and Huffington, 1953). Hopefully, the detailed studies still in progress in the area may shed further light on this question, which is a critical one in evaluating the origin and history of these important deposits.[1]

[1] Discussion at the Salt Symposium with Mr. S. S. Adams of International Mining and Chemical Company, Carlsbad, who is completing a very detailed study of bromide distribution in their mine, confirms low and irregular nature of the bromide in the potash zone.

The second profile is even more problematical. Evaporites of the Devonian Elk Point group extend across the western interior all the way from North Dakota to the Northwest Territories (Pearson, 1963; Grayston, Sherwin, and Allen, 1964); a part of this section is shown schematically in Fig. 16. The most widespread salt formation is the Prairie evaporite which, in Saskatchewan, also includes very extensive bodies of potash minerals. The bromide profile of Fig. 17 through the Prairie evaporite in Anglo-Canadian No. 1 Elk Point, east-central Alberta, is typical for this formation. A profile through Vermillion Consolidated Oil No. 15, south above the Prairie Plateau, as well as profiles in Saskatchewan published by Schwerdtner and Wardlaw (1963) and Wardlaw (1964) are only qualitatively similar to this one. The profile in the Elk Point well is of

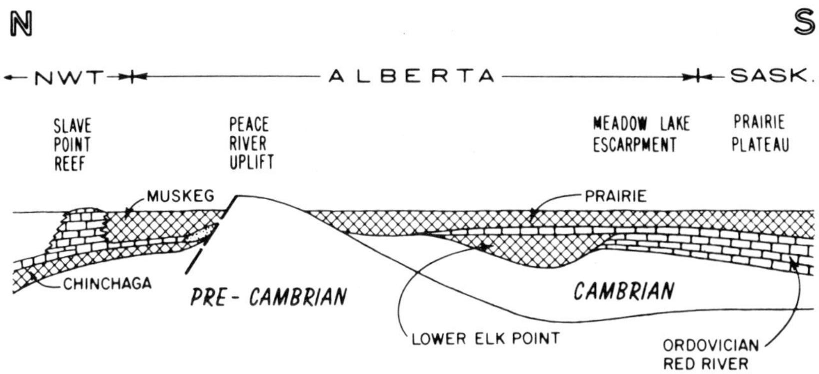

Figure 16. Schematic cross section through the lower Paleozoic of western Canada, referred to the top of the Prairie Formation.

the "commonly irregular" sort, although in the slightly higher range 60-100 ppm, with a number of spikes up to 150 ppm. This is rather reminiscent of the Salado profile, although here in Alberta only traces of potash minerals are present. The large variations are found even in the rather fine sampling shown by the inset on Fig. 17. W.M. Schwerdtner (1964, p. 1114) found low bromide in the Prairie evaporite potash zone in Saskatchewan, although not so low as I found in the Salado.

The startling feature of the Elk Point evaporites, however, is the two earlier salts, which only fill in the Alberta basin (actually two connected basins). These salts have an extraordinarily low bromide content, in the range 2-6 ppm, that is continuous almost without exception through the entire salt section. Again, no regular trends are evident within this low range of values. The purity of this salt can be appreciated when I tell you that a typical analytical-grade sodium chloride has 15 ppm bromide. Two or more cycles of crystallization are certainly necessary for the purification of salt to this low a level. Furthermore, the amount of bromide in sea water prevents reduction below 7 ppm no matter how many cycles of fresh sea water are involved. Consequently, recrystallization by meteoric waters is required for the formation of this salt. Possibly this is some sort of playa-lake deposit deriving its halides from evaporites of the Ordovician Red River Formation on the Prairie plateau to the south. Red bed character of the surrounding and interbedded sediments and nearly complete absence of anhydrite, support this model, but fragmentary marine fossils in the salt argue against it. It is also possible that recrystallization of these salt beds by the action of meteoric water took place in their present position after initial marine deposition. One other profile, in thin salts deposited far to the north in the Mackenzie basin, correlative with the Lower Elk Point, showed a more common irregular profile in the range 40-70 ppm

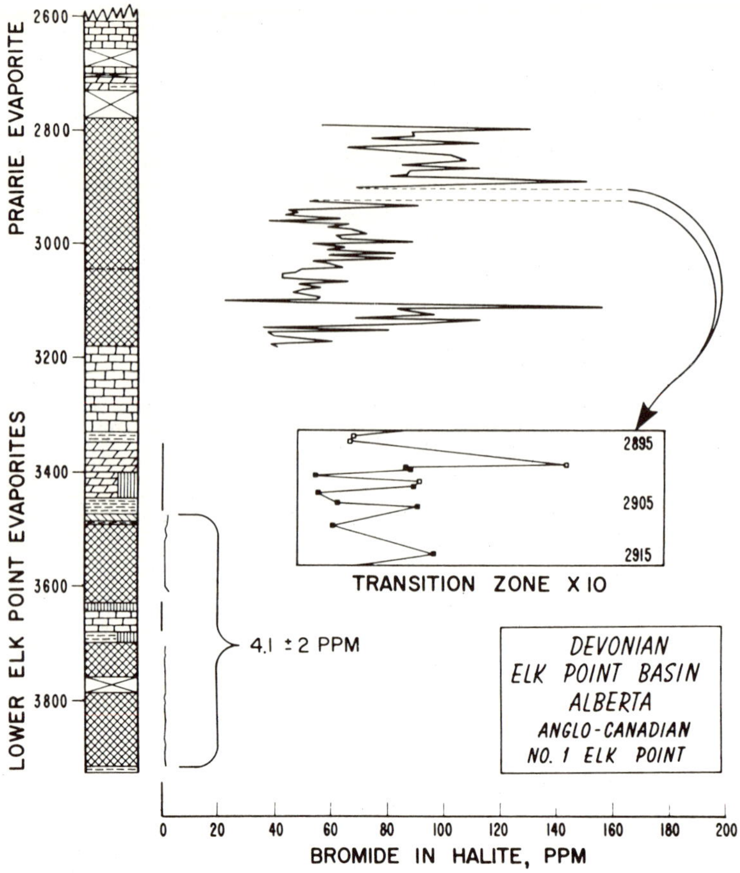

Figure 17. Bromide profile through the Elk Point Group in the Alberta basin, just northwest of the Meadow Lake escarpment (Figure 16). Detailed analysis for one 10 foot zone is shown on an enlarged scale, with dark symbols indicating gray, cloudy, possibly primary salt and open symbols indicating clear, white, possibly recrystallized salt.

bromide. It will be interesting to determine the extent of the extremely low bromide content in the Alberta basin, as more sample material becomes available in the future.[2]

I might mention in passing a problematical profile at the other end of the scale, from the depths of Siberia. The thick (up to 1000 m) and extensive lower Cambuan evaporites of this area were reviewed by Ivanov and Levitskii (1960, pp. 8-31) and by Yanshin (1962). Bromide profiles published by Ogienko (1959, pp. 895-898) and one which I did on a set of core samples (obtained for another purpose) show 120 to 200 ppm Br at the base, rising irregularly to 300-400 ppm Br at 600 meters above the base in a 700-meter section. Most of the halites have enough bromide to be

[2] Discussion at the Symposium revealed that D. W. Watson, working with N. C. Wardlaw at the University of Saskatchewan, had sampled the same core in Anglo-Canadian No. 1 Elk Point. They also found an extremely low bromide content, although it was below the sensitivity of their conventional van der Meulen procedure so that an actual profile was not obtained. They got similar results in Anglo-Canadian No. 2 Elk Point, about seven miles to the northwest. We are publishing a joint paper discussing the details of this low-bromide profile and its paleogeographical implications (Holser, Watson, and Wardlaw, in preparation).

in the potash facies, yet no sylvite was evident in the samples, and core analysis showed only traces of potassium. As of 1962 (Yanshin, 1962, pp. 2137-2140) no substantial deposits of potash had been discovered in the basin, although they certainly are expected. It is difficult to visualize a mechanism by which bromide can be so highly concentrated in halite without also depositing potash salts. Possibly the latter were diagenetically leached by solutions that did not stay around long enough to come equilibrate with the remaining halite. Or is it possible that K/Br was actually lower in the Cambrian sea than today in spite of the occurrence of high-K glauconite in Cambrian strata (Yanshin, p. 2138)?

SOME PROBLEMS IN BROMIDE GEOCHEMISTRY

I would like to go back over the details of some particular aspects of bromide geochemistry that could only be touched upon incidentally in the general geological descriptions of bromide profiles.

Origin of common bromide profiles. Most of the common type of bromide profiles in salt rock share some or all of the following characteristics:

1. The profile begins with a basal bromide content in the range 30-50 ppm, comparable with that found on evaporation of modern sea water both in the laboratory and in a Recent salt deposit, but far below the 75 ppm predicted by laboratory experiments on distribution coefficients.
2. The profile rises regularly for a few meters above the base.
3. The profile is very irregular through most of its duration, commonly in the range 30-80 ppm, with occasional spikes up to 150 ppm.
4. These relatively low and irregular bromide levels may continue into the potash zone, if there is one.
5. Limited sampling indicates homogeneity over several feet vertically, and at least that far horizontally.

In some cases the irregularity could be explained as simply due to repeated incursions of new sea water with efficient flushing of the old bromide-rich bitterns. It could also be explained as a result of one or more subsequent cycles of crystallization, which would also serve to explain point (4). Furthermore, such an explanation conforms with the recycling that is certainly required by the extremely low bromide content of the Lower Elk Point subgroup in Alberta (Fig. 17), and of the top of the Hutchinson salt of Kansas (Fig. 9). Although common bromide profiles show effects of multiple cycles of crystallization, the bromide analyses themselves cannot determine whether the new cycle of crystallization occurred by the circulation of new waters (sea water, meteoric waters, or metamorphic water from gypsum dehydration) through solid but porous salt, or whether the salt was actually dissolved in one place and redeposited elsewhere as a "descendent" salt. More needs to be known about the three-dimensional distribution of bromide on a large scale, and its relation to bedding and other structures, before this question can be settled. In the Carlsbad potash zone (and in Zechstein 3 and 4) good horizontal uniformity over short distances and at least rough correlations over hundreds of feet (horizontally) suggest that the salts were sedimented in their present condition on the surface, but in the Prairie evaporite potash zone of Saskatchewan Schwerdtner (1964, pp. 1111-1113) finds considerable inhomogeneity of bromide content even on a small scale, as well as petrographic evidence of secondary alteration in place. This question must be considered still open.

Recrystallization of halite. Dellwig (1955), Gottesman (1963), and Wardlaw (1964) have described patterns of fluid inclusions that are indicative of various growth forms in halite crystals, and are consequently firm evidence for primary crystallization of these particular rocks. These patterns of inclusions are in marked contrast to other salts that have apparently undergone some form of recrystallization, where inclusions are segrated to grain boundaries. Dome salts that have undergone extensive plastic deformation also show inclusions along (110) or irrational intercrystalline shear planes. There are a number of other characteristics of certain salt rocks that may be correlative with this distinction between "recrystallized" salt and "probably primary" salt: clear vs. cloudy color, coarse (2-200 cm) vs. medium- to fine-grained crystallinity, friable vs. compact texture, and crosscutting vs. bedding structure. Even casual observation will soon make it evident that not all of these characteristics go together, yet collectively they represent a suite of attributes that possibly indicate growth of new crystals in halite at some time after original crystallization. The friable salt, for example, I have so far seen only in mines in two Gulf Coast salt domes. This salt has been so well recrystallized that each single crystal is bounded by a set of plane or slightly convex polygons, unrelated to crystal faces, apparently in surface tension equilibrium (Smith, 1951, pp. 1-20). A large chunk of such rock easily disaggregates into the single crystals if you kick it. Crosscutting structures commonly transect the bedding in flat-lying formations of many evaporite basins, as well as occasionally being evident in salt domes. Figure 18 illustrates one of these bodies in the Salina salt of the Michigan basin. It may be useful to point out that in a core such salt would probably be logged as a differently textured salt bed, when actually it is not a bed at all.

Figure 18. Mass of white, coarsely crystalline, recrystallized salt crosscutting well-defined bedding of Salina unit F in the Ojibway Mine, Ontario. Area is about 8 feet wide.

It was rather interesting to discover that some of these clear-to-white, coarsely crystalline, crosscutting bodies, which seem to have been formed by some kind of recrystallization, actually had a higher rather than a lower bromide content compared with the surrounding bedded salt. In Table 1, I have collected a number of pairs, from a wide variety of occurrences, and find that in nearly all of these bromide increased in the "recrystallized" salt by a factor of up to 3. Figure 19 shows one situation studied in detail. The salt was homogeneous in bromide content throughout both the horizontal cross section of a 3-inch core, and for distances of a foot or more vertically, transecting several varve lamellae. Furthermore, grain size showed good negative correlation with bromide content. Considering the diversity of the pairs in Table 1, and the probability that they do not all represent the same process, their consistency is surprising. I have no explanation to offer for this relationship.[3]

[3] Discussion at the Symposium revealed that similarly high bromide in "recrystallized" halite had been found in the Prairie evaporite of Saskatchewan-Alberta by N. C. Wardlaw and in the Windsor evaporites of the Canadian Maritime provinces by C. A. Barr.

TABLE 1

BROMIDE IN RECRYSTALLIZED HALITE

Formation	Location	Bromide in ppm "Normal"	"Rextall"	Rextall Normal
Salado	Carlsbad, N. Mex.	18	42	2.3
Wellington	Hutchinson, Kans.	30-65	65	1-3
Prairie	Alberta	22-90	70-155	1.0-2.9
Salina	Goderich, Ont.	36-46	88-131	1.9-3.6
Salina	Windsor, Ont.	66-68	85	1.3
Salina	Fairport, Ohio	69	113-121	1.6-1.8
Louann	Grand Saline, Tex.	39-51	39	0.8-1.0

Basal halite. Presumably the first salt to crystallize in an evaporite basin should have a bromide content that was determined only by the bromide/chloride ratio in the water coming into the basin. If the salt had not been recrystallized or otherwise altered, and if the halides were predominately marine, this ratio should now allow us to calculate (equation 1) the bromide/chloride ratio in the ocean of that time. The problem is that we do not yet know the reason why evaporation of modern sea water, both natural and experimental, gives substantially lower bromide than the 75 ppm calculated from measured distribution coefficients. But it is interesting to see whether basal halites even lie within this range between 35 and 75 ppm.

Figure 20 collects data on basal halites, both from my own profiles and a few published ones. Formation names and locations refer to profiles published elsewhere in this paper, except for the following: Cheshire, England (Haslam, Allberry, and Moses, 1950; also letter of November 19, 1959 from J. Haslam); Hall, Austria: Kühn (1955b, p. 65); Germany: Schulze (1960), d'Ans and Kühn (1940), Barr (1954), Kühn (1962); Prairie, Saskatchewan: Schwerdtner and Wardlaw (1963), Wardlaw (1964). The scatter is very wide, even allowing for the fact that most of these "basal" salts, although they are all from the base of the salt section in that profile, are not usually from the deepest known part of the basin. The basal values for the Leine series of the Zechstein were extrapolated (Schulze and Seyfert, 1959; Kühn, 1959; Schulze, 1960, p. 44) through a sharp high at the base inherited from the underlying Stassfurt potash zone. If we eliminate the Lower Elk Point subgroup of Alberta, which must be a second cycle salt involving meteoric water, there does seem to be a general tendency for basal salts to lie in the range 30-50 ppm. I suggest that whatever process made the modern Baja California salts lower than the theoretical 75 ppm, could also have operated, sometimes to a greater extent, to cause the series of values displayed in Fig. 20.

There are two reasons that we might expect some variations in the bromide/chloride ratio of ancient sea waters. Both elements are "excess volatiles" in the sense that the present amounts in the hydrosphere and sediments cannot have been contributed by the weathering of igneous rocks, but must have been a product of volcanic degassing of the crust and mantle throughout their history. Modern vulcanism and hot springs emanate halites with a substantially higher chloride/bromide ratio than modern sea water (Fig. 1), so one might suppose either that a large slug of halite of low Cl/Br ratio was contributed by a primeval ocean (Ellis and Anderson, 1961, p. 427), or that the composition of volcanic emanations had been changing with time. This might be a reason to look for some regular trend in the halite ratios of basal salts through geological time, although the amount of time even back to the Cambrian represents only a rather small fraction of the total

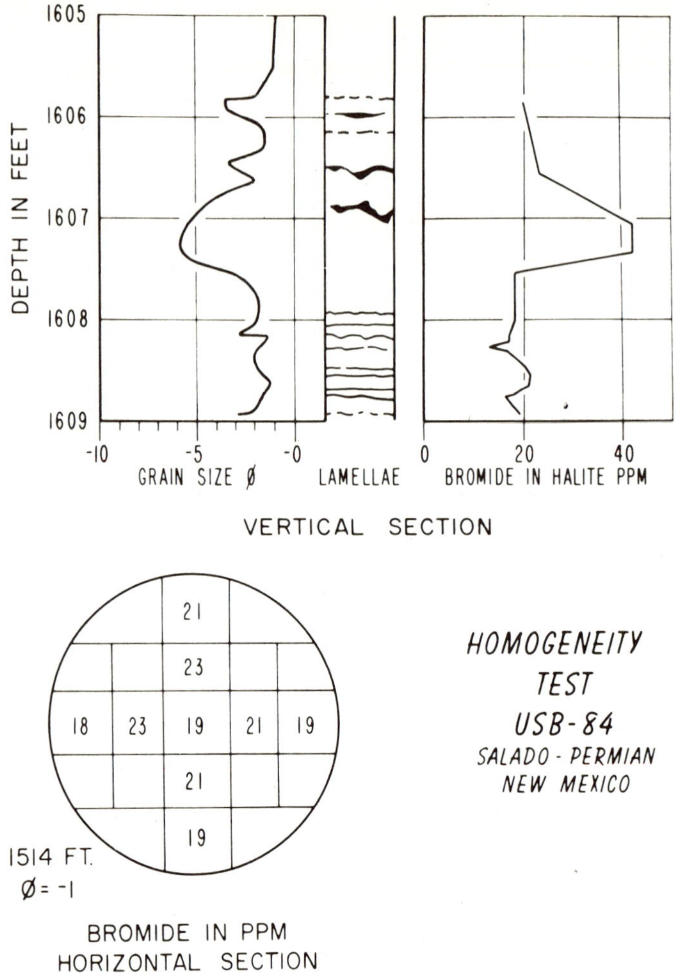

Figure 19. Vertical and horizontal homogeneity of bromide concentration, shown by detailed horizontal and vertical sampling of a core in the lower part of the Salado Formation. Note that the bromide content is relatively constant across the interbedded lamellae (mainly carbonate), but increase sharply with increasing grain size (decreasing ϕ). This homogeneity in a short section of core is only an episode in very large variations seen on the smaller scale display of Figure 15, both above and below this level in the evaporite.

time during which volcanic emanations have probably been accumulating. Certainly no regular trend can be demonstrated by the scatter of data presently available, as shown in Fig. 20.

The other viewpoint, adopted earlier in this paper, is that perhaps volcanic halides, whether they are primeval or simply recycled crustal material, may represent some kind of an average for the crust. If so, the volume of chloride stored up in evaporite rocks is at least as great, and probably greater, than the volume presently in the sea. This, then, leads to the interesting conclusion that as the proportion of evaporite and sea water chloride shifted throughout geological

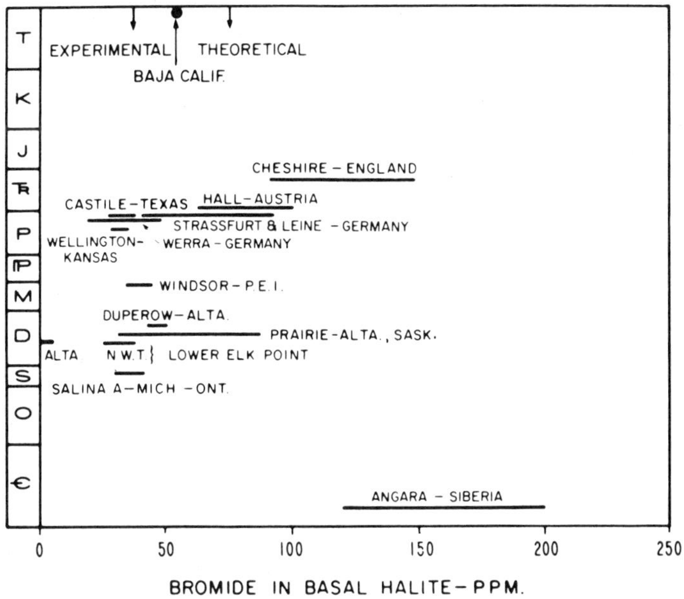

Figure 20. Bromide in basal halite plotted against geological age.

time, through the ever-changing relations of evaporite formation and salt rock erosion, both the chloride content and the chloride/bromide ratio should have been changing in the sea. That is, during periods when more salt rock was being eroded than was being deposited in evaporites, the Cl concentration and the Cl/Br ratio in sea water should have risen, and consequently the bromide content at the base of a salt profile should have begun at a smaller value. Periods of increasing evaporite deposition throughout the world, relative to rock salt erosion, should have increasing values in the basal salt bromide. This situation is strictly analogous to the sulfur isotope ratio in anhydrites, for which an age curve has now been fairly well established (Holser, Kaplan, and Silverman, 1963; Thode and Monster, 1964). The scatter of Fig. 20 gives no evidence that the sea was any richer in bromide at the end of the Permian, certainly a time when a very large amount of evaporites had been accumulated, than it is today. However, this possibility is something to look for in the future as we accumulate more data from profiles in a variety of basins, and with perhaps better petrographic control of the primary nature of the rocks.

At present, the results on basal salts are too scattered to support arguments for any kind of change in the chloride/bromide ratio of sea water since the Cambrian. Valiashko (1956b) came to the same conclusion but it is not clear what data he used.

Initial slope. Analysis of equations (2) to (5) for the general model of evaporite trace element deposition shows that, other conditions being uniform, the slope should be proportional to the mass of chloride into which the excess bromide is being rejected. In other words, if the fluid reservoir is large, the amount of bromide rejected during the crystallization of one meter of salt will not be of much consequence, and the bromide content of succeeding layers of salt will rise only slowly; whereas if only a thin layer of brine is present, this rejection of bromide will cause the brine composition, and consequently later salt layers, to be richer in bromide at a faster rate.

This leads to the concept of fictive depth, defined above, which is simply the starting mass, m_0 of equations (2) to (5), in different units. By the natural geometry of salt basins, and lateral variations in crystallization rate, the fictive depth may be somewhat greater than the average depth of brine.

270

**Footnote added 15 June 1972:* Later analyses of 25 samples of halite from cores in this area indicate a range of 100–195 ppm Br, which represents, in general, a smaller discrepancy from theoretical values than the earlier analyses (the single earlier analysis at 50 ppm may have been an isolated case of recrystallized halite).

Using a similar concept, Kühn (1955a, 1962) has calculated depths of the Zechstein sea from bromide profiles. Braitsch (1962, pp. 157-158) has criticized Kühn's approach on the basis that Kühn's model takes no account of the reflux of bromide-rich brines back to the sea, and concludes that the latter effect is so indeterminate that depths cannot be calculated. However, the generation of my equation (4) takes into account any amount and concentration of reflux by weighting it with the inflow to give a net inflow. Now if we differentiate equation (4) at the base, we find for the slope

$$\left[\frac{dr_c}{dm_c}\right]_{m_c = 0} = \frac{(g - K)\, r_{co} + (1 - g)\, K r_f}{m_o} \qquad (6)$$

In order to apply these equations to calculate a fictive depth from the ratio of basal bromide content to initial slope, we need only assume what I will call concordant inflow: that the regime of evaporation and inflow was constant just before and through the period taken for the initial slope. Under conditions like these, where crystallization is just starting in the basin, the trace element ratio in the inflow, that at the point of crystallization, and that in the outflow are all the same,

$$r_f = r_{co} / K \qquad (7)$$

It does not matter what the conditions (g) are, so long as they remain the same just after the beginning of salt crystallization in the basin, as they were before the crystallization started. Later, after the brine builds up an appreciable excess of trace element (assuming $K < 1$), equation (7) may no longer be a good approximation and in fact $r_f < r_{co}/K$. Now substituting (7) in the general equation (6)

$$\left[\frac{dr_c}{dm_c}\right]_{m_c = 0} = \frac{(1 - K)\, r_{co}}{m_o} \qquad (8)$$

Thus this assumption of concordant inflow allows an explicit solution for the fictive depth under any initially constant conditions, by measuring the initial value and initial slope of the trace element, and knowing the distribution coefficient for bromide in halite (including its variation with $MgCl_2$ concentration), and convenient units may be taken into account by determining the initial slope graphically in Fig. 7, whence

$$\text{Fictive depth in meters} = \frac{4.8 \times (\text{Initial value Br/NaCl in ppm})}{(\text{Initial slope Br/NaCl in ppm/meter of NaCl})} \qquad (9)$$

Falling back on the sometimes shaky assumption that the salts are primary, fictive depths were calculated from some of the bromide profiles in this and other published papers (Stassfurt: Schulze, 1960, p. 99; Prairie Industrial Minerals #3: Wardlaw, 1964, p. 272), as shown in Table 2. These must be considered doubly tentative in view of the scarcity of data and its wide variation, as a scan of the profiles will demonstrate. The recycling of salts discussed in previous sections would also complicate this model, although if the recycling were a sufficiently uniform process it would still come under equations (7) to (9). Even for the regular profiles of Zechstein 2 Braitsch (1962, p. 152) suggests that Schulze should have looked for petrographic evidence of diagnetic alteration in the lowermost section, because the bromide content was below 75 ppm. The shallow depths of 25-35 meters calculated here are well within the wide range of 2 to 130 meters (mean 20 meters) depth calculated by Kühn (1955a, pp. 659-662) using the bromide oscillations within single yearly varves of the upper part of the Stassfurt salt (various areas), and assuming no reflux. If I assume that the lower segment of the Stassfurt profiles is secondary, and that the very gradual rise of the middle segment can be extrapolated to the base of the section as a concordant flow, the depths calculated by equation (9) are over a thousand meters. If nothing else, this discussion demonstrates the sensitivity of such calculations to purely geological information and assumptions.

TABLE 2

FICTIVE BRINE DEPTHS FROM INITIAL SLOPES

Formation	Location	Basal Halite ppm	Initial Slope ppm/m	Fictive Depth m
Castile	Sotex #1 Allen	27	1.6	80
Stassfurt	Schierstadt 1/55	30	6.7	20
	Schierstadt 2/55	25	5.3	25
	Schierstadt 3/56	60	1.4	210
	Bernberger	35	5.0	35
Wellington	NAS Hutchinson	30	1.3	110
Prairie	I.M. #3 Ft. McMurray	32	0.62	250
	A.C. #1 Elk Point	40	2.	100
Salina	Pure #1 Rinehardt	42	4.	50

It should be further remarked that the fictive depths calculated in Table 2 are less than the thickness of the overlying salt formation (except the Prairie Formation at Ft. McMurray), so that if they bear any relation to reality, subsidence of the basin would seem to be required to keep it from filling up. Richter-Bernburg (1955a, p. 622) has emphasized that the very fast rate of salt deposition, say 0.05 m/yr, would on the other hand require a previously formed basin.

The principle of concordant inflow may be found useful when better data can be assembled.

CONCLUSIONS

If this paper has demonstrated anything, I think it is that bromide geochemistry of salt rocks is a lot more complicated than it may have seemed from the first research work on the Zechstein evaporates. Some very interesting questions have been raised, and only a few tentative answers have come out of this work. But there is no question that bromide has a unique place in geochemistry as an indicator for a history of halides in salt rock, potash, and brines. A great deal more work needs to be done before the full story can be untangled.

REFERENCES

Adams, J. E., 1963, Permian salt deposits of west Texas and eastern New Mexico, in Symposium on Salt: Northern Ohio Geol. Soc., pp. 124-130.

Alling, H. L., and Briggs, L. I., 1961, Stratigraphy of upper Silurian Cayugan evaporites: Am. Assoc. Petroleum Geologists Bull., vol. 45, pp. 515-547.

d'Ans, Jean, and Kühn, Robert, 1940, Über den Bromgehalt von Salzgesteinen der Kalisalzlagerstatten. I: Kali, 1940, no. 4, pp. 42-46.

__________, and __________, 1960, Bemerkungen zur Bildung und zur Umbildungen ozeanischer Salzlagerstatten: Kali u. Steinsalz, vol. 3, no. 3, pp. 69-84.

Baar, Auguste, 1954, Investigation of the bromine content in Zechstein salt: Bergbautechnik, vol. 4, pp. 284-288.

____________, and Kühn, Robert, 1962, Der Werdegang der Kalisalzlagerstätten am Oberrhein: Neues Jb. Mineralog. Abh., vol. 97, pp. 289-336.

Balatre, P., 1936, Colorimetric determination of small quantities of bromine in the presence of a large excess of chlorine: Jour. de pharmacie et de chimie (Ser. 2), vol. 24, pp. 409-413.

Bloch, M. R., and Scherb, J., 1953, On the Cl^-/Br^--ratio and the distribution of Br-ions in liquids and solids during evaporation of bromide-containing chloride solutions: Bull. Res. Council Israel, vol. 3, pp. 151-158.

Boecke, H. E., 1908, Uber das Kristallisationsschema der Chloride, Bromide, Jodide von Natrium, Kalium und Magnesium, sowie über das Vorkommen des Broms und das Fehlen von Jod in den Kalisalzlaggerstätten: Zeitschr. Kristallographie, vol. 45, pp. 346-391.

Braitsch, Otto, 1962, Entstehung und Stoffbestand der Salzlagerstatten: Berlin, Springer-Verlag, 232 p.

____________, and Herrmann, A. G., 1963, Zur Geochemie des Broms in Salinaren Sedimenten, Teil I: Experimentelle Bestimmung der Br-Verteilung in verschiedenen naturlichen Salzsystemen: Geochim. et Cosmochim. Acta, vol. 27, pp. 361-391.

Dellwig, L. F., 1955, Origin of the Salina salt of Michigan: Jour. Sed. Petrology, vol. 25, pp. 83-110,

____________, 1963, Environment and mechanics of deposition of the Permian Hutchinson salt member of the Wellington shale, in Symposium on Salt: Northern Ohio Geol. Soc., pp. 74-85.

Ellis, A. J., and Anderson, D. W., 1961, The geochemistry of bromine and iodine in New Zealand thermal waters: New Zealand Jour. Sci. and Technology, vol. 4, pp. 415-430.

Gottesman, W., 1963, Eine haufig auftretende Struktur des Halits im Kaliflöz Stassfurt: Geologie, vol. 12, pp. 576-581.

Grayston, L. D., Sherwin, D. F., and Allen, J. F., 1964, Devonian, in Geological History of Western Canada: Alberta Soc. Petroleum Geologists, Calgary, Alta., p. 49.

Haslam, J., Allberry, E. C., and Moses, G., 1950, The bromine content of the Cheshire salt deposit and of some borehole and other brines: Analyst, vol. 75, pp. 352-356.

Hewitt, D. F., 1962, Salt in Ontario: Ontario Dept. of Mines, Indus. Mineral Rept. no. 6, 38 p.

Holser, W. T., 1962, Diagnetic polyhalite in recent salt from Baja California (abs.): Geol. Soc. America Spec. Paper 73, p. 173 (to be published in Amer. Mineral).

____________, 1963, Chemistry of brine inclusion in Permian salt from Hutchinson, Kansas, in Symposium on Salt: Northern Ohio Geol. Soc., pp. 86-95.

____________, Kaplan, I. R. and Silverman, S. R., 1963, Isotope geochemistry of salt rocks (abs.): Geol. Soc. America Spec. Paper 76, p. 82 (to be published in Geoch. et Cosmoch. Acta).

Ivanov, A. A., and Levitskii, IU. F., 1960, Geology of halide deposits of the USSR: SSSR, Minist. Geol., Vsesoiuznyi geol. inst., Trudy, vol. 35, 424 p.

Jones, C. L., 1954, The occurrence and distribution of potassium minerals in southeastern New Mexico: New Mexico Geol. Soc., 5th Field Conf. Guidebook, pp. 107-112.

____________, 1962, Some geologic and petrographic features of upper Permian evaporites in southeastern New Mexico (abs.): Intern. Conf. on Saline Deposits, Houston, Texas, p. 22.

____________, 1965, Petrography of evaporites from the Wellington Formation near Hutchinson, Kansas: U. S. Geol. Survey Bull. 1201-A, 69 p.

Kühn, Robert, 1955a, Tiefenberechnung des Zechsteinmeeres nach dem Bromgehalt der Salze: Deutsche geol. Gessel. Zeitschr., vol. 105, pp. 646-663.

Kühn, Robert, 1955b, Mineralogische Fragen der in den Kalisalzlagerstatten vorkommenden Salze, in Kalium Symposium: Intern. Kali-Inst., Bern, pp. 51-105.

__________, and Schwerdtner, W.M., 1959, Nachweis deszendenter Vorgange wahrend der Entstehung der Leine-Serie des deutschen Zechsteinsalzes: Kali und Steinsalz, vol. 2, no. 11, pp. 380-383.

__________, 1962, Geochemistry of the German potash deposits (abs.): Intern. Conf. on Saline Deposits, Houston, Tex., (to be published in Geol. Soc. Amer. Spec. Paper).

Kulstad, R.O., 1959, Thickness and salt percentage of the Hutchinson salt: Kansas Geol. Survey Bull. 137, pp. 241-247.

Kupfer, D.H., 1963, Structure of salt in Gulf Coast domes, in Symposium on Salt: Northern Ohio Geol. Soc., pp. 104-123.

__________, 1965, The relationship between internal and external structures in Gulf Coast salt domes (abs.): Am. Assoc. Petroleum Geologists Bull. 49, p. 347.

Landes, K.K., 1945, The Salina and Bass Island rocks of the Michigan basin: U.S. Geol. Survey Map OM-40.

__________, 1951, Detroit River group in the Michigan basin: U.S. Geol. Survey Circ. 133, 23 p.

Lang, W.B., 1954, Relations of potassium, bromine, and iodine in marine saline sedimentation: Geol. Soc. America Bull., vol. 65, p. 1276.

Lindberg, M.L., 1946, Occurrence of bromine in carnallite and sylvite from Utah and New Mexico: Am. Mineralogist, vol. 31, pp. 486-494.

Maley, V.C., and Huffington, R.M., 1953, Cenozoic fill and evaporite solution in the Delaware Basin, Texas, and New Mexico: Geol. Soc. America Bull., vol. 64, pp. 539-546.

Muehlberger, W.R., 1959, Internal structure of the Grand Saline Salt Dome, Van Zandt County, Texas: Texas Univ., Bur. Econ. Geology, Rept. Inv., no. 38, 24 p.

__________, and Clabaugh, P.S., 1965, Internal structure and growth of salt domes (abs.): Amer. Assoc. Petroleum Geol. Bull., vol. 59, pp. 352-353.

Murray, G.E., 1961, Geology of the Atlantic and Gulf Coastal province of North America: New York, Harper & Bros., 692 p.

__________, 1965, Geology of Gulf Coast salt domes (abs.): Amer. Assoc. Petroleum Geologists Bull., vol. 49, p. 353.

Ogienko, V.S., 1959, Distribution of bromine in the rock salt of the Angara-Lena Salt Basin and the possibility of finding potassium salts: Geochemistry, 1959, pp. 893-900.

Pearson, W.J., 1963, Salt deposits of Canada, in Symposium on salt: Northern Ohio Geol. Soc., pp. 223-235.

Phleger, F.B., and Ewing, G.C., 1962, Sedimentology and oceanography of coastal lagoons in Baja California, Mexico: Geol. Soc. America Bull., vol. 73, pp. 145-182.

Raup, O.B., 1965, Bromine distribution in some halite rocks of the Paradox member, Hermosa Formation, Utah (abs.) in Second Symposium on Salt, (to be published by Northern Ohio Geol. Soc., Cleveland, Ohio).

Richter-Bernburg, Gerhard, 1955, Über salinare sedimentation: Deutsche Geol. Gesell., Zeitschr., vol. 105, pp. 593-645.

Schulze, Gunter, 1960, Stratigraphische und genetische Deutung der Bromverteilung in den mitteldeutschen Steinsalzlagern des Zechsteins: Freiberger Forschungshefte, vol. C83, 114 p

__________, and Seyfert, H., 1959, Der Bromgehalt an der Basis des Leinesteinsalzes und seine Beziehung zur Faziesdifferenzierung im Flöz Stabfurt: Zeitschr. angew. Geologie, 1959, no. 2, pp. 62-69.

Schwerdtner, W.M., 1964, Genesis of potash rocks in Middle Devonian prairie evaporite formation of Saskatchewan: Amer. Assoc. Petroleum Geologists Bull., vol. 48, pp. 1108-1115.

________________, and Wardlaw, N.C., 1963, Geochemistry of bromine in some salt rocks of the Prairie evaporite formation of Saskatchewan, in Symposium on Salt: Northern Ohio Geol. Soc., pp. 240-246.

Smith, C.S., 1951, Grain shapes and other metallurgical applications of topology: Amer. Soc. for Metals, 33rd Annual Conv., Detroit, Preprint no. 37, 42 p.

Thode, H.G., and Monster, J., 1964, The sulphur isotope abundances in evaporites and in the ancient oceans, in Khimiia zemloi kori, vol. 2, Izdatel'stvo Nauka, Moscow, pp. 589-600.

Valiashko, M.G., 1956a, The geochemistry of bromine in halogenesis processes and the use of bromine content as a genetic and prospecting criterion: Geochemistry, no. 6, pp. 33-48.

________________, 1956b, Geochemistry of deposits of potassium salts, in Voprosy Geologii Agronomicheskikh Rud.: Akad. Nauk SSSR, pp. 182-207 (Transl. deposited Univ. Microfilms, Ann Arbor, Mich.).

________________, 1958, Die wichtigsten geochemischen Parameter fur die Bildung ker Kalisalzlagerstatten: Freiberger Forschungshefte, vol. A123, pp. 197-235.

Wardlaw, N.C., 1964, Bromide in some Middle Devonian salt rocks of Alberta and Saskatchewan: International Williston Basin Symp., 3rd, pp. 270-273.

Yanshin, A.L., 1962, Potassium salt prospects in Siberia: Akad. Nauk SSSR, Sibir. Otd., Geol. Geofiz., 1962, no. 10, pp. 3-22 (Transl. Intern. Geol. Rev., vol. 6, pp. 2132-2149, 1964).

Editors' Comments on Papers 30 and 31

30 Thode and Monster: *Sulfur-Isotope Geochemistry of Petroleum, Evaporites and Ancient Seas*

31 Holser and Kaplan: *Isotope Geochemistry of Sedimentary Sulfates*

Isotopes of Sulfur

More than 60 million tons of sulfur have been produced from the cap rock of Boling Dome, Texas. Originally a mass of pure anhydrite, much of the cap rock has been altered to calcite and sulfur in a reaction involving calcium sulfate and petroleum. The mystery surrounding the cause of this reaction has been solved by studies of the isotopic composition of sulfur and carbon from cap rock (Thode et al., 1954; Jones and Starkey, 1957, Feely and Kulp, 1957) and bedded evaporites elsewhere (Dessau et al., 1962; Davis and Kirkland, 1970). It is difficult to believe that geologists would ever have established that bacteria were responsible for this reaction had not the causes of the fractionation of the isotopes of sulfur been understood and applied to this problem.

In the elucidation of this process alone, the geology of the isotopes of sulfur has made a significant contribution to our understanding of the diagenesis of evaporites. Recently, however, research on these isotopes has taken another direction and promises to reveal much about the composition of seawater during Phanerozoic time.

As values for the ratio of sulfur-34 to sulfur-32 began to accumulate from investigations of gypsum and anhydrite, it became apparent that there was a distinctive value for deposits of the same age, irrespective of their location. In 1959 Ault and Kulp constructed a curve showing the fluctuation of the value of $^{34}S/^{32}S$ throughout the Phanerozoic, and despite the paucity of values on some sections of the curve, a cyclic trend was evident. Since then, many more data have accumulated and the curve has been so revised and refined that it may now be used to identify the age of some deposits (Thode and Monster, 1964, 1965; Nielsen and Ricke, 1964; Nielsen, 1966, 1972; Holser and Kaplan, 1966; Claypool et al., 1972).

The theory behind the application of these studies to problems of evaporite sedimentation is summarized in the two papers included in this part. Of particular importance are the conclusions of Thode and Monster (1965):

1. The isotopic composition of combined sulfur in unaltered evaporitic sulfates represents the isotopic composition of sulfur in the ocean of the period when those sulfates were formed.

2. The various factors that altered the isotopic composition of sulfur in ancient oceans can be isolated and explained.

We, therefore, can accumulate data that will tell us much about the temporal variations in the medium from which evaporites are derived. Holser and Kaplan, elaborating on the work of Thode and Monster, consider that the fluctuations in the isotopic composition of sulfur in the oceans are a function of the balance between the *addition* of sulfur rich in sulfur-32 ("light sulfur") derived from the weathering of shale, and the *removal* of "light sulfur" from the oceans by the action of sulfate-reducing bacteria in muds on the sea floor.

Holser and Kaplan base their work on an examination of the geochemical inventory of sulfur, an exercise which also has led them to the conclusion that the concentration of sulfate in seawater may have varied throughout time. They estimate, for example, that the content of sulfate in the ocean may have been 45 percent greater in the Permian than at present. Perhaps at last we have an explanation for the abundance of Permian sulfate deposits.

H. G. Thode has been active in sulfur-isotope studies for the past 20 years, and after an 11-year absence during which he was President and Vice-Chancellor of McMaster University, he has now returned to his research interests in the Department of Chemistry. Jan Monster was born and educated in the Netherlands and since 1954 has collaborated with Thode on a number of geochemical studies.

W. T. Holser, now at the University of Oregon, is the author of two papers in this volume. A. I. Kaplan is originally from Australia, where he worked with both L. G. M. Baas-Becking and T. A. Rafter. He is unusual among earth scientists in that he is trained in bacteriology as well as in geology and the chemistry of isotopes.

Reprinted from *Fluids in Subsurface Environments—A Symposium,* Mem. 4, 367–377 (1965)

SULFUR-ISOTOPE GEOCHEMISTRY OF PETROLEUM, EVAPORITES, AND ANCIENT SEAS[1]

30

H.G. THODE[2] AND J. MONSTER[2]
Hamilton, Ontario

ABSTRACT

It has now been established that considerable sulfur-isotope fractionation occurs in the biological sulfur cycle and that the bacterial reduction of sulfate, which leads to the enrichment of S^{34} in sulfate and its depletion in sulfide, is largely responsible for the wide fluctuations in isotope ratio which occur in marine sediments.

In this regard, present-day ocean-water sulfate is remarkably uniform in sulfur-isotope content, both in depth and in geographical location at a value of $\delta = S^{34} = +20$ (20 parts per mil enriched S^{34} with respect to sulfur in meteorites), and provides a base level in isotope ratio from which fractionation can be reckoned. However, in dealing with ancient sediments and petroleum, we need to know the S^{34} content of the ancient oceans or seas.

Recently (Thode and Monster, 1963) a study of the sulfur-isotope distribution in the marine evaporites of some ten sedimentary basins of several continents was carried out. From this study it has been possible to estimate the sulfur-isotope ratio for the various ancient oceans and to establish the pattern of change throughout geological time.

The pattern of change for petroleum sulfur appears to be parallel to that for the evaporites and ancient seas. However, the petroleum sulfur is, in general depleted in S^{34} by about 15 °/₀₀ with respect to the contemporaneous gypsum and anhydrite deposits. This displacement of 15 °/₀₀ in the S^{34} content, which is about the isotope fractionation expected in the bacterial reduction of sulfate, is strong evidence that sea-water sulfate is the original source of petroleum sulfur and that it is first reduced by bacterial action in the shallow muds before being incorporated into petroleum. The lack of any sulfur-isotope fractionation in the plant metabolism of sulfate would seem to rule out plant sulfur as a major source of petroleum sulfur.

Inasmuch as the δS^{34} values for petroleum pools in a given horizon, for example, Devonian (D-2), are fairly uniform over a large sedimentary basin, and as these values vary from one horizon to another depending on the S^{34} content of the contemporaneous seas, sulfur isotope studies should be useful in solving migration problems.

Sulfur-isotope geochemistry has been a fruitful field of investigation, and already studies in this field have led to the solution of a number of geochemical problems. Large numbers of samples of sulfur-containing substances have been investigated, and the general distribution pattern of the sulfur isotopes is now well known. Finally, the story of how the sulfur isotopes have been fractionated in natural processes is beginning to unfold.

The interpretation of the sulfur-isotope distribution data requires a knowledge of certain base levels from which isotope fractionation began in any process. The most important of these is the primordial ratio of the sulfur isotopes, or the ratio of the sulfur isotopes at the time the elements were formed. It turns out that the sulfur-isotope ratios in meteorites are remarkably constant. Further, this constant level is almost identical with that found for sulfur in the earth's subcrust or mantle, as estimated from studies of large basic sills (Shima, Gross, and Thode, 1963).

The sulfur-isotope ratio for meteorites is therefore used by all investigators as the standard of comparison. The S^{32}/S^{34} ratio for any sample is expressed in terms of δS^{34} defined as follows:

$$\delta S^{34}\ ^{\circ}/_{\circ\circ} = \frac{(S^{34}/S^{32})_{\text{sample}} - (S^{34}/S^{32})_{\text{standard}}}{(S^{34}/S^{32})_{\text{standard}}} \times 1000$$

where $(S^{34}/S^{32})_{\text{standard}}$ is the ratio for meteorites.

Samples with positive and negative δ-values are therefore enriched and depleted, respectively, in S^{34} with respect to meteoritic sulfur.

The sulfur-isotope ratio for ocean-water sulfate is also of considerable importance in the interpretation of isotope-distribution data for sediments and materials of marine origin. Although the sulfate of present-day oceans is highly en-

[1] Read before the 6th Annual Meeting of the Southwestern Federation of Geological Societies and Southwestern Regional Meeting of The American Association of Petroleum Geologists, at Midland, Texas, February 1, 1964. Manuscript received, January 20, 1964.

[2] McMaster University.

We wish to thank the American Chemical Society for a Petroleum Research Fund grant (P.R.F. 107-A_2). We also thank the National Research Council of Canada for support.

riched in S^{34} with respect to meteorites ($\delta S^{34} = +20$), it is nevertheless very uniform in isotope ratio throughout its breadth and depth. It, therefore, provides a large reservoir of sulfur, uniform in isotope ratio, from which isotope fractionation in present-day marine processes may be reckoned.

For example, Thode and others (1951), Jones and others (1956), Jones and Starkey (1957), and Harrison and Thode (1958) have shown that fractionation of the sulfur isotopes occurs in the bacterial reduction of sulfate, which process takes place on a large scale in the shallow muds at the bottom of the sea. In any process where sea-water sulfate is reduced, isotope fractionation may occur, and the extent of this fractionation may be determined by comparing the S^{32}/S^{34} ratio of the reduced sulfide with that of sea water. All laboratory studies of recent and ancient sediments indicate that the biological sulfur cycle, and particularly the bacterial reduction of sulfate, is the major cause of sulfur-isotope fractionation. Because the ocean provides, in effect, an infinite reservoir of sulfate, its sulfur-isotope ratio will not be changed perceptibly as bacterial reduction takes place, and the $S^{=}$ produced will, in general, be depleted in S^{34} by one simple process factor of $\sim$15 ‰. Sulfides in recent sediments laid down off the coast of Venezuela in association with ocean-water sulfate are, indeed, depleted in S^{34} by about 15‰ (Thode, Monster, and Dunford, 1961).

If we are to interpret the sulfur-isotope distribution data for petroleum and related materials in ancient sediments, then we must have information as to the isotope content of the ancient seas from which they were derived. In this regard there is no reason to assume that in some point in the geological past ocean-water sulfate was less uniform in isotopic content than it is at present. An obvious approach to the problem of determining the sulfur isotope content of an ancient ocean is to investigate the isotopic content of the contemporaneous marine evaporites.

Thode and Monster (1963) proposed a method whereby the isotope ratio of an ancient sea could be determined from such an investigation, even though a spread in S^{34} δ-values is indicated for evaporites of the same geological period. They reasoned that in large, partially enclosed, shallow seas with intermittent or continuous contact with the oceans, the sulfate evaporites formed would either have the same S^{34} content as the ocean at the time of deposition, as in the case of some evaporites forming today (Thode, Monster, and Dunford, 1961; Ault and Kulp, 1959), or they would show S^{34} enrichment due to bacterial action in the basin. According to this reasoning, the evaporite with the lowest S^{34} content for a particular period should give the closest approach to the ocean-water isotope level at that time. Making this assumption, Thode and Monster (1963) were able to estimate the sulfur-isotope ratio for the various ancient oceans and to establish the pattern of change from a detailed study of the distribution of the sulfur isotopes in the evaporites of some ten sedimentary basins of several continents.

Mississippian evaporites.—Results showed that in a single evaporite bed—specifically the lower Ratcliffe of the upper Mississippian in the Williston basin of Saskatchewan—there was a remarkable uniformity in the isotopic ratio over a 5,000-square-mile area. A similar uniformity was found for an evaporite bed in the Devonian, covering an area of 200,000 sq mi in Western Canada. However, the S^{34} distribution data for the evaporites in a vertical section represented by a unit or formation showed a variation in the S^{34} content of 1 to 6 parts per mil (Table I). Their complete summary of isotope distribution data for gypsum and anhydrite evaporites of the upper and middle Mississippian for six different sedimentary basins is given in Table II. Finally, the spread in S^{34} δ-values obtained for the different evaporite beds of the upper and middle Mississippian for these six sedimentary basins is illustrated in Figure 1 on an age correlation chart.

According to Thode and Monster (1963), these fluctuations reflect the S^{34} level in the basin from which evaporites were deposited and provide information concerning changing conditions in the basin. For example, the Williston basin evaporites of the upper Mississippian are more enriched in S^{34} than are the contemporaneous marine evaporites from the Windsor basin, suggesting that there was more extensive reducing and bacterial action in the Williston basin with conditions favoring the enrichment of S^{34} in the depositing evaporites.

TABLE I. S^{34} DISTRIBUTION IN ANHYDRITE OF UPPER MISSISSIPPIAN SECTION OF WILLISTON BASIN

Unit	*Number of Samples Analyzed (different locations)*	δS^{34} ‰ *Range of Values*	δS^{34} ‰ *Minimum Value*
Upper Evaporite, Middle Evaporite, "B" Zone	9	+17.0 - +18.2	+17.0
Lower Evaporites	15	+17.7 - +19.3	+17.7
Upper Ratcliffe	8	+18.0 - +19.3	+18.0
Middle Ratcliffe	16	+18.3 - +19.7	+18.3
Lower Ratcliffe	22	+18.5 - +20.6	+18.5
Midale	21	+16.5 - +20.6	+16.5
Mission Canyon	5	+17.0 - +18.9	+17.0

After Thode and Monster (1963).

The degree of S^{34} enrichment in these evaporites depends on a number of factors. The most important is the overall content of biological activity in the basin. We know that in the bacterial reduction of sulfate the H_2S produced is depleted in S^{34}, leaving the residual sulfate enriched in this isotope. In an enclosed basin, this process will continue—the sulfate becoming ever more enriched as larger and larger fractions of the sulfate are reduced. Gypsum deposited in its last stages might, therefore, be highly enriched in S^{34}.

However, in a basin with contact, or intermittent contact, with the oceans over bars and reefs (Fig. 2), the extent of S^{34} enrichment in the depositing evaporites also depends on (1) the ratio of inflow of ocean water to outflow of brine, (2) the way and extent of mixing of the S^{34}-enriched bottom brines and incoming ocean sulfate, and (3) the rate of water evaporation and deposition of sulfate relative to ocean-basin flow interchange over the barrier.

Upper Mississippian oceans.—According to

TABLE II. S^{34} DISTRIBUTION IN MARINE EVAPORITES OF DIFFERENT SEDIMENTARY BASINS

MISSISSIPPIAN

Basin	*Formation*	*Number of Samples Different Locations*	δS^{34} ‰ *Range of Values*	δS^{34} ‰ *Minimum Value*
Windsor (Canadian Maritimes)	Windsor	56	+13.7 - +16.5	+13.7
Williston basin (Western Canada)	Upper Charles to Mission Canyon MC-5	107	+16.8 - +20.6	+16.8
Alberta basin (Western Canada)	Mt. Head	2	+17.9	+17.9
	Shunda	2	+20.0 - +22.6	+20.0
Illinois basin (North-Central U.S.A.)	St. Louis (Illinois)	11	+14.3 - +17.0	+14.3
Illinois basin (North-Central U.S.A.)	St. Louis (Iowa)	1	+15.2	+15.2
Illinois basin (North-Central U.S.A.)	St. Louis (Indiana)	5	+13.6 - +17.3	+13.6
Virginia-West Virginia	Maccrady	2	+15.0 - +15.3	+15.0
Moncton basin (Nova Scotia)	Albert	5	+13.9 - +19.4	+13.9

After Thode and Monster (1963).

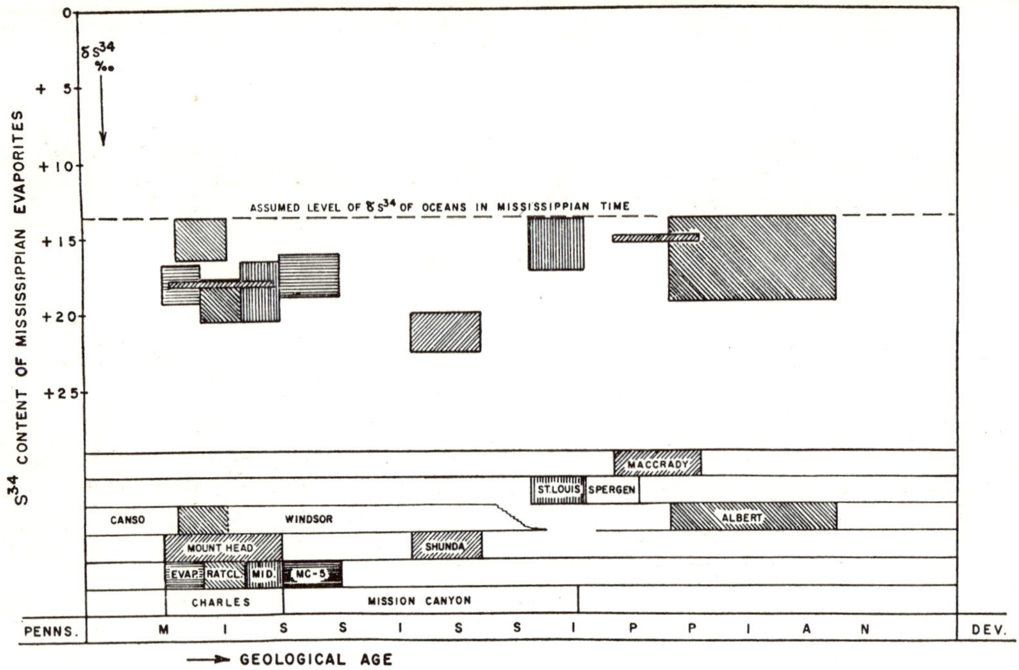

FIG. 1.—S^{34} content of Mississippian evaporites.

Thode and others (1961), the evaporite samples of any period with the least S^{34} enrichment, have S^{34} δ-values closest to that of the ancient oceans from which they are derived. Sulfur-isotope studies of such evaporites now being deposited from present-day oceans would seem to confirm this hypothesis (Thode and Monster, 1963). Applying this principle, it is seen from Table II and Figure 1 that $\delta S^{34} = +14$ is the best value for the ancient Mississippian oceans. The results also show that the sulfur-isotope ratio for ocean sulfate probably remained fairly constant throughout Mississippian time.

Permian evaporites and Permian oceans.—The Permian evaporites are of particular interest in that the ranges of S^{34} δ-values are narrow for each group of samples and in that the ranges are almost identical for each group taken from widely different locations in the great sedimentary basins of Western Europe and Western U.S.A. (Table III). The least enriched samples from the different groups have S^{34} δ-values ranging from +9.6 to +10.2 (Table III, Column 5), indicating a best value for the S^{34} content of upper Permian oceans of about +10 ‰.

Ancient oceans (Precambrian to Recent).—Fi-

TABLE III. S^{34} DISTRIBUTION IN EVAPORITES FROM WESTERN EUROPE, TEXAS, AND NEW MEXICO

UPPER PERMIAN

Sedimentary Basin	*Region*	*Number of Samples Different Locations*	*δS^{34} ‰ Range of Values*	*δS^{34} ‰ Minimum Value*
Western Europe	England	6	+10.1 – +11.2	+10.1
Western Europe	Netherlands	16	+ 9.7 – +11.1	+ 9.7
Western Europe	Germany	9	+10.2 – +12.0	+10.2
Western U.S.A.	West Texas	8	+ 9.6 – +11.5	+ 9.6

After Thode and Monster (1963).

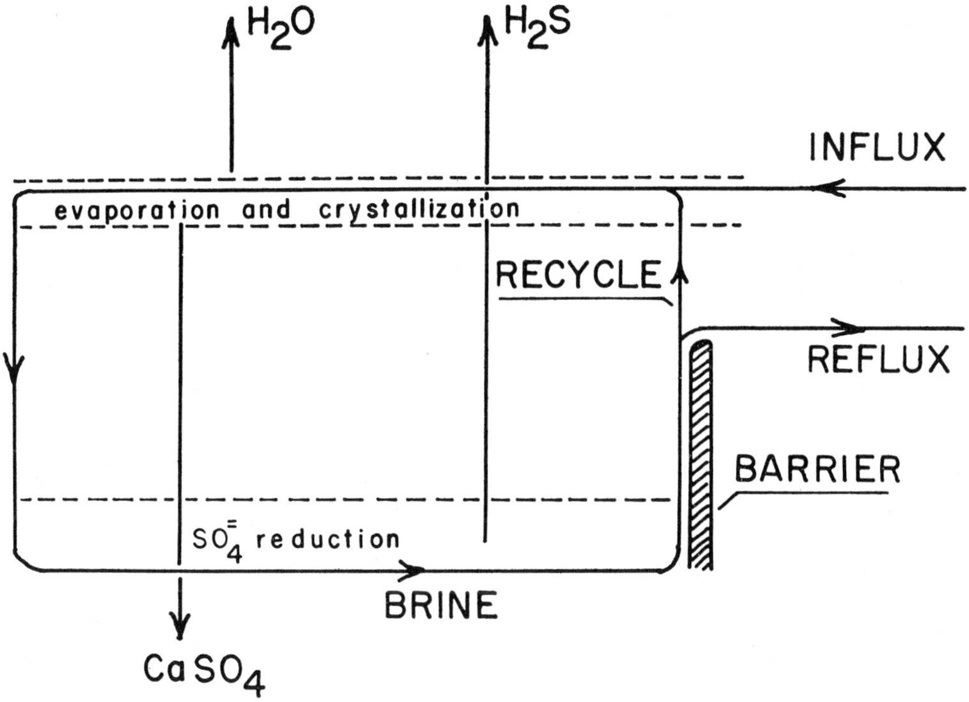

FIG. 2.—Simple diagram of an evaporite basin.

nally, the results of some 450 gypsum and anhydrite evaporite samples from sedimentary rocks ranging in age from Precambrian to Tertiary and Recent obtained by Thode and Monster (1963) are given in summary form in Table IV. In the case of each group of contemporaneous evaporites, the number of samples investigated, the range of S^{34} δ-values, and the minimum S^{34} enrichments are given. These latter minimum S^{34} δ-values are again taken as the best values for the contemporaneous ancient oceans. Figure 3 gives a plot of the S^{34} content of the ancient oceans so obtained, *versus* the geological age. Some possible fluctuations within geological periods are indicated by the broken lines. The results indicate a marked variation in the S^{34} content of the ancient oceans between Proterozoic and Recent time. The gross picture suggests a maximum enrichment in S^{34} during the Cambrian Period ($\delta S^{34} = +28$) and a minimum enrichment for the Permian oceans ($\delta S^{34} = +9.7$).

This cyclic nature of the changes in the S^{34} content of the oceans with time indicates that several factors must be involved. The following processes would have the net effect of increasing the S^{34} content of the oceans.

1. Re-dissolution of older evaporites with a composition heavier than that of ocean water.
2. A process whereby an older evaporite as in (1) would re-enter the sulfur cycle through magmatic metamorphosis.
3. Bacterial reduction. In the bacterial reduction of sulfate, the H_2S which is depleted in S^{34} generally becomes fixed as a sulfide. This means that the residual sulfate of the ocean and inland seas and evaporite basins will become ever richer in S^{34} as bacterial reduction continues.

The following are some processes which would result in a net decrease in the S^{34} content of the oceans.

1. Addition of "magmatic" sulfur with an isotopic content very close to that of meteorites ($\delta S^{34}\,^{\circ}/_{\circ\circ} = 0$).

TABLE IV. COMPLETE SUMMARY OF S^{34} DISTRIBUTION IN EVAPORITES

GYPSUM AND ANHYDRITE, PRECAMBRIAN TO RECENT

Geological Age	*Area and Description*	*Number of Samples*	δS^{34} ‰ *Range of Values*	δS^{34} ‰ *Minimum Value*
Precambrian Grenville Series, ~1,000 ×10⁶ yrs.	Balmat, New York	3	+14.5 – +28.6	+14.5
Precambrian Proterozoic >635×10⁶ yrs.	Victoria Island, N.W.T., Canada	2	+14.6 – +14.7	+14.6
Cambrian, Middle	Franklin Mts., N.W.T., Canada	3	+27.6 – +31.1	+27.6
Cambrian, Middle	Devon Island, Arctic Archipelago	2	+29.0 – +30.0	+29.0
Cambrian	South Australia, embedded in Cambrian Limestone	2	+29.7 – +30.0	+29.7
Ordovician, Knox Formation	Illinois basin	1	+23.6	+23.6
Ordovician, Joachim Formation	Illinois basin	1	+25.5	+25.5
Ordovician, Beekmantown Formation	Ontario, Canada	6	+29.2 – +35.8	+29.2
Ordovician, Red River Formation	Saskatchewan	4	+24.5 – +27.6	+24.5
Silurian (Cayugan) Salina Formation (or equiv.)	Ontario, Ohio, Mich., Pa., N. Y., W. Va.	31	+24.2 – +28.8	+24.2
Lower Devonian (Elk Point, Detroit River, Wapsipinicon, Dawson Bay, Prairie Evaporites, Winnipegosis)	Saskatchewan, Ontario	19	+16.7 – +19.9	+16.7
Middle Devonian (Woodbend, Duperow, Beaverhill Lake, Souris River, Leduc)	Williston basin and Alberta	34	+22.4 – +30.9	+22.4
Upper Devonian (Three Forks, Nisku, Potlatch)	Williston basin, Canada	4	+22.3 – +25.7	+22.3
Mississippian (St. Louis, Spergen, Maccrady Formations)	Iowa, Ill., Ind.	20	+13.9 – +18.7	+13.9
Mississippian (Albert Formation)	Nova Scotia	5	+13.9 – +19.4	+13.9
Mississippian (Upper) Charles to Mission Canyon, MC-5	Williston basin	107	+16.8 – +20.6	+16.8
Mississippian (Upper)	Windsor basin, Canadian Maritimes	56	+13.7 – +16.5	+13.7
Pennsylvanian (Lower) (Amsden Formation)	Montana (Lima and Heath)	3	+17.6 – +18.6	+17.6
Pennsylvanian (Middle) (Des Moines Formation)	New Mexico	1	+13.7	+13.7
Permian (middle) (Guadalupe and Leonard)	Tex., N. Mex.	14	+10.6 – +13.0	+10.6
Permian (upper) Ochoa and equiv.	Western Europe, U.S.A. (England, Netherlands, Germany, Tex., N. Mex.)	47	+ 9.6 – +12.0	+ 9.6
Triassic (Lower) (Upper Bunter, Rot)	Netherlands	4	+25.1 – +27.6	+25.1
Triassic (Upper) (Keuper)	England	4	+13.1 – +13.7	+13.1
Jurassic (Watrous Formation)	Saskatchewan	3	+19.3 – +21.0	+19.3
Jurassic (Gypsum Springs Formation)	Wyoming	1	+15.4	+15.4
Jurassic (Hith Formation)	Kuwait	1	+18.6	+18.6
Cretaceous (Lower)	Mississippi (Ferry Lake Anhydrite)	4	+15.8 – +16.5	+15.8
Cretaceous (Upper) (Tayarat Formation)	Kuwait	1	+22.7	+22.7
Tertiary-Paleocene (Cedar Keys Form.)	Florida (Fort Myers on W. Coast)	10	+19.0 – +20.3	+19.0
Tertiary-Eocene (Radhuma Form.)	Kuwait	1	+22.1	+22.1
Tertiary-Eocene (Rus Form., Middle Eocene, Florida)	Kuwait; Florida	11	+18.4 – +24.3	+18.4
Modern evaporites	Laguna Madre, Texas; Bocana Virrila, Peru	9	+20.2 – +24.3	+20.2
	Present-day oceans (average)			+20.1

After Thode and Monster (1963).

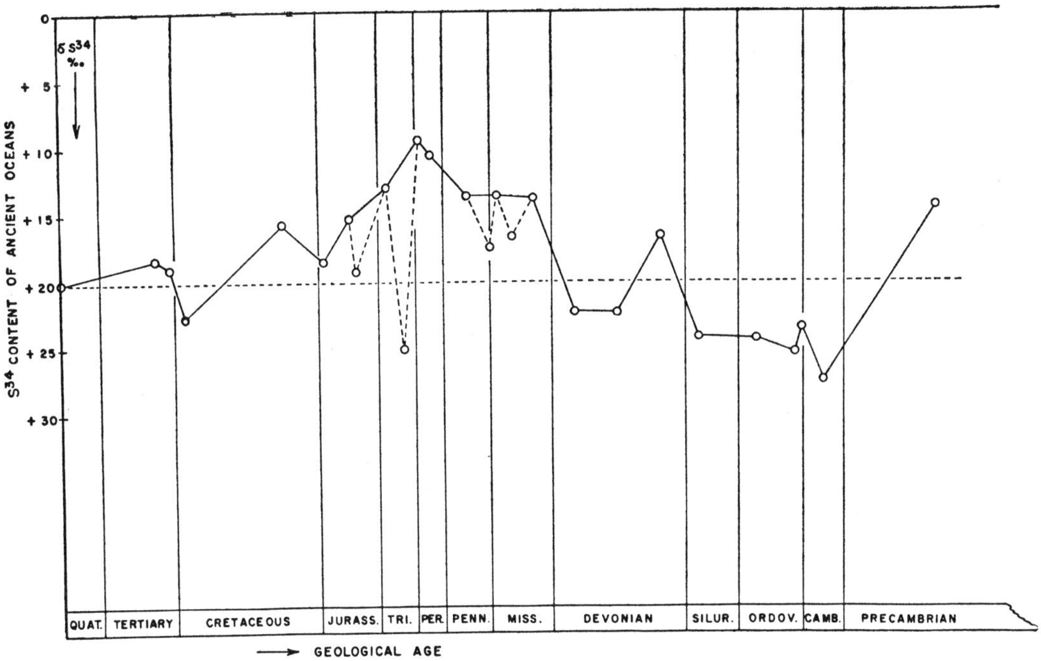

FIG. 3.—S^{34} content of ancient oceans from Precambrian to Recent.

2. Re-dissolution of older evaporites with a composition lighter than that of ocean water. According to Ricke (1960) and Krumbein (1951), there is far more sulfur fixed as sedimentary sulfates (evaporites) than had been previously thought.
3. Weathering of rocks and the inflow of fresh water into the oceans *via* rivers. Fresh-water sulfate in the Great Lakes, for example, has a S^{34} δ-value of $+ 10\ ^\circ/_{\circ\circ}$ as compared to that of present-day ocean water which is $+ 20\ ^\circ/_{\circ\circ}$. Sulfate in rain water returning to the oceans has a S^{34} δ-value of about $+5\ ^\circ/_{\circ\circ}$ in regions remote from industrial areas.

Isotope effects in crystallization.—In the interpretation of sulfur-isotope-distribution data of evaporites, the assumption has been made that little or no isotope fractionation occurs in the crystallization of $CaSO_4 \cdot 2H_2O$ from sea-water brines. However, the possibility of some fractionation must be considered. Harrison (1956) studied the equilibrium:

$$(1)\quad \underset{\text{(Solid)}}{CaS^{32}O_4 \cdot 2H_2O} + \underset{\text{(Soln)}}{S^{34}O_4^{=}} \leftrightarrows \underset{\text{(Solid)}}{CaS^{34}O_4 \cdot 2H_2O} + \underset{\text{(Soln)}}{S^{32}O_4^{=}}$$

From the experimental vibrational frequencies obtained for sulfate ion in solution and for sulfate ion in the crystal of $CaSO_4 \cdot 2H_2O$, and assuming the same isotopic shift for S^{34} substitution in $CaSO_4 \cdot 2H_2O$ as for sulfate ion in solution (Tudge and Thode, 1950), he obtained an isotope exchange constant K for reaction (1), above, of 1.001, indicating a $1^\circ/_{\circ\circ}$ enrichment of S^{34} in the $CaSO_4 \cdot 2H_2O$ crystals over that of the brine.

Experimentally he obtained a value of 1.002 for K by mixing a large excess of $CaSO_4 \cdot 2H_2O$ with water and measuring the sulfur isotopic composition of both solid and solution after a period of two weeks. Feely and Kulp (1957) also conducted laboratory experiments to measure the fractionation of sulfur isotopes in the crystallization of calcium sulfate from solution and concluded that fractionation was absent.

In the following experiment we attempted to approach isotopic equilibrium between crystalline $CaSO_4 \cdot 2H_2O$ and sulfate in solution in another way. The method involved the slow evaporation of a saturated solution of $CaSO_4$ under reduced pressure (16–20 mm Hg) at room temperature.

TABLE V. SULFUR-ISOTOPE EQUILIBRIUM STUDIES IN CRYSTALLIZATION OF $CaSO_4 \cdot 2H_2O$ FROM BRINE

Sample	δs^{34} °/oo	δs^{34} °/oo *Average*
Solid phase gypsum	+1.52 +1.60 +1.39	+1.49
Liquid phase ($SO_4^{=}$ in solution)	−0.12 −0.21 −0.16	−0.16

Isotope effect = 1.49 °/oo − (−0.16 °/oo) = 1.65 °/oo.
∴ Equilibrium constant K =

$$\frac{S^{32}/S^{34}\ CaSO^4\ \text{sol'n}}{S^{32}/S^{34}\ CaSO_4 \cdot 2H_2O\ \text{crystals}} = 1.00165$$

Evaporation was allowed to proceed for about 12 days, at which time the amount of $CaSO_4 \cdot 2H_2O$ that had crystallized out was approximately equal to the amount of $CaSO_4$ still in solution. At this point, the solution and crystals were allowed to stand for 24 months with occasional shaking. Solids and solution were then separated by filtration and the sulfur isotope ratios in both phases determined. The results are given in Table V.

These results, which are in good agreement with those of Harrison, indicate that if gypsum is deposited from a brine under equilibrium conditions, the crystals will be slightly enriched in S^{34} (1.6 °/oo) with respect to the brine. However, even this small fractionation would seldom be realized because the deposition of evaporites in a marine basin generally takes place with a certain amount of mixing of inflow ocean water and outflow brine which flow countercurrent across barriers or reefs. It would seem reasonable, therefore, to assume that little or no sulfur-isotope fractionation occurs in the deposition of $CaSO_4 \cdot 2H_2O$ from an evaporite basin.

Petroleum and related materials.—The sulfur-isotope distribution in a large number of samples of petroleum and related materials from the United States and Canada has been studied (Thode, Monster, and Dunford, 1958). Also, similar studies have been made of samples of petroleum and related materials from the Volga region of the Soviet Union (Eremenko, 1960; Eremenko and Pankina, 1963). The studies of Thode and others (1958) revealed the following.

1. Some petroleum samples are enriched and some depleted in S^{34} as compared to meteoritic sulfur, and the total spread in isotope ratio is more than 40 °/oo.
2. Sulfur-isotope distribution patterns for petroleum samples from marine and nonmarine sediments are completely different (Harrison and Thode, 1958b).
3. Petroleum samples from a single oil pool (Leduc, Alberta), which covers an area of 75 sq mi, are remarkably constant in their sulfur-isotope content.
4. Petroleum samples from reservoir rocks of the same age widely distributed over the western plains have very nearly the same sulfur-isotope ratios, but these ratios may vary from one stratigraphic level to another. For example, the Upper Devonian, Lower Cretaceous (heavy oils), and Upper Cretaceous oils have average S^{34} δ-values of +12, +8 and −2.5, respectively. Because these oils found at the different horizons have different levels of sulfur-isotope ratio, it may be possible to use the ratios in solving problems about oil migration.
5. H_2S gas intimately associated with petroleum appears to have the same sulfur-isotope content as that of the petroleum. This suggests that the two are related and that there is little or no isotope fractionation in the splitting off of H_2S in the maturation process.
6. Petroleum samples from widely separated pools in the same reservoir rocks—for example, Devonian—have very nearly the same S^{34} content but differ greatly in total sulfur content. Thode, Monster, and Dunford (1958) conclude that the sulfur-isotope ratio of petroleum in many large deposits does not change materially with time but is indicative of the nature of the source sulfur and the environment in which the petroleum was formed.

Their results indicated that the source sulfur and the fractionation of the sulfur isotopes in the reduction of this source sulfur, which occurs before or during petroleum formation, are the two main factors which determine the sulfur-isotope content of many present-day crude oils.

Inasmuch as most oils are formed in a marine environment, sea-water sulfate would seem to be the main original source of sulfur in petroleum. However, this sulfate is rapidly reduced by bacteria in the shallow muds in contact with the sea, and the isotopes of sulfur are fractionated to a considerable extent in the process. According to the laboratory experiments of Harrison and Thode (1958a) and of Jones and Starkey (1957), we can expect the reduced sulfur in the sediments to be depleted in S^{34} by $\sim$15°/₀₀ with respect to the contemporaneous sea-water sulfate.

If it is this reduced sulfur, present in the sediments in various forms, that is finally incorporated in the petroleum, and this seems likely, then we can expect a relationship between the sulfur-isotope ratios in petroleum and in the contemporaneous seas and evaporites. In general, the S^{34} δ-value for petroleum would be displaced by about 15°/₀₀ from that of the associated evaporites.

Apparent parallelism between sulfur-isotope composition of sea water and that of contemporaneous petroleum.—There is considerable evidence to show that, at least in many cases, these are the main factors involved and that there should be a relationship between the sulfur-isotope ratios in petroleum and those in the contemporaneous oceans and evaporites. Thode, Harrison, and Monster (1960) showed that the reduced sulfur in its various forms, including organically bound sulfur in the Recent sediments off the coast of Venezuela at depths of 20, 80, and 160 ft, is indeed depleted in S^{34} by about 15°/₀₀ with respect to present-day ocean sulfate.

Early in our studies of petroleum samples it was observed that a plot of the sulfur-isotope ratios in petroleum *versus* the geological age of the reservoir rocks gave a most interesting cyclic curve. We now know from the gypsum and anhydrite evaporite studies that there is a parallel variation in the sulfur isotope ratios for the ancient seas. However, the extensive evaporite studies indicate that the evaporites ($CaSO_4 \cdot 2H_2O$) in general reflect the sulfur-isotope ratio of the inland marine basin at the time of deposition rather than of the contemporaneous oceans. This suggests a closer parallelism between petroleum and associated evaporites than between petroleum and contemporaneous seas.

In Table VI the average sulfur-isotope ratios for petroleum of a given horizon or geological age are compared with those of the contemporaneous evaporites. It is clear from Table VI that for four major oil-forming periods of the Western Hemisphere, there is a parallelism between the sulfur-isotope ratios in petroleum and in the associated evaporites. In general, the displacement in the S^{34} content between the two is about 15°/₀₀ (column 5), the oil being depleted in S^{34} in each case relative to the associated evaporite.

Some oils do not seem to fit this pattern of parallelism. However, in most of these cases, we do not have sulfur-isotope data for the contemporaneous evaporites. It should be pointed out that the oils of a given period as well as the evaporites of the same period give a range of S^{34} δ-values, and we have compared the averages. We cannot, of course, pinpoint the exact S^{34} content of â basin at the time the source organic matter for the petroleum was laid down. Furthermore, the isotope fractionation in the bacterial reduction of sulfate which gives rise to the displacement factor can vary to some extent, depending on the metabolic rate (Harrison and Thode, 1958a). Considering these factors, the parallelism exhibited between the average sulfur-isotope ratios for the oils and for the associated evaporites is very good.

There are other factors which could disturb this pattern of parallelism and, in fact, some oils might not fit the pattern at all. For example, the oil might have migrated over a considerable vertical distance and be associated with evaporites of a quite different period. Furthermore, some of the oils might not have been of marine origin, and the petroleum sulfur might have been derived from a source other than the sea.

Finally, there is always the possibility that large quantities of sulfur are introduced into petroleum after its formation and during its migration through the reservoir rock at some later date. The results of studies indicate, however, that this does not happen often. In most of the oil horizons studied, the sulfur-isotope ratio in a single horizon is remarkably constant over large areas, even though the sulfur content may vary by a factor

TABLE VI. SULFUR-ISOTOPE RATIOS OF CRUDE OILS AND CONTEMPORANEOUS EVAPORITES ($CaSO_4$)

Geological Age	*Range in δ-values*	$\delta_1 = \delta S^{34}$ *Oil (average)*	$\delta_2 = \delta S^{34}$ *Evaporite (average)*	$\Delta = \delta_2 - \delta_1$
PERMIAN				
Oils from Wyoming and Texas	− 4.6 to + 4.5	− 1.3		
Upper Permian anhydrite N.E. Netherlands basin (16 samples)	+ 9.7 to +11.5		+10.7	+12.0
Upper Permian Delaware basin, W. Texas (8 samples)	+ 9.6 to +11.5			
MISSISSIPPIAN				
Charles and Mission Canyon oils, Sask.	+ 1.3 to + 6.8	+ 3.7		
Charles and Mission Canyon anhydrite, Sask. (large number of samples)	+16.3 to +21.0		+18.8	+15.1
DEVONIAN				
D-2 and D-3 oils, Leduc Oil field, Alta.	+11.7 to +12.8	+12.1		
D-2 anhydrite	+22.4 to +23.5		+23.0	+10.9
SILURIAN				
Oils from S.W. Ontario Oil from W. Texas	+ 8.4 to + 9.7 +7.7	+ 8.7		
Silurian evaporites (Salina Formation) Ont.-Ohio-Mich.–N. Y.-Pa.-W. Va.	+24.7 to +29.6		+26.9	+18.2

After Thode and Monster (1963).

of 10 (Thode, Monster, and Dunford, 1958). This would suggest a large, uniform source of sulfur such as occurs in a large inland sea or sedimentary basin. The sulfate in the connate waters associated with petroleum reservoir rocks is a very limited source, and the sulfur isotope ratio varies markedly as bacterial action takes place. For example, the sulfate in the connate waters of the Devonian D_2 reservoir rocks of Western Canada have sulfur-isotope ratios ranging from $\delta S^{34} = +22$ to $+37$. Even wider variations are reported for the elemental sulfur deposits in the salt domes of Louisiana and Texas (Thode, Wanless, and Wallouch, 1954; Feely and Kulp, 1957).

The uniformity of the sulfur-isotope ratios of oils formed in a given period in a particular sedimentary basin, the relationship which exists between these ratios and the contemporaneous evaporite deposits in the same basin, and the displacement between the sets of ratios being of the same magnitude as the fractionation factor in the bacterial reduction of sulfate—all these things point to the sea sulfate of the sedimentary basin as the original source of petroleum sulfur. The actual S^{34} concentration in the basin at any one time is either the same as that of the contemporaneous oceans or enriched through bacterial action in the basin. The important role of the bacterial reduction of sulfate in evaporitic basins has been recognized by Borchert (1959), Dietrich and Kalle (1957), Müller and Schwartz (1955), and Schwartz (1958).

It is clear that sulfur-isotope studies can give us a great deal of information about the nature of, and conditions in, a sedimentary basin at the time certain deposits were laid down. Such factors as the nature of the barrier between the basin and the open oceans, the ratio of inflow of ocean water to outflow of brine from the basin, the extent of evaporation in the basin relative to these flows, and finally, the extent of bacterial reduction of sulfate—all these factors have a bearing on the

S^{34} content of the sulfate in the basin at any time, and how it changes with time.

REFERENCES

Ault, W.U., and Kulp, J.L., 1959, Isotope geochemistry of sulphur: Geochim. et Cosmochim. Acta, v. 16, no. 4, p. 201–235.

Borchert, H., 1959, Ozeane Salzlagerstatten: Berlin, Gebrüder Borntraeger.

Dietrich, G., and Kalle, K., 1957, Allgemeine Meereskunde: Berlin, Gebrüder Borntraeger.

Eremenko, N.A., 1960, Variations of the sulfur isotope composition in U.S.S.R. crude oils according to stratigraphic sequence: Geol. Nefti i Gaza, v. 4, no. 11, p. 9–10.

——— and Pankina, 1963, The variation in the isotopic composition of sulfur in oils and gases depending on the age of deposits containing them: Conf., Moscow, March, 1963, pub. in Vernadsky Memorial Vol. I.

Feely, H.W., and Kulp, J.L., 1957, Origin of Gulf Coast salt-dome sulfur deposits: Am. Assoc. Petroleum Geologists Bull., v. 41, no. 8, p. 1802–1853.

Harrison, A.G., 1956, Isotope effects in the chemical and bacterial reduction of sulphur compounds: unpub. Ph.D. thesis, McMaster Univ.

——— and Thode, H.G., 1958a, Mechanism of the bacterial reduction of sulphate from isotope fractionation studies: Faraday Soc. Trans., v. 54, pt. 1, p. 84.

——— 1958b, Sulphur isotope abundances in hydrocarbons and source rocks of Uinta basin, Utah: Am. Assoc. Petroleum Geologists Bull., v. 42, no. 11, p. 2642–2649.

Jones, G.E., and Starkey, R.L., 1957, Fractionation of stable isotope of S by microorganisms and their role in deposition of native sulfur: Appl. Microbiology, v. 5, p. 111–118.

Jones, G.E., Starkey, R.L., Feely, H.W., and Kulp, J.L., 1956, Biological origin of native sulphur in salt domes of Texas and Louisiana: Science, v. 123, p. 1124–1125.

Krumbein, W.C., 1951, Occurrence and lithologic associations of evaporites in the United States: Jour. Sed. Petrology, v. 21, no. 2, p. 63–81.

Müller, A., and Schwartz, W., 1955, Über das Vorkommen von Mikroorganismen in Salzlagerstätten (Geomikrobiologische Untersuchungen III): Z. Deut. Geol. Ges., v. 105, p. 789–802.

Ricke, W., 1960, Ein Beitrag zur Geochemie des Schwefels: Geochim. et Cosmochim. Acta, v. 21, no. 2, p. 35–80.

Schwartz, W., 1958, Die Bakterien des Schwefelkreislaufs und ihre Lebensbedingungen: Freiburger Forschungsh, v. C-44, p. 5–14.

Shima, M., Gross, W.H., and Thode, H.G., 1963, Sulfur isotope abundances in basic sills, differentiated granites and meteorites: Jour. Geophys. Research, v. 68, no. 9, p. 2835–2847.

Thode, H.G., Harrison, A.G., and Monster, J., 1960, Sulphur isotope fractionation in early diagenesis of Recent sediments of northeast Venezuela: Am. Assoc. Petroleum Geologists Bull., v. 44, no. 11, p. 1809–1817.

Thode, H.G., Kleerekoper, H., and McElcheran, D., 1951, Isotope fractionation in the bacterial reduction of sulphate: Research, v. 4, p. 581–582.

Thode, H.G., and Monster, J., 1963, The sulphur isotope abundances in evaporites and in the ancient oceans: Conf., Moscow, March, 1963, pub. in Vernadsky Memorial Vol. II.

Thode, H.G., Monster, J., and Dunford, H.B., 1958, Sulphur isotope abundances in petroleum and associated materials: Am. Assoc. Petroleum Geologists Bull., v. 42, no. 11, p. 2619–2641.

——— 1961, Sulphur isotope geochemistry: Geochim. et Cosmochim. Acta, v. 25, p. 159–174.

Thode, H.G., Wanless, R.K., and Wallouch, R., 1954, The origin of native sulphur deposits from isotope fractionation studies: Geochim. et Cosmochim. Acta v. 5, p. 286–298.

Tudge, A.P., and Thode, H.G., 1950, Thermodynamic properties of isotopic compounds of sulphur: Canadian Jour. Research, v. 28, p. 567–578.

Reprinted from *Chem. Geol.*, **1**, 93–96, 114–135 (1966)

31

ISOTOPE GEOCHEMISTRY OF SEDIMENTARY SULFATES[1]

W.T. HOLSER and I.R. KAPLAN

Chevron Research Company, La Habra, Calif. (U.S.A.);
Department of Geology and Institute of Geophysics, University of California, Calif. (U.S.A.)

(Received April 6, 1966)

SUMMARY

Anhydrite, gypsum and other sulfates were sampled from a wide range of origins, differing in: geological age, mineralogy, stage of evaporation, inland basin or oceanic facies, and primary or replacement crystallization. For sulfates of oceanic origin, variation with geological age is so large that it nearly masks all other factors. Tertiary, Recent and synthetic gypsum have $\delta^{34}S$ within ± 2‰ of the +20‰ of modern sea water. A large number of Late Permian samples from Germany and the United States are all in the range $\delta^{34}S$ = +9 to +11‰ irrespective of mineralogy or other geological factors. Mesozoic samples are intermediate between the Permian and Tertiary levels, and Early Paleozoic sulfates have up to $\delta^{34}S$ = +30‰. These results confirm the general curve found by other workers, in which the isotopic composition of oceanic sulfate varies with time, on a worldwide scale.

These variations can only be accomplished by large-scale transfer of sulfur to or from the shale reservoir. Inflow of sulfate exceeded sedimentation in the Late Paleozoic, to increase the sulfate content of the sea by 45% with a corresponding enrichment in ^{32}S. In the Mesozoic, sedimentation of sulfide exceeded inflow removing the light sulfur preferentially from the sea. Review of the geochemical balance and cycle of sulfur indicates that an excess inflow of sulfate was derived from shale sulfide by an increase of either weathering or metamorphism-volcanism, along with an even larger amount of sulfur from old evaporites.

Alternatively, a decrease of sulfide sedimentation is directly related to biological activity, which in turn may have been controlled by one of several geomorphic changes. The isotope evidence thus indicates substantial changes in the sulfate concentration of sea water, but is symptomatic of even more important changes in the tectonic and paleogeographical regime in the Late Paleozoic and Early Mesozoic.

INTRODUCTION

This study was begun in 1961, by which time the general outlines of sulfur-isotope geochemistry had been published by Ault and Kulp (1959) and Thode

[1] Publication No.512, Institute of Geophysics, University of California.

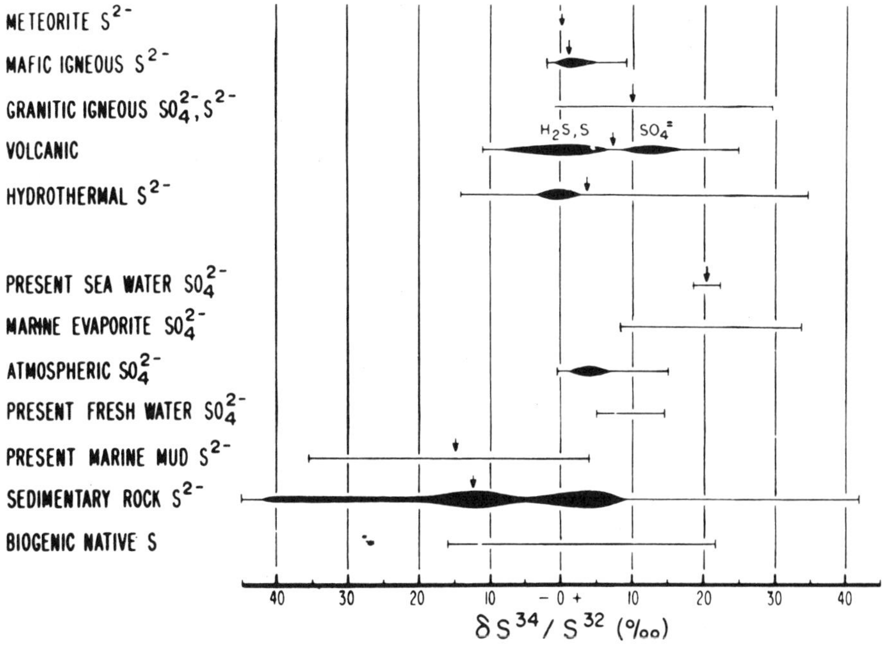

Fig.1. Isotopic geochemistry of sulfur. Arrows indicate mean values. Abstracted from all published data.

et al. (1961a) among others. A very general distribution of sulfur isotopes in nature is shown in Fig.1. Sulfides or sulfur produced by biological reduction are enriched in the light isotope, with a $\delta^{34}S$* that varies widely around an average of about $\delta^{34}S = -12$‰. Other "primary" sulfur - igneous, volcanic and hydrothermal - is only slightly heavier than meteoritic sulfur, and the residue of unreduced sulfate in the ocean and evaporites is variably heavy in the range $\delta^{34}S = +10$ to $+30$‰.

The objective of our study was a general survey of the distribution of sulfur isotopes in evaporite rocks representing a variety of geological situations: (1) various stages of evaporation of the brine from which the sulfate was precipitated; (2) marine or continental origin of the brine; (3) sulfate crystallized as gypsum, anhydrite or other sulfates; (4) post-depositional transformations; and (5) different geological time of deposition. Our results show that the last factor overrides all others. Any of these geological parameters with the exception of the geological age could best be studied in areas with good geological control, and for this purpose we used a large number of samples from four locales: (1) a synthetic evaporation of modern sea water; (2) tidal flats in Baja California; (3) Permian Basin of New Mexico–Texas; and (4) the Permian Zechstein Series of Germany. Supplemental samples were analyzed from a dozen other localities.

A preliminary report of our work was presented in 1963 (Holser et al.,

---

$$* \delta^{34}S \text{‰} = \frac{{}^{34}S/{}^{32}S \text{ sample} - {}^{34}S/{}^{32}S \text{ std}}{{}^{34}S/{}^{32}S \text{ std}} \times 1{,}000$$

The standard is troilite from Canyon Diablo Meteorite.

1963). At about the same time detailed work on sulfates in evaporites was discussed by Thode and Monster (1964, 1965), and by the Göttingen group (Buschendorf et al., 1963; Nielsen and Ricke, 1964; Nielsen, 1966). These contributions are considered in detail below in relation to our results.

METHODS

Sample selection and purification

Purity of samples was estimated visually, by petrographic microscope and by X-ray diffraction. Samples were chosen from localities relatively free of sulfide or sulfur for the most part, to minimize possible effects of local reduction of sulfate by late solutions. If clay or pyrite was present, sulfate was first separated mechanically and subsequently extracted with a 10% hydrochloric acid solution. The sulfate was then precipitated as barium sulfate. Elemental sulfur, where present, was extracted with benzene and purified by recrystallization. Pyrite and elemental sulfur were oxidized with bromine and aqua regia to sulfate, which was then removed from solution as barium sulfate. Special treatment of the Searles Lake brine sample is discussed later.

Isotope measurements

Samples separated from solution, or purified by extraction and reprecipitation were stored as barium sulfate. All barium sulfates, anhydrites and gypsums were reduced to hydrogen sulfide with a hypophosphorous-hydriodic acid mixture and converted to silver sulfide (Thode et al., 1961a).

Sulfur dioxide was prepared for mass spectrometer by an adaptation of the method described by Rafter (1957a,b), in which the silver sulfide was burnt at 1,200°C in a stream of oxygen.

The sulfur-dioxide samples were analyzed in a dual-collector, Consolidated Electrodynamics Corp. mass spectrometer, Model 21-401, in the Laboratories of Economic Geology, Yale University.

A few analyses are listed in Table III and IV (marked by an asterisk) from a later series of determinations in which recrystallized gypsum was reduced by metallic iron at 850°C for 30 min, converted to silver sulfide, then oxidized. These sulfur-dioxide samples were analyzed in the Department of Chemistry, McMaster University.

GEOCHEMISTRY OF MARINE SULFATES

Detailed reviews of the geochemistry of marine evaporites have been published recently by Stewart (1963), Braitsch (1962) and Borchert and Muir (1964, pp.64–125). Certain facts and interpretations should be mentioned here as particularly pertinent to the distribution of sulfur isotopes in these rocks:

(1) Although many Cenozoic evaporites have been deposited in closed interior basins, nearly all evaporites earlier in the geological section were formed by the evaporation of sea water in restricted basins.

(2) The sedimentary texture of many evaporites, particularly the well-

preserved laminae, each of an approximate thickness about that expected for a year's evaporation, indicates they have undergone only slight alteration, if any. Other evaporites have more massive or complex textures superimposed by diagenetic or later recrystallization and alteration.

(3) Primary evaporite facies are a function mainly of the overall stage of evaporation of the sea water, regardless of the extent of refilling of the basin by new sea water. Each facies is characterized by the appearance of an additional mineral phase. Present-day sea water will first precipitate carbonates, then at a concentration beyond 5× will precipitate calcium sulfate, beyond 11× halite, and beyond 63× various potash–magnesia salts[1].

(4) Many evaporite basins show a regular succession of these facies in space and time, but only in a very general way - the detailed variation of mineralogy is usually complex. Thus the calcium-sulfate facies in many basins predominates at the base of an evaporite and near the edge of the basin.

(5) In the calcium-sulfate facies gypsum is generally the stable phase in equilibrium with the brine, and it is usually but not always the one phase precipitated even under conditions of higher temperature or salinity where anhydrite may be thermodynamically stable.

(6) With increasing temperature during burial to about 300 m, gypsum becomes unstable and transforms to anhydrite.

(7) With high potash–magnesia concentrations in brines the stable form of calcium sulfate is polyhalite, $K_2Ca_2Mg(SO_4)_4 \cdot 2H_2O$; the brines in this facies have the capability of depositing polyhalite from newly introduced sea water, or of altering new or old gypsum to polyhalite.

(8) In the potash–magnesia facies, the magnesium-sulfate minerals expected from the equilibrium diagram are rather uncommon, although replacement by kainite, $KMgClSO_4 \cdot 3H_2O$, kieserite, $MgSO_4 \cdot H_2O$, and langbeinite, $K_2Mg_2(SO_4)_3$ are locally important.

(9) As an evaporite section is eroded and brought within a few hundred meters of the surface, cool ground waters transform anhydrite and polyhalite to gypsum and/or dissolve them.

* * * * * * *

MECHANISMS OF FRACTIONATION

The results indicate that $\delta^{34}S$ values within a single deposit are generally consistent. Small changes do occur in vertical sequences and these may relate to the stage of crystallization, as in the Zechstein deposits (Table III). There are also small differences between various deposits of any one geological age. These variations appear to be superimposed on a more radical variation that is related to the age of the deposits as seen later in Fig.5. We will consider in turn each of the several factors that can affect the sulfur-isotope ratio, particularly on a local scale within a single basin. These considerations will then be used to derive the best value of the isotopic composition of the ocean for a given geological time, in order to reconstruct and interpret the sulfur-isotope age curve.

The principle process by which the sulfur-isotope ratio of a mineral may differ from that of the brine in which it crystallized, is fractionation during crystallization. In addition, the brine may differ from the composition of sea water for one of several reasons: (1) progressive changes in sulfur-isotope composition of the brine due to the above fractionation; (2) input of non-marine sulfate; and (3) local or basin-wide biological reduction of sulfate. These will be discussed in turn.

Fractionation during crystallization

It has been generally considered that fractionation of sulfur isotopes was insignificant during the crystallization of gypsum from brine. As this paper was being prepared for publication, Thode and Monster (1965) released data from both new experiments and old calculations that indicate a fractionation:

$$K = \frac{^{34}S/^{32}S \text{ gypsum}}{^{34}S/^{32}S \text{ solution}} = 1.00165 \pm 0.00012$$

or:

$$\Delta^{34}S_c = \delta^{34}S_{gypsum} - \delta^{34}S_{brine} = +1.65 \pm 0.12‰ \qquad (1)$$

They suggested that ". . . Even this small fractionation would seldom be realized since the deposition of evaporites in a marine basin usually takes place with a certain amount of mixing of inflow ocean water and outflow brine. . .". Such mixing will only delay the attainment of a particular facies or stage of evaporation in the total brine (Holser, 1966b), but should not affect any correlation of isotope fractionation and stage of evaporation.

As shown in Table I, our experimental data on a closed constant-volume tank (equivalent to complete mixing of brine and inflow) show a mean sulfur-isotope ratio for the gypsum slightly (perhaps not significantly) heavier than that of the mean for brines. Our data for gypsum from modern brines (Table I) at Laguna Ojo de Liebre, Baja California, with an unknown amount of reflux of the brine, suggest a fractionation of $\Delta^{34}S_c = +2.3 \pm 0.7‰$. This may be compared with the inconclusive previously published values for Laguna Madre, Texas: $\Delta^{34}S_c = +0.5 \pm 0.4‰$ (Ault and Kulp, 1959, p.215) and +4.2‰ (Thode, 1964, p.258); and for Bacana de Virrila, Peru: $\Delta^{34}S_c = +0.6 \pm 0.2‰$ (Thode, 1964, p.258) and +2.4‰ (Thode et al., 1961a, p.171). All of these results suggest that gypsum precipitated from sea water will be slightly heavier than the sea water, but the scattering does not lend much authority to any quantitative value for the difference.

No examples of recently crystallized anhydrite were available for this study. One anhydrite whose petrographic relation suggested primary crystallization is heavier than our gypsum samples from the same locality, but not significantly. Replacement of one sulfate mineral by another after deposition also does not have any clear-cut change in sulfur-isotope ratio. Gypsum forming from anhydrite by weathering of the Zechstein 1 is unchanged (Table III). Polyhalite formed by the alteration of Recent gypsum (Table I) and Permian anhydrite (Table III) are generally equivalent to the original sulfate within the scatter of ±0.7‰ commonly found in these analyses. Even solutions of anhydrite and its redeposition as cement in sandstone show no consistent change (Table III).

Within these limits, sulfate minerals should be only slightly fractionated from the sulfate in the brine from which they crystallize.

Sulfate in highly evaporated facies

Since our work was completed, Nielsen and Ricke have described an apparent trend toward slightly lighter sulfate in evaporites of the later potash and magnesia facies in the Zechstein 1 and the Zechstein 2 (Nielsen and Ricke, 1964, p.582). However, our data do give some support to a lighter weight for late sulfates. Referring to Table III, one finds a difference $\Delta^{34}S_K = \delta^{34}S$ (potash–magnesia facies sulfates) – $\delta^{34}S$ (gypsum-anhydrite facies) = $-0.8 \pm 0.5‰$ in the Delaware Basin and $\Delta^{34}S_K = -1.0 \pm 0.8‰$ in the Zechstein basin. Furthermore, in both the Michigan Basin and the Saskatchewan Basin (Table IV) anhydrites that occur as thin single lamellae during the regular deposition of the halite facies, are lighter than anhydrites from the same formation that occur in thicker successions of anhydrite lamella which undoubtedly represent a temporary period of return to anhydrite facies conditions.

Although these differences may not be quantitatively convincing, the reason for such a possible trend should be examined. One possibility is that fractionation during crystallization, as described above, has gradually depleted late brines of the heavy isotope. The *maximum* extent to which this might be reflected in a change of isotope concentration in sulfate minerals of the potash facies can be estimated, by assuming, (1) no reflux of sulfate back to the sea, (2) an equilibrium sequence of salt minerals and (3) isotopic equilibrium during crystallization but no later re-equilibration. We have to make the further assumption that the fractionation is the same for all sulfate minerals as measured for gypsum (eq.1). The problem can be treated by applying the principles used for calculating trace element distribution in a closed system (McIntire, 1963, p.1250; Holser, 1966b), in which the isotope ratio $r_c = {}^{34}S/{}^{32}S$ changes from an initial value r_o according to the equation

$$\ln \frac{r_c}{r_o} = (K - 1) \ln \left[1 - \frac{m_c}{m_o}\right] \tag{2}$$

and:

$$\Delta^{34}S_c = \frac{r_c - r_o}{r_o} \times 1{,}000 \tag{3}$$

The weight fraction of sulfate crystallized is given by the ratio of final to initial sulfate m_c/m_o. Fig.4 gives a plot of $\delta^{34}S$ fractionation against the amount of sulfate crystallized (m_c/m_o) from the salt body, assuming K = 1.00165 for the initial gypsum that is formed. It can be seen that the $\delta^{34}S$ for this sulfate is at first more positive than the soluble sulfate from which it forms ($\delta^{34}S = 0$ in Fig.4) and then logarithmically declines according to eq.2. At the end of the halite facies with 31% of the original sulfate precipitated (Braitsch, 1962, p.66), $\Delta^{34}S_c = -0.6‰$ and at the beginning of carnallite deposition, with 90% of the sulfate precipitated, it has dropped $\Delta^{34}S = -3.8‰$. Relative to the original sea water, the fractionation would

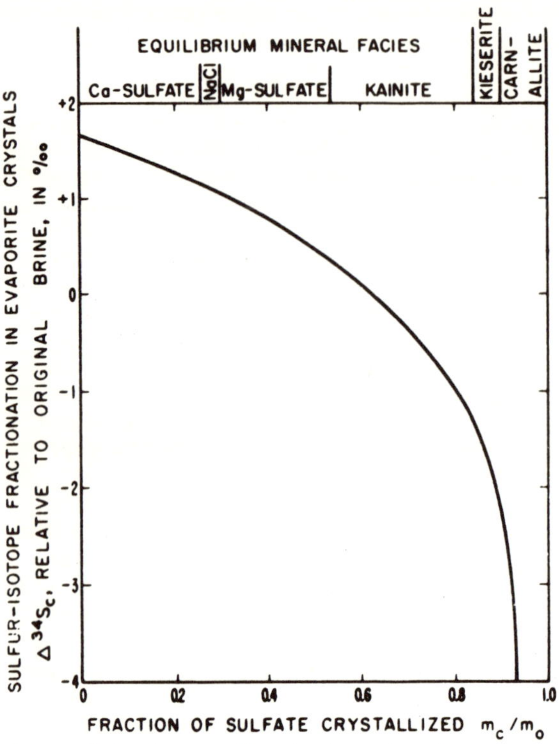

Fig.4. Calculated sulfur-isotope fractionation in evaporite crystals during equilibrium crystallization.

be $\Delta^{34}S_c$ = +1.0‰ and −2.2‰ respectively. The small maximum effects calculated here may be overwhelmed by other processes discussed later. However, our analyses in combination with those of Nielsen and Ricke (1964) suggest that late-stage sulfate may be lighter by an amount in line with the above calculation of fractionation. Detailed measurements of other late-stage evaporites are needed to verify this as a general effect.

Nielsen and Ricke (1964, p.587) considered crystal fractionation to be too small an effect to explain the lightness of late-stage evaporites, and proposed instead that diffusion in a salinity-stratified basin was a more likely explanation. However, it is well known that ionic diffusion is overwhelmed by mixing (eddy diffusion) in most large bodies of water (Hutchinson, 1957, p.587). Such diffusion would also have to involve hydrated sulfate ions with a mass ratio of less than 1.02, making the processes one of unlikely importance.

Sulfate inflow

Most pre-Tertiary evaporites are considered to be of marine origin, as mentioned above. By this it is only meant, however, that the general chemistry of the principle phases precipitated are those expected from modern sea water, in contrast to the variety of minerals precipitated in present interior basins. By their very nature, however, evaporites will tend to be formed near the borders of continents, and may be expected to receive some inflow from rivers. The association of many ancient evaporites with "red beds" of possible continental origin has often been emphasized (Krumbein, 1951; Ivanov and Levitskii, 1960, p.390.).

Where sulfate inflow from rivers may be important, it will generally involve the addition of light sulfur. The possible sources of sulfur in river water are: (1) solution of sulfate from earlier evaporites ($\delta^{34}S$ = +10 to +30‰); (2) springs having their source in earlier connate brines, ($\delta^{34}S$ = +10 to +30‰); (3) springs and fumaroles of volcanic origin ($\delta^{34}S$ = –10 to +15‰); (4) oxidation of pyrite during the weathering of shale, hydrothermol veins, and igneous rocks ($\delta^{34}S$ = –40 to +5‰); (5) precipitation and dry fallout of sulfur from the atmosphere ($\delta^{34}S$ = +3 to +15‰). All except (1) and (2) have lighter sulfur than sea water of any geological age (see below), while the weathered evaporite may be either slightly heavier or slightly lighter than sea water. This general relation is confirmed by the few results on sulfates from interior basins, as described above for the Quaternary Searles Lake Basin and by Harrison and Thode (1958) for the Eocene Uinta Basin, Utah.

Although the river waters have a highly variable chemistry, most of them (Durum and Heidel, 1961; Livingstone, 1963) are high in calcium, bicarbonate and sulfate relative to sodium, magnesium, chloride and other elements. On evaporation, such waters soon precipitate calcium carbonate, then calcium sulfate, leaving sodium, sulfate and chloride in the brine residue (Hutchinson, 1957, pp.564–571). When such rivers flow into an evaporite basin near the sea, they enrich the residual brine in ^{32}S sulfate relative to sea-water sulfate.

The sulfate concentration of rivers in arid regions is about 0.1 that of sea water, and of rivers in humid regions is about 0.01 that of sea water. The corresponding values for carbonate are 10× and 4× sea water, respectively (Livingstone, 1963, pp.10–17). These chemical relations should be evident as a larger ratio of carbonate to sulfate in yearly varves of evaporites with substantial river inflow, especially where total yearly evaporation was sufficient to keep the basin in the halite facies or some more saline facies despite the larger influx of dilute water. Such anhydrites should be depleted in ^{34}S. Quantitative textural measurements are not yet available to test a relation of this kind, but should be sought.

Sulfate reduction

The well-known fractionation of sulfur during biological reduction to sulfide, with a large but variably negative shift of at least $\Delta^{34}S_b = \delta^{34}S_{S^{2-}} - \delta^{34}S_{SO_4^{2-}}$ = –30‰ in the sulfide phase (Fig.1), is the most obvious candidate to explain various changes in sulfur-isotope composition. In marine muds (Kaplan et al., 1963) most of the sulfate reduction takes place by biological reduction of sulfate at the sediment–water interface, or in the interstitial water just below the interface, through diffusion of sulfate from the overlying sea water. The amount of sulfide formed and its isotopic composition are highly variable with both location and depth in the sediment, depending on details of biological activity and diffusion. In an evaporite basin, you might expect similar processes to go on wherever sufficient organic material was available to feed biological activities. You might also expect a certain amount of reduction of the sulfate in interstitial water, including that from previously deposited and redissolved gypsum or anhydrite.

The ultimate fate of the reduced sulfur would probably be one of the following:

(1) If sufficient iron were present in the clay minerals, the sulfide might be fixed in the sediment, eventually as pyrite.

(2) In the absence of iron, but with intermittent exposure of the sediment on mud flats, hydrogen sulfide might escape to the atmosphere, where it would be re-oxidized and returned to the rivers and sea as sulfate.

(3) Hydrogen sulfide might diffuse into a deep layer of the water that has been stabilized by high salinity, and will gradually mix with the overlying aerobic water and re-oxidize to sulfate.

(4) Deep-layer hydrogen sulfide formed as in (3) might be refluxed out of the basin to the ocean, where it would be gradually mixed and oxidized as in (3).

The efficiency of some much process is attested by the sulfate and sulfur-isotope profiles of fresh-water lakes (Deevey et al., 1963), and research in progress on the Dead Sea. All of these processes could lead to short-time changes in the sulfur-isotope ratio of the sulfate in the surface layer of an evaporite basin, where gypsum might be crystallizing. All except (3) might be cumulative within a given basin over a long period of time, leading to a vertical sequence of increasing isotope ratios in that basin. Only the first process could lead to long term changes in the sulfur-isotope ratio of the ocean, because only this process effectively removes ^{32}S from the hydrosphere (see below).

Pyrite is widespread in evaporite rocks, particularly as a constituent of dark bituminous carbonate lamellae in anhydrite. But its amount is small: a section of the Zechstein 1 that has been analyzed chemically shows only 0.15–0.30% iron, corresponding to 0.09–0.19% sulfide (Jung and Knitzchke, 1960, p.71). Mass balance calculations, assuming that the overlying reservoir of ^{34}S-rich sulfate is 3.2× the sulfate already deposited (halite facies), indicate that the sulfur in pyrite represents reduction of only about 1% of the total sulfur in the basin, which would mean a cumulative change in sulfur-isotope ratio of only 0.3‰. Evidently the process (1) above, pyrite formation, is not significant in this evaporite basin, although it may be effective in some lower-salinity pre-concentration zone between the sea and the evaporite basin. In most thick evaporite sequences sulfide formation may be small relative to the sulfate content, but a true estimate of the importance of the process of sulfate reduction may not be possible if hydrogen sulfide is not captured by the sediment.

VARIATION OF THE SULFUR-ISOTOPE RATIO WITH THE GEOLOGIC AGE

The age curve

In Fig.5 we have constructed a sulfur isotope age curve. All data published

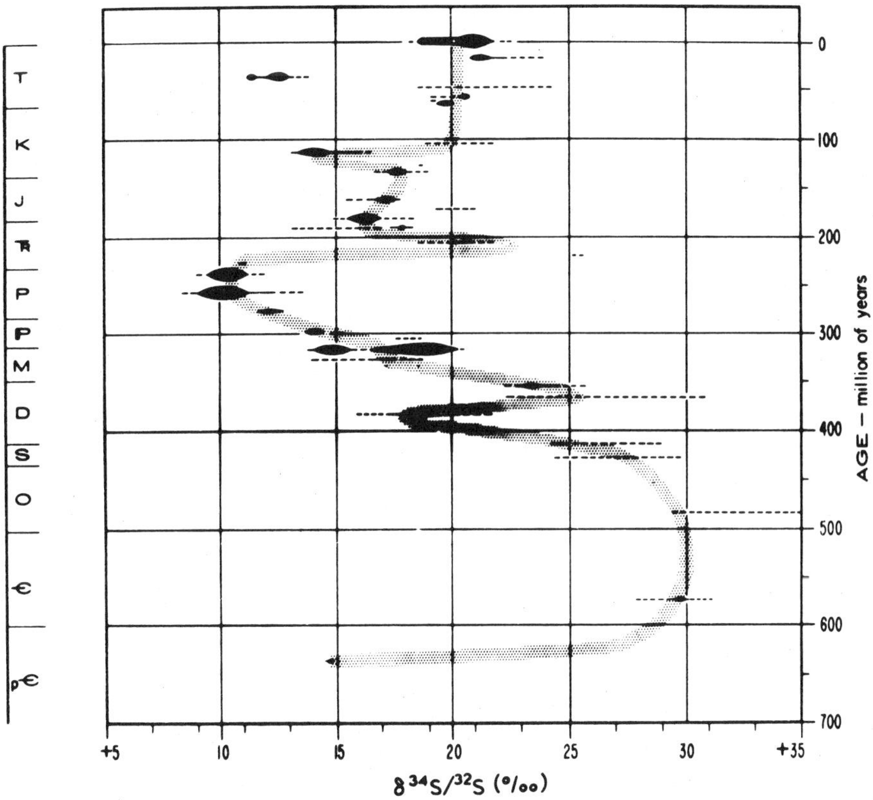

Fig. 5. Sulfur-isotope age curve, from all present and previous data. Thickness of line suggests number of analyses, and dashed lines indicate ranges where only a few analyses or insufficient details have been published.

through 1965 have been considered, including the determinations of Ault and Kulp (1959), Buschendorf et al. (1963), Nielsen and Ricke (1964), Thode and Monster (1964, 1965), our analyses, and several smaller collections. The data are now so numerous that it is impossible to show the individual analyses, their mineralogy, geological source, or analyst on a single-page diagram. In constructing this chart, the geological age of the samples was independently determined as closely as possible, from literature or correspondence. The data are plotted on a linear time scale, using the age assignments of Kulp (1961), with a few modifications from more recent age determinations.

Thode and Monster (1964), have proposed that the key to the isotope composition of the world ocean, at any one geological time, is given by the lightest sulfur-isotope ratio found in evaporite rocks of that age, because biological reduction within a single evaporite basin may push the sulfate of that basin toward a *heavier* average. While agreeing to this as a possibility, we also propose variations toward *light* sulfate in a given evaporite basin for reasons discussed in the above section: fractionation in late-stage evaporites, and in-flow of probable light sulfate from rivers. In the absence of any direct evidence on this question, we have assigned a most probable sulfur-isotope ratio to the world ocean at or a little below the mean for all analyses of that geological age.

The curve is similar in its broad aspects to those published by Ault and Kulp (1959, p.230), Thode and Monster (1964), and Nielsen and Ricke (1964, p.589). In general, a well-defined minimum in sulfur weight for the Permian is flanked by intermediate values in the Carboniferous and for much of the Mesozoic. Most of the Tertiary is similar to present sulfate, and most of the Early Paleozoic is heavier than the present sulfate. Within this general framework a number of details call for further discussion.

The most striking evidence of the change of sulfur-isotope ratio with time is provided by the contrast between the modern sulfates, described above, and the Permian evaporite rocks, as summarized in Fig.6, with an average throughout the Late Permian of around $\delta^{34}S$ = +10‰. As previously discussed,

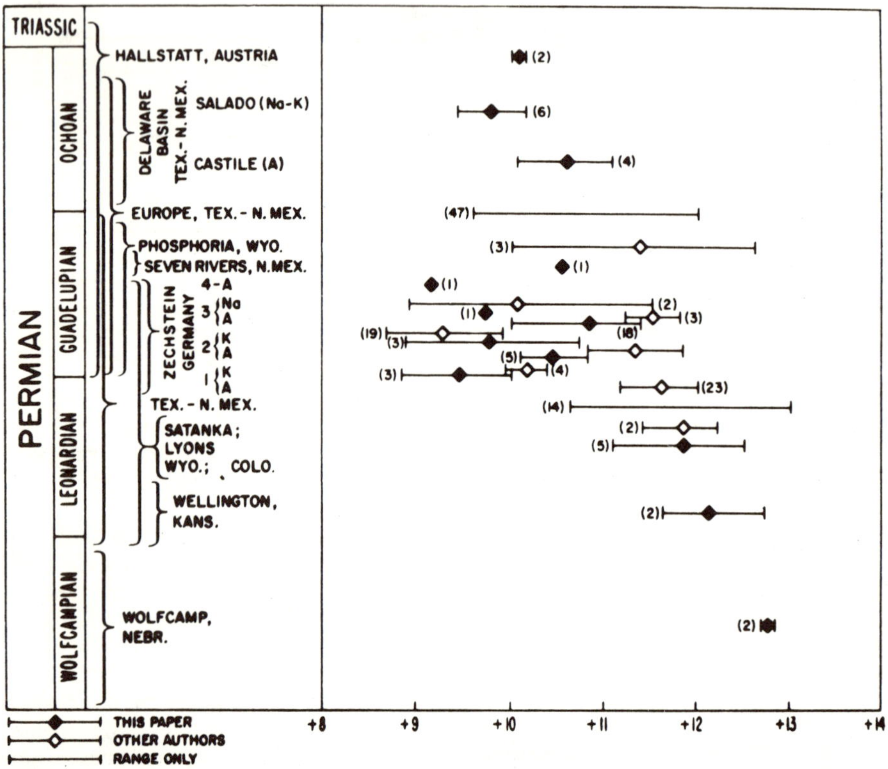

Fig.6. Summary of sulfur-isotope analyses of Permian evaporites. Means and standard deviations are shown for groups of samples of various ages and facies, as analyzed for this paper and as abstracted from other authors: Undifferentiated Upper Permian of Europe, Texas and New Mexico; and Leonardian–Guadelupian of Texas and New Mexico (Thode and Monster, 1964; see also Thode, 1964, pp.260–261); Phosphoria and Lyons Formations, Wyoming (Ault and Kulp, 1959); Zechstein 1–3 (Nielsen and Ricke, 1964). Brackets indicate possible range of age of each evaporite series relative to the standard Permian section.

these analyses represent a complete spectrum of evaporite mineralogy from around the world. They therefore indicate a different isotope ratio in the world ocean during Permian time than is present today.

On this basis, the ocean seems to have had a $\delta^{34}S$ = +10.3‰ during the most extensively sampled Late Permian (Guadelupian–Ochoan). The data collected in Fig.6 suggest a slightly heavier ocean sulfate in earlier Permian time, perhaps $\delta^{34}S$ = +11.7‰, in the Leonardian, and $\delta^{34}S$ = +12.7 in the Wolfcampian transitional from $\delta^{34}S$ = +13 in the Pennsylvanian (Fig.5). These data are only from the western United States, and must be confirmed by analyses from other areas. In this respect, analyses from the Soviet Union, which has extensive evaporite deposits throughout the Permian era, would be of particular interest. Even without this, the various data on the Permian are the most consistent, from the most wide-spread localities, and from a large variety of laboratories, to make this the best determination of the sulfur-isotope ratio in ancient seas.

Tertiary marine evaporites from various parts of the world show that the present sulfur-isotope ratio in sea water, $\delta^{34}S$ = +20‰, began somewhere in the Cretaceous period. The major exception is the series of twelve samples from the Upper Oligocene evaporites of the Rhine Valley analyzed by Nielsen and Ricke (1964, p.584), around $\delta^{34}S$ = +12‰, which is quite out of line with other Tertiary evaporites. We found $\delta^{34}S$ = +15.7 in a single sample 10 m below the potash section that they analyzed (Table IV). Part of the sedimentary section above and below the Rhine evaporites is non-marine, and there has been speculation in the past that the evaporites, or some of their components, were of continental origin (Wagner, 1953). More recent evidence from high bromide content in the salt tends to disprove this theory. In any case, the only obvious continental source for the sulfate would be the Zechstein evaporite basin, because of its light sulfur-isotope ratio, and this is over 300 km away, as Nielsen and Ricke (1964, p.586) have pointed out. These analyses suggest the importance of further work in the area, both in analyzing the several hundred meters of evaporites that still further underlie the Rhine Basin potash deposits, as well as analysis of other Lower Tertiary evaporites recently discovered in grabens of eastern France (Fraisse, 1965).

A number of widely separated samplings in the Jurassic and Late Triassic are in the range $\delta^{34}S$ = +16 to +18‰. The break between this value and the higher level in the Tertiary apparently happens rather quickly sometime in the Lower Cretaceous. Analyses by Ault and Kulp (1959, p.215) showed this break even within 2,000 ft. of section in a single well, between heavy sulfate that can be dated (E.L. Russell, Chevron Oil Co., personal communication, 1963) as Late Albian and light sulfate that can be dated near the Aptian-Albian boundary. This is confirmed by our analyses from Cay Sal, farther south, as shown in Table IV. However, a study in progress on other evaporites in the Caribbean indicates that the break may not be contemporaneous everywhere within this region.

Previous measurements of Devonian evaporites, particularly the numerous samples from Canadian evaporites analyzed by Buschendorf et al. (1963, p.513) show a very wide spread, from $\delta^{34}S$ = +17 to over +30‰, and our two analyses from the Prairie evaporites lie within their lower range. Buschendorf et al. (1963, p.520) ascribe the wide variation either to reduction to sulfide within the basin, or to reaction of the salt rocks with other sulfate-bearing solutions. An alternate possibility is shown in Fig.7, where we have

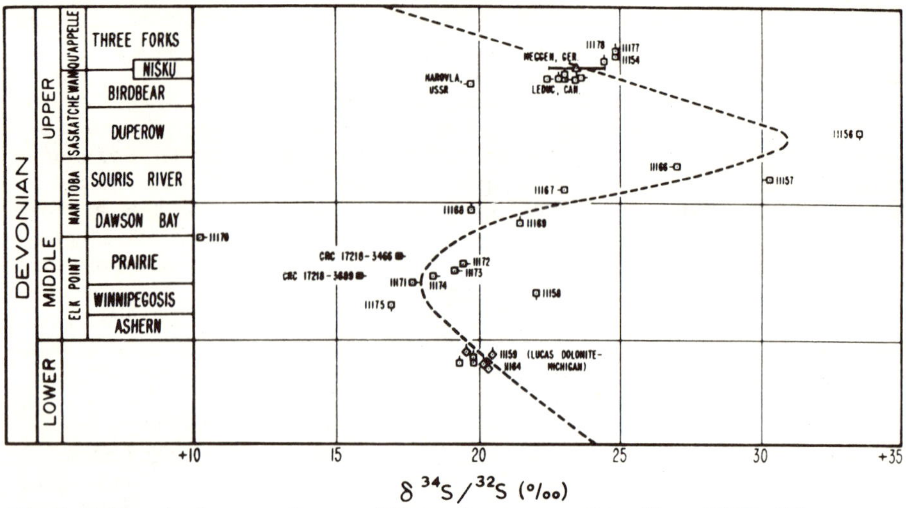

Fig.7. Sulfur-isotope analyses of Devonian evaporites. Key: CRC: this paper; Leduc: Thode et al. (1961b); Mengen, and 11177, etc.: Buschendorf et al. (1963); Narovla: Vinogradov et al. (1956).

taken the published descriptions of localities of all of the Devonian samples, analyzed their stratigraphic correlations (according to information from W.S. Campbell, Chevron Standard, Ltd., Calgary, Alta., and T.B. Howes, Chevron Oil Company, Billings, Mont.) and replotted them on a larger geological time scale. This replotting suggests that all of the Middle Devonian samples are relatively light in the range δ^{34} S = +16 to +22 ‰, and that the Late Devonian peaks at over δ^{34}S = +30‰. Uppermost Devonian evaporites in both Canada and Germany are down to δ^{34}S = +23‰, in line with the even lower values found in the Carboniferous period (Fig.5). Thode and Monster (1964) also suggest this threefold division of the Devonian, but we did not plot their data on Fig.7 inasmuch as only ranges of δ^{34}S values were given. It will be important to analyze Devonian evaporites from other parts of the world, with close stratigraphic control, to determine whether such large changes are indeed world-wide.

Geochemical cycle of sulfur

The ocean is a very large reservoir of sulfur, $1.3 \cdot 10^{15}$ metric tons, so that to change its isotopic composition requires an inverse change in another large reservoir. As Ault and Kulp (1959, p.231) have pointed out, "The major process of fractionation takes place in the oceans, ocean margins and recent sediments. . . where sulphate-reducing bacteria produce light sulphide and heavy sulphate. The actual $^{32}S/^{34}S$ ratio of sulphate in ocean water is regulated by the following factors: (a) influx of sulphate from rivers. . .; (b) deposition of ocean sulphate; and (c) bacterial activity. Since the ocean is mixed fairly rapidly (few hundred years) compared with the mean residence time of a sulphate ion (millions of years?) the bacteria are essentially operating on an infinite pool of sulphate ions of constant isotopic composition.

TABLE VII

Geochemical inventory of sulfur

1 Rock type	2 Mass of rock[1] (10^{15} T)		3 Mean S content (%) and standard deviation of the mean[2]	
Deep oceanic rocks				
Sediments	300	± 75	0.025	± 0.003
Mafic rocks (to Moho)	4,400	± 1,100	0.053	± 0.013
Sedimentary rocks				
Sandstone	280	± 70	0.090	± 0.01
Shale	750	± 190	0.27	± 0.04
Limestone	290	± 80	0.13	± 0.01
Evaporites	30	± 12	17	± 4
Volcanics	120	± 30	0.04	± 0.01
Connate water	140	± 30	0.019	± 0.02
Total evaporites and connate water	170	± 30		
Total other sediments	1,440	± 220	0.19	± 0.05
Total all sediments	1,600	± 220	0.49	± 0.13
Fresh water	0.3	± 0.2	0.0011	± 0.0001
Ice	35		0.00003	± 0.00001
Atmosphere	5.3			
Sea	1,420	± 40	0.090	± 0.003
Total in sedimentary rocks and sea				
Continental igneous and metamorphic rocks (to Moho)				
Granitic	10,500	± 2,600	0.021	± 0.003
Mafic	8,700	± 2,200	9.053	± 0.013
Total	21,000	± 3,000	0.032	± 0.005
Total in the crust				

[1] Masses from Poldervaart (1955, pp.127–313) except:

(1) Wedge of sediments underlying the present hemi-pelagic zone is considered to have average composition of young folded belts and shelf sediments, rather than that of modern hemi-pelagic sediments. This hemi-pelagic wedge represents 46% of the total sediments.

(2) The above estimates considered to have ± 25% precision.

(3) Mass of evaporites calculated by difference between: (a) sulfur analyses of combined limestones and anhydrites of the Russian platform (Vinogradov and Ronov, 1956); (b) all other limestone analyses.

(4) Mass of fresh water from Hutchinson (1957, p.225).

(5) Mass of ice from Gould (1965).

[2] Most analyses as abstracted by Ricke (1960; 1963) and Pettijohn (1963) plus Branner (1941), relisted to calculate mean and standard deviation of the mean, weighted by numbers of samples represented (Worthing and Geffner, 1943, p.199). For deep oceanic sediments assumed all S as SO_4^{2-} in porosity sea water (Ricke, 1960, pp.69–70). Average connate water and evaporite analyses from published and unpublished data collected by Holser. Ice composition from Cl content of Antarctic ice (Brocas and Delwiche, 1963) and average SO_4/Cl ratio of precipitation (Junge and Werby, 1958, p.424).

* Estimated from Fig.1.

TABLE VII (continued)

4 Mass of sulfur (10^{12} T) [=(2) × (3)]	5 Mean $\delta^{34}S$* (‰)	6 ^{34}S excess (10^{12} T) [=(4) × (5)]	1 Rock type
			Deep oceanic rocks
75 ± 20	+20 ± 1	+1.5 ± 0.4	Sediments
2,300 ± 800	+1 ± 2	+2.3 ± 5	Mafic rocks (to Moho)
			Sedimentary rocks
250 ± 60	−12 ± 5	−3.0 ± 1.5	Sandstone
2,000 ± 580	−12 ± 5	−24 ± 13	Shale
380 ± 110	−12 ± 5	−4.6 ± 2.3	Limestone
5,100 ± 1,600	+17 ± 2	+87 ± 30	Evaporites
50 ± 18	+5 ± 5	0.2 ± 0.2	Volcanics
27 ± 5	+17 ± 2	+0.5 ± 0.1	Connate water
5,100 ± 1,600	+17 ± 2	+87 ± 30	Total evaporites and connate water
2,700 ± 600 →	−11.8 ± 5.5	← −32 ± 13	Total other sediments
7,800 ± 1,700 →	+7.1 ± 4.5	← +55 ± 33	Total all sediments
0.003 ± 0.002	+10 ± 5	0.00	Fresh water
0.006 ± 0.002	+5 ± 5	0.00	Ice
($3.6 \cdot 10^6$)	+5 ± 5	0.00	Atmosphere
1,280 ± 55	+20 ± 0.5	+26 ± 1.3	Sea
9,100 ± 1,700	+8.5 ± 4.0	+81 ± 33	Total in sedimentary rocks and sea
			Continental igneous and metamorphic rocks (to Moho)
2,200 ± 600	+10 ± 5	+22 ± 12	Granitic
4,600 ± 1,600	+1 ± 2	+4.6 ± 9	Mafic
6,800 ± 1,700 →	+4 ± 3	← +27 ± 15	Total
18,200 ± 2,500 →	+6.0 ± 2.3	← +110 ± 40	Total in the crust

The competition of processes (a)-(c) noted above may cause gradual changes in the $^{32}S/^{34}S$ ratio in dissolved sulphate in the oceans which will not be uni-directional with time." They might just as well have left out process (b) because its effect on the isotopic composition of the remaining oceanic sulfate cannot be very great, as the contrast in isotopic composition is small. More recently Thode and Monster (1965) have in effect subdivided process (a) into sources of river water that increase the ^{34}S content of the ocean: by weathering of some older evaporites (either before or after metamorphism); and sources that decrease the ^{34}S content of the ocean: by weathering of other older evaporites, weathering of other rocks, addition of volcanic sulfur, and return of rain water to the oceans. The implication of Thode and Monster, that this and most other sources of river-water sulfate are lower in ^{34}S content than the ocean is beside the point; the critical relation is (as Ault and Kulp, 1959, suggest) the *balance* between input of light sulfur into the ocean by the long-term river cycle (see below), and the deposition of other light sulfur from the ocean into new sediments by bacterial reduction.

Some idea of the relative importance of the various reservoirs is shown in Table VII, which inventories sulfur in various parts of the crust. The details of its construction are not pertinent to the present discussion, but in general it differs from previous charts of this type (Eriksson, 1959, 1960, 1963) in that it uses actual estimated volumes of rock types in the crust (Poldervaart, 1955) rather than volumes of weathered rock and sediments found by indirect geochemical calculations involving an assumed model of single-cycle sediments. This calculation also assigns estimated precisions to the measured values, in order to see their effect on subsequent geochemical calculations. The calculations of this table show in the fourth and fifth column two important reservoirs of heavy sulfur (anhydrite rocks and the sea), and two important reservoirs of relatively light sulfur (shales and the igneous-metamorphic complex of the continents). The last column lists the excess of ^{34}S in each reservoir, over the meteoritic standard.

Some of the sulfur in the last category may be juvenile, in the sense of never having been on the surface of the earth, but general geochemical balances (Barth, 1961, 1962) as well as abundant petrological evidence suggest that most of it was recycled from earlier sediments. The general process of metamorphism of the sedimentary rocks includes both magmatism and volcanism, as well as associated hydrothermal activity, and the low content of sulfur in igneous rocks indicates that most of the sulfur took part in the volcanic branch of this cycle. Although geochemical calculations of the classical type (Conway, 1943, Rubey, 1951, Eriksson, 1959, 1960) indicate that sulfur is one of the "excess volatile" elements that may have been added to the crust, even on that model this could only have taken place over a long period of geologic time. Therefore, although juvenile non-cyclic sulfur cannot have been a major contributor to the sea during the relatively short intervals of geological time that we are considering, volcanic sulfur derived by metamorphism from earlier sediments (both shales and evaporites) may have been an important contributor to the sulfur cycle at one time or another. An independent check of the calculations of Table VII and Fig.8 is that the overall sulfur-isotope composition of volcanic exhalations should approximate that of all sediments, or $\delta^{34}S = +7.1 \pm 4.5‰$ according to the fifth column of Table VII. The mean isotope content of total sulfur in Wairakei (New Zealand) hot springs is $\delta^{34}S = +10‰$ and of White Island fumaroles is $\delta^{34}S = +4‰$ (Rafter and Wilson, 1963, p.26); workers in most other areas have not determined a weighted average, but the distribution of sulfur-isotope ratios in sulfide and sulfate exhalations is similar to that in New Zealand (Sakai and Nagasawa, 1958; Vinogradov, 1964). These values agree with our calculation, which quantitatively confirms the conclusions of Vinogradov (1964; see also Ault and Kulp, 1959, p.232).

The main flows among these various reservoirs are shown diagrammatically in Fig.8. Some idea of the long-term importance of the various flows is suggested by the thickness of the solid lines; these were estimated in the following way. The relative constancy of the sulfur-isotope ratio in the sea since Late Cretaceous time indicates that sulfur is neither accumulating nor decreasing in the sea at present. Such constancy is most likely if all of the long-term flows are now relatively constant. We make the assumption that evaporite and shale sulfur are being deposited in the proportion of 2:1 to their present accumulated masses, although the time necessary to establish this ratio was evidently very long. We then assume that both weathering and metamorphism of the shales and the evaporites are volume processes that affect their respective sulfur contents in this same proportion. But we assume that weathering of the total sedimentary volume and the igneous-

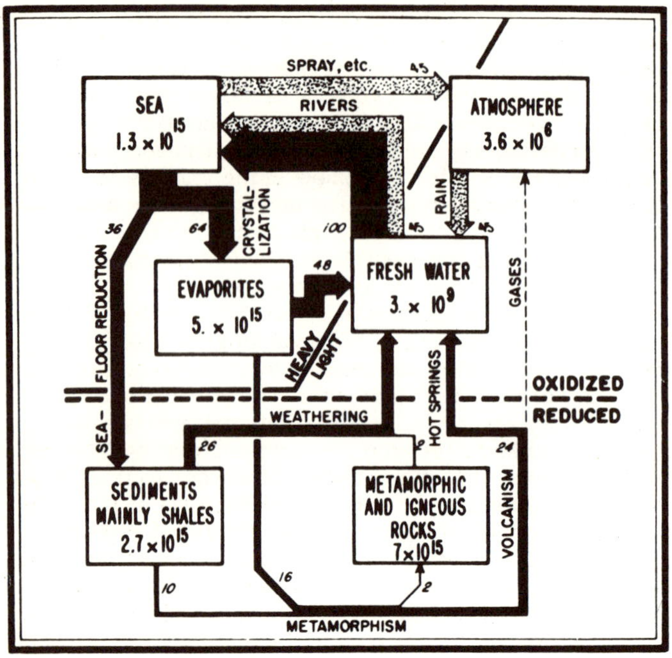

Fig.8. Geochemical cycle of sulfur. Masses are in metric tons. Most material above the dashed line is oxidized to sulfate, most below this line is reduced to sulfide; above the solid line heavy sulfur predominates ($\delta > +10‰$), below this line light sulfur predominates ($\delta < +5‰$). Long-term (dark) and short-term fluxes of sulfur between reservoirs are indicated on the basis of 100 for the long-term component of fresh-water sulfate flowing to the sea.

metamorphic complex takes place in proportion to their relative areas of exposure, presently about 3:1, which taken with the average sulfur contents of sediments and igneous-metamorphic rocks, 0.49:0.032 (Table VII, third column) gives a weathering ratio of 46:1 for sulfur in sediments and crystallines. Finally we note that the process of metamorphism must have released (0.49–0.032)/0.49 = 93% S on the average from every volume of total (shale plus evaporite) sediment that was metamorphosed. These numbers, applied to the solid closed circuits of Fig.8, give the single solution indicated by the line widths.

Contributions of these long-term cycles to the fresh-water reservoir decrease slightly in importance in the order evaporite weathering–shale weathering–hot spring activity. However, the various cycles involved are not directly tied to one another, and as they happen to change with the paleogeographic and tectonic environment, their relative importance may easily shift, and sulfur may be accumulated or depleted in any of the reservoirs. When these flow and storage variations are coupled with the accompanying isotopic fractionations, changes in the isotopic composition of the sea are a natural result. And the isotopic composition of the sea is symptomatic of the state of the flow-storage system at any given time. It is of course most sensitive to cycles that involve large isotopic fractionation, that is, to the relative importance at any time of the amount of biological reduction of sulfur into shales, the amount of weathering of shales, and the amount of volcanic generation of sulfur from deep sediments. Weathering of evaporites on the whole contributes more sulfur, but because the contrast of its isotope ratio is less, it can only slightly modify the isotope ratio of sea water.

The various flows converge in the fresh-water reservoir, and that should be the best place to sort out the relative importance of these flows in the present regime. However, the detection in fresh water of the long-term flows important for the history of sea water is rendered very difficult by the confluence of the very large short-term atmospheric precipitation cycle, shown as an open pattern in Fig.8. To this we must add in the immediate present a very large contribution from human "weathering" by mineral production. Junge and Werby (1958, p.424) estimated an addition of $4 \cdot 10^7$ tons/year to the atmosphere in 1957 from the burning of coal, petroleum, and sulfide ores, to which we may add $1.3 \cdot 10^7$ tons/year in native sulfur (Fogarty and Mollison, 1960) and $8 \cdot 10^6$ tons/year in gypsum, (Havard, 1960) most of which may end up in the fresh water reservoir. These figures have become much higher since 1958, which makes it very difficult to untangle the relative contributions of the various flows in today's fresh water. However, a detailed comparison of rainfall (Junge and Werby, 1958) and river flows (Livingstone, 1963) shows that the amount of river sulfate not accounted for by rainfall sulfate (exclusive of dry fallout and gas reactions) varies considerably among the various basins of the United States, with a mean of about 68%. If this same fraction is applied to the present world river flow of $1.2 \cdot 10^8$ tons/year of sulfur (Livingstone, 1963, p.41) we find $0.68 \cdot 1.2 \cdot 10^8 - 1.3 \cdot 10^7 - 8 \cdot 10^6 = 6.1 \cdot 10^7$ tons/year of sulfur left as a maximum for the present "long-terme" flows of weathering and hot springs. In Fig.8 the bulk of volcanic sulfur was somewhat arbitrarily assigned to continental hot spring activity, although of course some of this volcanic sulfur may reach the sea directly or through the atmosphere. However, the rough estimates of Rubey (1951, pp.1137–1139) of hot spring flow of $4.2 \cdot 10^{10}$ tons H_2O/year (on the continents) at an average concentration of 330 p.p.m. figures out to $1.2 \cdot 10^7$ tons/year of sulfur. According to the proportions of flows calculated for Fig.8, this corresponds to $5.4 \cdot 10^7$ tons/year for hot springs plus weathering. Considering all the approximations involved, this agrees much better than it should with the value of $6.1 \cdot 10^7$ tons/year calculated above. These same proportions should give a mean isotope composition for fresh water of about $\delta^{34}S = +8‰$, which is commensurate with the scanty data available. These checks suggest that the main features of our model for sulfur circulation are valid. The present relations of long and short-term flows are diagrammed on this basis in Fig.8

Changes in ocean chemistry related to the geochemical cycle

During the above discussion we have already indicated our view of the origin of the sulfur-isotope age curve for the sea: a redistribution of sulfur between the sea and shale reservoirs, in which biological reduction of sulfur into sea floor muds is off balance from the fresh water flow of sulfur from shales, via the sum of weathering and hot spring routes. We now consider both the magnitude and rate of these shifts, and their relation to the geochemical balance sheet.

A shift of the sulfur-isotope ratio from $\delta^{34}S$ = +10‰ in the Late Permian sea to $\delta^{34}S$ = +20‰ in the Tertiary can be explained by a net transfer of about $5.6 \cdot 10^{14}$ tons of sulfur (about 45% of the present amount) from the sea into the shale reservoir. Similarly the shift of sulfur-isotope ratio from $\delta^{34}S$ = +30‰ in the Silurian or Late Devonian to +10‰ in the Late Permian could be explained by a net transfer of $8.7 \cdot 10^{14}$ tons of sulfur from the shale reservoir into the sea. This is a net difference of about 25% less in the Early Paleozoic sea than at present.

These compositional changes of sulfur inferred from the isotope data are in addition to any transfers involving the evaporite reservoir, which might not be reflected in any marked changes of isotope ratio.

Other "excess volatile" elements, like chlorine and water, that are involved in the evaporite or metamorphic-volcanic cycles, may also have shifted their contents in the sea. Assuming that the amount of water in the sea did not change, the calculated transfers of sulfur would have required corresponding differences of sulfate concentration in the sea: 45% more in the Permian and 25% less in the Early Paleozoic.

The decline of the sulfur-isotope age curve (Fig. 5), from Devonian to Permian, indicates a rate of change in the sea of about 0.4‰ $\delta^{34}S/10^6$ years, and increases in the Cretaceous and Early Triassic seem to be similarly rapid. It is significant that the changes can be equally rapid in both directions, which suggests that the curve is the result of a single more or less symmetrical mechanism, rather than the result of an increase by one mechanism and decrease by another.

The present approximate flow of sulfur from the sediments to the sea (both through weathering and volcanism) was calculated above as about $6.1 \cdot 10^7$ tons/year, which gives a residence time in the sea for this long-cycle sulfur (solid flows in Fig.8) of about 21 million years. At this rate the shifts of sea water concentration and isotope composition are feasible. The corresponding residence time for average sulfur in average sediment is around 100 million years, but this reservoir is not very well mixed and the concept of residence time is not directly applicable.

Geological consequences of geochemical shifts

The sulfur-isotope data and the geochemical calculations require large shifts in the concentration of the total sulfur in the sea. Although it is usually assumed that the composition of sea water has been substantially constant since at least the Cambrian (Rubey, 1951; Chave, 1960; Kramer, 1965), such changes of sulfate concentration may have been overlooked.

If the source of an increase in sulfate content of the ocean (decrease in $\delta^{34}S$) was either excessive erosion or unusual volcanic activity, then probably other soluble components also were dumped into the ocean in comparable quantities. The extent to which the various soluble components were generated independently (e.g., by erosion of evaporites vs. shales) would shift ionic ratios. Increased erosion would also unload more detritus into the ocean, with probable increases in both minor inorganic and organic components.

Another possibility is that ^{32}S accumulated in the ocean due to a slowdown of sulfate reduction by bacteria in the bottom sediments. Such a phenomenon could be brought about in a number of ways: (a) by a marked cooling of the ocean waters; (b) by mixing and oxygenation of bottom ocean currents preventing the formulation of "stagnant" areas; or (c) removal of a readily available organic carbon food supply that is necessary for heterotrophic bacterial metabolism.

The possible implications of any such changes reach far beyond the significance of the change in sulfate itself. Biological systems in the sea may be profoundly affected by input from the continent that changes the ionic ratios of main components, amount of essential minor elements, amount of poisonous minor elements, or amount of organic food material. Temperature and circulation are also obviously important for the overall life system of the sea. Any one of these changes may led to the extinction of organic types that had low intrinsic or evolutionary adaptability. Such effects have been suggested in the past, particularly to explain changes in the biological pattern in the Permian–Triassic interval. Now the sulfur-isotope age curve provides the first external evidence of a variation in the nature of sea water, with which such suggestions may be tested. These implications will be discussed in detail elsewhere.

ACKNOWLEDGEMENTS

We are particularly indebted to Dr. Robert Kuhn of the Kaliforschungsinstitut, Hannover, Germany, for most of the samples from the Zechstein. Samples from other areas were contributed by A.L. Lyth (Denver, Colorado), G.A. Arrhenius (La Jolla, California), L.A. Briggs (Ann Arbor, Michigan), D.E. Garrett and the American Potash Co., (Trona, California), D.W. Graham (Los Angeles, California), D.W. Levandowski (La Habra, California), Maynard Nichols (Los Angeles, California), E. Middendorf (Hildesheim, Germany) and Leo Ogniben (Catania, Italy).

The sea-water evaporation experiment at Chevron Research Co. was conceived by G.A. Risley and carried out by J.O. Gardner, with the cooperation of Gale Baker, K.E. Chave, E.D. Hanson, R.G. Smalley, E.D. Sproul and others.

We are grateful to E.D. Cheney for carrying out most of the mass-spectrometer analyses, and to M.L. Jensen for laboratory facilities at Yale University. As mentioned above, a few of the analyses were made by Jan Monster, using the facilities of H.G. Thode at McMaster University.

Part of this work was done by I.R. Kaplan under a Ford Foundation Fellowship at the California Institute of Technology.

REFERENCES

Adams, J.E., 1944. Upper Permian Ochoa Series of Delaware Basin, west Texas and southeastern New Mexico. Bull. Am. Assoc. Petrol. Geologists, 28: 1596–1625.

Adams, J.E., 1963. Permian salt deposits of west Texas and eastern New Mexico. Symp. Salt, 1st, Columbus, Ohio, 1962, pp.124–130.

Ancarani-Rossiello, N., Bettinali, C. and Ferraresso, G., 1963. Sul rapporto isotopico $^{32}S/^{34}S$ nello zolfo elementare e nello zolfo dei solfati associati di alcuni giacimenti Italiani. Accad. Ital. Classe Sci. Fis. Mat. Nat., Rend., 34: 246–249.

D'Ans, J. and Kuhn, R., 1960. Bemerkungen zur Bildung und zur Umbildungen ozeanischer Salzlagerstatten. Kali Steinsalz, 3 (3): 69–84.

Applin, P.L. and Applin, E.R., 1965. The Comanche Series and associated rocks in the subsurface in central and south Florida. U.S. Geol. Surv., Prof. Papers, 447.
Arrhenius, G., 1963. Pelagic sediments. In: M.N. Hill (General Editor), The Sea. 3. The Earth Beneath the Sea. Interscience, New York, N.Y., pp.655–727.
Arrhenius, G., 1965. Neptunism and vulcanism in the ocean. In: M. Sears (Editor), Progress in Oceanography. Pergamon, London, pp.1–20.
Arrhenius, G., 1966. Sedimentary record of long-period phenomena. In: P.M. Hurley (Editor), Advances in Earth Sciences, M.I.T. Press, Cambridge, Mass., in press.
Ault, W.U. and Kulp, J.L., 1959. Isotopic geochemistry of sulphur. Geochim. Cosmochim. Acta, 16: 201–235.
Banks, J.E., 1964. The geologic setting of southern Florida. Geol. Soc. Am., Guidebook, Miami Meet., 1964, pt.1: 3–14.
Barr, A. and Kuhn, R., 1962. Der Werdegang der Kalisalzlagerstätten am Oberrhein. Neues Jahrb. Mineral. Abhandl., 97: 289–336.
Barth, T.F.W., 1961. Abundance of the elements, areal averages and geochemical cycles. Geochim. Cosmochim. Acta, 23: 1–8.
Barth, T.F.W., 1962. Die Menge der Kontinentalsedimente und ihre Beziehung zu den Eruptivgesteinen. Neues Jahrb. Mineral., Monatsh., 1962: 59–67.
Borchert, H. and Muir, R.O., 1964. Salt Deposits. Van Nostrand, London, 338 pp.
Braitsch, O., 1962. Entstehung und Stoffbestand der Salzlagerstätten. Springer, Berlin–Heidelberg–Göttingen, 232 S.
Branner, G.C., 1941. Limestones of northern Arkansas. Arkansas Geol. Surv., Little Rock, Ark., 24 pp.
Brocas, J. and Delwiche, R., 1963. Cl, K and Na concentrations in Antarctic snow and ice. J. Geophys. Res., 68: 3999–4000.
Buschendorf, F., Nielsen, H., Puchelt, H. and Ricke, W., 1963. Schwefel-Isotopen-Untersuchungen am Pyrit–Sphalerit–Baryt-Lager Meggen/Lenne (Deutschland) und an verschiednen Devon-Evaporiten. Geochim. Cosmochim. Acta, 27: 501–523.
Chave, K.E., 1960. Evidence on history of sea water from chemistry of deeper subsurface waters of ancient basins. Bull. Am. Assoc. Petrol. Geologists, 44: 357–370.
Conway, E.J., 1943. Chemical evolution of the ocean. Proc. Roy. Irish Acad., 43B: 161–212.
Deevey, E.S., Nakai, N. and Stuiver, M., 1963. Fractionation of sulfur and carbon isotopes in a meromictic lake. Science, 139: 407–408.
Dellwig, L.F., 1955. Origin of the Salina salt of Michigan. J. Sed. Petrol., 25: 83–110.
Dessau, G., Gonfiantini, R. and Tongiorgi, E., 1960. L'origine die giacimenti solfiferi Italiani alla luce delle indagini isotopiche sui carbonati della Serie Gessoso–Solfifera della Sicilia. Boll. Serv. Geol. Italia, 81: 313–348.
Dessau, G., Jensen, M.L. and Nakai, N., 1962. Geology and isotopic studies of Sicilian sulfur deposits. Econ. Geol., 57: 410–438.
Droste, J.B., 1961. Clay minerals in sediments of Owens, China, Searles, Panamint, Bristol, Cadiz, and Danby Lake Basins, California. Bull. Geol. Soc. Am., 72: 1713–1722.
Durum, W.H. and Heidel, S.G., 1961. The influence of continental water quality on shallow marine waters. Natl. Coastal Shallow Water Res. Conf., Tallahassee, Fla., 1961, pp.281–287.
Eriksson, E., 1959, 1960. The yearly circulation of chloride and sulfur in nature; meteorological, geochemical and pedological implications. Tellus, 11: 375–403; 12: 63–109.
Eriksson, E., 1963. The yearly circulation of sulfur in nature. J. Geophys. Res., 68: 4001–4008.
Feely, H.W. and Kulp, J.L., 1957. Gulf Coast salt-dome sulphur deposits. Bull. Am. Assoc. Petrol. Geologists, 41: 1802–1853.
Feth, J.H., Robertson, C.E. and Polzer, W.L., 1964a. Sources of mineral constituents in water from granitic rocks of the Sierra Nevada, California and Nevada. U.S. Geol. Surv., Water Supply Papers, 1535–I, 70 pp.
Feth, J.H., Rogers, S.M. and Roberson, C.E., 1964b. Chemical composition of snow

in the northern Sierra Nevada and other areas. U.S., Geol. Surv., Water Supply Papers, 1535-J, 39 pp.
Flint, R.F. and Gale, W.A., 1958. Stratigraphy and radiocarbon dates at Searles Lake, California. Am. J. Sci., 256: 689–714.
Fogarty, C.F. and Mollison, R.D., 1960. Sulfur and pyrites. In: J.L. Gillson (Editor), Industrial Minerals and Rocks, 3rd ed. Am. Inst. Mining Met. Petrol. Engrs., New York, N.Y., pp.819–833.
Fraisse, H., 1966. Salt basins of Bresse and Valence, France. Symp. Salt, 2nd, Cleveland, Ohio, 1965, in press.
Friedrich, K., Kuhn, R. and Strunz, H., 1961. Identität von Wathlingenit mit Kieserit. Kali Steinsalz, 3 (7): 221–227.
Füchtbauer, H. and Goldschmidt, H., 1959. Die Tonminerale der Zechsteinformation. Beitr. Mineral. Petrog., 6: 320–345.
Furrazola-Bermudez, G., Judoley, C.M., Mijailovskaya, M.S., Miroliubov, Y.S.,
Novojatsky, I.P., Jimenez, A.N. and Solsona, J.B., 1964. Geologia de Cuba. Editora Consejo Nacl. Univ., Havana, 239 pp.
Gale, H.S., 1914. Owens, Searles, and Panamint Basins, southeastern California. U.S. Geol. Surv., Bull., 580-L: 251–323.
Garrett, D.E., 1960. Borax processing at Searles Lake. In: J.L. Gillson (Editor), Industrial Minerals and Rocks, 3rd ed. Am. Inst. Mining Met. Petrol. Engrs., New York, N.Y., pp.119–122.
Goldman, M.I., 1952. Deformation, metamorphism, and mineralization in gypsum-anhydrite cap rock, Sulphur-salt Dome, Louisiana. Geol. Soc. Am., Mem., 50, 163 pp.
Gorham, E., 1961. Factors influencing supply of major ions to inland waters, with special reference to the atmosphere. Bull. Geol. Soc. Am., 795–840.
Gould, L.M., 1965. Antarctica - continent of international science. Science, 150: 1775–1781.
Haines, D.V., 1959. Core logs from Searles Lake, San Bernardino County, California. U.S. Geol. Surv., Bull., 1045-E: 139–317.
Harrison, A.G. and Thode, H.G., 1958. Sulphur-isotope abundances in hydrocarbons and source rocks of Uinta Basin, Utah. Bull. Am. Assoc. Petrol. Geologists, 42: 2642–2649.
Havard, J.G., 1960. Gypsum. In: J.L. Gillson (Editor), Industrial Minerals and Rocks, 3rd ed. Am. Inst. Mining, Met. Petrol. Engrs., New York, N.Y., pp.471–486.
Hay, R.L. and Moiola, R.J., 1963. Authigenic silicate minerals in Searles Lake, California. Sedimentology, 2: 312–332.
Holser, W.T., 1966a. Diagenetic polyhalite in Recent salt from Baja California. Am. Minera logist, 51: 99–109.
Holser, W.T., 1966b. Bromide geochemistry of salt rocks. Symp. Salt, 2nd, Cleveland, Ohio, 1965, in press.
Holser, W.T., Kaplan, I.R. and Silverman, S.R., 1963. Isotope geochemistry of salt rocks. Geol. Soc. Am., Spec. Papers, 76: 82 (abstract).
Hoy, R.B., Foose, R.M. and O'Neill, B.J., 1962. Structure of Winnfield Salt Dome, Winn Parish, Louisiana. Bull. Am. Assoc. Petrol. Geologists, 46: 1444–1459.
Hoyt, J.H., 1963. Permo-Pennsylvanian correlations and isopach studies in the northern Denver Basin. Rocky Mt. Assoc. Petrol. Geologists, Field Conf., 1963, Guidebook pp.68-83.
Hutchinson, G.E., 1957. A treatise on Limnology, Wiley, New York, N.Y., 1015 pp.
Ivanov, A.A. and Levitskii, Yu.F., 1960. Geology of halide deposits of the U.S.S.R. Tr. Vses. Nauchn. Issled. Geol. Inst., 35 (1960); 424 pp. (In Russian.)
Jones, B.F., 1966. Geochemical evolution of closed basin water in the western co-terminous United States. Symp. Salt, Cleveland, 2nd, Ohio, 1965, in press.
Jones, C.L., 1954. The occurrence and distribution of potassium minerals in south-eastern New Mexico. New Mex. Geol. Soc., Field Conf., 5th, Guidebook, pp.107–11
Jones, C.L., 1962. Some geologic and petrographic features of Upper Permian evaporite in southeastern New Mexico. Intern. Conf. Saline Deposits, Houston, Texas, p.22 (abstract).

Jung, W. and Knitzchke, G., 1960. Kombiniert-feinstratigraphisch-geochemische Untersuchungen der Anhydrite des Zechstein 1 in SE-Harzvorland. Geologie, 9: 58–72.
Junge, C.E. and Werby, R.T., 1958. The concentration of chloride, sodium, potassium, calcium and sulfate in rain water over the United States. J. Meteorol., 15: 417–425.
Jux, U., 1961. The palynologic age of diapiric and bedded salt in the Gulf Coastal province. Louisiana, Dept. Conserv., Geol. Bull., 38: 46 pp.
Kaplan, I.R., Rafter, T.A. and Hulston, J.R., 1960. Sulfur-isotope variations in nature. 8. Application to some biogeochemical problems. New Zealand J. Sci., 3: 338–361.
Kaplan, I.R., Emery, K.O. and Rittenberg, S.C., 1963. The distribution and isotopic abundance of sulphur in recent marine sediments off southern California. Geochim. Cosmochim. Acta, 27: 297–331.
King, R.H., 1947. Sedimentation in Permian Castile Sea. Bull. Am. Assoc. Petrol. Geologists, 31: 470–477.
Klaus, W., 1955. Über die Sporendiagnose des deutschen Zechsteinsalzes und des alpinen Salzgebirges. Z. Deut. Geol. Ges., 105: 776–788.
Knopf, A., 1918. A geologic reconnaissance of the Inyo Range and the eastern slope of the southern Sierra Nevada, Calif. U.S. Geol. Surv., Prof. Papers, 110, 130 pp.
Kramer, J.R., 1965. History of sea water. Constant temperature–pressure equilibrum models compared to liquid inclusion analyses. Geochim. Cosmochim. Acta, 29: 921–945.
Krumbein, W.C., 1951. Occurrence and lithologic associations of evaporites in the United States. J. Sed. Petrol., 21: 63–81.
Kühn, R., 1935. Violet kainite containing hydrogen sulfide. Zentr. Mineral. Geol., 1935A: 193–194.
Kulp, J.L., 1961. Geologic time scale. Science, 133: 1105–1114.
Lang, W.B., 1942. Basal beds of Salado Formation in Fletcher potash core test, near Carlsbad, New Mexico. Bull. Am. Assoc. Petrol. Geologists, 26: 63–79.
Linn, K.O. and Adams, S.S., 1966. Halite zones in potash deposits, Carlsbad, New Mexico. Symp. Salt, 2nd, Cleveland, Ohio, 1965, in press.
Livingstone, D.A., 1963. Chemical composition of rivers and lakes. U.S. Geol. Surv., Prof. Papers, 440-G, 64 pp.
McIntire, W.L., 1963. Trace element partition coefficients - a review of theory and applications to geology. Geochim. Cosmochim. Acta, 27: 1209–1264.
Momper, J.A., 1963. Nomenclature, lithofacies and genesis of Permo–Pennsylvanian Rocks - Northern Denver Basin. Rocky Mt. Assoc. Geologists, Field Conf., 1963, Guidebook, pp.41–67.
Murray, G.E., 1961. Geology of the Atlantic and Gulf Coastal Province of North America. Harper, New York, N.Y., 692 pp.
Newell, N.D. and Rigby, J.K., 1957. Geological studies on the great Bahama bank. Soc. Econ. Paleontologists Mineralogists, Spec. Publ., 5: 15–72.
Nichols, M.M., 1965. Composition and Environment of Recent Transitional Sediments on the Sonoran Coast, Mexico. Thesis, Univ. Calif., Los Angeles, 440 pp.
Nielsen, H., 1966. Schwefelisotope im marinen Kreislauf und das $\delta^{34}S$ der früheren Meere. Geol. Rundschau, 55: 160–172.
Nielsen, H. und Ricke, W., 1964. Schwefel-Isotopenverhältnisse von Evaporiten aus Deutschland; ein Beitrag zur Kenntnis von S im Meerwasser-Sulfat. Geochim. Cosmochim. Acta, 28: 577–591.
Niemann, H., 1960. Untersuchungen am Grauen Salzton der Grube Konigshall–Hindenburg. Reyershausen bei Göttingen. Beitr. Mineral. Petrog., 7: 137–165.
Ogniben, L., 1955. Inverse graded bedding in primary gypsum of chemical deposition. J. Sed. Petrol., 25: 273–281.
Ogniben, L., 1957a. Petrografia della Serie Solfifera Siciliana e considerazioni geologiche relative. Serv. Geol. Ital., Mem. Descrittive Carta Geol. Ital., 33, 215 pp.
Ogniben, L., 1957b. Secondary gypsum of the sulphur series, Sicily, and the so-called integration. J. Sed. Petrol., 27: 64–79, 469–470.
Östlund, H.G. and Alexander, J., 1963. Oxidation rate of sulfide in sea water, a preliminary study. J. Geophys. Res., 68: 3995–3997.

Pettijohn, F.J., 1963. Chemical composition of sandstones, excluding carbonate and volcanic sands. U.S. Geol. Surv., Prof. Papers, 440S, 19 pp.
Phleger, F.B. and Ewing, G.C., 1962. Sedimentology and oceanography of coastal lagoons in Baja California, Mexico. Bull. Geol. Soc. Am., 73: 145–182.
Poldervaart, A., 1955. Chemistry of the earth's crust. In: A. Poldervaart (Editor), Crust of The Earth. Geol. Soc. Am., Spec. Papers, 62: 119–144.
Rafter, T.A., 1957a. Sulphur-isotopic variations in nature. 2. A quantitative study of the reduction of barium sulphate by graphite for recovery of sulphide - sulphur for sulphur-isotopic measurements. New Zealand J. Sci., B, 38: 955–968.
Rafter, T.A., 1957b. Sulphur-isotopic variations in nature. 3. A study of the combustion characteristics of silver sulphide and lead sulphide for sulphur-isotopic measurements. New Zealand J. Sci., B, 38: 969–981.
Rafter, T.A. and Wilson, S.H., 1963. The examination of sulphur-isotope ratios in the geothermal and volcanic environment. New Zealand Dept. Sci. Ind. Res., Inst. Nucl. Studies, Rept., INS-R-25; 45 pp.
Richter-Bernburg, G., 1955. Stratigraphische Gliederung des deutschen Zechsteins. Z. Deut. Geol. Ges., 105: 843–854.
Richter-Bernburg, G., 1957a. Isochrone Warven im Anhydrit des Zechstein 2. Geol. Jahrb., 74: 601–610.
Richter-Bernburg, G., 1957b. Zur Paläogeographie des Zechsteins. In: Atti del Convegno di Milano su I Giacimenti Gassiferi dell'Europa Occidentale, 1. Accad. Nazl. Lincei e dall'Ente Nazl. Idrocarburi, Rome, pp.87–99.
Ricke, W., 1960. Ein Beitrag zur Geochemie des Schwefels. Geochim. Cosmochim. Acta, 21: 35–80.
Ricke, W., 1963. Geochemie des Schwefels und ihre Anwendung auf Faziesprobleme. Fortschr. Geol. Rheinland Westfalen, 110: 271–278.
Roberson, C.E., 1961. Geographic distribution of major constituents in stream waters of the western conterminous United States. U.S. Geol. Surv., Prof. Papers, 424: D-334–D-336.
Rubey, W.W., 1951. Geologic history of sea water. Bull. Geol. Soc. Am., 62: 1111–1148.
Sakai, H. and Nagasawa, H., 1958. Fractionation of sulphur isotopes in volcanic gases. Geochim. Cosmochim. Acta, 15: 32–59.
Schauberger, O., 1955. Zur Genese des alpinen Haselgebirges. Z. Deut. Geol. Ges., 105: 736–751.
Smith, G.I., 1962. Subsurface stratigraphy of Late Quaternary deposits, Searles Lake, California, a summary. U.S. Geol. Surv., Prof. Papers, 450C: 65–69.
Smith, G.I. and Haines, D.V., 1964. Character and distribution of nonclastic minerals in the Searles Lake evaporite deposit, California. U.S. Geol. Surv., Bull., 1181-P: 58 pp.
Stewart, F.H., 1963. Marine evaporites. U.S. Geol. Surv., Prof. Papers, 440-Y: 58 pp.
Story, J.A. and Howell, D.H., 1963. Gypsum deposits of the northern Colorado Front Range. Rocky Mt. Assoc. Geol., Field Conf., 1963, Guidebook, pp.258–263.
Stuiver, M., 1964. Carbon-isotopic distribution and correlated chronology of Searles Lake sediments. Am. J. Sci., 262: 377–392.
Thode, H.G., 1964. Stable isotopes a key to our understanding of natural processes. Bull. Can. Petrol. Geologists, 12: 242–262.
Thode, H.G. and Monster, J., 1964. The sulphur-isotope abundances in evaporites and in the ancient oceans. In: Vinogradov, A.V. (Editor), Khimiia Zemloi Kori, 2. Izd. Akad. Nauk S.S.S.R., Moscow, pp.589–600.
Thode, H.G. and Monster, J., 1965. Sulfur-isotope geochemistry of petroleum, evaporites and ancient seas. Am. Assoc. Petrol. Geologists, Mem., 4: 367–377.
Thode, H.G., Monster, J. and Dunford, H.B., 1961a. Sulphur-isotope geochemistry. Geochim. Cosmochim. Acta, 25: 159–174.
Thode, H.G., Monster, J. and Dunford, H.B., 1961b. Sulphur-isotope abundances in petroleum and associated materials. Bull. Am. Assoc. Petrol. Geologists, 42: 2619–2641.
Vinogradov, V.I., 1964. Isotopic composition and origin of volcanic sulfur. Geol. Rudn. Mestorozhd., 6 (3): 7–15 (In Russian).

Vinogradov, A.P. and Ronov, A.B., 1956. Composition of sedimentary rocks of the Russian platform in relation to the history of its tectonic movements. Geokhimiya, 1956 (6): 3–24. (In Russian; Transl.: Geochemistry, 1956: 533–559.)

Vinogradov, A.P., Chupakhin, M.S., Grinenko, V.A. and Trofimov, A.V., 1956. The isotopic composition of sulfur in connection with the growth of pyrites of sedimentary origin. Geokhimiya, 1956 (1): 96–105. (In Russian; Transl.: Geochemistry, 1956: 97–108.)

Wagner, W., 1953. Die tertiären Salzlagerstätten im Oberrheintal-Graben. Z. Deut. Geol. Ges., 105: 706–728.

White, D.E., Hem, J.D. and Waring, G.A., 1963. Chemical composition of subsurface waters. U.S. Geol. Surv., Prof. Papers, 440F: 67 pp.

Whitehead, H.C. and Feth, J.H., 1961. Recent chemical analyses of waters from several closed-basin lakes and their tributaries in the western United States. Bull. Geol. Soc. Am., 72: 1421–1426.

Worthing, A.G. and Geffner, J., 1943. Treatment of Experimental Data. Wiley, New York, N.Y., 338 pp.

Wright, J.C. and Dickey, D.D., 1963. Block diagram of the San Rafael Group and underlying strata in Utah and part of Colorado. U.S. Geol. Surv., Oil–Gas Invest., Chart OC-63.

References
Cited by Editors

Adams, J. E., and M. L. Rhodes, 1960, Dolomitization by seepage refluxion: *Am. Assoc. Petroleum Geologists Bull.*, **44**, 1912–1920.

Adams, S. S., 1969, *Bromine in the Salado Formation, Carlsbad Potash District, New Mexico:* Socorro, New Mex. State Bur. of Mines and Min. Res., Bull. 93, 122p.

Agricola, G., 1558, *De Natura Fossilium,* Basel, Hieron Frobenium et Nicolaum Episcopium: 2nd ed., 270 p.

Alberti, F. von, 1852, *Halurgische Geologie:* Stuttgart and Tübingen, J. B. Cotta, v. 1, 570p; v. 2, 414p.

Alderman, A. R., and H. C. W. Skinner, 1957, Dolomite sedimentation in the south-east of South Australia: *Am. Jour. Sci.,* **255,** 561–567.

Anderson, R. Y., and L. H. Koopmans, 1963, Harmonic analysis of varve time series: *Jour. of Geophysical Res.,* **68,** 3, 877–893.

Anderson, R. Y., and D. W. Kirkland, 1966, Intrabasin varve correlation: *Geol. Soc. Am. Bull.,* **77,** 241–256.

Anderson, R. Y., W. E. Dean, Jr., D. W. Kirkland, and H. I. Snider, 1972, Permian Castile varved evaporite sequence, West Texas and New Mexico: *Geol. Soc. Am. Bull.,* **83,** 59–86.

Arkhangel'skaya, N. A., and V. N. Grigor'yev, 1960, Formation of halogenic zones in marine basins illustrated by the example of the Lower Cambrian evaporite basin of the Siberian Platform: *Acad. of Sci. U.S.S.R., Geol. Ser.,* **4** (Eng. trans.), 41–54.

Ault, W. U., and J. L. Kulp, 1959, Isotope geochemistry of sulphur: *Geochim. Cosmochim. Acta,* **16**, 201–235.

Baar, A., 1954, Untersuchungen des Bromgehalts im Zechsteinsalz. Auswertung des "Bromtests" bei Aufschlussarbeiten im Kalibergbau: *Bergbautechnik,* **4,** 284.

———, 1963, Der Bromgehalt als stratigraphischer Indikator in Steinsalzlagerstatten: *N. Jb. Miner. Mh.,* **7**, 145–153.

Baar, A., and R. Kühn, 1962, Der Werdegang der Kalisalzlagerstätten am Oberrheim: *N. Jb. Miner., Abh.,* **97,** 289.

Bathurst, R. G. C., 1971, *Carbonate sediments and their diagenesis:* Amsterdam, Elsevier Publishing Company, 620p.

Bischof, K. G. C., 1854, 1855, 1859, *Elements of Chemical and Physical Geology:* London, Harrison and Sons, trans. by B. H. Paul, v. 1 (1854) 455p., v. 2 (1855) 523p., v. 3 (1859) 566p. (Eng. trans. of 1st Ger. ed., 1847).

Bock, E., 1961, On the solubilities of anhydrous calcium sulphate and of gypsum in concentrated solutions of sodium chloride at 25°C, 30°C, 40°C, and 50°C: *Can Jour. Chem.,* **39,** 1746–1751.

Boecke, H. E., 1908, Uber das Kristallisationsschema der Chloride, Bromide, Jodide von Natrium, Kalium, und Magnesium, sowie über das Vorkommen des Broms und das Fehlen von Jod in den Kalisalzlaggerstätten: *Zeitschr. Kristallographie,* **45,** 346–391.

Borchert, H., and R. O. Muir, 1964, *Salt Deposits: The Origin, Metamorphism and Deformation of Evaporites:* New York, Van Nostrand Reinhold Company, 338p.

Branson, E. B., 1915, Origin of thick gypsum and salt deposits: *Geol. Soc. Am. Bull.,* **26,** 231–242.

Brill, K. G., Jr., 1952, Stratigraphy in the Permo–Pennsylvanian Zeugogeosyncline of Colorado and northern New Mexico: *Geol. Soc. Am. Bull.,* **63**, 809–880.

Brongersma-Sanders, M., and P. Groen, 1970, Wind and water depth and their bearing on the circulation in evaporite basins: in J. L. Rau and L. F. Dellwig (eds.), *Third Symposium on Salt,* v. 1, Cleveland, Northern Ohio Geological Society, Inc., p. 3–7.

Bryson, R. A., and J. A. Dutton, 1961, Some aspects of the variance spectra of tree rings and varves: *Ann. N.Y. Acad. Sci.,* **95,** 580–604.

Bundy, W. M., 1956, Petrology of gypsum-anhydrite deposits in southwestern Indiana: *Jour. Sed. Petrology,* **26**, 240–252.

Butler, G. P., 1969, Modern evaporite deposition and geochemistry of coexisting brines, the *Sabkha,* Trucial Coast, Arabian Gulf: *Jour. Sed. Petrol.,* **39,** 70–89.

———, 1970a, Holocene gypsum and anhydrite of the Abu Dhabi sabkha, Trucial Coast: an alternative explanation of origin: in J. L. Rau and L. F. Dellwig (eds.), *Third Symposium on Salt,* v. 1., Cleveland, Northern Ohio Geol. Soc., Inc., p. 120–152.

———, 1970b, Secondary anhydrite from a sabkha, northwest Gulf of California, Mexico: in J. L. Rau and L. F. Dellwig (eds.), *Third Symposium on Salt,* v. 1, Cleveland, Northern Ohio Geol. Soc., Inc., p. 153–155.

Carpelan, L. H., 1957, Hydrobiology of the Alviso Salt Ponds: *Ecology,* **38,** 375–390.

Caspers, H., 1952, Untersuchungen über die Tierwelt von Meeressalinen an der bulgarischen Küste des Schwarzen Meeres: *Zool. Anz.,* **148,** no. 5/8, 243–259.

Claypool, G. E., W. T. Holser, I. R. Kaplan, H. Sakai, and I. Zak, 1972, Sulfur and oxygen isotope geochemistry of evaporite sulfates: *Geol. Soc. Am. abs. with program, 1972 annual meeting,* **4,** no. 7, 473.

Cruft, E. F., and Chao, Pao-Chin, 1970, Nucleation kinetics of the gypsum–anhydrite system: in J. L. Rau and L. F. Dellwig (eds.), *Third Symposium on Salt,* v. 1., Cleveland, Northern Ohio Geol. Soc., Inc., p. 109–118.

Curtis, R., G. Evans, D. J. J. Kinsman, and D. J. Shearman, 1963, Association of dolomite and anhydrite in the Recent sediments of the Persian Gulf: *Nature,* **197,** 679–680.

D'Ans, J., D. Bredtschneider, H. Eick, and H. E. Freund, 1955, Untersuchungen über die Calciumsulfate: *Kali u. Steinsalz,* **9** (Ser. 1), 17–38.

D'Ans, J., and R. Kühn, 1940, Uber den Bromgehalt von Salzgesteinen der Kalisalzlagerstätten: *Kali,* **34,** 42.

Davis, J. B., and D. W. Kirkland, 1970, Native sulfur deposition in the Castile Formation, Culberson County, Texas: *Econ. Geol.,* **65,** 107–121.

Deffeyes, K. S., F. J. Lucia, and P. K. Weyl, 1965, Dolomitization of Recent and Plio-Pleistocene sediments by marine evaporite waters on Bonaire, Netherlands Antilles: in *Dolomitization and Limestone Diagenesis,* Soc. Econ. Paleon. and Miner. Spec. Pub. no. 13, p. 71–88.

Dellwig, L. F., 1955, Origin of the Salina Salt of Michigan: *Jour. of Sed. Petrol.,* **25,** 83–110.

Dellwig, L. F., and R. Evans, 1969, Depositional processes in salina salt of Michigan, Ohio, and New York: *Am. Assoc. Petroleum Geologists Bull.,* **53**, 949–956.

Dessau. G., M. L. Jensen, and N. Nakai, 1962, Geology and isotopic studies of Sicilian sulfur deposits: *Econ. Geol.,* **57,** 410–438.

Dickey, P. A., 1968, Contemporary nonmarine sedimentation in Soviet Central Asia: *Am. Assoc.*

of Petroleum Geologists Bull., **52**, 2396–2421.
Dittrich, E., 1962, Zur Gliederung der Werra-Serie (Zechstein 1) im Werra-Kaligebiet: *Ber. Geol. Ges. DDR 6*, p. 296.
Dorheim, F. H., and R. B. Campbell, 1958, Recent gypsum exploration in Iowa: *Iowa Acad. Sci. Proc.*, **65,** 246–253.
Douglas, G. V., and N. R. Goodman, 1957, The deposition of gypsum and anhydrite: *Econ. Geol.*, **52,** 831–837.
Dzens-Litovskiy, A. I., 1967, The problem of Kara-Bogaz-Gol: *Lithol. Miner. Resour.*, **1**, 70–76.
Dzens-Litovskiy, A. I., and G. V. Vasil'yev, 1962, Geologic conditions of formation of bottom sediments in Karabogaz-Gol in connection with fluctuations of the Caspian Sea level: *Acad. Sci. U.S.S.R., Geol. Ser.*, no. 3 (Eng. trans.), 79–86.
Eugster, H. P., 1971, Review of *Salt Deposits, Their Origin and Composition* by Otto Braitsch: *Science*, **173**, 713–714.
Fairbridge, R. W., 1968, Kara-Bogaz Gulf: in R. W. Fairbridge (ed.), *The Encyclopedia of Geomorphology:* Encyclopedia of Earth Sciences Series, v. **3**, p. 579–581.
Feely, H. W., and J. L. Kulp, 1957, Origin of Gulf Coast salt dome sulfur deposits: *Am. Assoc. Petroleum Geologists Bull.*, **41,** 1802–1853.
Fisk, H. N., 1959, Padre Island and the Laguna Madre Flats, Coastal South Texas: *Nat. Acad. Sci.–Nat. Res. Counc., 2nd Coastal Geog. Conf.*, 103–151.
Friedman, G. M., and J. E. Sanders, 1967, Origin and occurrence of dolostones: in *Carbonate Rocks,* Amsterdam, G. V. Chilingar (ed.), Elsevier Publishing Company, p. 267–348.
Fulda, Ernst, 1927, Der Assalsee im Somaliland und seine Bedeutung für die Erklärung der Entstehung mächtiger Salzlager: *Z. deutsch. geol. Gesellsch.*, **79,** Monatsber., 70–75.
———, 1938, Steinsalz und Kalisalze: in F. Beyschlag, P. Krusch, und J. H. L. Vogt (eds.), *Die Lagerstätten der nutzbaren Mineralien und Gesteine,* **III,** Teil 2, Stuttgart, 240p.
Garbell, M. A., 1963, The sea that spills into the desert: *Sci. Am.*, **209,** no. 2, 94–100.
Grabau, A. W., 1920, *Geology of the non-metallic mineral deposits other than silicates:* v. 1, *Principles of Salt Deposition:* New York, McGraw-Hill Book Company, 435p.
Hardie, L. A., 1967, The gypsum–anhydrite equilibrium at one atmosphere pressure: *Am. Mineral.*, **52,** 172–200.
Hedgpeth, J. W., 1947, The Laguna Madre of Texas: *Trans. 12th North Am. Wildl. Conf.*, 364–380.
Hill, A. E., 1934, Ternary systems, XIX. Calcium sulfate, potassium sulfate and water: *Jour. Am. Chem. Soc.*, **56,** 1071.
Hite, R. J., 1960, Stratigraphy of the saline facies of the Paradox member of the Hermosa Formation of southeastern Utah and southwestern Colorado: in *Four Corners Geol. Soc. Guidebook Field Conf.*, no. 3, *Geology of the Paradox basin fold and fault belt,* 86–89.
———, 1961, Potash-bearing evaporite cycles in the Salt Anticlines of the Paradox basin, Colorado and Utah: in *U.S. Geol. Sur. Prof. Paper 424-D,* p. D135–D138.
———, 1968, Salt deposits of the Paradox basin, southeastern Utah and southwestern Colorado: in *Saline Deposits:* Geol. Soc. Am. Spec. Paper 88, p. 319–330.
———, 1970, Shelf carbonate sedimentation controlled by salinity in the Paradox basin Southeast Utah: in J. L. Rau and L. F. Dellwig (eds.), *Third Symposium on Salt,* v. 1, Cleveland, Northern Ohio Geological Society, Inc., p. 48–66.
Hite, R. J., and W. C. Gere, 1958, Potash deposits of the Paradox basin: in *Guidebook to the Geology of the Paradox Basin,* Intermountain Assoc. Petroleum Geologists 9th Ann. Field Conf., p. 221–225.
van't Hoff, J. H., E. F. Armstrong, W. Hinrichsen, F. Weigert, and G. Just, 1903, Gips und anhydrit: *Z. Phys. Chem.*, **45,** 257–306.
Holliday, D. W., 1968, Early diagenesis in Middle Carboniferous nodular anhydrite of Spitsbergen: *Proc. Yorks. Geol. Soc.*, **36**, 277–292.
Holser, W. T., 1966, Bromide geochemistry of salt rocks: in J. L. Rau (ed.), *Second Symposium on Salt,* Cleveland, Northern Ohio Geol. Soc., v. 1, p. 248–275.
———, 1966, Diagenetic polyhalite in Recent salt from Baja California: *Am. Mineral.*, **51,** 99–109.
Holser, W. T., and I. R. Kaplan, 1966, Isotope geochemistry of sedimentary sulfates: *Chem. Geol.*, **1,** 93–135.

Hsü, K. J., 1972, When the Mediterranean dried up: *Sci. Am.*, **227,** 6, 26–36.

Illing, L. V., and Wells, A. J., 1964, Present day precipitation of calcium carbonate in the Persian Gulf: in M. J. U. van Straaten (ed.), *Deltaic and Shallow Marine Deposits,* Amsterdam, Elsevier Publishing Company, p. 429–435.

Illing, L. V., A. J. Wells, and J. C. M. Taylor, 1965, Penecontemporary dolomite in the Persian Gulf: in *Dolomitization and Limestone Diagenesis,* Soc. Econ. Paleon. and Miner. Spec. Pub. no. 13, p. 89–111.

Ivanov, A. A., 1967, Depths of halogenic basins of the geologic past: *Lithol. Miner. Resour.,* **2,** 161–171.

Jones, V., 1935, Origin of the gypsum deposits near Sandusky, Ohio: *Econ. Geol.,* **30,** 493–501.

Jones, G. E., and R. L. Starkey, 1957, Fractionation of stable isotopes of sulfur by microorganisms and their role in deposition of native sulfur: *Applied Microbiology,* **5,** 111–118.

Kelley, K. K., J. C. Southard, and C. T. Anderson, 1941, *Thermodynamic Properties of Gypsum and Its Dehydration Products:* U.S. Bur. Mines Tech. Paper 625, 73p.

Kerr, S. D., and A. Thomson, 1963, Origin of nodular and bedded anhydrite in Permian shelf sediments, Texas and New Mexico: *Am. Assoc. Petroleum Geologists Bull.,* **47,** 1726–1732.

King, R. H., 1947, Sedimentation in Permian Castile Sea: *Am. Assoc. Petroleum Geologists Bull.,* **31,** 470–477.

Kinsman, D. J. J., 1964, *Recent carbonate sedimentation near Abu Dhabi, Trucial Coast, Persian Gulf:* unpublished Ph.D. thesis, Univ. of London, 320p.

———, 1966, Gypsum and anhydrite of Recent age, Trucial Coast, Persian Gulf: in J. L. Rau (ed.), *Second Symposium on Salt,* v. 1, Cleveland, Northern Ohio Geol. Soc., Inc., p. 302–326.

———, 1967, Huntite from a carbonate-evaporite environment: *Am. Mineral,* **52,** 1332–1340.

———, 1969, Modes of formation, sedimentary associations, and diagnostic features of shallow-water and supratidal evaporites: *Am. Assoc. Petroleum Geologists Bull.,* **53,** 830–840.

Kudryavtsev, N. A., 1971, Evaporites and petroleum: discussion: *Am. Assoc. Petroleum Geologists Bull.,* **55,** 2033–2061.

Kulstad, R. O., P. Fairchild, and D. McGregor, 1956, *Gypsum in Kansas:* Kansas State Geol. Survey Bull. 113, 110p.

Lichtenberger, T., 1927, Entstehung der Salzlager: *Kali,* **21,** 65.

Liebig, J. von, 1840, *Die Chemie in ihre Anwendung auf Agrikultur und Physiologie:* Braunschweig, Fr. Vieweg.

Lotze, F., 1938, Steinsalz und Kalisalze, Geologie: in *Die wichtigsten Lagerstätten der "Nicht-Erze":* Berlin, Gebrüder Borntraeger, Bd. III, Teil 1, 936p.

———, 1957, *Steinsalz und Kalisalze:* Berlin, Gebrüder Borntraeger, 465p.

Lucia, F. J., 1968, Recent sediments and diagenesis of South Bonaire, Netherlands Antilles: *Jour. Sed. Petrol.,* **38,** 845–858.

MacDonald, G. F. J., 1953, Anhydrite–gypsum equilibrium relations: *Am. Jour. Sci.,* **251,** 884–898.

McKee, E. D., 1938, *The Environment and History of the Toroweap and Kaibab Formations of Northern Arizona and Southern Utah:* Carnegie Inst. Washington Pub. 492, 268p.

Madgin, W. M., and D. A. Swales, 1956, Solubilities in the system $CaSO_4$–NaCl–H_2O at 25°C and 35°C: *Jour. Appl. Chem.,* **6,** 482–487.

Manger, G., 1961, Die praktische Anwendung der Bestimmung des Bromgehaltes im Steinsalz zur Horizontierung bei Tiefbohrungen und Untertage-Aufschlüssen der Gerwerkschaft Baden in Buggingen (Südbaden): *Jb. geol. Landesanst. Baden–Württemberg,* **5,** 279.

Masson, P. H., 1955, An occurrence of gypsum in southwest Texas: *Jour. of Sed. Petrol.,* **25,** 72–77.

Morris, R. C., and P. A. Dickey, 1957, Modern evaporite deposition in Peru: *Am. Assoc. Petroleum Geologists Bull.,* **41,** 2467–2474.

Muir, J. L., 1934, Anhydrite–gypsum problem of Blaine Formation, Oklahoma: *Am. Assoc. Petroleum Geologists Bull.,* **18,** 1297–1312.

Murray, R. C., 1964, Origin and diagenesis of gypsum and anhydrite: *Jour. Sed. Petrol.,* **34,** 512–523.

———, 1969, Hydrology of South Bonaire, N.A.—A rock selective dolomitization model: *Jour. Sed. Petrology,* **39**, 1007–1013.
Nielsen, H., 1966, Schwefelisotope im marine Kreislauf und das δ ^{34}S der Früheren Meere: *Geol. Rund.,* **55,** 160–172.
———, 1972, Sulfur isotopes and the formation of evaporite deposits: in *Geology of Saline Deposits,* Paris, UNESCO, p. 91–102.
Nielsen, H., and W. Ricke, 1964, Schwefel-Isotopenverhaltnisse von Evaporiten aus Deutschland; ein Beitrag zur Kenntnis von Sim Meerwasser-Sulfat: *Geochim. Cosmochim. Acta,* **28,** 577–591.
Ochsenius, Carl, 1877, *Die Bildung der Steinsalzlager und ihrer Mutterlaugensalze, unter spezieller Berücksichtung der Flötze von Douglashall im Egeln Mülde:* Halle, C. E. M. Pfeffer, 172p.
———, 1888, On the formation of rock-salt beds and mother liquor salts: *Acad. Nat. Sci. Philadelphia Proc.,* **40,** 181–187.
———, 1893, Uber die Bedeutung des orographischen Elementes Barre usw: *Zeitschr. Prakt. Geol.,* **1,** 189–201, 217–233.
Ogniben, Leo, 1957, Secondary gypsum of the sulfur series, Sicily and the so-called integration: *Jour. Sed. Petrology,* **27,** 64–79.
Pannekoek, A. J., 1965, Shallow-water and deep-water evaporite deposition: *Am. Jour. Sci.,* **263,** 284–285.
Partridge, E. P., and A. H. White, 1929, The solubility of calcium sulfate from 0° to 200°: *Jour. Am. Chem. Soc.,* **51,** 360–370.
Perthuisot, J. P., 1971, Recent polyhalite from Sebkha el Melah (Tunisia): *Nature,* **232,** 186–187.
Peterson, J. A., and R. J. Hite, 1969, Pennsylvanian evaporite-carbonate cycles and their relation to petroleum occurrence, southern Rocky Mountains: *Am. Assoc. Petroleum Geologists,* **53,** 884–908.
Phleger, F. B., 1969, A modern evaporite deposit in Mexico: *Am. Assoc. Petroleum Geologists Bull.,* **53,** 824–829.
Phleger, F. B., and G. C. Ewing, 1962, Sedimentology and oceanography of coastal lagoons in Baja California, Mexico: *Geol. Soc. Am. Bull.,* **73,** 145–182.
Posnjak, E., 1938, The system, $CaSO_4$–H_2O: *Am. Jour. Sci.,* **35A,** 247–272.
———, 1940, Deposition of calcium sulfate from sea water: *Am. Jour. Sci.,* **238,** 559–568.
Raup, O. B., 1966, Bromine distribution in some halite rocks of the Paradox Member, Hermosa Formation, in Utah: in J. L. Rau (ed.), *Second Symposium on Salt,* v. 1, Cleveland, Northern Ohio Geol. Soc., Inc., p. 236–247.
Richter-Bernburg, G., 1950, Zur Frage der absoluten Geschwindigkeit geologischer Vorgange: *Die Naturwissenschaften,* **37,** no. 1, 1–8.
———, 1955, Uber salinare Sedimentation: *Z. deutsch. geol. Gesellsch.,* **105,** 593–645.
———, 1957, Isochrone Warven im Anhydrit des Zechstein 2: Germany, Geol. Landesanst., *Geol. Jb.* Bd. 74, 601–610.
———, 1959, Die Korrelierung isochroner Warven im Anhydrit des Zechstein 2 (zweiter Beitrag): Germany, Geol. Landesanst, *Geol. Jb.* Bd. 75, 629–639 (1958).
———, 1960, Zeitmessung geologischer Vorgange nach Warven-Korrelation im Zechstein: *Geol. Rund.,* **49,** Heft 1, **49,** 132–148.
———, 1964, Solar cycle and other climatic periods in varvitic evaporites; in A. E. Nairn (ed.), *Problems in Palaeoclimatology,* New York, Wiley (Interscience Publishers), p. 510–521.
Rogers, A. F., 1910, Anhydrite and associated minerals from the salt mines of central Kansas: *Am. Jour. Sci.,* **29,** 4th ser., 258–261.
Schaller, W. T., and E. P. Henderson, 1932, *Mineralogy of Drill Cores from the Potash Field of New Mexico and Texas:* U.S. Geol. Survey Bull. 833, 124p.
Schulze, G., 1960, Stratigraphie und genetische Deutung der Bromverteilung in den mitteldeutschen Steinsalzlagern des Zechstein: *Freib. Forsch.-H.,* C83, p. 1–116.
Schwerdtner, W., and N. C. Wardlaw, 1963, Geochemistry of bromine in some salt rocks of the Prairie Evaporite Formation of Saskatchewan: in J. L. Rau (ed.), *First Symposium on Salt,* Cleveland, Northern Ohio Geol. Soc., Inc., p. 240–246.
Scruton, P. C., 1953, Deposition of evaporites: *Am. Assoc. Petroleum Geologists Bull.,* **37,** 2498–2512.

Sellards, E. H., and C. L. Baker, 1934, *The Geology of Texas:* Texas Univ. Bull. no. 3401, **2,** 623–631.

Shaw, A. B., 1964, *Time in Stratigraphy:* New York, McGraw-Hill Book Company, 365p.

Shearman, D. J., 1966, Origin of marine evaporites by diagenesis: *Inst. Mining Met., Trans.,* **75,** 208–215.

———, 1970, Recent halite rock, Baja California, Mexico: *Inst. Mining Met., Trans.,* **79,** 155–162.

Stewart, F. H., 1949, The petrology of the evaporites of the Eskdale no. 2 boring, east Yorkshire; pt. 1—The Lower Evaporite Bed: *Mineralog. Mag.,* **28,** 621–675.

———, 1951a, The petrology of the evaporites of the Eskdale no. 2 boring, east Yorkshire; pt. 2—The Middle Evaporite Bed: *Mineralog. Mag.,* **29,** 445–475.

———, 1951b, The petrology of the evaporites of the Eskdale no. 2 boring, east Yorkshire; pt. 3—The Upper Evaporite Bed: *Mineralog. Mag.,* **29,** 557–572.

———, 1963, Marine evaporites: in *Data of Geochemistry,* 6th ed., U.S. Geol. Sur. Prof. Paper 440-Y, 52p.

Thode, H. G., and J. Monster, 1964, The sulfur-isotope abundances in evaporites and in the ancient oceans: in A. V. Vinogradov (ed.), *Khimiia Zemloi Kori,* v. 2, Izd. Akad. Nauk. S.S.S.R., Moscow, p. 598–600.

———, 1965, Sulfur-isotope geochemistry of petroleum, evaporites and ancient seas: *Am. Assoc. Petroleum Geologists,* Mem. 4, 367–377.

Thode, H. G., R. K. Wanless, and R. Wallouch, 1954, The origin of native sulfur deposits from fractionation studies: *Geochim. Cosmochim. Acta,* **5,** 286–298.

Toriumi, T., and R. Hara, 1934, On the transition of calcium sulfate in water and concentrated sea water: trans. from *Jour. Chem. Soc. Japan,* **55,** 1051–1059.

Toriumi, T., T. Kuwahara, R. Hara, 1938, *On the Calcium Sulphate in Sea Water, II. Solubilities of Calcium Sulphate Hemihydrate in Sea Water of Various Concentrations at 65°–150°C:* Sendai Tohoku Univ. Technol. Rept. 12, 560–571.

Udden, J. A., 1924, Laminated anhydrite in Texas: *Geol. Soc. Am. Bull.,* **35,** 347–354.

Usiglio, M. J., 1849, Études sur la composition de l'eau de la Mediterranée et sur l'exploitation des sels qu'elle contient: *Ann. Chim. Phys.,* **27,** no. 3, 172–191.

Valyashko, M. G., 1962, *Geokhimicheskie Zakonomernosti Formirovaniya Mestorozhdenii Kaliinykh Solei (Geochemical Laws Governing the Formation of Salt Deposits):* Moscow, Izd. Moskov Univ., 397p.

Walther, Johannes, 1900, *Das Gesetz der Wüstenbildung in Gegenwart und Vorzeit:* Berlin, Dietrich Reimer.

———, 1903, Die Entstehung von Salz und Gyps durch topographische oder klimatische Ursachen: *Centr. Min.,* p. 211–217.

———, 1924, *Das Gesetz der Wüstenbildung in der Gegenwart und Vorzeit:* Berlin, Dietrich Reimer, 4th ed., 421p.

Wardlaw, N. C., and W. M. Schwerdtner, 1966, Halite–anhydrite seasonal layers in the Middle Devonian Prairie Evaporite Formation, Saskatchewan, Canada: *Geol. Soc. Am. Bull.,* **77,** 331–342.

Wells, A. J., 1962, Recent dolomite in the Persian Gulf: *Nature,* **194,** 274–275.

West, I. M., 1964, Evaporite diagenesis in the Lower Purbeck Beds of Dorset: *Proc. Yorks. Geol. Soc.,* **34,** 315–330.

Wilfarth, Martin, 1933, Sedimentationsprobleme in der Germanischen Senke zur Perm- und Triaszeit: *Geol. Rund.,* **24,** 349–377.

Wilson, J. L., W. C. Ward, and M. J. Brady, 1970, Northeast Yucatan, Mexico—A new area opens for study of carbonate-evaporite sediments: *Jour. Sed. Petrol.,* **40**. 745–749.

Wood, G. V., and M. J. Wolfe, 1969, Sabkha cycles in the Arab/Darb Formation off the Trucial Coast of Arabia: *Sedimentology,* **12,** 165–191.

Yanshin, A. L., 1961, The depth of salt bearing basins and some problems of salt series: *Geologiya i Geofizika,* no. 1, 3–15 (Eng. trans. available from Nat. Trans. Center, John Crerar Library, 35 West 33rd St., Chicago, Ill. 60616).

Zen, E-An, 1965, Solubility measurements in the system $CaSO_4$–NaCl–H_2O at 35°, 50°, and 70°C and one atmosphere pressure: *Jour. Petr.,* **6,** 124–164.

Author Citation Index

Subject Index